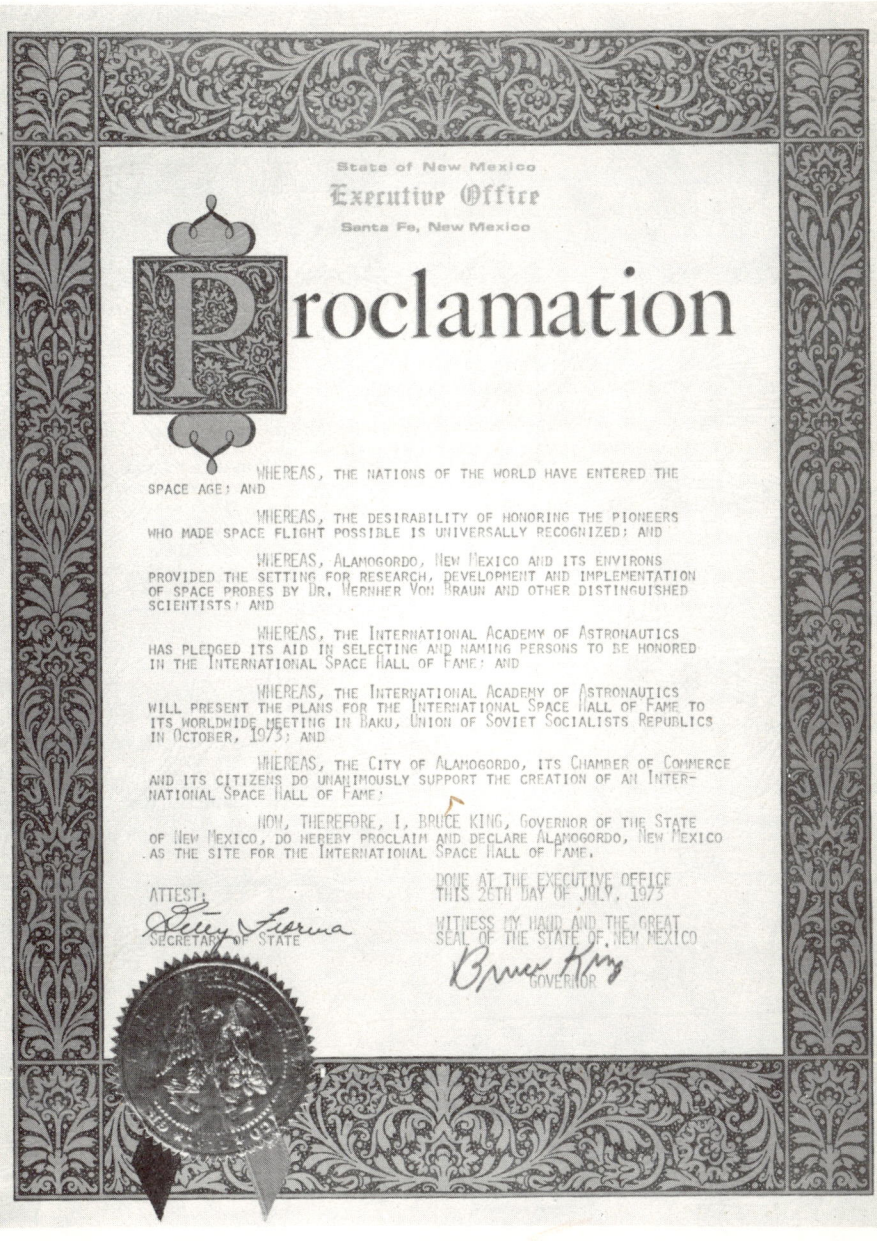

Upper Illustration: Black and White reproduction of the Space Hall Winterscape of William L. Taylor of Alamogordo, taken 30 minutes past Sunset in February 1977. The original color photo does not show the Moon. The Frontispiece color picture is printed on Fire-Orange stock and shows a sky coloring, often observed at the White Sands. While it is already dark for all practical purposes, infrared-sensitive film was used to photograph the Moon, which at the time of the colored picture original was about as far north of the right margin of the picture, as it is now north of the left margin of the color photo. The shadow of the Space Hall shows the incidence of the moonlight. The Moon image was copied from the infrared picture.

Center Illustration:

Daylight photo of the Space Hall taken on 5 Oct.76, showing reflection of the Mountains east and south of the Hall. The glass surface of the Hall shows colors from pure white to gold to dark black, depending on the incidence angle of the sunlight.

Bottom Illustration: Shows night shot of the Space Hall from the evening of the 5th Oct. 1976. While the lights shine bright, the internal details of the Space Hall are barely visible. Dedication Day pictures on page 134 have been taken by foreign visitors and provided to the New Mexico Rsearch Institue for illustration of the Proceedings.

THE EAGLE HAS RETURNED
SECOND PART

PUBLICATIONS OF THE AMERICAN ASTRONAUTICAL SOCIETY

Following are the principal publications of the American Astronautical Society.

JOURNAL OF THE ASTRONAUTICAL SOCIETY (1954-   )

Published quarterly and distributed by AAS Business Office, 6060 Duke St., Alexandria, Virginia 22304.

AAS NEWSLETTER (1962-   )

Published bimonthly and distributed by AAS Business Office, 6060 Duke St., Alexandria, Virginia 22304.

ADVANCES IN THE ASTRONAUTICAL SCIENCES (1957-   )

Proceedings of major AAS technical meetings. Published and distributed by Univelt Inc. (P.O. Box 28130, San Diego, California 92128) for the American Astronautical Society.

SCIENCE AND TECHNOLOGY SERIES (1964-   )

Supplement to Advances in the Astronautical Sciences. Proceedings and monographs, most of them based on AAS technical meetings. Published and distributed by Univelt Inc. (P.O. Box 28130, San Diego, California 92128) for the American Astronautical Society.

AAS MICROFICHE SERIES (1968-   )

Supplement to Advances in the Astronautical Sciences. Consists principally of technical papers not included in the hard-copy volumes. Published and distributed by Univelt Inc. (P.O. Box 28130, San Diego, California 92128) for the American Astronautical Society.

Subscriptions to the Journal and the AAS Newsletter should be ordered from the AAS Business Office. Back issues of the Journal and all books and microfiche should be ordered from Univelt Inc.

AN AMERICAN *Astronautical* SOCIETY PUBLICATION

The Eagle has Returned

SECOND PART

VOLUME 45, SCIENCE AND TECHNOLOGY

A SUPPLEMENT TO ADVANCES IN THE ASTRONAUTICAL SCIENCES

Edited by
Dr. Ernst A. Steinhoff

Proceedings of the Dedication Conference of the International Space Hall of Fame, held at Alamogordo, New Mexico, from 5 through 9 October 1976, as a tribute to 35 Space Pioneers, citizens of eight different nations, honored as the first inductees into the International Space Hall of Fame.

Copyright 1977

by

NEW MEXICO RESEARCH INSTITUTE

P.O. Box 454
Alamogordo, New Mexico 88310

ISBN 87703-092-8

Printed and Bound in the U.S.A.

An American Astronautical Society Publication

AAS President

    Robert L. Gervais   McDonnell Douglas
                                          Astronautics Co.

Vice President-Publications

    Dr. Peter M. Bainum   Howard University

Series Editor

    Dr. Horace Jacobs   Univelt, Inc.

Editor

    Dr. Ernst A. Steinhoff   New Mexico Research
                                           Institute

Front Cover and Title Sheet, courtesy of
    New Mexico Research Institute, Inc.

Front Cover Illustration

    Photograph by William L. Taylor
    Alamogordo, New Mexico

FOREWORD

This Volume at hand is the Part II of the Dedication Conference Proceedings, and contains those manuscripts for the book which were not on hand for submission in early September 1976, particulary those of very busy people, actively involved in one way or the other in Space Flight activities which were held to honor some 28 deceased Space Pioneers and seven living Pioneers. These Pioneers, by their achievements and support of the budding young Space Age, were found eligible for this induction by their peers - coming from eight different Nations. In this particular case, due to the availability of the names of five additional Pioneers selected for induction at the first Anniversay of the International Space Hall of Fame, will increase on the Third of November 1977 to 31 deceased, and 9 living Pioneers enshrined in the Space Hall, coming from nine different Nations, on the publication date of this volume.

The purpose of this second volume is to provide the international scientific community with those presentations and addresses, which due to their nature and resulting conference deliberations could not have been included at the publication deadline of Part I of the Conference Proceedings. They continue to provide an overview and summary of the past achievements, current state of the art and future near- and far-term achievement goals of international and national spaceflight efforts, expected from scientific, engineering, life sciences, space law, and managerial combinations of all these disciplines contributing toward the common objectives of spaceflight.

This conference featured four roundtable sessions which beyond the two regular International Academy of Astronautics Committee sessions on Spacestations and History of Spaceflight, and a Special Session, treating the contributions of the development of inertial guidance, flight control and navigation as well as the overall management of the Saturn-Apollo and Soyuz-Apollo toward its success.

Such a one-week conference cannot completely reflect the myriad of historical and contemporary contributions made by a legion of participants who succeeded, in advancing a dream of ours, for humanity to leave its earthly cradle. This venture in the realm of our solar system of the last two decades to eventually set foot on the first celestial bodies by unmanned and manned probes designated to report physical conditions, and environmental data found on and near such bodies and their moons, or asteroids and comets. However, two themes have been selected to illustrate the roadmap of efforts to the overall objective to achieve such exploration, namely the manned Saturn/Apollo project which culminated in man's first setting foot on the Moon, and the international manned joint project, the Soyuz/Apollo, during which two nations symbolically joined hands to mutually perform manned space research with the objective of providing valuable knowledge for the benefit of all Mankind.

The two efforts, and their milestones from concept acquisition to their realization have been narrated by living pioneers or their close associates, who have spent a substantial part of their lifetime toward the materialization and achievement of these objectives, of the early dreams of humanity. In 1976, seven pioneers, and in 1977, an additional three living pioneers have been selected from a large number of eligible persons who actively participated to realize these objectives, and who were chosen by the New Mexico Governor's Commission, to whose judgement it is entrusted to determine a small number, compared to the number of deceased selected space pioneers. This Commission holds the authority to determine such living persons worthy of such honor in the eyes of their peers and fellow scientists, judged on previous awards and honors, bestowed on them. The now living nine persons come from three nations which historically made major pioneering efforts to achieve the goals of Astronautics, the science and art of exploring the Universe by whatever means available to Man and his resourcefulness. The names and laudatia of the selected Space Pioneers are in this book.

We now present, with one or two exceptions, the balance of all keynotes and addresses in Sessions, Luncheon, Dinner and Banquet speakers, which are not already in the preliminary Proceedings, Vol.43 of the AAS Science and Technology Series, and therefore already have been covered. You may find that the strongest contributions on the average are in this Volume. It outnumbers Vol.I by about 20%, and contains some of the most interesting findings of the entire 1976 international space conferences. Regarding the Space Hall Dedication Conference Proceedings the so far unpublished contributions with two exceptions are in this volume. The missing contributions will be published and will become part of another volume when on hand. For financing reasons we did not publish Part II within the six months predicted which did not permit adequate time for the review of several contributions on account of lack of time to update them. We would like to note that one delay was caused by the arrival of the Space Hall first anniversary and the decision to include the announcement of the second part.

Ernst A. Steinhoff
Director, Dedication Conference
President, New Mexico Research Institute

SUMMARY AND CONCLUSIONS

In Part I of the Proceedings of the Dedication Conference of the International Space Hall of Fame, (Volume 43 of the Science and Technology Series of the American Astronautical Society titled "The Eagle Has Returned"), on page x is contained a "Note about this Book", which explains the experiment which was made with the publication of Volume 43. Since a number of national and foreign scientists of outstanding credentials in their field had committed themselves to participate in this experiment, in which much of the preparations for the Roundtables in Science in Spaceflight, Engineering in Spaceflight, Developments in Space Law and Life Sciences in Spaceflight were performed. At least two communications beetweem the participants after receiving an assessment of the current state of the art in each specific field were carried through including a summary of the achievements in each field, and an outlook on results from current and future work in near- and far-term future efforts. Part II of "The Eagle Has Returned" has the publication date of 3 November 1977, the day of the induction of five new Space Pioneers, selected by the New Mexico Governor's Commission on the International Space Hall of Fame, New Mexico's Bicentennial Project, funded by its Legislature in 1974, and completed on 5 October 1976.

The "Experiment" conducted with this book has been a full success, as each reader by reading Part I and Part II of the Proceedings will recognize. The two volumes are stimulating books, particularly for our youth, to convey to it the full impact of all the worldwide space efforts on our lives, our economy and the future of Homo Sapiens. To achieve this result 10 chapters have been written reflecting the preparations and deliberations of the Dedication Conference from 5 October thru 9 October 1976, just prior to the 27th Congress of Astronautics held from 11 thru 17 October 1976 at Anaheim, Calif. Many of the foreign visitors to Alamogordo during the first week attended the Anaheim Congress the following week. The feedback received from those was that the Alamogordo Dedication Conference was an outstanding event of the year. The History Committee of the American Institute of Aeronautics and Astronautics, in its session on 30 January 1977, passed a resolution that the Dedication Conference at Alamogordo together with the opening of the National Space Museum were the two most outstanding space-related events of 1976.

Volume One of the book has been adopted on recommendation of the Superintendent of the Alamogordo Public schools as textbooks for the New Mexico Public School System. Efforts are in progress to have the Volume Two reviewed for the same objective and to also extend the adoption of this book to as many U.S. States as possible. The financing of the printing of the second volume has been secured in July 1977 and this book will be available in December 1977. This volume contains all the short biographies of all the 1976 and 1977 inducties, and 17 comprehensive full biographies including biographical summaries for the 12 persons who were still living at the time of their induction, presented by persons closely familiar with the curricula vitae of those pioneers.

We wish to thank NASA for the use of the pictures marked NASA; Fairchild Industries, Inc. for the von Braun pictures; William L. Taylor for the cover picture which is also on an inside page, and the Aerospace Industry for those which have been provided by the presenters.

TABLE OF CONTENTS

CHAPTER I   DEDICATION CEREMONIES

Official up-dated List of Inductees including five new 1977 Space Pioneers   2

Dedication Ceremony Address   5
    Dr. Manfred Lachs

Welcome Address of the Governor of New Mexico   8
    Honorable Jerry Apodaca

Keynote Address: "The Future of Space"   9
    Dr. H. Guyford Stever

Greetings of the International Academy of Astronautics and Opening Address   15
    Dr. Charles Stark Draper

Greetings from the Federal Republic of Germany   16
    Dr. Heinz Fuchs

CHAPTER II   SPECIAL SESSION

Opening Address of Special Session   19
    Dr. H. Guyford Stever

Address to the Special Session   21
    Dr. Robert Gilruth

On the Evolution of Accurate Inertial Guidance Instrumentation   23
    Dr. Charles Stark Draper

Reply to Dr. Stever   29
    Dr. Ernst A. Steinhoff

The Life and Achievements of Neil Armstrong   33
    Astronaut Dr. Harrison "Jack" H. Schmitt

John Herschel Glenn, Jr., A Biographical Sketch   36
    Donna Utsunomiya and Ernst A. Steinhoff

Biography of Michael Collins, Composed from NASA Data   39
    Donna Utsunomiya and Ernst A. Steinhoff

Harrison "Jack" H. Schmitt - Astronaut and Scientist   40
    Donna Utsunomiya and Ernst A. Steinhoff

Soyuz Crew Biographies:
    Commander Cosmonaut Allexey Arhipovich Leonov   42
    Flight Engineer Cosmonaut Valeriy Nikolayevich Kubasov   43

    Composed fron NASA and USSR Academy of Sciences Data by Liovy Chaves and Ernst A. Steinhoff

CHAPTER III   ENGINEERING IN SPACE FLIGHT ROUNDTABLE

Assessment for Engineering in Space Flight, Letter 2   46
    Teofilio Tabanera

CHAPTER III   ENGINEERING IN SPACE FLIGHT, continued

    Response letter No. 2 Engineering in Space Flight   47
      Leland Belew

    Point of View in Engineering in Space Flight   49
      Teofilio Tabanera

    Assessment for Engineering in Space Flight by   51
      Segismundo Sanz Aranguez

    Airbreathing and Oxygen Collecting HTOHL Launch Vehicle   53
      Richard A. Nau

    Questions and Answers for Engineering in Space Flight   63
      Rudi Beichel

    Questions and Answers   69
      Subrata Sarkar

CHAPTER IV   SCIENCE IN SPACE FLIGHT

    Overview of the Viking Project and Most Recent Conclusions   73
      Dr. James Martin

    PART I   - OPTICAL ASTRONOMY

    Assessment of Solar System by Optical Astronomy On-Site   90
    Observations and Crust Samples
      Professor Audouin Dollfus

    Assessment of Future Tasks of Astronomy   95
      Professor Arthur Code

    Optical Astronomy   98
      Dr. Lubos Perek

    Remarks on Professor's Code's Survey Paper, Amendment   100
      Dr. John Strong & E.V. Silvertooth

    Past Interpretation of Mars in Retrospect   107
      Clyde W. Tombaugh

    Jupiter - A Laboratory for Meteorologists   112
      Dr. Reta F. Beebe

    Unsolved Problems in the Solar System   115
      Dr. Jack Evans

    The Large Space Telescope, an Introduction into the Future   119
    of Spaceborne Optical Astronomy
      Maxwell Hunter II

    PART II   - RADIO ASTRONOMY

    Very Long Baseline Interferometry (VLBI) in Modern Radio   124
    Astronomy
      Dr. Berry Clark

    The Very Large Array at Magdalena, New Mexico   128
      Dr. Victor Herrero

CHAPTER V   LIFE SCIENCES IN SPACE FLIGHT, ROUNDTABLE

    Introduction to Life Sciences in Space   135
       Dr. Charles Berry

    First Response to Dr. Berry's Survey Paper   144
       Dr. Hasashi Saiki

    Life Sciences in Space Flight   150
       Dr. Harald J. von Beckh

    Second Responses to the papers of: Dr. Campbell; Dr. White;   153
    Dr. Luft; Dr. Graul
       Dr. Hasashi Saiki

    Comments   155
       Dr. Heinz S. Fuchs

    Comments   156
       Dr. Harold Sandler

CHAPTER VI   DEVELOPMENTS IN SPACE LAW, ROUNDTABLE

    Introductory Address   161
       Dr. Carl Q. Christol

    Review of Space Law Developments   165
       Dr. Stephen Gorove

    Comments   170
       Dr. Isabella Diederiks-Verschoor

    Comments   172
       General Martin Menter

    Issues Pending Before the United Nations Legal Subcommittee   176
       Mrs. Eileen Galloway

    The Utility of Morphology to Space Law   181
       Dr. Ernst Fasan

    Comments   184
       Dr. Stephen Gorove

    Comments   187
       Dr. Carl Christol

    Remote Sensing and Direct Broadcast with the United   192
    Nations' System
       Subrata K. Sarkar

    Comments   196
       Mrs. Eileen Galloway

    Comments   197
       Dr. Charles Stark Draper

    Space Law Developments, New Comments   198
       Dr. Istvan Herczeg

    Greetings From The American Bar Association   199
       Edward R. Finch, Jr.

    Comments   200
       General Martin Menter

CHAPTER VI   DEVELOPMENTS IN SPACE LAW, Continued

    Summary and Conclusion   202
      Dr. Stephen Gorove

CHAPTER VII   INTERNATIONAL ACADEMY OF ASTRONAUTICS HISTORY SYMPOSIUM

    Select Biographies of Outstanding Space Pioneers Introduction 207
      E.M. Emme, Organizing Co-Chairman

    The Creativity of Rober H. Goddard   209
      F.C. Durant, III

    William Randolph Lovelace II   216
      Jacqueline Cochran
      Presented by Brig. General Charles Yeager

    Robert Esnault-Pelterie   220
      Edmond A. Brun
      Presented by Michelle Pige

    Sir William Congreve   224
      Frank H. Winter

    Max Valier: Space Pioneer   238
      Mitchell Sharpe

    G. Arturo Crocco   246
      Luigi Crocco   Presented by E. M. Emme

    Johannes Winkler: Early Investigator in Liquid Propellants 255
      Mitchell Sharpe

    Theodore von Karman   262
      R. Cargill Hall

    Hugh Latimer Dryden   270
      Eugene M. Emme

    Willie Ley: Pioneer Publicist for Space Exploration   285
      Mitchell Sharpe

    Andrew Gallagher Haley: A Biographical Sketch   294
      Stephen E. Doyle
      Presented by Mrs. Eilene Galloway

    Fritz Zwicky   303
      R. Cargill Hall

    Dr. Eugene Saenger Biography   312
      Mrs. Irene Saenger-Bredt

    Wilhelm Theodor Unge   317
      A. Ingemar Skoog   Condensed by Donna Utsunomiya

    Clinton P. Anderson: Life as Statesman, Pioneer of U.S.   320
    Space Objectives, Goals and Legislation as well as Executive Manager.
      Frank C. Diluzio   Condensed by E.A. Steinhoff

    Wernher von Braun in Memoriam   321
      F.C. Durant III

CHAPTER VIII EVOLUTION OF FLIGHT CONTROL SYMPOSIUM

The Evolution of Flight Control of the Apollo Mission   324
   Dr. Maxime A. Faget

CHAPTER IX SPACE STATION SYMPOSIUM

Space Stations-Symbols and Tools of New Growth in an Open World   332
   Krafft A. Ehricke

Space Stations and Their Prologue   345
   J. F. Madewell

Astropolis and Androcell -- The Psychology and Technology of Space Utilization and Extraterrestrialization   373
   Krafft A. Ehricke

Contributions of Spaceflight and Space Stations to the Solution of Social and Environmental Problems   397
   Dr. Leonard Jaffe

About the Need for Social Institutions to Change to Accommodate Technological Advances and Change. Comments on Dr. Jaffe's and Dr. Ehricke's Keynotes   404
   Dr. Warren Armstrong

The European Space Agency and the Spacelab Programme   405
   W. J. Mellors, ESA

CHAPTER X LUNCHEON AND DINNER SPEAKERS AND BANQUET PROGRAM

PART I - LUNCHEON SPEAKERS

The Future of Space   409
   Dr. H. Guyford Stever   (see Dedication Ceremonies)

The Space Shuttle Orbiter   410
   Mr. George Merrick

Trends in Space Communications   416
   Mr. George Harter

From Ballistic Missiles to Space Operations - An Exercise in Perspectives   421
   Mr. Grant Hansen

The Role of Space Technology in Society   428
   Robert Gervais

PART II - BANQUET SPEAKER

The State and Promise of Space   433
   Dr. Thomas O. Paine

PART III - CLOSING CEREMONY

Closing Remarks   436
   Dr. Charles Stark Draper

Closing Remarks   437
   Dr. Ernst A. Steinhoff

INTERNATIONAL SPACE HALL OF FAME DEDICATION CONFERENCE

Alamogordo Conference Program Director
Rev. Dub Bryant

supported by

THE INTERNATIONAL ACADEMY OF ASTRONAUTICS

sponsors

Space Symposia & Roundtables

AT
ALAMOGORDO, N.M.
OCTOBER 5-9, 1976

"BIRTHPLACE OF AMERICAN SPACE FLIGHT"

in conjunction with
the dedication of

THE
INTERNATIONAL
SPACE
HALL OF FAME

Dr. Ernst A. Steinhoff
Conference Chairman

supported by

HOLLOMAN-ALAMOGORDO SECTIONS
AIAA and IEEE

TUESDAY, 5 OCTOBER

0800 OPENING OF THE ISHF DEDICATION CONFERENCE
Host: (IAA) Dr. Charles S. Draper, President International Academy of Astronautics
INVOCATION: Rev. Dub Bryant

0802 OPENING REMARKS: Dr. Charles S. Draper, Honorable Conference Chairman

0805 IAA HISTORY SYMPOSIUM
Co-Chairmen: Dr. E.M. Emme(USA); Mr. F. Durant III (USA)
Place: Alamogordo High School

1000 PAGEANTRIES
Host: Alamogordo Chamber of Commerce
Reviewer: Hon. Jerry Apodaca, Gov. of New Mexico
Place: Vicinity of Space Hall

1200 IAA LUNCHEON
Speaker: Dr. G. Stever, NSF, US President's Science Advisor
Place: Chaparral High

1400 DEDICATION CEREMONY
Chairman: Gov. Apodaca
Noted Participants: New Mexico Congressional Delegation
UN Address: Dr. Lubbs (Czechoslovakia) UN Outer Space Affairs
Mr. F. Durant III, National Space Museum
Place: Space Hall Site

1600 EVOLUTION OF FLIGHT CONTROL, GUIDANCE AND NAVIGATION SYSTEMS USED IN OUTSTANDING MANNED SPACE MISSIONS (IAA SYMPOSIUM)
Organizer: Col.L.R. Sugerman, (USA)
Presenters: Dr. Walter Haeussermann(NASA); Dr. D. Hoag (Draper Labs); Dr. H. Hoelzer (NASA)
Place: Alamogordo H.S.

1800 SPECIAL SESSION DINNER
Master of Ceremonies: Dr. Guyford Stever, Director NSF & U.S. President's Science Advisor
Invocation:
Rev. Dub Bryant
Introduction of Guests and Speakers:
Judge M. Lachs, International Court of Justice, The Hague, The Netherlands, Address to Special Session
Dr. C.S. Draper, President, International Academy of Astronautics

Introductory Address:
Dr. Guyford Stever
Place: Chaparral High

2000 Introduction of Special Session and Events by Dr. Stever
SATURN/APOLLO AND SOYUZ/APOLLO SPECIAL EVENTS:
I. Recount of Missions by Participants:
Dr. R. Gilruth
II. Homage to deceased Astronauts and Cosmonauts
Dr. C.S. Draper
III. Tribute to Members of the Saturn/Apollo and Soyuz/Apollo Teams
Addresses:
Dr. Harrison Schmitt
Dr. Ernst Stuhlinger
Dr. Maxime Faget
Dr. David Hoag
Dr. Ernst Steinhoff
(Prof. Sokolsky)

WEDNESDAY, 6 OCTOBER

0830 LIFE SCIENCES IN SPACE FLIGHT (Roundtable)
Organizer:
Co-Chairmen:
Dr: P. Campbell (USA)
Dr. H. Fuchs (FRG)
Prof. H. Saiki (Japan)
Dr. H. von Beckh (USA)
Dr. O. Luft (USA)
Dr. H. Sandler (USA)
Place: Alamogordo H.S.

1215 LUNCHEON
Master of Ceremonies:
Guest Speaker: Mr. G.B. Merrick, President, Rockwell International Space Division
Place: Chaparral High

1400 EXCURSION TO SACRAMENTO PEAK SOLAR OBSERVATORY

1900 DINNER
Inn of the Mountain Gods, Ruidoso, N.M. Mescalero Indian Reservation
Speaker to be announced

Thursday, 7 October

0830 SCIENCE IN SPACE FLIGHT (Roundtable)
Organizer: Dr. Steinhoff (USA)
Co-Chairmen: Dr. James, JPL; Dr. Thomas, NMSU (all USA)

I PIONEER SESSION
Dr. Wolfe, Dr. Atkins, Dr. B. Oliver, Dr. K. Champion, Dr. R. Beebe, Mr. J. Long, Dr. E. Stuhlinger

II VIKING SESSION
Dr. Tombaugh, Dr. Masursky, Dr. Soffen, Dr. Martin, Dr. Friedman
Place: Alamogordo H.S.

1030 EVOLUTION OF FLIGHT CONTROL, GUIDANCE AND NAVIGATION SYSTEMS USED IN OUTSTANDING MANNED SPACE MISSIONS II
Co-Chairmen: Dr. Faget, Col. Sugerman, USA
Presenters: Dr. Faget, Dr. Jaenke, NMRI
Place: Alamogordo H.S.

1215 LUNCHEON
Master of Ceremonies:
Guest Speaker:
Mr. G. Harter, VP & Gen Mgr, Electronics Systems Div, TRW, Inc.
Place: Chaparral H.S.

1400 SCIENCE IN SPACE FLIGHT (Continued)

III OPTICAL ASTRONOMY
Co-Chairmen: Dr. Doll, Roundtable: Dr. Pickering,USA
Dr. Strong, Mr. M. Hunter, Dr. Perek, Dr. H. Masursky, Dr. Zeilick
Moderator: Dr. Code

IV RADIO ASTRONOMY
Co-Chairmen: Dr. Drake, Roundtable: Dr. B Clark, Dr. V. Herrero, Dr. E. Stuhlinger, Dr. B. Oliver, Dr. A. Code
Moderator: Dr. F. Drake
Place: Alamogordo H.S.

1600 ENGINEERING IN SPACE FLIGHT (Roundtable)
Spaceflight Subsystems*
Co-Chairmen: Prof. S. Sanz Aranguez (Spain)
Dr. M. Faget (USA)
Roundtable: Dr. Santini, (Italy); Mr. R. Beichel, Mr. Belew, Dr. Kaplan, Dr. Saikeld (USA)
Dr. Karl O. Brauer (South-Africa)
Dr. Haeussermann (USA)

FRIDAY, 8 OCTOBER (PARALLEL SESSION)

0830 INTERNATIONAL SPACE STATIONS (IAA SYMPOSIUM)
to
1130
Organizer: Dr. Ehricke, USA
Co-Chairmen: Dr. Davis, UNM, Dr Steinhoff,USA
Keynote Paper: Dr. Ehricke, Dr. Chatel, France
Presenters: Mr Madewell, USA, Mrs Galloway,USA
Dr. Fasan, Austria
Amb. Finch, USA, Dr I. Herczeg, Hungary
Place: Alamogordo H.S.

1215 LUNCHEON
Master of Ceremonies:
Guest Speaker: Dr. G. Hansen, VP & Gem Mgr, Convair Div., General Dynamics, Inc.

1400 INTERNATIONAL SPACE STATIONS (Continued)
to
1615
Contributions of Spaceflight and Space Stations to the Solution of Social and Environmental Problems
Co-Chairmen:
Dr. Armstrong, President ENMU, USA
Dr. E.A.Steinhoff,USA
Keynote: Mr.L.Chaffe USA
Presenters:
Dr. L. Perek UN
Dr. K. Ehricke USA
Place: Alamogordo H.S.

1700 PICNIC AND PAGEANTRIES
Sponsor: Chamber of Commerce & Tularosa Lions Club
Place: White Sands National Monument

FRIDAY, 8 OCTOBER

0930 SPACE LAW DEVELOPMENT (Roundtable)
to
1130
Organizer: Prof.Gorove USA
Co-Chairmen: Dr. Diederiks-Verschoor, Dr. Manfred Lachs,Holland
Moderator: Dr. Carl Christol, USA
Roundtable: BGen Menter USA, Mrs Galloway,USA
Dr. Fasan, Austria
Amb. Finch, USA, Dr I. Herczeg, Hungary
Place: Alamogordo, H.S.

1215 LUNCHEON
Master of Ceremonies:
Guest Speaker: Dr. G. Hansen, VP & Gem Mgr, Convair Div., General Dynamics, Inc.

1400 SPACE LAW DEVELOPMENT (Roundtable) (Continued)
to
1615
Organizer: Prof. Gorove USA
Co-Chairmen: Dr. Diederiks-Verschoor, Dr. Manfred Lachs,Holland
Moderator: Dr. Carl Christol, USA
Roundtable: BGen Menter USA, Mrs Galloway,USA
Prof. H. Lay,USA, Hon. Amb. Finch, USA, Dr I. Herczeg, Hungary
Place: Alamogordo, H.S.

1700 PICNIC AND PAGEANTRIES
Sponsor: Chamber of Commerce & Tularosa Lions Club
Place: White Sands National Monument

SATURDAY, 9 OCTOBER

0830 ENGINEERING IN SPACE FLIGHT (Roundtable) continued)
Spaceflight Systems *)
Co-Chairmen:
Dr Bisplinghoff - USA
Dr Maxime Faget USA
Roundtable: Dr A.P. Mitra, Prof. N. Munjal (India); Dr. M. Kaplan, Mr R. Beichel, Dr. W,Haeussermann (USA)
Mr. R. Beichel, Prof M. Kaplan USA, Dr S. Sarkar (Switzerland and India)
Place: Alamogordo H.S.

1030 HISTORICAL SKETCHES OF OUTSTANDING SPACE PIONEERS (Part II)
Co-Chairmen: Dr. Emme, Mr. Fred Durant, III
Introduction Dr. Emme
Presenters: Mr. F. C. Durant,III; Brig. F.C. Gen. Charles Yeager, Michelle Pige; Frank H. Winter; Mitchell Sharpe; E. M. Emme; W Carl Hall; Mrs. Eilene Galloway; Mrs Irene Saenger-Bredt, A. Ingemar Skoog; F. C. Diluzio

1215 LUNCHEON
Master of Ceremonies:
Dr. R. Bisplinghoff
Guest Speaker: Mr. Robert Gervais, Pres. American Astronautical Society
Place: Chaparral H.S.

1400 HISTORICAL SKETCHES (Part III)
Co-Chairmen: Dr. Emme, Mr. Fred Durant, III
Place: Alamogordo H.S.

1600 CLOSING CEREMONY
Honorary Chairman: Dr. Draper
Program Chairman: Dr. Steinhoff
Presenters: Reports by Session Chairmen
Place: Alamogordo H.S.

1900 BANQUET
Master of Ceremonies: Dr. R. Bisplinghoff
Guest Speaker: Dr. Thomas Paine, President, Northrup Corporation

xiii

List of Paricipants on Space Hall Dedication Conference and Authors

A. Dr. Abbott, Shirley ARBA
   Hon.Apodaca, Jerry Gov.
   Dr. Armstrong, Warren
   Prof. Sanz Aranguez,Spain
   Dr.Atkins, Kenneth L.

B. Bagley,Charles S.
   Bagley,M.
   Bakel,
   Dr.Beebe, Reta F.
   Dr. Beebe, Stephen
   Beichel, Rudolf
   Beichel,
   Dr. Bechthold, Ira C.
   Dr. Berry. Charles
   Dr. Bisplinghoff,Raymond
   Bolger, Philip,
   Black, Maureene
   Dr.Brauer, Karl O.S-Afr.
   Brady,J.F. Jr.
   Brown, Richard R.
   Brown,
   Belew,
   Dr. Bradey, Hugh
24. Bryant, Dub. Rev.

C. Dr. Campbell, Paul
   Campbell,
   Dr. Carlson, Robert L.
   Castenholz,
   Castenholz,
   Dr.Chatel, Bertrand UN
   Dr. Champion,C
   Dr. Christol, Carl
   Chavez, Fabian
   Dr. Clark, Barry
   Dr. Clark, James
   Cochran, Jackie,Aviatrix
   Dr. Code, Arthur
   Col. Clark, Richard E
   Clark,
40. Clark, Howard G.

D. Dr. Davis,L.I., Lt.Gen
   Dr. Davis, Prs. UNM
   Dr. Demetriades,, Srge T.
   Dr. Diederichs-Vershoor
       The Netherlands
   Doom, Richard C.
   Prof. Dollfus, Audouin
        France
   Durant, Frederic C.III
           Ntl. Space Museum
   Dunn, Aubrey, State Sen.
   Dr. Draper, Charles S.IAA
51. Davis,

E. Dr. Eber, Gerhard R.
   Eber, H.
   Dr. Ehricke, Krafft
   Elms, James
   Dr. Emme, Eugene, NASA
   Emme,
   Engel, Rolf
   Engel,
60. Dr. Evans, Jack

F. Dr. Faget, Maxime A.
   Faget,
   Dr. Fásan, Ernst, Austria
   Fleissig, Ross
   Fleissig,
   Fichtner,
   Hon. Finch, Edward R. ABA
                U.S. Ambassad.
   Dr. Fong, James
   Frahnert, Walter   FRG
   Frahnert, Traute   "
   Frahnert, Helga   "
   Dr. Francisco, E.
   Francisco,
   Dr. Friedmann, Lou
   Friedman,
   Fthenakis,Emanuel
78. Dr. Fuchs, Heinz,Surg.Gen
            RFG.
G. Galloway, Eilene U.S.Sen.
   Dr. Gardenhire, Lawrence R.
   Gardenhire,
   Gervais, Robert,Pres. AAS
   Dr. Gilruth, Robert

G. Gilruth,
   Dr. Gievers, John
   Gievers,
   Dr. Giffoni
   Dr. Glazer, Henry
   Glazer,
   Dr. Gorove, Stephen
   Gorove,
   Greinacher, Richard &
93. Greinacher, Switzerl.

H. Dr. Haeussermann, Walter
   Hall, Cargill R.
   Hall, George A.
   Hannover, Carl, IEEE
   Hannover
   Dr. Harter, George TRW
   Hansen, Grant, GDA-A
   Dr. Hermann, Rudolf
   Hermann
   Dr. Herczeg,I, Hungary
   Dr. Herrero, Victor VLA
   Dr. Hjertstrand,Ake,
                  Sweden
   Dr. Hnatiuk,Bhodan T.
   Dr. Hoag, David, MIT
   Hoag,
   Dr. Hoelzer, Helmut
   Hoelzer, Christel
   Dr. Hoff, Nicholas
   Hunter, Maxwell III.
112
I. Dr. Ingold, Norman

J. Dr. Jaenke, Martin
   Jaenke, Eva
   Jaffe, Leonard, Dep.Ast.Adm
                  NASA
   Dr.James, Jesse, JPL
   Dr. James,Jack
   Dr. Johnson, R.
119
K. Dr. Kaplan, Marshall
   Dr. Kershner,Richard B. ABL
   Kershner,
   Dr. Kline, Richard L.
   Dr. Knausenberger, George
124
L. Dr. Manfred Lachs, Chief
       Justice,Int.Ct.ofJustice
       Den Hague, Netherlands.
   Prof.Lay, H.
   Long,James
   Dr. Ulrich Luft

M. Prof. Maeda, Hiroshi, Japan
   Madewell, J.F. Rockwell Int.
   Mapes,, Joe, CIGTF
   Dr.Masurski, H.USGS
   Maxwell, Robert, Pergamon G.B.
   Dr. Martin, J.S. NASA Langley
   Brig.Gen. (ret) Menger, Martin
   Col. Mellors, W.   ESA
   Mc. Murdo
   Mc. Natt, Muriel
   Dr. Mitra, A.P., India
   Dr. Murphy.James S.
   Murphy
   Dr. Murthy,H.G.S. UN
   Mc.Murdo
   Merrick, George B. Pres.Rockwell
          International,Space Div.
   Robert Munson
145
N. Nau, Richard, GDA

O. Ohlinger, Dwight,Chm. N.M. ISHFC
   Dr. Oberth, Adolph, Aerojet Gen.
   Dr. Oliver, Bernie, VP.Hewlett Pd
   Overton,
150
P. Dr. Paine, Thomas, Pres.Northrop
   Dr. Pattermann, Embassy of FRG
   Dr. Patterson, John L.
   Patterson,
   Dr, Perek, Luos, UN
   Pige, Michelle Staff IAA'
   Dr. Pope, Alan, Sandia Corp.
157
R. Rasmussen, Hans AFSC-Track HAFB
   Dr. Donald Rathbun, El Paso Med.
           Center.

R. Cont.
   Rauschenbach,G.J. Dir.COMSAT
   Robinson, Marvin W. UN
   Dr. Rodenberger, C.A
162.
S. Dr. Saiki, Hisashi, Japan
   Dr. Shaver,
   Dr. Salkeld, Robert
   Dr. Sandler, Harold, AMES
   Savage, Thea, El Paso, TX
   Dr. Sarkar. Subrata, India
   Dr. Silvertooth, E. W.
   Silvertoooth, Muriel
   Dr. Strong, John
   Strong,
   Dr. Strughold, Hubertus
   Strughold, Mary Webb
   Dr. Stuhlinger, Ernst
   Sugarman, Leonhard
   Dr. Soffen, Gerald
   Salvador, Galdino
   Dr. Harrison Schmitt,U.S.Senate
   Dr. Spurk, T.H. Darmstadt-FRG
   Stungis, John
   Dr. Stapp, John Paul
   Dr. Steinhoff, Ernst A.
   Steinhoff, Hildegard
   Steinhoff, Monika
   Dr. Steinhoff, John
   Steinhoff, Beverly
   Dr. Stever, Guyford, U.S. Pre-
188   sidents Science 7 Techn.Adv
T. Tabanera, Teofilio Argentina
   Prof. Clyde Tombaugh, NMSU
   Tombaugh,
   Dr. Thomas, Gerald, Pres. NMSU
   Thomas, LOvell
   Thomas, Richard
   Tribbets
195
V. Dr. Vega, GDA
   Miss Van Gelder, Helene IAA
   Vick, Austin WSMR
   Dr. von Beckh, Harald
199
W. Dr. Wolfe, NASA Ames
   Dr. White, Clayton S. Oklahoma
            Medical Foundation, Pres.

Y. B.Gen.USAF (ret) Yeager,"Chuck
   Yates, Jerry   NMRI
   Young, Tom

Z. Zagone, Peter, CIGTF
205
125 NM Officila Visitors and
       Governors Staff
 25 Press
  2 NMRI Staff
257 Total participants

   Dr. Ingomar Skoog-Sweden-Author
259 Stungis, WSMR

xiv

CHAPTER I

DEDICATION CEREMONIES

"It is difficult to say what is
impossible, for the dream of yesterday
is the hope of today and the reality
of tomorrow."
    Dr. Robert H. Goddard, 1904, (U.S.A.)

INTERNATIONAL SPACE HALL OF FAME

OFFICIAL LIST OF SPACE PIONEERS INDUCTED DURING THE DEDICATION CEREMONY, 5 OCTOBER 1976

(Deceased Candidates)

Anderson, Clinton Presba (1895-1975), inducted in 1977; U.S. Senator; Chairman of the Committee on Aeronautical and Space Sciences; leader in promoting cooperative efforts between NASA and other agencies; encouraged Project Rover, SNAP, the manned Orbiting Laboratory, scientific probes and international cooperation in Space; Senate's prime spokesman in Space. (USA)

Blagonravov, Anatoli A. (1894-1975), Soviet scientist, specialized in applied mechanics. Head of Engineering Research Institute and the Commission for Exploration and Use of Outer Space of USSR Academy of Sciences. (USSR)

Crocco, Arturo Gaetano (1877-1968), Italian pioneer of rocketry and flight mechanics. Initiated experiments on solid and liquid-propellant rockets, ramjet propulsion; founder of Italian Rocket Society; vice-president of International Astronautical Federation 1955-56; Member of founding committee of International Academy of Astronautics. (Italy)

Damblanc, Louis (1889-1969), French rocketry pioner, sometimes called "The Father of Multi-Stage Rockets". Patented and conducted experiments on multi-stage rockets, 1930's; his patent later purchased by Americans and incorporated in Terrier and other missiles (tandem, dual-stage system); undertook scientific investigation of combustion of solid fuel rockets on test stand and pendulum apparatus, which paper won REP-Hirsch Prize. (France)

Dryden, Hugh L. (1898-1965), American aerodynamicist, Director of Research, NACA; while Chairman of the Air Force-Navy-NACA Research Airplane Committee, the X-15 and other high altitude aircraft were developed; helped establish NASA in 1958, and was named Deputy Administrator; played major role in organization of Project Mercury and in planning Projects Gemini and Apollo. (USA)

Esnault-Pelterie, Robert (1881-1957), French astronautical and aerodynamical pioneer. Began writing mathematical theories of space flight from 1912 and later published his findings in L'Astronautique, 1930-35; founded the REP-Hirsch Prize for astronautical achievements; constructed his own test stand, and conducted rocket experimentation. (France)

Gagarin, Yuri A. (1934-1968), USSR cosmonaut pilot; first man to carry out a space flight orbiting the Earth in the Vostok spaceship (12 April 1961); killed in a crash of aircraft training flight. (USSR)

Ganswindt, Hermann (1856-1934), German rocket pioneer, engineer; one of the first to propose a manned spaceship (1891) based on his theories of reactive motion. (Germany)

Goddard, Dr. Robert H. (1882-1945), American rocket pioneer; began experimenting with rockets in 1907; developed and flew (1926) world's first liquid-propellant rocket; granted over 100 patents in rocketry; his Method of reaching extreme Altitudes (1919) a classic in field on early rocketry and astronautics; speculated upon space travel from early childhood. (USA)

Haley, Andrew G. (1904-1966), American pioneer in space law; co-founder and first president of Aerojet Engineer Corp. (now Aerojet General); one of the first advocates of space law from 1930's; Chairman of First Colloquium on the Law of Outer Space 1958; General Counsel, Vice-President (1952-54); and President of the International Astronautical Federation (1957-59); member of the founding committee of International Academy of Astronautics (1960) and Legal Counsel of Academy (1960-66). (USA)

Hohmann, Walter (1880-1945), German pioneer in theory of space flight; published first book (1925) treating problem of interplanetary flight with mathematical accuracy, with concepts basic to modern trajectory studies, "Hohmann transfer ellipse" early member German Rocket Society (VfR). (Germany)

Kibalchich, N. I. (1854-1881), Russian inventor, author of plan for rocket-type flying apparatus for manned flight (1881). Plan included coverage of such questions as arrangement of a powder rocket engine, programmed burning conditions, flight control, provision of stability to the apparatus, etc. (USSR)

Korolyev, Sergei P. (1906-1966), USSR rocket pioneer; co-founder and first head of Moscow Group for Study of Reactive Motion (MosGIRD), 1931; in 1952 one of organizers and later head of the Group for Study of Reactive Motion (GIRD), which created and launched first Soviet rocket plane; worked on rocket boosters for aircraft in WWII, later directed development of ballistic, geophysical, and space rockets, including Vostok and Voskhod. (USSR)

Ley, Willy (1906-1969), German-American astronautical pioneer, writer; co-founder and active member of German Rocket Society (VfR), 1927; experimenter on early liquid-fueled Repulsor rockets, 1930-31; prolific writer on interplanetary flight and rocketry; did much to popularize and educate public on possibilities of space flight. (Germany)

Lovelace, William R. II (1907-1965), American astronautical pioneer, physician; Director of Space Medicine, NASA; developer of high-altitude oxygen equipment for pilots; research in space medicine and physiology for manned flights. Member of founding committee (1960) and of Board of Trustees of International Academy of Astronautics (1961-65). (USA)

Low, Prof. Archibald M. (1888-1956), British astronautical, rocket pioneer; designed and constructed first electrically guided rocket missile, 1917, and radio-controlled monoplanes; later, 1930's, became leading member of British Interplantary Society; wrote prophetic books on future of space flight. (England)

Riedel, Klaus (1907-1944), German rocket pioneer; worked with Wernher von Braun and others on liquid-fuel rockets, 1930's; tested first oxygen-gasoline rocket engines in Germany; credited with building first water-cooled rocket engine; from 1937 was Director for Ground Equipment in rocket development center of Peenemünde. (Germany)

Saenger, Eugen (1905-1964), German-American rocket pioneer; wrote classic work on theory and technology of rocket propulsion, 1930; founder (1936) of Rocket Research Center at Trauen, near Hannover; research on high-altitude aerodynamics, rocket-powered skip plane, or antipodal bomber; ramjet experiments. President of International Astronautical Federation (1951-53). (Germany)

Thiel, Walter (1910-1943), German rocket pioneer; in charge of development of V-2 (A-4) engine, a pump-fed, liquid-cooled, double-walled rocket motor of 56,000 lbs. thrust, and its production after first successful test in 1940; killed August 1943 in bombing of Peenemünde. (Germany)

Tikhonravov, M. K. (1900-1974), Soviet scientist, pioneer in rocketry. Under his supervision, first Soviet liquid-propellant rocket, the 09, was launched successfully (17 August 1933). Author of number of works on theory of rocket motion, on problems of designing a LPRE (liquid-propellant rocket engine), and on problems of cosmonautics. (USSR)

Tsander, Friedrich A. (1887-1933), USSR rocket pioneer; began lecturing on possibilities of interplanetary travel, 1921, afterwards publishing the conclusions of his theoretical researchs; constructed, 1930-31, compressed air and gasoline engine; first liquid-propellant rocket engine tested 1933; from 1931 took active part in Moscow Group for Study of Reactive Motion (MosGIRD); wrote Problem of Flight by Means of Reaction Vehicles, 1932, in which he proposed high energy propellants as oxygen and flourine and combustion of metals. (USSR)

Tsiolkovsky, Konstantin E. (1857-1935), Russian/Soviet astronautical pioneer; founder of modern cosmonautics and rocket technology; began these studies in 1883 and wrote prolifically on his mathematical theories of interplanetary travel, rocket dynamics, including effects of air drag and gravitation, etc.; theory of multi-stage space rockets, 1929; evolved fundamental theories of liquid-fuel rockets, independently considered liquid oxygen/liquid hydrogen and other optimal performance combinations; suggested artificial satellites, space stations; designed space suits, life-support systems, etc. "Tsiolkovsky Formula"; "Tsiolkovsky Number". (USSR)

Unge, Wilhelm Theodor (1845-1915), inducted in 1977; Swedish Technologist-Inventor in solid-propellant field; invented all the following innovations needed to make solid rockets qualified for later spaceflight use: introduction of double-base smokeless powder, a stabilizer to prevent chemical decompostion in storage, used a plasticizer to increase propellant shock-insensitivity, combined with case-bonding to increase mechanical propellant grain strength; worked with life-saving rockets. (Sweden)

Valier, Max (1895-1930), German rocket pioneer; wrote Der Vorstoss in den Weltenraum (The Advance Into Space), 1924, promoting the idea of interplanetary travel; by 1928 conducted numerous rocket experiments, including rocket cars, rocket planes, sled, to further promote the cause of rockets and to demonstrate their possible and future application; killed in an explosion of a test rocket. (Germany)

Von Braun, Wernher Freiherr (1912-1977), (See full biography, page 26, Part I). Memorial by Fred Durant, page 321. (USA)

Von Hoefft, Dr. Franz (1882-1954), Austrian astronautical pioneer; founded the Austrian Society for High-Altitude Exploration, 1928, the first rocket society in Europe (after a Soviet Air Force group which lasted only a year, in 1924); the Society was later superseded by the Austrian Society for Rocket Technology; also influential in the German Rocket Society (VfR) and wrote numerous theoretical rockets designated as the RH (Rakete Hoefft) series; considered interplanetary mail and photographic missions. (Austria)

Von Karman, Theodore (1881-1963), Hungarian-American astronautical pioneer; Director, Guggenheim Lab; Director, Jet Propulsion Lab; one of the founders of Aerojet Engineering Corp. (later, Aerojet General); Chairman of the Founding Committee (1960) and first President of the International Academy of Astronautics (1960-63); pioneer in development of high speed aircraft and missiles; development of supersonic wind tunnel; research in thermodynamics as applied to hight speed flight, hydrodynamics, etc. (USA-Hungary)

Von Pirquet, Baron Guido (1880-1966), Austrian astronautical pioneer, early member, Austrian Society for High-Altitude Exploration and Austrian Society for Rocket Technology; calculated spacecraft trajectories, early 1920's, including orbit and escape paths; advocate of manned space stations, 1930's; later active member, International Astronautical Federation; "Pirquet's Paradox". (Austria)

Winkler, Johannes (1898-1947). German rocket pioneer; launched first successful liquid-fuel rocket in Europe, 1931; the rocket burned oxygen and methane and attained an altitude of 1000 ft.; one of the founders of the German Rocket Society. (Germany)

Wyld, James H. (1913-1953), American rocket pioneer, early member and experimenter in the American Rocket Society; in 1938 he built and tested first successful American regeneratively cooled rocket motor; design was basis of the formulation of Reactive Motors and the concept became widely used thereafter. (USA)

Zwicky, Fritz (1898-1974), Swiss astrophysicist; jet propulsion, astronomical pioneer; director of research of Aerojet Engineering Corp, 1943-49, worked on the aeropulse, hydropulse, hydroturbojet, monopropellants for rockets,; catalogued and defined supernovae; studied 4000 galaxies, found scores of remnants of exploding stars; headed a committee of the International Astronomical Union; actively promoted space travel; helped conceive (1957) Air Force project on firing tiny artificial meteroids into space; author of many papers and books on astrophysics and reaction propulsion. Vice-President of the International Academy of Astronautics (1965-74). (Switzerland)

(Living Candidates)

Armstrong, Neil A. (b. 1930), American astronaut, engineer; command pilot of the Apollo 11 spacecraft; was the first man to pilot a spacecraft to a landing on the Moon and step on its surface (20 July 1969; currently Professor of Aerospace Engineering, University of Cincinnati, Ohio. (USA)

Collins, Michael (b. 1930), inducted in 1977; American astronaut; pilot 3-day Gemini 10 mission; command module pilot on Apollo 11; currently Director of the National Air and Space Museum at the Smithsonian Institute. (USA)

Draper, Charles Stark (b. 1901), (See full biography, page 8, Part I). (USA)

Gilruth, Robert Rowe (b. 1913), (See full biography, page 16, Part I). (USA)

Glenn, John Herschel Jr. (b. 1921), inducted in 1977: American astronaut; one of the 7 original Project Mercury astronauts; was first American into space (20 February 1962) in Mercury-Atlas 6; currently U. S. Senator from Ohio. (USA)

Kubasov, Valeriy Nikolayevich (b. 1935), Soviet cosmonaut, civilian; graduated as mechanical engineer for aircraft building from the Moscow Aviation School (1958), received master of science degree (1968); joined cosmonaut corps in 1967; backup technical scienctist for Soyuz 5 and flight engineer on Soyuz 6; was flight engineer for the Apollo Soyuz International Project (Soyuz 19). (USSR)

Leonov, Aleksey Arkhipovich (b. 1934), Lt. Colonel of the USSR Air Force, graduated from Chuguyev Air Force School, the Zhukovskiy Air Force Engineering Academy; became a cosmonaut in 1960; was co-pilot of Voskhod 2 (first man to perform extravehicular activity in space); command pilot of the Apollo/Soyuz Project. (USSR)

Oberth, Hermann Julius (b. 1893), (See full biography, page 20, Part I). (Germany)

Schmitt, Harrison H. (b. 1935), inducted in 1977; American astronaut, geologist; lunar module pilot on final Apollo lunar landing, Apollo 17; Chief of Scientist-Astronauts at LBJ Space Center, Houston, Texas; Assistant Administrator of the Office of Energy Programs at NASA Headquarters in Washington D.C.; currently U.S. Senator from New Mexico. (USA)

ADDRESS TO THE DEDICATION CEREMONY ASSEMBLY

Judge Manfred Lachs*

This is a great occasion, and I feel honoured to be with you today. Nineteen years ago, the first man-made instrument reached outer space - thus man's great venture began; man embarked on making his own stars. This has been the outcome of a long process, of penetrating the world, of discovering the secrets of nature - as long as human history.

"The main key to wisdom is studious and frequent questioning", said Abelard, that scholastic philosopher of the 12th century, but he received a most enigmatic answer from his great opponent, St. Bernard of Clairvaux, who ironically replied:"The seminarist learned and learned until he died." Seven centuries later Mark Pattison, that famed Oxford scholar claimed:"Learning is a peculiar compound of memory, imagination, scientific habit, accurate observation, all concentrated for a long period, on the analysis of the remains of literature. The result of this sustained mental endeavour is not a book but a man." Today we celebrate the triumph of man.

We take pride in achieving what our ancestors may have called miracles. Notwithstanding the claim of a Michel Foucault that "we are bred in the mythology of our culture to prefer one conception to another"-our achievements have been tremendous. There is little room for skeptics.

But, who am I to speak of Science in the presence of such great authorities - you know much better than I, and you have visions going beyond anything I can possibly say. It is to your great predecessors that we pay tribute today. Mine is only an all too modest contribution to the ceremony of today. It is one of a jurist. Yet you recall, as I do, Durrenmatt's words in his Physicists:"Every effort of an individual to resolve in his own way what concerns all men must be doomed to failure." For the great issue encompasses the social implications of the application of Science. There is no reason to go the whole way with such pessimists like Wells, when we face man's centuries-old yearning for learning and above all for greater control over the forces of nature. The basic and fundamental issue is that of using knowledge, science and technology in a way that would not harm -"the fragile cargo of life"- more to use it for its benefit. Centuries ago Plato explained the link between philosophy and technology; d'Alembert linked Archimedes with Homer; the great Leonardo personified philosophy and painting, mechanics and mathematics; Aristotle had his horse and cart; and Galileo thought of shipbuilding and artillery. But these individual illustrations apart,

----

*Chief Justice, International Court of Justice, The Hague, The Netherlands. Paper presented by Dr. Lubos Perek, United Nations Outer Space Affairs Division.

it has been wisely said that "the mental processes involved in observation, verification and use of the imagination in producing ideas are fundamentally the same in poets, artists, musicians and scientists". For there are no frontiers between the various branches of human sciences - each of them is part of the whole - integrated as what we may call social reality. As Aldous Huxley so rightly said: "For Science in its totality, the ultimate goal is the creation of a monistic system in which the world's enormous multiplicity is reduced to something like unity."

All this requires a social commitment of Science, of the scientist, to society - to humanity; it calls for social responsibility. This in turn leads to the need to establish rules of conduct, of restraint - it leads to the need to establish the rule of law.

I recalled on another occasion that Zeus dispatched Hermes to give man a sense of law, so that the gifts of Prometheus, wisdom and fire, should be supplemented - by rules of cooperation.

This is the great task we face: we jurists, like you,are engaged in the study of physical phenomena; but our field is to see that the movements of electrons, atoms, energy, whatever its form, be brought under control so that men can wisely use them in the interest of men. While you move into ever new spheres, States and Governments should be able to use the power of new instruments for weal and edification, not woe and destruction. We cannot lag behind. Let me recall that the discovery of this continent, as well as others, gave rise to a series of problems of international character, and law had to follow in the footsteps of a Columbus as it later did in those of Picard, Cook, Perry, Shakleton and Burke - and recently of Tsiolkowsky, Goddard, Oberth and their followers.

Thus law is bound to take account of the effects on humanity of the scientific revolutions. It is one of the most fortunate phenomena of our age that jurists became very quickly aware of the need for legal regulation of man's activity in Outer Space. Work on the Law of Outer Space began soon after the first man embarked upon his journey among the stars. Only five years later the first principles of the law were laid down in an historic Resolution of the General Assembly. A year later - in December 1966 - the Treaty on Principles Governing the Activities of States in the Exploration and Use of Outer Space - was born. Other rules followed. Today, we can take pride in the existence of three more Conventions: the Agreement on the Rescue of Astronauts and Return of Objects Launched into Outer Space, the Convention on International Liability for Damage Caused by Space Objects, and the Convention for Registration of Space Objects. Some principles on Direct Television Broadcasting from Satellites on the Remote Sensing of the Natural Resources of the Earth have been elaborated; others are in the process of codification. A whole network of agreements has come into being: such as concerning tracking facilities, launching assistance, telecommunications and weather observation. Earlier, in 1963, the important Treaty Banning Nuclear Weapons Tests, <u>inter alia</u> - in Outer Space was signed. The work is slow and painstaking, yet if you compare it with the development of rules of other chapters of International Law, it has been unusually speedy. Imperfect as it might be, law is trying to keep pace with scientific developments, more perhaps than in other areas.

This is so important. It is encouraging, and it may serve as both a lesson and warning. For in this, as in other fields, the jurist should,as a witness of events in the international arena, keep in constant touch with developments in Science and technology; as a thinker, he should reflect on ways and means of making International Law work more effectively in the interests of all States and Nations; as a writer, he should convey his ideas to the legal community, to scientists and to public opinion and thus help in shaping the latter; as a practitioner, he is called upon to devise and perfect the tools on international cooperation and bring about progress in law; as an adviser, he should forewarn of the dangers resulting from disregarding law or the interests of other States and help to forge new links of international cooperation. As a judge, he must interpret the law and decide upon its application in the spirit of its progressive development and ultimate objective; he should administer law in a forward-looking spirit, relying on words not as instruments of manipulation or reflection of a remote past.

Though we may sometimes despair that the great principles in International Law remain a dead letter, that many of our efforts have failed in view of its persistent fragility, yet these thoughts come only in moments of frustration and despair for, just like Science and technology, International Law is bound to develop and to move forward, helping States and Nations to enhance their common destiny through cooperation and mutual understanding. Faced with the great strides of Science, the jurist should constantly ask himself: Whom do they threaten? What rights do they actualize, create or infringe? "The legal response to a given social or technological problem is, therefore, in itself a major social action which may aggravate a given problem or alleviate and help to solve it." (1)

I have perhaps been guilty of over-stressing the role of the public international lawyer in this process. However, I have done so, because in all this we need your help. Some centuries ago it was said: "The proud doctors disputed and philosophized and the medieval world paused to listen." This, of course, is not the case today. You, members of the great and distinguished family of scientists move with supersonic speed and we, jurists, have to follow you.

Today, we honour, at the Opening of the International Space Hall of Fame, the most famous amongst you who ousted the dreams of a Cyrano de Bergerac, Jules Verne or H. G. Wells by the reality of their scientific achievements; in fact, those who have done more: the reality they have created is much bigger than the dream.

They can be proud of their record - and so are all men. Should this lead us to greater conceit, to the strengthening of our conviction that man is unique in the Universe? Whether life does or ceased to exist elsewhere, we are those who challenge nature and make it serve our ends. This should lead us to greater humility. This in turn should be translated into rules of conduct among nations - find its expression in man-made law.

As one who had the privilege to preside over the preparation and drafting of the first principles and rules of the Law of Outer Space, I take particular comfort in its speedy and successful development. Presenting the first Space Treaty to the United Nations, I called it "a milestone in the developing of international relations." And so it has remained. But our thoughts should be future oriented. What we have built is a bare scaffolding for the law of tomorrow. We are bound to accept the sacred obligation to make Science serve man, social justice, the betterment of mankind, to do away with starvation and mysery, with the threat of annihilation. Unfortunately the gaps remain wide. Hence, what is called the Janus face, remains. Centuries ago, Thucydides told us: "True wisdom is shown by those who make careful use of their advantages in the knowledge that things will change."

Thus, the great changes brought about in Science and technology must lead to adequate changes in relations between man and nations.

This is an occasion in which we all rejoice; it is also one for reflection on the value and use of knowledge and technology; on the danger that the most innocent discovery could produce a threat to our future. In order to assure their beneficial use the law-makers need the help of the scientists, in domains of which you are unsurpassed masters. By joining hands we not only honour the great, but help in making their achievements serve all men, all races - those who populate our small planet and are in dire need of that help.

Reference: Friedman, Changing Role of Law in the Interdependent Society

OPENING ADDRESS
INTERNATIONAL SPACE HALL OF FAME DEDICATION

Hon. Jerry Apodaca*

Thank you very much, Fabian Chavez, Senator Bird. It's a pleasure to have you in our beautiful state, and I am glad that not only are we having a great day here in Alamogordo, but that we are having such beautiful weather to make you feel at home. Mayor Glover, it's nice to be in your city. Senator Montoya and Congressman Runnels, it again is a pleasure to share the platform with you; our many visiting dignitaries, ladies and gentlemen.

Today is certainly a day that all New Mexicans can be proud of, because as we dedicate this great facility, not only do we have to think about the past and the many accomplishments of this country, and the many large ways in which our State has contributed to it, but we can more importantly think about the future. Because New Mexico is a state of the future, with so much to look forward to, and so much hope, and so many things to contribute to mankind, and so the dedication of this facility is indicative not only of our past, and not only of the great accomplishments of this country, but hopefully a living example of what the future means to each of us. So I am very glad I am able to participate in some small way in this very memorable day here in Alamogordo, in the southern part of New Mexico, and of course, a memorable day to the people of New Mexico as well. We appreciate the many long hours of dedication that many men and women have spent in the last three years or so to make this a reality. We hope that not only this community gives this Hall of Fame the support that it needs, but that the people of New Mexico do likewise.

So again, my tribute to all of you for this great accomplishment. We wish you well, and as Governor of New Mexico, I can only say on behalf of all the people in New Mexico, we are extremely proud to see this day come true, and again we hope we can continue to participate, not only in the future of this state, but in the future of the world as well. Thank you very much.

----

*Governor of New Mexico

THE FUTURE OF SPACE

Keynote Address

Dr. H. Guyford Stever*

We are here in the City of Alamogordo and State of New Mexico to celebrate the history of our venture into space by the dedication of a hall to its heroes of the past and present and by this symposium. The candidates for the International Space Hall of Fame are from some 14 nations. They go back in time to the days of ancient Greece - - to Ptolomy (ca A.D. 90-168) and Lucian of Samosota (ca A.D. 125-190). Not all the pioneers whose work paved the way for our entrance into this Space Age were honored in their own time. Perhaps recognition in this Hall of Fame will in some way serve to right that injustice.

For thousands of years only the minds of men and women could enter space, primarily through their intellect and imagination. Within the past century or so advances in the sciences of astronomy and astrophysics have allowed us to penetrate more of the mysteries of that majestic realm. Now, in just the last few decades we have begun a much more extensive probing of space. And through our space technology, our astronauts, our cosmonauts, and our unmanned missions, humanity has taken its first faltering and firm steps away from its mother planet.

This age of flight and space has descended on us with amazing rapidity. There are many people alive today - - and perhaps some in this room - - who can recall the news from Kitty Hawk, who may have known Robert Goddard or Hermann Oberth, who have looked at White Sands, Cape Canaveral, or Houston, and who witnessed the success of Sputnik and Neil Armstrong's First step onto the Moon. These and other great milestones have occurred in the span of only some 73 years.

But the Hall of Fame is not to be simply a monument to commemorate the past. It is to live into the future. With these and other significant accomplishments behind us, we are at the threshold of a new and different stage of the Space Age - - one that raises complex and difficult questions concerning the future of space. For our entrance into this stage coincides with a time when mankind faces many other transitions - - transitions that will test its thinking and its technologies and are bound to affect its activities in space.

It is a time of great expectation and demand for more than half of the human race who are living close to the subsistence level but who have mounted a concerted effort to raise their standard of living.

It is a time of growing sensitivity and concern over the global environment and resources that support all life on this planet, and particularly the technological civilization we have built over the past century.

----
*Science and Technology Adviser to the President and Director, Office of Science and Technology Policy

And it is a time when all these concerns and demands have combined to put great pressure for change and priority setting on our economic and social institutions.

In the light of all this it seems quite worthwhile at this time to examine some of the possibilities for future space activities, to see how they might fit in with other aspects of our global outlook, and to consider their possible roles in the resolution of some of humanity's most urgent problems. This is what I will attempt to do, based on current U. S. space activities and studies for the outlook for space.

In an examination of this nature perhaps it is best to start closest to Earth -- with a discussion of the remarkable results of Earth-orbiting satellite activities and the promises they hold.

Almost every high school science student is familiar with the claim of Archimedes that if he had a lever long enough and a fulcrum in space he could move the Earth. Figuratively speaking, our Earth-orbiting satellites are levers in space. In near-Earth orbits and in synchronous orbits at about 22,000 miles from the Earth's surface, they provide the means to accomplish prodigious tasks we cannot do alone from down here. From those vantage points, and with the variety of sophisticated technologies that have been advanced with them, these satellites have become new eyes, ears, and voices of tremendous perception and range for all of mankind. And we have only begun to exploit their many existing uses and to think about new ways they can be employed.

As an example, let us first take a look at the communications satellite. The growth in this field has been remarkable. We usually think of a decade as a reasonable time span for the development of the first generation of a complex technological system. But during such a ten-year interval, four generations of commercial communications satellites have been introduced, and each one offering major advances. We have gone from the first Intelsat craft orbited in 1965, weighing 40 kilograms and with a capacity of 240 telephone circuits, to Intelsat IV, a 740-kilogram satellite with a capacity of 4,000 simultaneous circuits. In this time we have gone from one link across the Atlantic to a system of seven satellites supporting 380 links around the globe. This system is used at 115 Earth terminals in 65 countries. It provides 6,500 full-time telephone circuits plus television, data, and facsimile service.

What is perhaps even more remarkable than the technological accomplishments of this system is that it involves the cooperation of the 91 member nations of the International Telecommunications Satellite (Intelsat) Organization, and that cooperation from countries of different racial, religious, and political background has led to a most successful and financially profitable operation. Revenues from this global investment of over $1 billion now exceed $500 million annually.

Aside from the international aspects of communications satellites, some individual countries have benefited remarkably through the use of a satellite system operating within their own borders. Indonesia is an excellent example as it is a nation composed of 13,000 islands. Through the use of a single satellite and 40 ground stations it has been able within a period of 17 months to provide telephone, telegraph, TV, and radio service to all of its 26 provinces. Through that same system Indonesia has been able to extend its elementary education to 85 percent of its youths.

India is another nation that may benefit immensely by the use of satellites to provide education. A major portion of India's more than 400 million people live in some 560,000 villages. Only some 71,000 of these are electrified, therefore, a principal way of reaching a great number of India's people with vital communication and education is via a powerful synchronous satellite capable of beaming programs directly to battery-or generator-powered television sets in these many villages. The U. S. National Aeronautics and Space Administration's Advanced Technology Satellite -6 (ATS-6) has been used experimentally for this purpose. Last year in a cooperative program with the Government of India this satellite transmitted educational programs to some 2,400 villages. It brought important information on hygiene, agriculture, family planning, nutrition, and other useful subjects. The programs were produced by the Government of India which supplied each village with a television set and antenna system costing about $1,000.

With this experiment successfully completed in India, the ATS-6 satellite was moved to a different orbital location. Over the past few months it has been used in a demonstration program co-sponsored by the U.S. National Aeronautics and Space Administration (NASA) and the Agency for International Development (AID). In this program, beamed to 11 developing nations in Africa and Asia, a series of broadcasts showed how Earth-orbiting satellites, such as Landsats 1 and 2, can be used for environmental monitoring, mapping, soil surveys, crop and forest inventories, and other studies relating to natural resources. The ATS-6 will begin next week to beam a similar series of programs to countries of Latin America.

These programs are in effect a preview of what can be done and is being planned to use space systems to come to grips with some of the world's major problems -- in both the industrialized and developing nations. Let me mention briefly some of the programs, plans, and long-range thinking along these lines.

One of the principal problems the international community will face in the years ahead is the adequate production and distribution of renewable resources -- food, fiber, and forest products. Space activities can play a vital role in our monitoring and management of such resources. Satellites are already surveying and reporting on crop conditions in some parts of the world, indicating blighted areas, pest infestation, drought conditions, and so forth. Under consideration for the future is the concept of Global Crop Production Forecasting. The objective of this idea would be to provide, via satellite systems surveying a variety of agricultural conditions, a bi-weekly forecast of the global production of major crops. This, of course, would have an economic as well as a humanitarian focus. Such a system would provide government agricultural managers with information that would allow them better control of food inventories. This information would give them early warnings of possible crop failures. It would provide information pertinent to trade agreements, potential market changes, and transportation requirements. One result would be that agricultural producers and nations could plan their production and distribution of food in ways that would eliminate much of the waste our food system suffers today.

We already have some elements of this program under way today in the Large Area Crop Inventory Experiment, which is based on the work of existing satellite systems. This experiment will provide insight into the problems and possibilities of global crop estimation. A recent NASA report indicates that by 1982 we could have an operational system for global wheat forecasting, and that by the late 1980's an improved system could be operational for forecasting the production of all major crops.

Methods of developing timber inventories in the United States by satellite surveys are also under way, as are methods for rangeland assessment. Range feeding of cattle is now on the increase because of the rising cost of grain and the growing demand for it for direct human consumption. Therefore, the use of spacecraft data becomes more effective in large areas, such as the western United States where marginal grazing conditions exist and the cattle need to be moved to optimum locations. It has been estimated that the use of satellite range surveys could offer economic benefits here from $4 million to $30 million annually. They will be equally effective in other parts of the world, particularly where overgrazing not only causes economic hardships but results in environmental problems.

Closely related to the problems of agriculture and forestry are, of course, those of weather, climate, and water. And space technologies are playing an increasingly more important role in improving our understanding of these. An ability to make accurate and timely weather forecasts provides great economic advantages, in addition to its many other benefits. It has been projected that in the United States alone total economic losses due to weather problems run about $12 billion annually, of which approximately $5 billion are preventable. Better long-range weather forecasts -- up to 30 days -- could save an estimated $500 million per year. On a global basis the savings in money and human suffering are far greater.

As you may know, the international science community has been pursuing a cooperative large-scale weather forecasting program. One experiment in this program, an experiment confined to the tropical Atlantic Ocean area, has been completed. It employed surface vessels, aircraft, and land stations as well as satellites. The First Global Experiment is scheduled to begin in 1978. Its goal is to extend forecasting accuracy from three days to between 10 to 20 days, which is considered by weather experts to be the ultimate limit of predictability. The success of this experiment will be based primarily on obtaining global atmospheric observations at least once every 12 hours. Plans to accomplish this will depend on the observations of two to four Sun-synchronous polar-orbiting satellites carrying instrumentation that can measure atmospheric temperature and humidity and sea surface temperatures, and five geosynchronous satellites capable of returning images that will derive winds, cloud amounts, and altitudes. The satellites will also track data from balloons, ocean buoys, and remote monitoring stations.

Based upon the success of this system, even more complex operational systems could be implemented in the mid-1980's and early 1990's. These weather forecasting systems could help support weather modification programs that some day may save many lives and the enormous losses of property around the world.

Closely related to weather is climatic change, and this is a subject in which there is currently a great deal of interest. Our interest in climate change and climate prediction has been stirred by some indications that our energy and chemical technologies could affect the world's climate, and by the fact that even a very small degree of climatic change can result in weather variability producing drought or shifts in monsoons that can play havoc with global agricultural activity. Enormous numbers of scientific observations and analyses are needed to understand and possibly predict climatic change. They include such observations as those of solar radiation, the surface radiation of the Earth -- over land, sea, and ice -- hydrological change, snow/sea ice change, and atmospheric gases and particles. Satellites, rockets, and other space equipment will play a major role in such observations and in the development of climate predictive models.

I could go on to describe several other uses of satellites that will play a vital role in our future, including its important applications in global hydrology, energy and mineral exploration, the world-wide control of harmful insects, and earthquake prediction. But I want to conclude with some brief comments on the benefits of other aspects of space activities, particularly the Space Shuttle and the Space Lab.

On September 17, I attended the roll out ceremony for the U. S. Space Shuttle in Palmdale, California. This reusable spacecraft will introduce a new era in space activity. It will allow for regular flights into Earth orbit and return carrying several men and women and tons of equipment. It is envisioned that during the 1980's when fully operational, the shuttle will be going into orbit routinely and frequently, carrying out a number of important missions. What will some of these missions be? Among other things the space shuttle will allow for experimentation in the production of certain drugs, metals, glass, and electronic crystals. Production of these in a gravity-free environment will offer some considerable economic advantages as well as allowing for us to produce items of unusual purity and strength.

Functions such as these will make the shuttle a space transportation system and space work horse of the future. With its cargo bay of 15 by 60 feet it will be able to carry several satellites or other technological gear and scientific experiments into orbit at one time. It will be capable of delivering a payload of some 65,000 pounds at about half the cost of the expendable rockets now in use.

While the Space Shuttle is a U. S. system, it will be serving the International space and scientific communities in a number of ways. In addition to the launching of satellites for Comsat, the shuttle will be made available to carry the Space Lab being planned by the European Space Agency in which 11 nations participate. The Space Lab, a $400 million project, will be constructed to fit into the cargo bay of the shuttle and flown for periods up to 30 days. Among the experiments being planned for Space Lab, for which equipment is already in the selection and design stage, are some in solar physics, infrared astronomy, ultraviolet/visible astronomy, high energy astrophysics, space biology, and atmospheric physics.

Space Lab flights will carry into orbit a crew of two pilots and five scientists. Studies are now being conducted to examine the possibilities of constructing a permanent 12-person space lab from modules carried into orbit by the shuttle. If such a lab is built -- possibly in the late 1980's -- it would be supplied and serviced by the shuttle, which would periodically bring its personnel back to Earth. Once such a working lab is established in orbit, it would not be difficult to add modules to it to expand its activities. Among these could be some important space manufacturing, particularly of biological products such as those vaccines and serums that can be produced in a weightless environment ten times faster than on Earth.

From this point, when we have developed the capability and confidence to work in space to this extent, it is not difficult to imagine a future in space that not too many years ago was only the subject of science fiction and fantasy. Today some people are giving serious thought to such projects as large solar power stations in orbit and space colonies capable of supporting thousands of people located between the Earth and the Moon. Detailed studies have been made on how such projects could be carried out, and in the case of the Space Colony how it might be constructed with materials mined from the Moon and processed and assembled in orbit.

The men and women being recruited today for the shuttle and skylab programs may be the pioneers of these new space frontiers. Surely they, and all the heroes and heroines we honor here in our International Space Hall of Fame, will have blazed the way for the next generation in space.

In the meantime our unmanned probes explore the planets and solar system, and the giant eyes and ears of our astronomy and radioastronomy observatories search the universe, the galaxies, and beyond. They look for signs of life, for new knowledge of the physical forces that form the basis of all existence, and for clues to the eternal mysteries that have challenged the minds and souls of all humanity since its eyes first turned in awe and reverence to the splendor of the heavens.

The Viking missions have been outstanding successes, providing us with much important information of the surface of Mars and its atmosphere. Great advances are being made toward our probing of the universe and toward the possibility of what Carl Sagan calls "the Cosmic Connection" -- contact with any other intelligent life in the cosmos. Radio signals to the farthest reaches of space have been beamed out from Arecibo and other radioastronomy centers. In this very State of New Mexico the world's largest radioastronomy facility, the Very Large Array (VLA) is now being constructed. The VLA, which will be ten times as powerful as any existing radioastronomy facility in the world today, should be completed in 1980. But as early as next year it should begin a program of limited scientific observations which will eventually lead to its search for signals from the most distant sources in space.

The future also holds the possibility of improved optical and radio searches through the use of orbiting observatories, and radio telescopes on the Moon. These would offer great advantages due to the absence of an atmosphere, the slow rotation rate of the Moon, and no interference from human activity. The low gravity of the Moon and the lack of wind would facilitate the construction and use of very large structures. And a lunar science base might also be useful in X-ray and cosmic ray observation and as a site for detecting gravity waves.

All this is part of our basic quest in space -- to discover from whence we came and where we might be going. Its pursuit will always continue to challenge us, to excite us, to move us into new and unknown frontiers. This is what the future of space is all about -- and we stand only at the threshold of that future.

OPENING ADDRESS

Dr. Charles Stark Draper*

Members of the Conference, distinguished guests; it is a real pleasure to see you all here. My first duty is to say a few words in starting the whole proceedings. Humanity has been developing, evolving, whatever you want to call it, for a long, long time. If you follow the developments of archeology, you find out that, sure enough, this wonderful capability that the human species developed of being able to think rationally....not everybody thinks rationally....but having the capability to think rationally, is a unique capability, and yet out of any given population of the earth, there, at any time, are only a very, very few individuals who make contributions to lead everybody on into higher and higher grounds. If you look, for example, at the stories in legends, religions, all kinds of books, you see there has been a lot of talk about going up to the stars for many thousands of years, and yet it has been only very recently that this has occurred.

I, at one time as the President of the International Academy of Astronautics, had an audience with the Pope in Rome to present to him an international astronautical dictionary. He was very cordial. He said to me, "We're both in the same business." I said, "Oh?" He said, "Yes, we are both navigating the Heavens."

Anyway, the situation that we have here is that the human beings, the very few human beings that make extraordinary contributions to the race, these people who make noteworthy contributions, are being honored by what they did in space. It is a great thing, a tremendous thing, and in the next few days you will see this spelled out in more detail. The detail of how great achievements are made becomes very complex as you all know, and we have heard in our recent days much unhappiness that people have about technology. They generate unhappiness and wish to lay it at the doorstep of technology, that is, the changing of the environment, into situations more suited to our needs and desires in a natural state. We have heard that this endeavor is bad, and various people are trying to stop any further advances. I, myself, have experienced some adventures because I'm regarded as sort of an exponent of technology.

Now I, looking backwards at the whole thing over a considerable period of time, would say that technology has been a very good thing for the human race because we know that even as recently as 125 years ago that slavery for human beings was not extraordinary. It was a common thing and had been a common thing for all of history and probably before that, because in building things like the pyramids, the only way it could be done was to use many, many people that you had to grind into the dirt like low animals. So technology comes along and changes all that by making machines that will do this work, and that I think, was the thing which finally set people free, not only from actual work, but politically also, and from the days of the Renaissance when people began to really think and to study nature in terms of science and to apply this to terms of technology to the endeavors of today, has been a very rapid advance, and an advance that gets faster all the time.
----

*President of the International Academy of Astronautics.
Director Emeritus MIT (Charles Stark Draper Laboratories)

Here, we are dealing with the cutting edge of this advance. We are working towards things that humanity has imagined but has never really touched actually until, well, until the last 10 years or so. And this technology of actually going into space, not in legends, but in actual pictures you can see on your own television, has given a lift to the human spirit that has hardly been equalled in all time.

Here we are gathered for the purpose of honoring those people, who, out of all human beings, led the way, and I think that this is a fine, fine thing because it does make a mark that will be permanent and will call attention to the fact that if a mind has capability and produces something that is truly great, that his fellows will recognize it, and I, for one, am going to be very interested in listening to the stories that we will hear of how this occurred for the distinguished people that are going to be inducted into the Space Hall of Fame. I thank you.

GREETINGS FROM FRG

Dr. Heinz Fuchs*

Distinguished guests, ladies and gentlemen. In my capacity as President of German Society of Aviation and Space Medicine, as member Executive Council of German Aerospace Society, and as an honorary member of the Hermann Oberth Society, which of course succeeded the former German Rocket Society, it's my privilege and great pleasure to extend to all those in attendance of the International Space Hall of Fame Conference the most cordial regards and very best wishes of these three German aerospace devoted societies. Even more than that, the German Ambassador to the United States, his Excellency Helmut Matthias, has kindly asked me to convey to all those present his best personal regards for this special event. We Germans are feeling highly honored that among those space pioneers and scientists being inducted into the Space Hall of Fame today, there are ten of our countrymen. They indeed contributed to the conquest of space, to the exploration of its mysteries and to the development of ideas, techniques and technologies. The impact of which in Earth-bound applications, we have only started to comprehend these days.

The Federal Republic of Germany, being very much aware of those pioneers and their devoted successful work, expanded into very remarkable efforts in the arena of the peaceful use of space. It is very well known to you all that my country belongs to those whose expenditures for space applications and technologies are the highest ones all over the world. In a very close cooperation with the United States, we successfully completed a highly sophisticated solar probe as Helius A and B. We also pushed forward the creation of the European Space Agency (ESA). We contribute to more than 50% of the budget and development of the forthcoming Spacelab, and Spacelab, of course, will be the European contribution to the highly ambitious U.S. Space Shuttle Program. I strongly feel that these efforts of our two countries are within the best traditions of all those who internationally pioneered on space exploration since many years ago, and whom we gratefully honored today. Thank you very much.

----

*Surgeon General of the Military Forces of the Federal Republic of Germany; President of the German Association of Biomedicine

CHAPTER II

SPECIAL SESSION

"The Earth is the cradle of mankind, but one does not stay in the cradle forever."
    Konstantin E. Tsiolkovski, 1903
    (U.S.S.R.)

Armstrong sets the flag after landing on the moon   NASA

U.S.A./U.S.S.R. joint Apollo-Soyuz mission, July 1975; standing from left Tom P. Stafford, Commander of U.S. crew; Aleksey Leonov, Commander of U.S.S.R. crew; seated from left, Donald K. Slayton, docking pilot; Vance D. Brand, command module pilot; Valeriy N. Kubasov, engineer-scientist, USSR   NASA

OPENING ADDRESS

Dr. H. Guyford Stever*

We're going to reserve Dr. Gilruth's reply for a moment and hear from two others who will speak - well Stark Draper. I've often found myself in my scientific and engineering career, in the midst of a battle. When I was a young physics graduate and World War II was coming on, I went to MIT to get involved in the radar program and we had a very successful radar program there and in this country. The associates that I've built up over those many years were electronics oriented to the nth degree. Right after the war, I discovered myself as an Assistant Professor in the Department of then Aeronautics which later became the Department of Aeronautics and Astronautics. A little while after that, one of the leading professors was Dr. Draper and he later became head and my boss. He felt that some things could be done with some non-electronic devices in the field of guidance of missiles and for part of my life I discovered that I was caught amongst my old friends and my new friends trying to explain each side to the other.

Well, all I have to say is that Dr. Draper won out but when he was explaining the intricacies of inertial guidance, etc. to me, he always talked about some of the people in his Instrumentation Laboratory in the laboratory -- an unbelievable laboratory in its accomplishments. I didn't know all of them but there was one fellow he was always talking about. Now, Davey Hoag can tell you such, and such, and such. Well I'm delighted - we're now going to hear about Dr. Draper and Davey.

Thank you Davey and will you thank Dr. Duffy for that and I'm sure that you could have done -- well don't tell him this -- I'm sure you could have done exactly as well yourself.

One way that you know that you're getting old is when you begin to put all of your old connections and I have to tell you that amongst those inducted in the Space Hall of Fame today, I had many friends and I met them under some unusual circumstances and this story does relate eventually to Dr. Draper and also to Dr. Steinhoff who will speak about Dr. Draper when I'm finished. One of them was, I was in radar but as you know the United States and the British did not have large scale rockets in World War II. I was assigned as Dr. Bush's man in radar in the London Embassy. Since we didn't have experts in guided missiles, when one problem came up, I was always brought in to those problems and I'm going to tell a story which I don't think Ernst knows about but, one time a scurry occurred in London and I went to a meeting and a young acquaintance of mine, Flight Lt. Wilkinson, I recall the name, had just flown in a mosquito bomber to Sweden to bring back the parts of a test V-2 which somehow had been misguided and had landed in a farmer's field in Sweden, and the British were notified, and came and picked it up. It
----

*Science and Technology Adviser to the President and Director, Office of Science and Technology Policy.

was reconstructed like a dinosaur is reconstructed in a museum, you know, with little bits and pieces hanging on strings from the ceiling so that the parts would approximately fit, and you had to imagine most of the parts. By the way, Ernst, we reconstructed it very well. I hope in your discussion in a few minutes, you will tell me what happened to that guidance system that got it off and landed it in Sweden instead of the Baltic. That is a story I've never known the answer to in these years. But that does relate Steinhoff to Draper in guidance. But my relationship with Steinhoff came after the war, and in fact, again because we didn't have missile guidance experts -- the United States did not have guidance missile experts in London -- now a few of them were coming over like Frank Malina and others, but Dr. van Karman took a mission as the war was ending to a place called Rechlin, Meklenburg -- is that right? -- where we discovered that the wings of the B-47 should be swept back and other great pieces of aeronautics and astronautics. On that same trip, Dr. Van Karman went with Dr. Dryden -- both of them inductees -- to Goettingen where he had once been to meet with his old master, Dr. Prandel. I was present at Dr. Prandel's home at that meeting, and I began to get acquainted with some of the German scientists who had been in aeronautics, which was my later field, and astronautics and also on that mission, some first steps were taken to make sure that some of the German rocket group did come to the United States where I grew to know them very, very well over the years. So that was kind of the beginning of my acquaintance with Dr. Steinhoff, who will now speak about Dr. Draper, and, I hope will also answer that question that has mystified me for these years.

ADDRESS TO THE SPECIAL SESSION

Dr. Robert R. Gilruth*

Ladies and gentlemen, this has been a pretty wonderful day. I want to thank Max Faget for his very kind remarks about my career. They mean a great deal to me coming from him, who himself has done so much to make the program successful. It seems at this time it is best for me simply to make a few general remarks about the space program and the things that I feel have been important to making it successful. I can't help but think about President Kennedy's definition of the task of Apollo. He said, "Let's fly Americans to the Moon, land them, and bring them back safely." He went on to say, "in this decade." That definition of the task was so clearly stated that you knew exactly what you had to do. First, it eliminated the controversy of should you fly an instrument instead of a man. You could have one man, two men, or ten, but it had to be manned.

Flying to the Moon was so difficult in that it required that we not only build and design new hardware, but we needed new concepts as well. So while our competitors had been beating us to the punch almost every other week, it seemed this was so great a task that none of their hardware was <u>adequate to do it either</u>. And so we both had to sort of start from scratch and it took us off the hook for awhile.

The President also said "in this decade" which meant that the pace of the program was defined. This was really a big help--a much bigger help than we thought at the time. I think that was a key factor in the success of the program.

The other factor was that the people of this country were solidly behind us. I don't know how many of you will remember it, but the vote of Congress for the Moon program was virtually unanimous.

Can you imagine anything like that today? While we didn't have any real hard plan of just how we'd go to the Moon, we did have a lot of very good people in this country. Not only were there good people in government and the universities, but we had a very strong aerospace industry. We also had the full cooperation of the military services. Mr. Kennedy said, "Let's go to the Moon" in a speech given in May. One year later we had the whole plan put together. We had decided how the rockets and the spacecraft would be designed. The Saturn V consisted of a lower or basement stage as Wernher would call it, with five huge kerosene/oxygen engines, and the second stage with the 5 hydrogen oxygen engines and the upper stage with one hydrogen oxygen engine. We had decided on the Command Module, the Service Module, and the Lunar Module. This was all done in one year, and when we started that year, we were on opposite ends of the spectrum. The Huntsville Group wanted to use Earth-orbit rendezvous or possibly direct ascent. Those of us at Houston weren't quite sure what we wanted at the very start, but in about one year we had zeroed in on the lunar orbit mode. And further more, we had come to agreement on this mode throughout the whole program. We had gotten to know each other well enough so that we could work together effectively. We were also lucky that the plan of using the Lunar Module worked out so well with the exisiting Command and Service module designs. You have to have good luck as well as good engineering, I believe, to make such complex systems as these work effectively. We were lucky that one Saturn V was powerful enough to take a whole spacecraft to the Moon--that we didn't need two of them and we didn't have to rendezvous with them. We were lucky that the rendezvous in lunar orbit proved as straightforward as it was and, here again, we were lucky that we had the Gemini program earlier and had a chance to work out all different ways of rendezvousing and to train our flight and ground crews. We had assembled a team of very well trained men by the time Gemini was over and Apollo was ready to begin. We also had a chance for Dr. Berry, our Head of Medical Operations, and the astronauts to learn to work together. Also the Mission Control Center was proven and shaken down in the Gemini program and was ready when the Apollo missions started.

We were fortunate that the rockets worked as well as they did. The Saturn V was a tremendous accomplishment, and it had relatively few problems. The only one I can think of was the Pogo and that was solved by some brilliant engineering on the part of the Marshall people. They were smart enough to keep something in their back pocket in the way of extra weight. They figured we'd never bring in the Lunar Module and the Command and Service Module for the weight that we said we would and that extra margin allowed us to put the Rover onboard so that the astronauts didn't have to plod around the Moon on foot. They could drive an automobile in the American tradition. We were fortunate that the blunt body solution to the reentry problem worked so well even in returning from the Moon. There never was a qualm about the way the heat shield and the afterbody and the controlled gliding reentry brought the astronauts to a pinpoit landing by the recovery carrier. We were fortunate that weightlessness was not a debilitating thing because at the time when Mr. Kennedy said "Let's fly man to the Moon," we had only fifteen minutes of flight time in space and we knew nothing about a mission where you'd fly for two weeks, have to back down on the Moon and takeoff and rendezvous and come back to Earth. And we were lucky that weightlessness really wasn't a problem.

We couldn't have gone to the Moon without hydrogen, titanium, high speed computing, or inertial guidance that has been pointed out here many times tonight.

We were lucky that our training plans were so sound--that the men had all really experienced all the different modes of flight, rendezvous, landing site, everything in their trainers before they got into the Apollo spacecraft. It wouldn't have been possible if the trainers hadn't been so good.

All in all, it was a great experience. I was very fortunate to work with so many wonderful people, and I think we were all lucky to have this great opportunity. Thank you.

ON THE EVOLUTION OF ACCURATE INERTIAL GUIDANCE INSTRUMENTS

Dr. Charles Stark Draper*

Thank you, I'm a bit overwhelmed by the many kind words that have come my way today. My remarks will be devoted to an overall glance at some of the events that occurred along a path starting with conventional aircraft instruments during the early 1930's, to guidance for the final water landing of the Apollo Command Module at the end of the last Lunar flight of the 1970's. The functions required from the system involved were quite different from those provided by equipment available before the Apollo project was conceived, so that new arrangements had to be brought into operational existence.

Creative technology was definitely necessary and the Apollo situations were faced by the difficulties and possibilities that affect the course of all attempts to carry any developments into regions involving unmapped areas of activity. Before any new results can be realized, creativity, emphasised by scientific and engineering imagination, assisted by understanding management and made possible by sympathetic administration, supported by adequate funding, must be present and cooperative. Well educated, capable and strongly motivated engineers in adequate numbers are always essential for effectively realizing revolutionary developments of technology.

My own positions at the Massachusetts Institute of Technology starting at the beginning of the 1930's as a faculty member of the Department of Aeronautical Engineering and Director of the group that became the Instrumentation Laboratory and later the Charles Stark Draper Laboratory, Inc., gave me effectively understanding management and administration for some forty years, while funding from various government agencies in return for contracturally specified technological results provided money and generally understanding management.

My very fortunate starting position as Professor in charge of instrument courses made it possible for me to venture into fields that had never before been given significant attention, led me into control, then into navigation, and finally by practical combinations of control and navigation, into guidance. Very strong motivation for this sequence of developments was put into my teaching and research efforts by some perilously near catastrophic adventures that I, as a private pilot for aircraft, experienced during the late 1920's and early 1930's. Under conditions of zero visibility with solid earth known to be near, lack of information on aircraft location is likely to be a fatal circumstance. Radiation aids, radio and radar in regions with proper ground installations have in later years reduced "no visibility" difficulties to a considerable extent, but in the 1930's and still in the 1970's for many practical situations, the realization of self-contained, no radiation-contact-required navigation equipment remains of great importance. During the 1930's and the 1940's self-contained systems for navigation seemed very important to me. Because of this I was strongly motivated to slant my courses toward this subject. I developed the physical theory, the geometrical relationships involved, and technological patterns for designing and bringing self-contained navigation systems to realization.

----

*President of the International Academy of Astronautics
 Director Emeritus MIT (Charles Stark Draper Laboratories)

Because the arrangements that seemed to be practical, depended on the gyroscopic principle of the spinning gyro rotor changing the direction of its spin axis with respect to inertial space toward the axis of an applied torque, while the gyro itself depends on Newton's laws of inertia, the devices that attracted attention were called inertial systems. As I worked to complete, during 1938, the requirements for my Doctor's Degree in physics with a minor in mathematics, while I developed and taught the theory and engineering of self-contained systems for navigation, I was able to attract a number of very able thesis students who carried out tests of experimental instruments in my personal airplane while I served as pilot. As a result of all these influences my motivation toward realization of the development of practical self-contained systems for control, navigation and guidance became stronger and stronger.

In the years of the 1930 decade no funding was available for inertial system developments. However, I did have complete freedom for composing and presenting course material in any way that matched my desires. I was able to find metal scraps and other not-desired-by-anybody-else items that made it possible to conceive, design and build thesis arrangements and special project devices. A most favorable circumstance during the 1930's and 1940's was that my classes included a considerable number of able and interested students who took pertinent academic subjects and who were strongly motivated to carry out graduate thesis on materials covered by my lectures. In addition, many students found it worthwhile to accept part time and later full time working positions in the Laboratory. This situation came about from personally developed convictions that a few years experience in actually carrying out successful work in pioneering technology provided very valuable continuing engineering education beyond the requirements associated with academic degrees.

Two gentlemen who are with us in this room have been associated with the education and practical output patterns of the Laboratory. General Leighton I. Davis, a former Commander of Holloman Air Force Base, is certainly high on the list of distinguished former students and collaborators with the Laboratory. General Davis was a student first concerned with aircraft engine instruments and later was a strong contributor to gunsights for machine guns and fire control instruments for fixed gun fighters. David Hoag who is also present here, was a student during the 1940's, has remained with the Laboratory, and has been in charge of several large and important projects.

As commanding officer of the Armanent Laboratory at Wright Field, General Davis certainly provided the Laboratory with creative, understanding and inspiring Administration and Funding for the Laboratory during and after the years of World War II.

During one of the autumn months in 1945 Lee Davis and I were returning to Wright Field from Flint, Michigan where we had looked over some General Motors manufactured gunsights that the Laboratory, with Lee's direction and collaboration, had developed for an Air Force Fighter Group he was soon to lead into the Pacific. As we flew in the C45 airplane from Flint to Dayton, the radio informed us that Japan had surrendered. As we discussed this momentous event, it appeared that some funds that were assigned to purchase bomb shackles might not have to be used for this purpose. I suggested that a most important and worthy use for the funds that might become available, would be to initiate and support an inertial bombing system development. Although the basic function involved certainly had to be navigation, this area of endeavour at Wright Field was assigned, not under Lee Davis's responsibility but to another organization. So as we discussed the matter it was judged that working arrangements would be most favorable if a project was directed toward the objective of bombing without radiation contacts. If this could be arranged, the requirements could be written by and the funding provided from the Armament Laboratory. Because of good past performances of the Armament Laboratory-Instrumentation Laboratory combination in various developments we judged that good results could be expected in the future from its continuation.

The Laboratory had a significant record of achievements for applications of inertial principles to important operational problems. During World War II after naval vessels, notably the Repulse and the Prince of Wales, had been sunk by multiple attacks of aircraft, the Laboratory had constructed experimental devices, had made prototype production designs and had supervised production of Gunsights Mark XIV and XV for the United States Navy. By using single-degree-of-freedom gyro units as the sensing elements for angular velocity, and generating indications by effective computations carried out with respect to inertial space, fire control was made rapid enough to deal with close-in flying aircraft, so light weight that it could be mounted on existing machine guns and other available weapons and so cheap that it was practical to provide all the guns of various sizes on ships with fire control. In practice, the Naval Gunsights were so effective that the balance of power in the Pacific shifted from multiple aircraft attacks to defense by anti-aircraft guns.

The Laboratory experience included the creation of Gunsights XIV and XV, units that were produced by the thousands during World War II, and several manually powered Directors Mark 51, Mark 52 and Mark 63 for Naval anti-aircraft guns. After this development was carried through for the X-1, a fire control system capable of generating indications throughout all of the upper hemisphere. Because of the lack of guns capable of covering the required angular ranges of operation and the decision of the United States Navy to substantially abandon guns for anti-aircraft purposes, the X-1 project was never carried beyond the stage of an engineering model mounted on a rolling platform to track and generate fire control solutions against close-aboard flying jet aircraft. During the latter part of the 1940's anti-aircraft work in the Laboratory was substantially dead. Dave Hoag, a very able individual who read General Bob Duffy's paper on Stark Draper to you, was strongly involved in the Navy anti-aircraft work. The demise of the X-1 project during the middle 1940's was something of an embarrassment to Dave Hoag who had been a leader in Naval fire control for some years. His background in this field is important to remember as, in later years, he became leader for projects of primary significance to the Laboratory.

Lee Davis and I landed at Wright Field on VJ Day and in due time with the assistance of Dr. John Clemens, the Instrumentation Laboratory received a contract to start development of a self-contained bombing system with, as I remember it, not more than two miles error after ten hours of flight. The funds involved were less than one million dollars and we realized that we had to have something significant in the way of results during a period of not more than two years. For the first system, assistance from solar and stellar observations were permitted as a means of evaluating the inertial phases of operation.

With the contractural arrangements in a form that allowed work to start, Dr. John Hutzenlaub, Dr. Don Atwood and Roger Woodbury organized the inertial guidance group and started work on the FEBE System. It was obvious that technical performance in almost every area of involved technology had to be improved, not moderately, but by substantial factors. For example, gyro elements commonly used in the orientation-indicating aircraft instruments that had been developed by Elmer Sperry Jr. and Jimmy Doolittle as the Artificial Horizon and Directional Gyro had drift rate specifications of about one earth's rate. With one minute of arc between local verticals equivalent to about one nautical mile of earth surface distance, and earth's rate of 15 degrees per hour, it was obvious that one mile error buildup in ten hours of flight correspond to one ten thousandth of earth's rate. This meant that for inertial system performance to approximate specified goals, gyro behavior had to be improved by about four orders of magnitude. It also appeared that accelerometer performance had to be made better by about the same factor if operational goals were to be satisfied. Some components such as synchros approximated the necessary behavior but had been designed for service on ships so that both weight and size were excessive for aircraft. Amplifiers and other electronic components were large and heavy but worked fairly well.

It was realized that the fundamentally important matter was to give Lee Davis performance from a principle-demonstrating inertial system that would both justify and stimulate further developments of inertial technology. The first system was given the name of FEBE, a contraction of PHOEBUS, who as god of the sun might be called on to help in flight operations by means of a photocell sun tracker that was incorporated in the system.

FEBE began to work in time for engineering tests to be made during 1948. The flights were between Boston and Wright Field. Results could not be regarded as excellent but their quality made it certain that navigation systems based completely on inertial principles could be regarded as technologically feasible. After FEBE, the Laboratory went on to design and test the completely inertial SPIRE and SPIRE JR systems with improved performance over trips, during the 1950's, between Boston and Holloman Air Force Base and between Boston and Los Angeles, with some flights longer than ten hours. The Laboratory had demonstrated the feasibility of inertial navigation and passed the knowledge generated along to manufacturers designated by the sponsors

During 1948, results from two areas of technology taken together suggested a pattern of activity that has proved to have great significance for the Laboratory. The decision of the Navy to abandon fire control developments and the demonstration of practical usefulness for inertial systems suggested that the group concerned with fire control applied its efforts toward the development of inertial navigation for ships and submarines. Dave Hoag was a leader in the first steps that applied the now generally known and accepted theory to a combination of gyro units constructed for the Air Force FEBE to produce the MAST system which was, in effect, the marine gyro compass and the marine stable vertical put together in a single unit. Overall performance was better for both heading and the vertical than that commonly achieved from the separate instruments. Tests at sea showed that MAST was particularly effective for high seas in small craft.

With encouraging results from the MAST experiments it was an obvious step to follow the pattern of SPIRE and add navigational indications to the directional and vertical outputs of the MAST system. The result was SINS, the Submarine Inertial Navigation System that was put together as a working entity and tested during trips in a truck-pulled trailer traveling over the eastern seaboard of the United States. These landborne journeys were supplemented by ocean trips up and down the east coast of the United States and across the Atlantic Ocean. Results were generally good enough to be useful although certainly not of ultimate quality. The SINS system, in its trailer along with a final report, was delivered to the Navy in 1955.

Conferences with officers and civilians during the 1952-1954 period indicated that the commonly accepted opinion of the Navy, held that inertial navigation could not be useful for its purposes. The Laboratory was "dead in the water" as far as inertial navigation work for the Navy was concerned. Because of this situation I was surprised when in mid-1955, I received an invitation from Admiral Bergen, who had been the officer in charge of our X-1 fire control work, to attend a meeting in Washington. A long table was surrounded by a very distinguished company of admirals, other officers and civilians. I was told that they were thinking of a fleet ballistic missile. It was concluded that guns were no longer useful and that a new kind of vessel would be required. These were to be submarines carrying ballistic missiles. The meeting wished to know if I thought that the submarine could be guided well enough under the water, so that a missile could be shot from 100 feet below the surface. They also wished to know if something could be put in the missile to give it accurate guidance. In addition, they asked me if I thought that I could make the guidance equipment. I answered, "Yeah, sure we can do it." The next question was 'how do you know it can be done?' I answered, "You have Instrumentation Laboratory Report Number so and so..." I actually knew the number at the time and also that the report was in Bureau of Ship files. There was more discussion at the meeting but the essential matter was that the Laboratory received contracts to work on an Inertial Navigation System, and to design, build, test, and generate manufacturing information for the POLARIS guidance system. Dave Hoag carried great responsibility in this development, and certainly deserves the highest possible praise for his efforts and their results. He is a most extraordinary fellow.

Many events have occurred in my life and in progress for the Laboratory. In a talk already too long I will pass over with a bare mention of our involvement with inertial guidance for Ballistic Missiles of the Air Force and for Air Force Cruise Missiles. Because it has a very direct association with matters discussed in the meetings of this International Space Hall of Fame Conference, I will tell briefly of another highly improbable sequence of events that led to important consequences for the Laboratory.

Milton Tragaser who had written a report on technology for unmanned trips to Mars, and I had, during the last years of the 1950 decade, been discussing possible trips into space and technology for feasible guidance arrangements with friends in NASA. Much discussion had occurred among people holding many different opinions, but no definite and approved plans had emerged when I was invited to Washington by Mr. Jim Webb, the Administrator of NASA. Milt and I attended a meeting in Mr. Webb's office that included Hugh Dryden and Bob Seamans. These people got me in a room, invited me to sit down and told me how they were going to make manned flights to the moon and return. That is, go to Earth Orbit, make a trip out to Moon Orbit, leave a vehicle in this orbit and make a trip to the moon surface in a landing module and after exploration activities return to lunar orbit, with reunion to the orbiting vehicle followed by return to earth and splashdown in the ocean. They asked me if I thought I could provide the necessary control, navigation and guidance for the command module and the lunar landing module. I answered, "Yeah, I can do it." They asked me, "How long will it take?" I said, "You'll have it before you need it.".

Of course the thing I had in mind for Apollo, as a first model anyway, was the equipment we had already used for guidance of the POLARIS missile. I personally would have liked to have seen it redesigned. If I could have achieved this, NASA would have had something of smaller size with better performance. Anyway I tried to get approval for some redesign without any luck. They were skeptical about operation, so I told them I would go along and run the equipment. Whether this offer had any effect I do not know, but history says that NASA gave the Control, Navigation and Guidance jobs for the Command Module and the Lunar Landing Module to the Laboratory.

Guess who carried full responsibility for the project, it was Dave Hoag over there. So Davey, you have to take full credit for many great things that have been done. Lee Davis started the whole process of inertial navigation development, and Davey Hoag carried on until final successful splashdowns in the water.

Ernst Steinhoff here had very much to do with my life starting long before I came to know him in person. One day during World War II, John Clemens, who worked for Lee Davis at Wright Field, came to see me with an instrument that was about a double hand full in size, that included ball bearings and gyro rotors among other things. If you got the wheel spinning and you pushed it in one direction it would go around and around an axis at right angles. You see, it was a gyro with an unbalanced wheel, so it would be an integrating accelerometer. I didn't know who had designed and made it. Later I got suspicious as to who had done it. But anyway, from the beginning I thought it was a hell of a good idea. I cleaned the device up a little bit and as I knew that ball bearings were not exactly the most perfect of all of man's mechanical devices, there were a few changes. I remembered that in connection with work on gunsights for ships, we had been responsible for using some 3 million ball bearings that had gone into gyro wheels. We didn't have data on all of them, but we had information enough to be very sure of facts in talking about friction and ball bearings. If you got ball bearing friction down to a few tenths of dynecentimeters, you were doing real good. And any fool could figure out that with the size of rotor you could use, you had to have the friction level down to about 10,000 times less than that of ball bearings used directly. So I made a little gyro with ball bearings or gas bearings in the rotors and I floated the damn things in heavy and viscous liquid, and I used servos to drive all moving parts. With these design principles applied, friction troubles were reduced to levels that permitted very high quality performance. As a matter of fact, you can find companies that are doing respectable business quantities by building devices incorporating small servos to eliminate friction effects. Some time after I first saw a pendulous gyro accelerometer, I found out that Ernst had much to do with their design and applications. So in the end it appears that I stole ideas from Ernst even before I knew him. He is a great man and a friend for whom I have much affection.

Thank you very much.

COMMENTS ON DR. STEVERS INTRODUCTION

Dr. Ernst A. Steinhoff*

I sure knew that Guy would sneak up something on me! I think I'm prepared to give the answer -- it was a first! It was a first in several respects, and I think I will give you little bits of the story. I've not written it up here. We had a project at Peenemuende which was a kind of a side issue, -- a side kick of the V-2, and its objective was to find out how one could use the "Wasserfall - Missile" as anti-aircraft missile, and possibly a ballistic missile defense system, eventually in the long-run.

I had one of my associates instructed to properly plan a demonstration that one could do it, and all our instrumentation could follow, and demonstrate its capability. For this purpose we used a standard A-4 missile (V-2) with its complete radio control system and modified the lateral displacement control such that line of sight angles from the target, and line of sight angles from the missile could be measured independently by two optical cinetheodolites and used to generate an error signal and its derivative, and the integral of displacement error such that the lateral control system would reduce the displacement error towards zero and also eliminated drift and wind errors by keeping the integral of the displacement error zero (isodrome). In this test, precomputed flight path data were used to introduce a schedule of error signals, which would by tracking the missile used to track down a fictive target. The project engineer had to memorize this control pattern such that the A-4 would fly an s-curve in the lateral plane such that its impact would be outside Swedish or Finnish territorial waters. In order to be able to follow the missile response to these artificial signals, I memorized the signals and their duration myself.

In the prelaunch tests of the control system, every item of the used equipment checked out well. We decided to go on with the count-down. The missile lifted off and turned into the direction it was to fly and gained speed. After passing the transsonic speed regime, my colleague started applying the memorized signals with the proper durations, we had worked out in advance. The missile response was as expected and well damped, as the original A-4 flight control system would do it. Watching the Azimuth angle I noticed that the missile was drifting farther north than the Azimuth schedule was calling for, and when a turn to the right should have come, it turned even farther north, with the error signals despite of the deviation again going towards zero. Then, when a return to left turn signal should be applied the missile started turning right at a time it should have established a course roughly parallel to the coastline between Danzig and St. Petersburg. This clearly indicated that while the signal duration was as planned and also signal variations followed the proper time sequence, their sign was reversed and pointed the missile in the direction towards Karlskrona, Southeast Sweden. No control malfunction could by accident produce the pattern with a negative sign of the signal direction, unless it was in response to the signal inputs. Since the missile had reached by that time its burn-out time and was no longer propelled, I asked my colleague, one of my most outstanding theoretical staff members, whether he had inadvertently given

---

*President, New Mexico Research Institute

the signals with the wrong sign, since the missile had appeared to be heading towards a point some 40 to 50 kilometers left of the intended flight path, and everytime when we had planned to have it turn right, it turned in a more left heading, so that there was a good chance that the impact could be on land in Sweden or in Swedish territorial waters. Each case could have diplomatic repercussions. He was very surprised by my statement, and tried to recollect his signal motions again, and admitted that it could have happened. We agreed that an immediate analysis of the tracking and doppler data be performed to insure that my observations were correct and necessary steps would be taken to prevent any diplomatic involvement of our secret work at Peenemuende. One of our crucial cine-theodolite stations was located at an island 15 km north of Peenemuende. In order to obtain the records of this station, I flew with a STOL aircraft, a 5-seater Fieseler Storch, to this island to review the tracking records, used to back up our electronic data when I was met by the military Commander of the Peenemuende Army Rocket Research Center on the airstrip at my arrival, asking me "Ernst, what happened?" There is a lot of commotion going on in Berlin. The supreme headquarters called whether we fired an A-4 into Sweden? I replied that we did not know that exactly. We would have the possibility of an impact on land or on sea I would think that the outmost position of an impact would be a location about 10 miles south of Karlskrona, but I would not say right now anything before we have the tracking results of our local station. We could otherwise give an estimate which could be incorrect and could later be proven to not be correct. I have already made all necessary arrangements to have the data analysis completed. My staff completed the analysis, and it turned out that in fact the missile had impacted on Swedish territory. If the colleague of mine had followed the instructions to memorize the command signals, but inadvertently in his excitation gave the inverse values of the precomputed signals, then there was just the possibility of the impact within a square of about five miles initial uncertainty in our quick-look data. This turned out to be the case. We could answer the question of the most probable impact location within about an other hour so that our government did know how to reply to inquiries from Sweden.

While Dr. Draper, in World War II, was working on one side of the international conflict, I was working on the other side for my home country. My task was to provide the V-2, or as we called it, the A-4, with guidance and control, some far reaching dreams for future spaceflight, but these were inactive during the war. However, we had a few items on the drawing board which were of great interest to us as future tools to demonstrate spaceflight, which history eventually proved to have been right. The Saturn-Apollo Project was the demonstration of the feasibility of our dreams. In the office of Ed Buckbee, the Director of the Huntsville Space Museum, is a picture of the Apollo II team with three dedications, if my memory serves me correct, saying:

"To Werner, the Initiator, Technocrat, Scientist, Pusher and Colleague, Neil Armstrong,

To Wernher with admiration and respect, Michael Collins,

To Wernher, whose dreams materialized this journey, Buzz Aldrin".

One of this trio was honored today, and I am sure at this point several of our astronauts as time progresses besides Neil Armstrong will join those who have been honored today. Before I change the subject, I would like to state with Neil Armstrong, his comment after setting his foot as the first human being on the Moon, that his merit in the great achievement of a millenial dream of humanity was possible only by the labors, efforts, and devotion to these tasks of dedicated people the

number of whom exceeds a large portion beyond 100,000 and Wernher in all
of his life has given more credit to the many people, on whose effort
and support he could "count on"!

Before I close my reply to Dr. Stever (I have in this case edited
and expanded my remark in response to his question to make it less impulsive and in greater detail than the time available to me on the program
would have permitted), I would like to make a few observations concerning our next honoree, I am introducing in the next few minutes.

Before 1946, I never had heard of Dr. Charles Stark Draper. At the
time I was part of the German Paperclip Scientists at Fort Bliss, who by
a very few of our American colleagues were considered spoils of the war,
while working on advanced concepts of spaceflight for our American Government, our work also was subject to regular reviews. In 1946, such a
review took place, and among the members of the U.S. National Academy
of Scinces and of the National Research Council, of a body of distinguished scientists, I now have several very close friends. I have had in
the past the honor to work with or for them, a friendship which is based
on mutual respect and admiration. With the team was a bespectacled distinguished looking member, who we quickly discovered spoke our language.
Not German, as could be interpreted, but what Theodore von Karman termed,
when asked what the scientific language of the time is, a question, he
answered with "Bad English", but what he meant is the full understanding
of the professional terms used in particular disciplines at discussion.
In this case, inertial guidance was the subject of review. It was Dr.
Charles Stark Draper, at that time a full professor at the MIT Aeronautics
Department, and on which Faculty I happened to be a full professor of
Astronautics during the academic year 1969-70, invited by two distinguished scientists of the MIT Faculty, who spent some time with us this
week. One of these is Dr. Draper, the other is Ray Bisplinghoff, now
Chancellor of the University of Missouri, at Rolla. It was not my first
invitation to teach there. In 1956 I received a similar invitation, but
I had to turn it down because I could not see, with seven children in all
levels of education, high and post secondary, how I could take my children out of school and move the whole family for nine months. I promised
at that time, however, that I would accept such an invitation in time,
the youngest of my children (I have two native Americans among these)
could be trusted to be left alone. This occasion came in late 1966,
when the Air Force decided that I should go on a sabbatical from the
Air Force and study any subject or problem I would consider worthwhile,
anywhere I wanted to, if necessary at a foreign University. Ray Bisplinghoff, and with him Stark Draper, as his friends call him, renewed their
invitation. At that time, I was on the Staff of the Air Force Systems
Command at Holloman, and on its Technical Management Council, and the
first one, who was accorded this honor, so far only given to outstanding
scientists of the A.F. Office of Scientific Research at Air Force
Headquarters. This new case needed policies and other pre requisites
before the administrative "red tape" can catch up with this innovation.

I replied to the invitation, that as much as I wanted to come right
away but could not be legally incorporated in the Fall Session of 1967, but
if I had to accept before 1 April 1967, I would caution them to expect me
to accept before 1 April 1968. To make the 1968-69 academic schedule, because by committing myself to that close of a deadline, I had expected that
one year later the Air Force Channels would have my orders out. They did,
but it was close. This period was for my family and me one of our happiest
events after my graduation. I left MIT after nine months with many friends,
among these students, and faculty, becoming the honorary lifetime president
of the MIT Soaring Association, and having among my 14 MIT students,
regularly attending my classes, three who received financial awards for
outstanding scholarship of ten, out of, if I remember right, 178 candidates of the Aero-Astro Department. Dr. Draper, who is not an average
teacher or educator, he is, by his friends considered to have disciples,

a point that Davey Hoag just convincingly made, without however saying it. I admire and fully back Stark Draper without a second thought. That day at Fort Bliss, where the Review Committee met, was the Day when I met Stark Draper, who could present in English our scientific terms according to which we solved guidance and navigation problems of the V-2 which till today in their fundamentals govern the theory behind the Saturn 5 and Apollo Flight control and navigation systems, and are for the first time in English presented in an Interrogation Report of the Dryden-Von Karman US Team, which eventually selected those German scientists, found worthy to be invited to continue their work in the United States, a Program initiated by General Eisenhower in Spring 1945, and code named "Paperclip". We were never prisoners of war, but felt that we earned our respect which was accorded to most of us since then, by the mutual respect, the larger majority of American Scientists in our fields brought towards us. This made it easy for us to become integrated into the American Nation, whom all of us owe very much. Now, with this of my conviction, I will augment Davey Hoag's story of Stark Draper, which he is not fully familiar with. Dr. Draper has not "only" disciples, he is a sage and a saint in the History of Science and Aerospace Technology. Dr. Charles Stark Draper, President of the International Academy of Astronautics, and President emeritus of the old MIT Instrumentation Laboratory, his life's real achievement; Stark.

With this I have fully answered Guy Stever's question, under whose control a V-2 impacted in Sweden, and just about a dozen years later another misguided missile impacted at a graveyard at Juarez, Mexico. In a few minutes after the second impact, enterprising citizens of Juarez had erected hamburger stands, and sold about twice the empty weight of the missile in souvenirs, made up from tin can metal. This is the first recorded economic benefit of a projectile re-entering from outer space (Ref). I can present my contribution to Dr. Draper's biography now.

Reference:

Bob Ward, "A funny thing happened on the way to the Moon" Facwett Publications Inc., Greenwich, Conn., 1969, page 47.

Dr. Kaplan   Dr. Steinhoff   Dr. Diederiks-Verschoor

THE LIFE AND ACHIEVEMENTS OF NEIL ARMSTRONG

Dr. Harrison Schmitt*

Thank you very much, Guy, Dr. Gilruth and Dr. Draper. I'm glad you set the theme of being among friends, because I certainly think that we are. There are so many familiar faces, and so many memories that come with those faces, that it's almost impossible not to recall a few of them such as the courses in the primary guidance and navigation for Apollo 15 and 17. Then, Bob Gilruth and I here one time sitting down and discussing (he may even have forgotten this) about whether or not the last mission to the Moon should go to the far side of the moon. As he said today, when I introduced him as the man who put me on the Moon, "No, I'm the man that got you back." That's certainly, although others will talk about Bob, I think, that will put his attitude toward our lives in perspective, and he certainly was a great friend to all of us.

I'd first like to tell you about a place that I've seen in the solar system and try to relate that to the man, Neil Armstrong, who we honor today. This valley on the Moon that became known as a valley of "Taurus-Littrow" was truly a magnificent place, and it was a name that wasn't chosen with poetry in mind, but I know in the minds of Gene Cernan and myself, there is a poetic quality to it. It is magnificent and stands with many of the valleys in the solar system, certainly comparable to the fjords of our North countries, the rocky valleys of the Rockies in the West, and even now more intriguing, the rifts of Mars. The bright slopes that rise from the pitted plains of Taurus-Littrow are tracked like snow, and rise to the rocky tops 7,000 ft. above.

The cratered floor contains the pages of history - a history of the sun, and in fact, the valley itself, records the history of our Earth through billions and billions of years in time. Other men have gone to other places at other times on this planet and now on the Moon. A return to the Moon is primarily the concern of men alive today. The man, Neil Armstrong, is very much concerned about Man and always has been. One of my most enlightening experiences about Neil was in the conduct of a conference that a few of us attended, that I organized at the California Institute of Technology, called the First Fairchild Conference on the Exploration of the Moon and Planets. We, Bob Gilruth was there and others, sat around for two days, and had what I think was the most stimulating discussion of my life, where five years, more or less, after the first lunar landing, we were asking ourselves, "What did we do? How well did we do it, and what would we do next time if we were called upon for similar endeavors?" Maybe more importantly as the second day drew to a close, we came to a very general conclusion, and I don't think there was a man among us who did not believe as Neil, and Mike Collins, Dr. Gilruth and many others said, that the essence of what has happened is that civilization is moving into space. Whether it's clearly there yet or not, it none-the-less is moving. Neil expressed that in his way, and he also, better than any of us, expressed what it took mankind and Americans in general, to achieve these great things. I think all of us sat in some awe when we heard the man that we had not listened to much since his flight, talk in these very philosophical terms. I think that we today have inducted more than just the first astronaut on the Moon, or a symbolic astronaut, if you will. We've inducted a man with very great insight into why Man does things and what Man believes.

----

*United States Senator of New Mexico.

As a matter of fact, our movement into space is such that I think we can predict with some confidence, although not in time, but certainly with confidence, that alive today on this planet Earth, there are the parents of the first Martians. The technical capability basically is understood and most of it exists. Some of the children you saw today, at the Hall of Fame dedication, will actually become the parents of the first Martians. It's very exciting, and I find as I talk to children around this state and around the country, that the idea is exciting to them. Those of you who are parents, and those of you who are teachers know that if you communicate with children, you are communicating with the future. It's about the only opportunity you have to do so.

Neil Armstrong was born in a place in Ohio that it took all of us about 2 or 3 years to learn to pronounce, Wapakoneta, Ohio, and I'm not sure I still pronounce it right, on August 5, 1930. As you heard today, just very briefly reviewing again, he received a Bachelor of Science Degree in Aeronautical Engineering from Purdue University, and a Master of Science Degree in Aerospace Engineering from the University of Southern California. He was an outstanding Naval aviator during the Korean conflict having flown 78 combat missions, then went back into civilian life primarily as an engineer and a test pilot, and as a manager for NASA. As a matter of fact, Guy, he was the first civilian on the Moon. I just happened to be the last.

Neil spent 17 years in NASA. I'm sure Neil would call them very exciting years, certainly very rewarding for the country. His X-15 flights, his Gemini flights, and his Apollo flights all contributed in very, very important ways to the progress of our nation's space effort. He is now, as most of you are aware, a professor of aerospace engineering at the University of Cincinnati. Professor and Mrs. Armstrong, and their two sons, reside near Cincinnati, Ohio. He also is probably, something you are not aware of, he is a very active, and he hopes to become a successful, breeder of beef cattle. He is as much a farmer now as he is a professor, and the idea behind this, as he and I talked, is to try to do something that contributes to the betterment of mankind, here on Earth, through specialized beef cattle.

Professor Armstrong is a Fellow of the Society of Experimental Test Pilots; Fellow of the American Astronautical Society; Honorary Fellow of the American Institute of Aeronautics and Astronautics; Honorary Member of the International Acadamcy of Astronautics of the International Astronautical Federation; U.S. National Delegate to the Advisory Group for Aerospace Research and Development (AGARD); and a member of the Soaring Society of America. He is a director of the Gates Learjet Corporation, the Cincinnati Gas and Electric Company, and the National Center for Resource Recovery.

Professor Armstrong has been decorated by 17 countries. He is the recipient of many special honors, including the Presidential Medal for Freedom; the NASA Exceptional Service Medal; the Explorers Club Medal; the U.S. Military Academy's Sylvanus Thayer Award; the National Geographic Society's Hubbard Gold Medal; the Arthur S. Fleming Award; the National Civil Service League's Career Service Award; the Robert H. Goddard Memorial Trophy; the Asa V. Call Achievement Award; the William J. Donovan Award; the NASA Distinguished Service Medal; the Harmon International Aviation Trophy; the Royal Geographic Society's Gold Medal; the Federation Aeroanutique Internationale' Gold Space Medal; the American Astronautical Society Flight Achievement Award; the Robert J. Collier Trophy; the Kitty Hawk Memorial Award; the Pere Marquette Medal; the General Thomas D. White USAF Space Trophy; the AIAA Astronautics Award; the Octave Chanute Award; and the John J. Montgomery Award.

I worked with Neil on many different occasions. I guess the high points for me and for most of us in the space program, were Apollo 8 and Apollo 11. Apollo 8, strangely enough, is not recognized outside of the space program with quite the importance, I think, it should be. That was, for many of us, the mission. It was the mission that was conceived and flown in four months. It really was something we didn't expect to do, but when they asked us to do, four months later, we had done it, done successfully, done it without any feeling, that I was aware of, that it could not be done. Neil was the backup Commander of Apollo 8 and contributed greatly in terms of his general stable nature of approach to problems, to the success of that mission.

With Apollo II, I personally believe and I said this after the selection and I will still say it, that we could not have found a more qualified and better representative for mankind than Neil to step on the Moon for the first time. For science, it was doubly rewarding, because I do not know a person who worked closely with Neil and with the results of Apollo 11 and the planning of that mission, who does not feel that we sent in Neil, the best observer we had available at the time. Neil would have made an outstanding geologist. There's just no question about it. We did not have a great deal of time to train him, but in the time that we had -- in the time that we had on the Moon -- we received per man-minute more information than we received from any other mission, maybe even including my own, and that's tough to say. All of us wish that he had much more time than he had in order to look around him, to observe and to sample the environment that he found. I know Neil feels the same way.

The importance of the space effort came to light in many, many ways. I think that Neil, and his future activities, will help us more than maybe he has in the past, articulate Man and Man's role in space, and the role of nations and particularly the United States in Space. I'm looking forward to seeing him much more actively involved in this future of ours. I guess part of that is wishful thinking, because I think that many of us feel that Armstrong is the Lindberg of our generation, and we hope that he will become as profound a thinker about the condition of mankind, and where it's headed, as Charles Lindberg was. Many of you may not be as familiar as some of us with Lindberg and his thoughts, but I got to know him coincidentally, at the same time that I really got to know Neil Armstrong - Apollo 8 and Apollo 11.

I think that as we view the accomplishments and the symbolism of Neil Armstrong's induction into the Hall of Fame, we are really viewing the daring, and the curiosity, and the achievements of mankind, and we're viewing again a window into the future in what mankind will do. Historians, when they write of the accomplishments of the Apollo explorations of the Moon, I think, will find that truly unique things were done in science, technology, and in the history of free men on this planet, and also in the overall history of mankind. In particular, I think they will say that there was a unique evolutionary event, that mankind for the first time had the confidence and the ability to evolve into the universe. We took ourselves, members of our species, and put them in orbit and then onto a planet completely foreign from the one on which they evolved, and with that, we took that very first step that meant mankind was, in fact, going to live in the solar system. As Neil said, "It was one small step for a man and a giant leap for mankind." Thank you.

JOHN HERSCHEL GLENN, JR
A BIOGRAPHICAL SKETCH
(1921 -   )

For John H. Glenn, Jr., Colonel, USMC (Ret.), his swearing in as Ohio's junior United States Senator on Christmas Eve, 1974, was another milestone in a career encompassing military service, private business, science, civic activities and politics.

John Glenn was born July 18, 1921, in Cambridge, Ohio. As a child, he moved with his parents to New Concord, Ohio, where he grew up and also attended public schools. While attending Muskingum College in New Concord, World War II broke out. He entered the Naval Aviation Cadet Program in March 1942, and was graduated from this program and commissioned in the Marine Corps in 1943. After advanced training, he joined Marine Fighter Squadron 155, and spent a year flying F8U fighters in the Marshall Islands.

During his World War II service, he flew 59 combat missions. After the war, he was a member of Fighter Squadron 218 on North China patrol, and had duty in Guam. From June 1948 to December 1950, Glenn was an instructor in advanced flight training at Corpus Christi, Texas. He then attended Amphibious Warfare Training at Quantico, Virginia. In Korea, he flew 63 missions with Marine Fighter Squadron 311, and 27 while an exchange pilot with the Air Force in F-86 Sabrejets. In the last nine days of fighting in Korea, Glenn downed three MIG's in combat along the Yalu River.

After Korea, Glenn attended Test Pilot School at the Naval Air Test Center, Patuxent River, Maryland. After graduation, he was project officer on a number of aircraft. He was assigned to the Fighter Design Branch of the Navy Bureau of Aeronautics (now Bureau of Naval Weapons) in Washington from November 1956 to April 1959, during which time he also attended the University of Maryland.

Two days before his 36th birthday, in July 1957, Glenn experienced his first taste of national publicity. While project officer of the F8U, he set a transcontinental speed record from Los Angeles to New York, spanning the country in 3 hours and 23 minutes. This was the first transcontinental flight to average supersonic speed.

More test pilot hours were amassed before he was assigned to the NASA Manned Spacecraft Center in April 1959, after his selection as one of the original seven Project Mercury Astronauts. On February 20. 1962, Glenn piloted the Mercury-Atlas 6 "Friendship 7" spacecraft on the first manned orbital mission of the United States. Launched from Cape Canaveral (renamed Cape Kennedy), Florida, he completed a successful three-orbit mission around the Earth, reaching a maximum altitude (apogee) of approximately 162 statute miles, and an orbital velocity of approximately 17,500 miles per hour. Glenn's "Friendship 7" Mercury spacecraft landed in an area in the Atlantic approximately 800 miles southeast of Cape Kennedy in the vicinity of Grand Turk Island. He landed 41 miles west and 19 miles north of the planned impact point. The time of the flight from launch to impact was 4 hours, 44 minutes, and 23 seconds. Prior to his flight, Glenn had served as backup pilot for Astronauts Shepard and Grissom. When the astronauts were given special assignments to ensure pilot input into the design and development of spacecraft and flight control systems in January 1963, Project Apollo became Glenn's specialty area.

Glenn resigned from the Manned Spacecraft Center on January 16, 1964. He was promoted to the rank of Colonel in October 1964, and after a military career spanning 23 years, retired from the Marine Corps on January 1, 1965.

Glenn began his business career in 1965. He became vice-president, corporate planning, with the Royal Crown Cola Company. Later he was named president of Royal Crown International, travelling extensively to open world-wide markets for RC.

In 1964, his initial bid for the United States Senate ended abruptly when a serious household injury hospitalized him. Following his 1964 campaign, Glenn continued his active involvement in Ohio Democratic politics. Glenn headed Citizens for Gilligan in the 1970 gubernatorial election, and has been an active fund raiser for the Party. He is a member of the state Democratic executive committee, and chaired the fund-raising Buckeye Executive Club. He also headed the Governor's Task Force on Environmental Protection, which was commended for its recommendations, many of which are now Ohio law. Glenn made a second try for the Senate in 1970, and was narrowly defeated in the primary by Clevelander Howard M. Metzenbaum. It was Metzenbaum, appointed in 1974 to a vacancy in the Senate, whom Glenn defeated in that year's primary.

Senator Glenn's general election victory by more than a two-to-one margin ranks as one of the largest in Ohio history, and demonstrated not only the unity of the Democratic Party behind his candidacy, but support by voters of Republican and Independent persuasion as well. He was the only candidate of either party this century to carry all 88 counties.

Since arriving in Washington, Glenn has worked systematically on an issue-by-issue basis to build a Senate record reflecting his concern for the nation. Our critically dwindling supply of energy, for instance, was a key issue in the Senator's campaign as he repeatedly stated that fuel availability would help determine whether America maintained its leadership in the world. On taking office, he set out to secure appointments to the two committees--Interior and Government Operations--which between them handle most energy legislation.

Senator Glenn has taken quickly to the legislative process, and has won approval of many far-reaching amendments in Committee and on the Senate floor. Senator Abraham Ribicoff (D-Conn), Government Operations Chairman, designated him _ad_ _hoc_ committee chairman for nuclear matters, and in that capacity Glenn has won national notice for eight days of hearings he conducted on the danger of nuclear proliferation, sessions that helped spur concern about world-wide nuclear safety, and which will to a large extent set the tone for future U.S. nuclear policy.

Glenn received his B. S. degree in Engineering from Muskingum College, New Concord, Ohio. Muskingum College also awarded him an honorary Doctor of Science degree in Engineering. He's received honorary Doctor of Engineering degrees from Nihon University in Tokyo, Japan, Wagner and New Hampshire Colleges.

Glenn has been awarded the Distinguished Flying Cross on five occasions, and holds the Air Medal with 18 Clusters for his service during World War II and Korea. He also holds the Navy Unit Commendation for service in Korea, the Asiatic-Pacific Campaign Medal, the American Campaign Medal, the World War II Victory Medal, the China Service Medal, the National Defense Service Medal, the Korean Service Medal, the United Nations Service Medal, the Korean Presidential Unit Citation, the Navy's Astronaut Wings, the Marine Corps' new insignia (an astronaut medal), and the NASA Distinguished Service Medal.

Glenn married his childhood sweetheart, Anna Margaret Castor (Annie) in 1943. They have two children. John David, born December 13, 1945, recently graduated from Case Western Reserve University Medical School in Cleveland. Carolyn Ann was born March 19 1947. Senator and Mrs Glenn reside in a suburb of Columbus.

JOHN HERSCHEL GLENN, JR

MICHAEL COLLINS

No. 1

No. 2

MICHAEL COLLINS
A BIOGRAPHICAL SKETCH
(1930 -   )

MICHAEL COLLINS, born Rome, Italy, on 31 October 1930. Command Module pilot of Apollo XI, and pilot of Gemini 10, was member of 3rd Astronaut pilot group, selected in October 1963. Graduated from Saint Alban's School, Washington D.C., recived BS Degree at West Point, and elected Air Force career, served as experimental flight test officer at Edwards AFB, testing flight performance, stability and control characteristics of AF jet fighter aircraft. has 4200 hrs flying time. Between 18 and 21 July 1966, on Gemini 10, Collins and Command Pilot John Young performed docking with an Agena, and changing orbit plane, redezvoused with a second passive Agena. During this flight, Collins conducted two EVA missions. 1st Agena provided power for the Gemini 10 plane change. Collins recovered meteorite detection equipment. Collins served as Apollo XI Command Module Pilot on July 16-24, 1969 in the "Columbia", on station in lunar orbit, while Armstrong and Aldrin made 1st lunar landing. After retiring from NASA and the Air Force, Collins served in the U.S. Dept. of State as Ass. Secretary of State for Public Affairs in November 1969, he became, in February 1971 Director of the National Air and Space Museum. Collins performed the final re-docking maneuver with the Eagle with Armstrong and Aldrin after their ascent from the lunar surface performed substantial extravehicular actions. During the lunar mission he monitored descent and ascent of the lunar module excursion, as well as voice communication with Houston. Examples of Houston - Columbia, Eagle and Traquility Base:

Columbia to Houston: As we ease around on the left side of the Moon, I marvel again at the precision of our path. We have missed hitting the Moon by a palty 300 n.miles, at a distance of nearly a quarter of a million miles from Earth, and don't forget that the Moon is a moving target and that we are racing thru the sky just ahead of its leading edge"....
...... " As we pass behind the Moon, we have just over eight minutes to go before the burn. We are super-careful now, checking and re-checking each step several times. When the moment finally arrives, the big engine instantly springs into action and reassuringly plasters us back into our seats......? I read out the results: Minus one, plus one, plus one". The accuracy of the overall system is phenomenal. Out of our 3000 feet per second, we have velocity errors in our body-axes coordinate system of only a tenth of one foot per second in each of the three directions....... even Neil acknowledges that fact...."

From Space to Ground tapes: Eagle: 540 feet, down 30 (feet per second), down at 15, 400 feet, down at 9, forward 350 feet, down at 4, 47 feet, 1.5 down, 13 forewart, 11 forewrd coming down nicely, 200 feet, down 4.5, down 5.5, 75 feet, 6 foreward, ... lights on, ... down 2.5, 40 feet, down 2.5, kicking up some dust, 30 feet, ... down 2.5, faint shadow 4 foreward 4 foreward- drifting to right a little... ok

Huston: 30 seconds (fuel remaining) ....
Eagle: Contact light, engine stop, engine command override off
Houston: We copy you down, Eagle ....
Eagle: Houston, Tranquility Base here. The Eagle has landed!
Houston: Roger, Tranquility Base, we copy you on the ground. You have got a bunch of guys turn blue! We are breathing again. Thanks a lot! ......
Tranquility: Thank you- That may have seemed like a very long final phase. The auto targeting was taking us right into a football-field-size Crater, with large boulders and rocks for about one to two crater diameters around it, and required flying manually over the rock field to find a reasonably good area!
Houston: Roger, Tranquility. It was beautiful from here, Tranquility!. Over.
Tranquility: We'll get you to the details of what's around here, but it looks like a collection of just about every variety of shape, angularity, granularity, about every variety of rock you could find.
Houston: Roger, Tranquility. Be advised there's lots of smiling faces in this room, and all over the world.
Tranquility: There are two of them up here.....
Columbia: And don't forget one in the Command Module....... ....

*) From " Apollo Expeditions to the Moon" U.S. Government Printing Office. NASA SP; 350
TL 789.8 U 6 A 513 629.45'4 75- 600071

Michael Collins, Colonel, USAF (Ret.), was born in Rome, Italy, on October 31, 1930. He graduated from Saint Albans School in Washington, D. C., and received a Bachelor of Science degree from the United States Military Academy at West Point, New York, in 1952.

Collins chose an Air Force career following graduation from West Point. He served as an experimental flight test officer at the Air Force Flight Test Center, Edwards Air Force Base, California, and in that capacity, tested performance and stability and control characteristics of Air Force aircraft--primarily jet fighters. He has logged more than 4200 hours flying time.

Colonel Collins was one of the third group of astronauts named by NASA in October 1963. He served as backup pilot for the Gemini 7 mission. As pilot on the 3-day Gemini 10 mission, launched July 18, 1966, Collins shares with command pilot John Young in the accomplishments of that record-setting flight. These accomplishments include a successful rendezvous and docking with a separately launched Agena target vehicle, and using the power of the Agena, maneuvering the Gemini spacecraft into another orbit for a rendezvous with a second, passive Agena. Collins' skillful performance in completing two periods of extravehicular activity included the recovery of a micrometeorite detection experiment from the passive Agena.

Gemini 10 attained an apogee of approximately 475 statute miles and traveled a distance of 1,275,091 statute miles--after which splashdown occurred in the West Atlantic, 529 miles east of Cape Kennedy. The spacecraft landed 2.6 miles from the USS Guadalcanal, and became the second spacecraft in the Gemini program to land within eye and camera range of the prime recovery ship.

Collins was assigned to Apollo 8, but was removed to undergo surgery. He served as command module pilot on Apollo XI, July 16-24, 1969--the first lunar landing mission. He remained aboard the command module, "Columbia", on station in lunar orbit while Neil Armstrong (spacecraft commander) and Edwin Aldrin (lunar module pilot) descended to the lunar surface in their lunar module, "Eagle". Collins performed the final re-docking maneuvers following a successful lunar orbit rendezvous, which was initiated by Armstrong and Aldrin from within the "Eagle" after their ascent from the lunar surface. Among the accomplishments of the Apollo XI mission were collection of lunar surface samples for return to Earth, deployment of lunar surface experiments, and an extensive evaluation of the life supporting extravehicular mobility unit worn by astronauts.

In November 1969, Collins resigned from NASA and the Air Force, after completing two space flights, and logging 266 hours in space--of which one hour and 27 minutes was spent in EVA. After he resigned from the Air Force, he was appointed Assistant Secretary of State for Public Affairs. In February 1971, he became Director of the National Air and Space Museum at the Smithsonian Institution. His book, <u>Carrying the Fire</u>, describes his selection into the astronaut program, and his experiences on Gemini 10 and Apollo 11.

Collins is married to the former Patricia M. Finnegan, and he has three children; Kathleen, born May 6, 1959; Ann S., born October 31, 1961; Michael L., born February 23, 1963.

HARRISON "JACK" HAGAN SCHMITT
A BIOGRAPHICAL SKETCH
(1935 -   )

HARRISON "Jack" HAGAN SCHMITT, born at Santa Rita, New Mexico, on 3rd July 1935. First U.S. Scientist-Astronaut and Geologist in Space and on the Moon. Education and Training: 1957 BS CALTECH, 1964 PhD Harvard U. Geology. 1965 Astronaut Corps, Science Option; 1957-58 Fullbright Fellow-U. of Norway; NSF Postdoctoral Fellow, 1964; Honorary Dr.Eng., Colorado School of Mines, 1973. Lunar Module Pilot, Apollo 17,(12th man to walk Moon). NASA Asst. Administrator for Energy Programs, 1974. U.S. Senator of New Mexico 1976. Instrumental in providing Apollo Flight Crews instructions in lunar navigation, geology and feature recognition. 6 Dec. 72 landed at Taurus-Littrow on SW edge of Mare Serenitatis- Longest lunar stay (301 hrs) and longest in lunar orbit (147 hrs 48 min), 22 hrs 4 min extravehicular activity on lunar surface. Largest sample return (115 kg). Former Chief of first Scientist-Astronaut Group. Major results of 6 Apollo landings evidence of six distinct lunar evolution phases: 1. Crust molten to 200 miles depth 4.6 to 4.4 billion years ago. 2. Major meteor impacts created lunar highlands 4.4 to 4.1 billion years ago. 3. The large lunar Mare evolved during next 200 million years. 4. The light colored plains formed during a short-term event of unknown nature 3.9 Billion years ago. Substantial volcanic activity with basaltic lava erupting in the Mares from 3.8 thru 3.1 billion years ago in phase 5. 6. Development of a quiet crust gradually over the past 3.0 billion years. While similar in basic element composition, Earth and Moon developed along lines of different thermal history, leading to different crust composition, with the Moon having more titanium, iron and heavy metals in its crust, while the Earth has more carbon, sodium and water in its upper crust basaltic lava. Of the Earth's history we know so far only five phases which correspond to the first five lunar phases. On Earth these phases developed at a slower rate.

Harrison H. Schmitt was born 3 July 1935, in Santa Rita, New Mexico. He graduated from Western High School, Silver City, New Mexico. A graduate of California Institute of Technology, and holder of a doctorate in geology from Harvard, he also studied at the University of Oslo in Norway during 1957-58.

Schmitt was a teaching fellow at Harvard in 1961, where he assisted in teaching a course in ore deposits. Prior to his teaching assignment, he did geological work for the Norwegian Geological Survey on the west coast of Norway, and for the U. S. Geological Survey in New Mexico and Montana. He also worked two summers as a geologist in southeastern Alaska. Before joining NASA, he was with the U. S. Geological Survey's Astrogeology Center at Flagstaff, Arizona. He was project chief for lunar field geological methods and participated in photo and telescopic mapping of the moon, and was among USGS astrogeologists instructing NASA astronauts during their geological field trips.

Dr. Schmitt was selected as a scientist-astronaut by NASA in June 1965. He later completed a 53-week course in flight training at Williams Air Force Base, Arizona. In addition to training for future manned space flight, he was instrumental in providing Apollo flight crews with detailed instruction in lunar navigation, geology and feature recognition. Schmitt also assisted in the integration of scientific activities into the Apollo lunar missions and participated in research activities requiring geologic petrographic, and stratigraphic analyses of samples returned from the moon by Apollo missions.

On his first journey into space, Dr. Schmitt occupied the lunar module pilot seat for Apollo 17--the last scheduled manned Apollo mission to the moon for the United States--which commenced at 11:33 p.m. (CST), December 6, 1972, and concluded on December 19, 1972. He was accompanied on the voyage of the command module "America" and the lunar module "Challenger" by Eugene Cernan (spacecraft commander) and Ronald Evans (command module pilot). In maneuvering

"Challenger" to a landing at Taurus-Littrow, which is loccated on the southeast edge of Mare Serenitatis, Schmitt and Cernan activated a base of operations facilitating their completion of three successful excursions to the nearby craters and the Taurus Mountains. This last Apollo mission to the moon for the United States broke several records set by previous flights and include: longest manned lunar landing flight (301 hours, 51 minutes); longest surface extra-vehicular activities (22 hours, 4 minutes); largest lunar sample return (an estimated 115 kg (249 lbs.); and longest time in lunar orbit (147 hours, 48 minutes).

Dr. Schmitt logged 301 hours and 51 minutes in space--of which 22 hours and 4 minutes were spent in extravehicular activity on the lunar surface.

In February 1974, Dr. Schmitt assumed addtional duties as Chief of Scientist-Astronauts at the LBJ Space Center in Houston, Texas. In May 1974, he accepted the position of Assistant Administrator of the Office of Energy Programs at NASA Headquarters in Washington, D. C. In 1976, he was elected to the United States Senate from New Mexico.

SENATOR HARRISON SCHMITT

Harrison Schmitt & Eugen Cernan on their traverse in front of Taurus-Littrow range.

SOYUZ CREW BIOGRAPHIES

Commander Aleksey Arkhipovich Leonov

Brigadier General Aleksey Arkhipovich Leonov, USSR space-pilot, Hero of the Soviet Union was born in 1934 in the village of Listvayanka, Altay Kray, Kemerovo region.

After World War II the Leonovs moved to the city of Kaliningrad. In 1953 a Komsomol member Aleksey Arkhipovich Leonov entered Chuguyev Air Force School. He graduated from it with an honors diploma and became a professional military pilot.

In 1957 he joined the Communist Party of the Soviet Union. The Headquarters highly appreciate Aleksey Arkhipovich's skill, self-control, discipline. He is a well-trained parachutist -- performed more than 100 parachute jumps of varying degrees of difficulty.

In 1960 with the first group of Soviet cosmonauts he began training for a space flight.

On March 18, 1965 he made a space flight in Voskhod-2 spacecraft (with Pavel Belyayev as the commander). On that flight Leonov went outside the ship. His walk into open space opened a new chapter in the history of space exploration.

All the subsequent years the cosmonaut continued his studies and training. He participated in preparation for all Soviet space flights. In 1968 he graduated from the Zhykovskiy Air Force Academcy.

Aleksey Arkhipovich Leonov is a member of the Young Communist League Central Committee (YCLCC), and a deputy to Moscow Regional Soviet. He is a vice-president of the USSR-Italy Friendship Society, the Chairman of the Constituent Council of the Press and News Agency. He is an honorary citizen of towns: Kemerovo, Kaliningrad, Vologda, Kremenchug, Nalchik, Belgorod, Drogobych, Sochi, Chuguev, Termez, Kapsukas, Sofia (PBR), Perm, Ust-na-Labe (Czechoslovakia), Altenburg, (GDR), Houston (USA).

Aleksey Arkhipovich Leonov has a passion for painting. He is a member of USSR painter's union. His paintings were exhibited in Moscow, Orel, Simferopol, Bratislava, Prague, Ottawa, Helsinki and in the Bjenal annual show. He is also keen in filming, hunting, water skiing, track and field athletics, and parachute sports.

Aleksey Arkhipovich Leonov has received the following rewards:
- Hero of the Soviet Union
- Hero of the DRV
- Hero of the PRB
- 18 orders and medals, including 8 orders and medals of other countries and:
- Tsiolkovskiy Gold Medal
- Great Gold Medal of FAI
- 1st Degree Diploma
- Great Gold Medal of Ch. SSR Academy of Sciences (for services to Mankind)
- Gold Damask Blade

Aleksey Arkhipovich Leonov's wife, Svetlana Petrovna, graduate from a Teacher's Institute. She is an editor. They have two daughters: Victoria, 13 and Oksana, 7.

SOYUZ CREW BIOGRAPHIES

Flight Engineer Valeriy Nikolayevich Kubasov

Valeriy Nikolaevich Kubasov, USSR space pilot, Hero of the Soviet Union was born on 7 January 1935, in Vyazniki, Vladimir region, North-East of Moscow.

In 1952 Valeriy graduated from secondary school with a silver medal which meant that he did not have to take the entrance examination for the Moscow Aviation Institue. He graduated in 1958 from the Institute with the diploma of engineer-mechanic (aircraft building). Upon graduating, he went to work in Sergei Korolyov's design office. He was assigned to the ballistics studies department, where he became acquainted with another future cosmonaut, Georgi Grechko. While at the design office, Kubasov wrote several books and papers on the theory of flying vehicles, and his works appeared in many scientific journals.

Kubasov was accepted into the Cosmonaut Corps in 1966, and during his training he studied for his M. Sc. degree in Engineering, which he received in 1968. His first assignment was as the back-up to Alexei Yeliseyev, the Soyuz 5 flight-engineer in January 1969. Kubasov made his first space flight as flight-engineer of Soyuz 6, launched on 11 October 1969. During the flight. Kubasov carried out the world's first Space-welding and metals melting experiments. The Vulkan equipment was located in the orbital module, and was remotely controlled by Kubasov from the descent module. In addition, Soyuz 6 carried out joint maneuvring experiments with the Soyuz 7 and 8 spacecraft. The Soyuz 6 descent module touched down after 4 days, 22 hours and 42 minutes.

His next assignment came in 1973, when he was named flight-engineer for the Soyuz to be launched for the Apollo/Soyuz Text Project (ASTP). From May 1973 he trained for the flight in both the United States and the USSR.

Kubasov has a passion for filming, hunting, fishing, skiing, and water skiing.

Valeriy Nikolayevich Kubasov is an honorary citizen of Karaganda, Kaluga, Vladimir, Vyazniki, Houston USA. He has received the following rewards:
Hero of the Soviet Union
Tsiolkovskiy Gold Medal (USSR Academcy of Science), medal "For
    valiant labour in commemoration of V.I. Lenin Centenary "
Emblem "For active participation in Komsomol activities"
Gold Medal for Yugoslavian Academcy of Science
Medal "For development of virgin lands", and a number of medals
    of other countries.

Kubasov's wife, Lyudmila Ivanovna, graduated from Moscow Aviation Institute; she is an engineer at a machine-building plant. The Kubasovs have two children: Katya, 8 and Dima, 3.

Aleksey A. Leonov (Born May 20, 1934, Lisvayanka, Altay Kray). Lt. Colonel of the USSR Air Force, graduated from Chuguyev Air Force School, the Zhukovskiy Air Force Engineering Academy, and became a cosmonaut in 1960. He was the co-pilot of Voskhod 2 (first man to perform extravehicular activity in space), and was command pilot of the Apollo-Soyuz International Project.

Valeriy N. Kubasov (Born January 7, 1935, Vyazniki). Civilian, graduated as mechanical engineer for aircraft building from the Moscow Aviation School (1958), received master of science degree (1968), and joined the cosmonaut corps in 1967. He was the backup technical scientist for Soyuz 5 and flight engineer on Soyuz 6. Mr. Kubasov was the flight engineer for the Apollo Soyuz International Project (Soyuz 19).

CHAPTER III

ENGINEERING IN SPACE FLIGHT ROUNDTABLE

ISHF

"There is no stronger force in this world than an idea whose time has come."
    Victor Hugo (1802-1885)

ASSESSMENT FOR ENGINEERING IN SPACE FLIGHT
ROUNDTABLE - LETTER II

Ing. Teofilo M. Tabanera*

Apparently we all agree on the fact that an effective transportation system is the key for a useful and great scale utilization. Most of the participants who have sent their assessments for engineering in space roundtable appear to put their emphasis on the short term solutions.

Nevertheless, I believe that near future advance in the propulsion systems technology will be based on the well known or proved schemes of the present and the results will not be spectacular, nor enough to solve our problems to enter decidedly inside the space transportation era.

The near term activity, the next 10 years, will show us some important progress as has been described by some colleagues, as the air breathing concept, or the 3000 to 4000 psi chamber pressure or even the dual mix-fuel solution (mixed-mode propulsion). But all of those systems will not bring us to the goal of less than 150 dollars per kilogram on near orbit. The medium term or long term period, the next 20 - 25 years, might be the one that brings the chance to solve our problem. Because for that period we must be prepared to satisfy the demand of a cheap space transport system to construct big space station structures for industrial uses, communication, education and other applications, or solar power stations, and many scientific activities.

Consequently, the short term solutions or immediate ones will not offer us a situation clearly decisive for this matter. It's true that we will attain the lowering of costs for the transport in an appreciable manner, perhaps arriving at the 300 dollars per kilogram; but this success will not be sufficient to enter efficiently in the new stage we are anxious to reach.

For the medium and long term period beyond the year 1985, we need to have a scheme based on an entirely new philosophy, by which we could reach a cost below 150 dollars per kilogram on near orbit. Such a goal compels us to look for other propulsion systems, based on fundamental change; this new concept could be one of such as to propose not to be obliged to bring with it all the "energy" you need.

This new philosophy brings us to look for schemes such as the possible utilization of propulsion system based on laser, with energy transport by microwave or light wave length. We must remember we are talking for the years after 1985, or better perhaps, for 1990 on; but we must concentrate from right now our studies on this scheme of transport of energy to the vehicles, solution which if for the moment looks with certain aspects of speculation, appears sufficiently promisory to merit a vigorous impulse of research. No matter how great efforts in basic research and development we will still need, but it will be necessary to put, from now on, large human and economic resources to be in situation to implement such an important and unavoidable step, if really we want to enter decisively in space transport era. On the other hand we need to make some considerations about the different basic factors which confront this space transport, and also the means to be utilized to specify which are the parameters that are more important.
----

*Buenos Aires, Argentina.

It will be necessary to decide from the very beginning on those essential factors so that we will be in a better position not to be obliged later to accept a sacrifice of one important parameter in favor of another one not so valuable. These parameters are: a) the time, or the time spent to reach the near satellite, and the far or sincronous satellite. b) payload and eventually volume minimum, desirable, to a near orbit. c) Operationability or operation mode preferred, maneuverability type of trajectory, type of launching, descent and recuperation, etc. d) cost per kilogram of payload to a near orbit.

Which of these factors has priority' Probably the cost, because this result is essential to reach the highest possibility of full utilization of this means of transport for the development of an ambitious plan of constructions of space stations for industrial activities, for applications and power stations, to deliver energy to the earth, apart from the more ambitious ones such as space colonization. Following the merit order we probably have the payload, because one particular system of construction of big size will need a minimum payload for each travel, to be in the near optimum solution. The other two factors, perhaps three, being important, they are however secondary with respect to the first two. We could say that it will be convenient to establish under which conditions we can sacrifice one parameter in favor of another one. This type of decisions must be made before we are in the research and development period, and we must establish the limits for this kind of exchange of factors.

The economic optimization perhaps will not coincide with the satisfaction of the minimum requisites previously fixed. If from now on we establish as an objective the maximum limit of $150 / kg we already have a convenient guide that will limit our speculations in the search of a preferred solution between the alternatives we surely will have. In the same way if we do not want to expend one year or one month to bring the payload to the near orbit or the total payload to final orbit, we already have an imperative rule for our studies.

All of this kind of decisions appear to be necessary if we want to limit the number of alternatives, variants and subvariants. I am in total agreement with the opinion of the majority of the participants expressed in their assessments received until today, when they pay attention to the near future activities based on new and even brilliant solutions already proposed and analized now, but without changing the basic philosophy and with which we could and surely we will continue the great and sustained progress that we are having already in the space activity. It is evident that the second generation Space Shuttle will be a necessity to continue on the plan to place very large payloads in orbit, between which one of the most promising will be the one to help to solve the world's energy problem. The 100% recoverable single stage to orbit appears to be one of the most promising solutions. Probably we need 10 years more to develop such technology in sight if we keep inside the concept of chemical propulsion. This means that we need urgently to start parallel investigations on a chemical propulsion concept that brings us to a better way to introduce an economic space transportation era.

I agree with the statement made that Laser propulsion "presently is well beyond the practical power (gigawatts) and technology (mirrors, etc.) requirements for booster systems", but it appears mandatory to solve the problems and to follow on this line of thinking. Or at least we should make strong efforts to solve them.

RESPONSE LETTER NO. 2
ENGINEERING IN SPACE FLIGHT

Leland F. Belew*

In reviewing the assessment papers, it became evident that the new challenge of defining, designing, building and operating an industrial base in space was not recognized as the new engineering challenge in space. Various concepts as related to space transportation were adequately covered.

Studies indicate that space industrialization activities such as Satellite Power Systems (SPS), Space Processing, Communications and other areas will be the basis of committing an effort that is vital to the future world needs. Preliminary study results further indicate that to develop an SPS capable of providing up to 20% of the electrical energy growth by 2025 in the U. S., payloads of 500,000 pounds must be launched some 500 times per year for a period exceeding 20 years. (The 2025 timeframe represents a desirable period for obtaining solar power and is not necessarily an optimum for developing the capability.) Included in the overall activities of establishing an SPS are challenging engineering activities of major proportion such as Heavy Lift Launch Vehicle, Large Space Station, orbital delivery systems and associated supporting systems.

Space industrialization on the scale of fulfilling a vital national need must evolve through successfully demonstrating the benefits and projected payoffs during the Shuttle. In order to achieve this a major effort must be devoted to Shuttle payloads. It is suggested that appropriate emphasis be placed on the future space industrialization objectives in defining the near-term payloads for Shuttle. Cost (low cost) will be the determining factor in most cases in future space flight programs. The big challenge in engineering lies in our ability to conceive and produce payloads and associated equipment and necessary operations that are competitive.

----

*Director, Science and Engineering Nasa Marshall Space Flight Center.

POINT OF VIEW IN ENGINEERING IN SPACE FLIGHT

Ing. Teofilo M. Tabanera*

From the different assessments made by participants to the round-table, we can conclude that there are two ways to see the matter:

I. From a pragmatical point of view, adapting the objective to the present situation. This is a political attitude, pointed to get an immediate technical result for today.

II. From a technical or socio-technical point of view. This is a technical idea and procedure to get an ambitious social result for a long period ahead.

The first one looks at the best strategy to get a result, perhaps modest but successfully adapted, to the political facts of operational constraints. It is an executive way of action.

The second way to focus the matter, looks at the optimum or near optimum technical solution for the not too immediate future. This has nothing to do with the operational or administrative difficulties existing at the present time, or for the next 5 to 10 years. We are learning from the past, now doing the present and shaping the future. We must keep together but not interfering those three factors: learning, doing and shaping, but never stop shaping the future because we are having difficulties at present.

With this mental approach we are looking to a future concept which will put us on a new way to solve the problem of space transport in such a way that space activities (industrial, energy, science and even colonization) would be a true activity of the entire society and not only a priviledged one for only a group of people or enterprises. But if we still need to think necessarily of the present and the near future - and we need - I would prefer:

a. Present or very near future: (up to 1985) give all our support to the present shuttle development and leave the "details" and operational problems (technical, economical) to the people engaged in the practical activity itself.

b. Near future (next 10 to 20 years): give some emphasis to the air breathing system also (or alternately to the 3000 psi chambers or the mixed-mode system) because this one seems the more natural or logical solution for the immediate period to come.

c. Future (for the period 20 years onward): approach the problem with a new philosophy (starting today) which could bring the great economic difference on the cost of transport which we need imperatively if we want to truly enter the space transport era. Laser or radiated energy is only but one way to follow.

----

*Buenos Aires, Argentina.

Then we could have one group of people and one group of investigators and engineers separately for each state - a., b., and c., concentrating each one in their specific matter or goal. Even if we have some people who can think at the same time in the immediate objectives and also in the future great goals, it would be better to give each one his own task separately and without interference.

ENGINEERING IN SPACE FLIGHT
ROUNDTABLE SESSION

Prof. Segismundo Sanz Aranguez*

By the following, I would like to point out, as a personal point of view some ideas about what the Space Shuttle System means to Europe and for the European technical and scientist environment.

From the transportation point of view, the Space Shuttle System can be considered as the pioneer of a new era. However, it is necessary to differentiate between "Routine Transportation" and "Special Transportation." The Shuttle, for "Routine Transportation," is very far from being, and it will not be for many years, a competitive means. Probably it will not be so until the middle of the 90's, because competitivity is mainly based on the economic factor. But saving, in the space transportation, can only be obtained when the current space transportation has been sufficiently developed; this will be when the Shuttle has been fully operated and tested as a means of "Special Transportation." Only then, will it be possible to have a clear idea about what the adequate techniques will be for the next Shuttle generation, enabling reduction of costs from $500 to $100 per pound in low orbit, which is normally given as the competitive figure to open up the space transport market.

On the other hand, the cost per pound in orbit will not essentially fall in the use of the Shuttle as a means of "Special Transportation" to place into orbit everything that represents an achievement from the scientist or technical point of view, a prototype, or the configuration of a station. This can be the case of the "Satellite Solar Power Station" SSPS); of "Manufacturing in orbit"; of the "Spacelab", or the placing into orbit of a "Space Tug" carrying nuclear reactors, or isotopes, to establish the transfer from a low orbit to a higher orbit, or geostationary, or another type, and vice versa.

From the logistic point of view, it would be desirable to use in any airport, coming to the conclusion that it should be rather convenient to use the take-off and landing horizontal system (HTOHL); but the main problem is the need of getting oxygen from the air in order to reach low weight during take-off. This increases costs due to the heavier weight of the propulsion system. It appears logical to use the vertical take-off system and the horizontal landing system for the orbiter (VTOHL), and the take-off and landing vertical system (VTOVL) for the booster using solid propellant.

As a mere European observer, I would like to point out that the American Program for the Space Shuttle System has been accepted, with a great interest within the European scientific and technical environments, until having the Spacelab concept. The Spacelab is an integral part of the Space Shuttle, and any change in it would affect the Spacelab. The Spacelab weight and diameter are adjusted to the Shuttle capability; the total Spacelab launch and landing weights
----

*Professor of Aeronautical Engineering, Madrid University, Madrid, Spain.

(29,500 Kg. and 14,500), with a diameter of 4.2 m. Spacelab's center of gravity and payload fall within narrow limits to satisfy the orbiter re-entry stability conditions.

The modular construction of the Spacelab allows flexibility to the different users. Spacelab has both autonomous and orbiter-dependent subsystems, searching for saving and the minimum interfaces between the Spacelab and the Shuttle.

Among the autonomous subsystems are the following:
- Command and Data Management System (CDMS).
- Humidity and thermal control.
- Air revitalization.

Some of the orbiter-dependent subsystems are as follows:
- Heat rejection.
- Primary electrical power.
- Fuel oxygen-hydrogen cell.
- Habitability and personal hygiene.
- Stabilization and control.

The European scientific and technical environments will be having experiences on board the preoperational Shuttle flights for Orbital Test Flight (OTF), and for the Long Duration Exposure Flight (LDEF). Many European proposals, in number exceeding 40, are known; about half of them are in the area of Astronomy and Cosmic Ray research. Among the rest, different technological experiences are represented; specifically the Science of Life and Materials. The Spacelab will carry crews with 4 members, made of American and European scientists and engineers; whereas the orbiter crew will consist of Astronauts. In this way, men will perform space activities for meteorological missions, earth resources, telecommunication, biology, biochemistry and the manufacturing of new materials.

Among the Spacelab technological experiences, there is one proposal from my own country's technicians covering the "Floating Liquid Zones in Zero Gravity". This experience has been proposed for its application to Metallurgy and composites, crystal growth, fluidphysics and electrophoresis.

The European budget for the Spacelab Program amounts to $420 million. This is very high for the European Economy, and it is my personal point of view, that Europe would be more interested and satisfied in obtaining the experience and maximum profit, with the greatest possible number of manufactured similar units, rather than changing the current model adapted to the Shuttle; at least, until reaching the necessary experience by means of the Shuttle exploitation to keep pace with the 2nd Shuttle generation.

AIRBREATHING AND OXYGEN COLLECTING HTOHL LAUNCH VEHICLES

Addendum to Assessment Paper for "Engineering in Space Flight"

Richard A. Nau*

This addendum is in response to the suggestions of Dr. Maxime A. Faget that I elaborate on the HTOHL schemes that use airbreathing engines, and oxygen collection launch vehicles referred to in my original assessment paper.

HTOHL AIRBREATHING BOOSTER COMPARISON

The following comparison of HTOHL airbreathers with both VTOHL and HTOHL is based on results of Aerospaceplane and reusable booster studies conducted in the 1960's (Ref. 1). The relative weights, sizes and costs of these vehicles are still valid today, although the weights should be approximately doubled, based on detail space shuttle design experience.

Eight fixed-wing reusable horizontal landing booster point design concepts are compared on the basis of relative weight and cost (Figure 1). All flight vehicles are fully recoverable and capable of flying back and landing at the launch site. Vehicle types are vertical take-off horizontal landing rockets, air breathing first stages, combined airbreathing and rocket first stages, oxidizer collection concepts, supersonic combustion ramjets, and in-flight refueling vehicles.

System non-recurring costs tend to be proportioned to the inert weight of the vehicle, and therefore the rocket propelled vehicles have lower investment costs than the airbreathing types. The turnaround or recurring costs are relatively independent of the launch vehicle types with the exception of the in-flight refueling system, which is about double those of the other systems because of the large weight of flight hardware to be maintained. The airbreathing first stage boosters, however, may offer more mission flexibility than the rocket first stages because of their capability to fly for longer periods within the earth's atmosphere.

The eight vehicles compared here are all designed to carry a 40,000 lb. payload to a 300 n.mi. low inclination orbit.

Previous studies have indicated that high energy cryogenic propellants such as liquid hydrogen/liquid oxygen are highly desirable for efficient upper stages of reusable launch vehicles. They have also indicated that high pressure rocket engines with combustion chamber pressure in the 2,000-3,000 psi range are desired. Therefore, all eight vehicles to be considered here are based on liquid hydrogen fuel, and liquid oxygen as rocket engine oxidizer.

---

*General Dynamics Convair Division, San Diego, California

For convenience in this discussion, most of the vehicles, with the exception of the in-flight refueling concept, have basically the same reusable lifting body second stage, with essentially identical staging additions of Mach 8 and 200,000 feet altitude. For the pure airbreathers, these staging conditions are generally close to the optimum, which gives the lightest total system weight; for the rocket-power systems, each will optimize at different staging conditions depending on staging mode. However, rocket vehicle size, costs, and development probably are not highly sensitive to staging conditions near the selected values.

The weights and weight fractions of the eight vehicle configurations are listed in Table 1. For purposes of comparison, the weights are also plotted in Figure 2, giving the gross liftoff or take-off weight as well as the hardware weight of each vehicle. As might be expected, the gross weights of the airbreathing vehicles are significantly less than the gross take-off weights of the rocket vehicles. However, the hardware weights of the rockets are lower than the hardware weights of the airbreathers because they do not require sophisticated airbreathing propulsion systems and have lower structural weight fractions because of their relatively simple configurations as compared to the airbreathers. The VTOHL rocket has the lightest inert weight of any of the configurations considered and also the simplest structure since its fuel tanks are simple, cylindrical shapes and its wing area is sized only for the flyback portion of the flight, whereas all of the other configurations have wings sized to support the entire weight of the vehicle at take-off. However, the airbreathers have offset range capability that may have some advantage in mission flexibility and they all are quite similar to aircraft in their abort characteristics. The acceleration loads of the rockets can be held to a maximum of 3 g's without significant penalties to vehicle size and weight, however, the maximum acceleration loads of the airbreathers are all below the 1 g level. The acceleration levels for both types are within the tolerance of non-astronaut trained personnel, however, and no problem is anticipated for either type if a large volume of non-astronaut traffic develops to support orbital space station operations.

Current airbreathing engine technology would support the development of airbreathing boosters with staging velocities up to Mach 3. For payloads comparable to the 40,000 pounds of the previous eight vehicles, or 65,000 to 100,000 pounds of interest today, these would be very large with take-off weights of 2,000,000 or 3,000,000 pounds. The large second stage (orbiter) velocity increment would result in a very large vehicle, probably not feasible at this time. Use of large subsonic aircraft such as C-5 or 747 derivatives as subsonic first stages has been suggested, but such vehicles would be limited to payloads of a few thousand pounds at best, with an expendable second stage.

The costs of the various launch vehicles considered here are shown for comparative purposes in Figure 3. As might be expected, the non-recurring costs, which are predominated by the research and development costs, are primarily a function of the inert weight of the vehicles involved, varying approximately as the 0.5 power of the vehicle hardware weight for vehicles of comparable complexity and technical difficulty. This trend is generally evident here with the exception of the air collection and enrichment airbreathing vehicle, which has the highest R & D costs of all of the configurations because of its complex propulsion and oxygen collection system. The R&D costs vary from approximately $9 billion for the VTOHL to a high of $15 billion for the air collection and enrichment vehicle. The development cost of the supersonic combustion ramjet vehicle is also high, approximately $14.5 billion, because of the lengthy development cycle foreseen for the scramjet engine and because of the sophisticated structure and heat shield required for the severe heating environment.

The investment in operational flight hardware and facilities, however, is highest for the in-flight refueling concept because two large airbreathing vehicles must be built and maintained. The airbreathing vehicles are inherently more costly to develop because they have all the structural problems of the rockets, rocket engines for second stages, airbreathing engine development engine development costs,stringent requirements for high aerodynamics performance, and major airframe/propulsion integration problems.

The operational or recurring costs per flight do not vary significantly across the range of vehicles, with the exception again of the in-flight refueling concept which requires the operation of twice as many large size vehicles.

OXYGEN COLLECTION AEROSPACEPLANE

Studies begun in 1959 indicated that it may be feasible to place a large single stage recoverable vehicle in orbit with a hydrogen fueled airbreathing propulsion system which collects oxidizer during flight in the atmosphere. This concept was designated Aerospaceplane by the USAF and was studied extensively from 1959 through 1964.(Ref. 2,3,4,5, 6,7.)

Aerospaceplane is a winged, horizontal take-off and landing, hydrogen fueled fully recoverable space vehicle which takes off in the conventional aircraft manner. The vehicle uses liquid air condensing engines of the LACE or air condensing turbojet types from take-off to about Mach 3, when it switches over to ramjet operation. Beginning at approximately Mach 4 and continuing to approximately Mach 5, the vehicle collects oxidizer from the air, condensing and storing it in tanks that have been previously emptied of hydrogen fuel. The single-stage-to-orbit vehicle more than doubles its weight during the oxidizer collection phase. This oxidizer is then used in hydrogen fueled rocket engines to continue the acceleration of the vehicle to orbital velocity. The unique features of this vehicle that make possible single-stage-to-orbit capability are the high specific impulse of the airbreathing engines in the early phases, and the collection of oxidizer in flight, that eliminates the need for carrying the large weight of oxidizer from the ground, thereby reducing engine weight, since the large oxidizer weight is not carried at lift-off.

A simplified weight history of an oxidizer collecting single-stage-to-orbit vehicle is shown in Figure 4. Early studies (Ref. 2) were concentrated on the single-stage-to-orbit concept, but payload was marginal with the structures technology available in the early 1960's. Emphasis was shifted to the two-stage concept in 1963 and 1964. The following is a brief summary of the point-design two-stage concept defined in detail in References 3 and 4.

An artist concept of this vehicle is shown in Figure 5. Specified requirements for the vehicle are shown in Figure 6, along with the values actually achieved in the study. Details of the vehicle are shown in Figure 7.

The over-all take-off weight including both stages was 700,000 lbs, with the first stage at 588,000 lbs, and second stage at 112,000 lbs.

Note that the second stage weight at launch (staging) is 351,057 lbs, with 239,057 lbs of oxidizer collected in flight. The second stage is carried amid ships on top of the first stage. The three men crew capsule is shown at the forward part of the vehicle. Propulsion system components are shown below the second stage, with the air separator being the dome-ended cylinder. The configuration is of the blended body type wherein the wing apex and the body apex are coincident, with no sharp demarkation between the wing and the body. One of the advantages of this type of configuration is low aerodynamic heating due to minimization of sharp discontinuities in the shape. The configuration also has a relatively low wetted area for minimization of structural weight. The inlets are seen on the under side of the vehicle in the high pressure region of the expanding fuselage and lifting wing. The nozzle is a low angle cone type, designed to fill most of the base of the configuration to minimize base drag. The leading edge sweep is 80 degrees in order to attain a low aerodynamic heating rate along with a low aerodynamic drag on the leading edges. The vertical fins of the first stage are located at the wing tip and are rolled out from the vertical plane for high effectiveness. The section of the first stage aft of the nose wheel bay and forward of the propulsion bay is the forward hydrogen tank. Other hydrogen tanks are located in the nozzle area and in the wings of the first stage.

The second stage, which stages vertically from the first stage well, has a lifting body type of configuration. This particular configuration has a maximum hypersonic lift to drag ratio of 1.35 and was designed to minimize the aerodynamic heating effects of the vehicle. The vertical fins for the second stage are located at the wing tips or sides of the vehicle, but are folded flat while the second stage is stored in the first stage well.

The second stage for the ACES Aerospaceplane Point Design is shown in Figure 8. The configuration is the lifting body type and the vertical fins are located at the sides of the vehicle. They are also rolled out for more effectiveness at high angles of attack. There are two sets of elevons as shown - one to control the flow from the top surface of the vehicle and the other to control the flow from the bottom surface of the vehicle. This is necessary, due to the large vehicle depth. The second stage has a three man crew with the cockpit also on escape capsule. The rocket nozzle has as large an exhaust area for the ED (external deflection) type nozzle as possible for this vehicle. The payload bay volume is 4000 cu. ct., with 23,150 lbs payload. Access to the bay is through two large doors that open upward on top of the vehicle. The landing gear uses wheels rather than skids to facilitate control during the landing run and maneuverability on the field.

The vehicle is equipped with two turbot jet engines that use hydrogen fuel to provide power for the landing and go-around capability in case of wave-off. This second stage contains liquid hydrogen at take-off and receives its oxidizer (liquid enriched air) during the air collection and enrichment phase from the first stage.

The ACES system (Figure 9) is the propulsion subsystem on which the complete system is predicated, and is closely integrated with the turboramjet and rocket subsystems. Thrust is provided by the ramjets when the ACES system is operating.

About one-third of the air entering the inlets is ducted to a heat exchanger system. In the heat exchangers it is cooled to a saturated vapor condition. The saturated vapor is admitted to a double-column distillation separator which is rotated about an axis to increase the g field within it. The air is separated into an oxygen-rich liquid, 90% oxygen and 10% nitrogen, and a nitrogen-rich vapor, 90% nitrogen and 2% oxygen.

The liquid is referred to as liquid enriched air of LEA, and the vapor as the waste product. Half the LEA is further cooled and stored at low pressure in the second stage, and half is stored at higher pressure in the first stage. The waste product has considerable refrigeration capacity as it leaves the separator. This capacity is utilized to cool the air to be separated.

The waste product is finally injected into the ramjet combustion chambers as the most efficient way of exhausting it overhead. In order to do so it must be compressed up to ramjet combustion chamber pressure. Hydrogen cools the LEA in the subcooler, operates the reflux condenser and shelf-cooler in the separator, and does the final cooling of the air entering the separator. It is then admitted to the ramjets for regenerative cooling and combustion.

The same heat exchangers are utilized to liquify air, without enrichment in the transonic speed range for rocket thrust augmentation.

The weight variation vs. Mach number up through staging for the Point Design Vehicle is shown in Figure 10. Superimposed on this plot are the propulsion modes involved in the acceleration to staging velocity. From take-off to Mach 3 the main propulsion is provided by six turbofan engines. At high subsonic velocities prior to accelerating through the transonic region, the air collection equipment is used to collect liquid air (not enriched) to be used for the transonic acceleration. This collection is made at about 900 fps at approximately 20,000 ft. altitude and is the first time during the mission that the heat exchanger or any of the air collection equipment is used. This liquid air is used as oxidizer in the hydrogen fueled rocket engines to aid the turbofan in accelerating through the high drag transonic region. From approximately Mach 3 to Mach 6 the propulsion is provided by six ramjets that are integrated with the turbofan engines. During a portion of this ramjet operation, from approximately Mach 4 through Mach 5.1, the air collection and enrichment takes place.

The increase in weight at Mach 5.1 is due to the vehicle taking on enriched air faster than it is using the hydrogen fuel, yielding 4.0 lbs of oxidizer per pound of hydrogen burned. Most of this collection is done at a loiter condition at Mach 5.1. From Mach 5.1 to approximately Mach 8.6 the rocket engines are again used for propulsion. There is some over-lap of the rocket propulsion with the ramjet for maximum efficiency. After ramjet cut-off the sole propulsion of the first stage are the rockets, using the stored liquid enriched air as oxidizer. At approximately Mach 8.6, the second stage is released and the first stage returns to base. The second stage rocket uses the liquid enriched air that was collected by the first stage air collection and enrichment equipment to propel it into 300 n. mi. orbit.

The point design trajectory is shown on an altitude vs. velocity plot in Figure 11. The first stage boost trajectory is shown by the heavy solid line and the first stage return trajectory from the staging point is shown by the dotted line. The second stage to boost trajectory is shown by the solid line and its return from orbit by the dashed line. The total time to reach the 300 n. mi. orbit is one hour and 29 minutes. The time for the second stage to return from orbit is slightly over one hour and the total flight time of the first stage during the mission is approximately one hour. The first stage takes off and accelerates at low altitude at approximately Mach .6. A subsonic climb is made to over 20,000 ft. The trajectory then levels off for acceleration through the transonic range. At about Mach 4, the internal pressure of the propulsion system reaches 200 psia, and the vehicle climbs more steeply to maintain this collection pressure. The vehicle completes the air collection and enrichment process 34-1/2 minutes after take-off. Above the collection Mach number, a pull-up is made using rocket power. After staging, the first stage coasts to an apogee and then re-enters the atmosphere. The peak temperatures are experienced during this re-entry and are 1800 degrees F on the wing leading edge and 1520 degrees F (Reynolds number transition) on the lower surface of the vehicle. The remainder of the first stage flight is a glide return to the main base, with the turboramjets providing powered approach and go-around capability.

After staging, the second stage accelerates to a low orbit at somewhat less than 300,000 ft. altitude. The acceleration is continued to make a Hohmann transfer to the 300 n. mi. orbit. The re-entry also uses the Hohmann transfer by initiating a retro-rocket and then descending on the elliptical path to re-enter the atmosphere. The second stage uses a high angle of attack pullout to zero flight path angle and then a constant altitude trajectory until the quasi steady glide line is reached. From there a quasi steady aerodynamic glide is made to the landing site. Airbreathing engines are extended to provide powered approach and go-around capability.

The maximum temperatures for the second stage occur during the initial pullout in the atmosphere and are 2,435 degrees F on the nose and 2,360 degrees F on the lower (Reynolds number transition) surface of the vehicle.

The basic mission performance for the ACES Aerospaceplane is shown in Figure 12. The combined vehicle takes off in 4400 ft. of ground run. The first stage accelerates to the staging point at 8600 ft. per sec., 176,000 ft. altitude. After staging, the first stage returns to base and lands with a ground run of 4700 ft. The second stage accelerates from the staging point to the 300 n. mi. orbit and is capable of delivering a payload of 23,150 lb. in a 300 n. mi. polar orbit. The second stage was designed to remain in orbit 72 hours and then to make a re-entry with or without the payload. After the aerodynamic glide through the atmosphere a horizontal landing with the use of hydrogen burning turbojets is performed. The second stage landing distance is over 2000 ft.

Figure 13 presents the surface and structural temperature histories of the first stage vehicle, at a point 120 ft. aft of the nose. These surface temperatures are for an abort case in which the first stage re-enters with the second stage on board but without the second stage oxidizer. The peak lower surface temperature is approximately 2000 degrees R, occurring during the first stage re-entry. The peak upper surface temperature of approximately 1400 degrees R occurs during the pull-up maneuver to the staging point. Neither the upper nor the lower surface exceeds the Hasteloy X temperature limit of 2260 degrees R. Many areas of the upper surface are below the 1460 degree R limit for 8-1-1 titanium material, and hence titanium is used on large portions of the upper surface. The lower surface is Hasteloy X.

The vehicle requires an insulation system. The upper and lower surface structural temperatures attained using the insulation system are shown in Figure 13. The structural temperature starts at an extremely low value due to the presence of the large amount of liquid hydrogen. The peak lower surface structural temperature obtained is approximately 700 degrees R, and the upper surface peak is approximately 400 degrees R, well within the 1060 degree R limit of the titanium structure.

The second stage temperature histories for the re-entry trajectory at a point 40 ft. aft of the nose are shown in Figure 14. The second stage re-entry trajectory determined the lower surface material and insulation thicknesses required for the second stage. The nominal second stage re-entry trajectory is a maximum lift re-entry at an angle of attack of approximately 50 degrees. The peak lower surface temperature is approximately 2800 degrees R and occurs during the initial pull-out maneuver. This temperature dictated the use of coated columbium as the lower surface material. The upper surface temperature during the re-entry is seen to reach a peak temperature of approximately 1100 degrees R. This temperature is below the peak temperatures reached on the upper surface of the second stage during the first stage boost. The upper surface material of the second stage is thus dictated by the first stage flight region. The lower surface structural temperature attained with 2.5 in. of insulation is also shown. It is seen that this amount of insulation keeps the lower surface structure temperature below the 1060 degree R limit of the titanium structure.

The structure and material concepts for the tank areas of the Point Design vehicle (both first and second stage) are shown in Figure 15. The frames are made with sinusoidal web members and are located inside the tank. The stringers running fore and aft also have sinusoidal webs. Between the stringers and frames is the hydrogen tight tank skin. This load carrying tank structure - the frames, skin, and stringers - is A110 titanium material. The very high strength to weight ratio capability of titanium at cryogenic temperatures is utilized in this concept. The thermal protection system is external to the stringers as shown. A titanium mesh over the stringers is covered with quartz fiber insulation and cover panels. The cover panels are made of various materials depending upon the temperature. The lower surface and leading edges of the first stage utilize the Hastelloy X, and the cooler upper surfaces are 811 titanium. Coated refractories are not required on the first stage of this vehicle.

On the second stage, the lower surface and leading edges utilize coated columbium cover panels and supports. Because the upper surface is cooler, it can utilize 811 titanium cover panels. The structural concept and the material are the same for all tank areas of both stages. The quartz fiber insulation will vary, depending upon the aerodynamic heating characteristics of the particular area.

CONCLUSION

There were many in-depth studies and technology demonstration programs carried out on the air collection and enrichment system (ACES) oxygen collection concept in the 1960's (Ref. 5, 6, 7). Technical feasibility and performance of the compact air-air separator system were experimentally verified by the Union Carbide Corporation Linde Division Cryogenic Development Laboratory, with a 1/20th flow scale prototype (Ref. 6). No technology work has been done on the hypersonic airbreathing propulsion system or the oxygen collection system since the mid-1960's, so an extensive propulsion development program would be required for this system, and it is not a candidate next generation shuttle.

NOTES

1. Nau, Richard A., " A Comparison of Fixed Wing Reusable Booster Concepts," SAE Paper 670384,SAE Space Technology Conference Proceedings, May 9-12, 1967.

2. Nau, R. A. and Campbell, S.A., "Oxygen Collection System for an Aerospaceplane." General Dynamics/Astronautics Report AE 62-0741, April 1962, Prepared for USAF and Aerospace Corporation 1962 Symposium on Ballistic Missile and Space Technology.

3. "Aerospaceplane Propulsion System/Vehicle Integration Study." Contract AF33(657)-10162, GD/A Report No. GD/A-63-1069 (Sixteen Volumes), October 1963.

4. "Air Collection and Enrichment/Chemical Rocket (ACES/CR) Aerospaceplane Development Planning Study Final Report." Contract Report No. GD/A-DCB-64-046 (Seven Volumes), 30 April 1964.

5. "Airframe Support of a Boilerplate Air Separator Air Enrichment Program." Contract AF33-(758)-8663, Report No. ASD-%DR-63-083, October 1963.

6. "Feasibility Study of a High Capacity Distillation Separator for an Air Enrichment System (Experimental and Analytical Program)." Contract AF339657-8722, Report ASD-TDR-63-665, Union Carbide Corporation, Linde Division, November 1963.

7. Nau, R. A. and Campbell, S.A., "Rotary Separator," United States Patent 3,779,452, December 18, 1973, Filed September 22, 1960.

POINT DESIGN VEHICLE

Figure 7

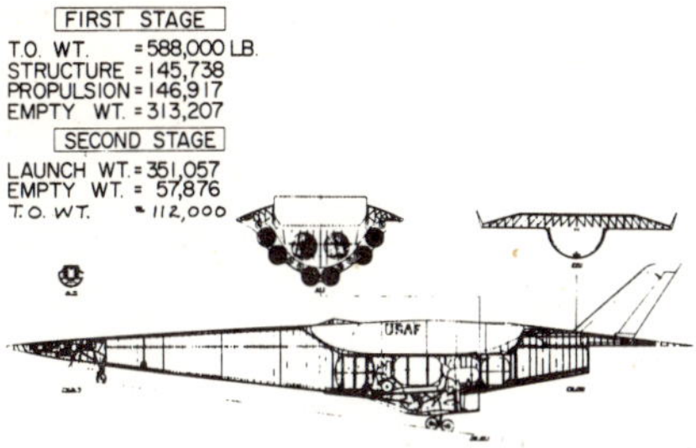

FIRST STAGE
T.O. WT. = 588,000 LB.
STRUCTURE = 145,738
PROPULSION = 146,917
EMPTY WT. = 313,207

SECOND STAGE
LAUNCH WT. = 351,057
EMPTY WT. = 57,876
T.O. WT. = 112,000

Figure 8

POINT DESIGN SECOND STAGE

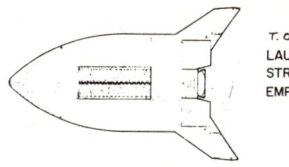

T.O. WT ≈ 112,000
LAUNCH WT. = 351,057 LB.
STRUCT WT. = 43,507
EMPTY WT. = 57,876

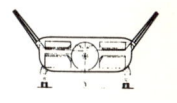

Figure 6

REQUIREMENTS

	ACHIEVED
VEHICLE	
• TAKE-OFF WT. = 700,000 LB.	✓
• PAYLOAD = 10,000 TO 40,000 LB. (300 N.MI.)	23,150 LB.
• 3 MAN CREW (BOTH STAGES)	✓
• RECOVERABLE PAYLOAD (4,000 FT³ BAY)	✓
OPERATIONS	
• HORIZONTAL TAKE-OFF (10,000' RUNWAY)	4,400 FT. - 200 Kn.
• HORIZONTAL POWERED LANDING (BOTH STGS)	✓
• DESIGN LIFE = 100 FLIGHTS	✓
• FERRY RANGE = 3,000 N.MI.	3,172 N.MI.
• MISSION DURATION = 72 HOURS	✓
• RETURN TO LAUNCH SITE (1ST STAGE)	WITH 315 N.MI. OFFSET

AEROSPACEPLANE

Figure 5

COST COMPARISON - 40,000 LB PAYLOAD REUSABLE BOOSTERS

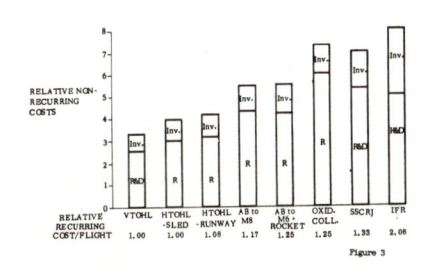

Figure 3

$$\eta = \frac{\text{Oxidizer Collected}}{\text{Fuel Used During Collection}} = \frac{W_{c_2} - W_{c_1} + W_{f_{1-2}}}{W_{f_{1-2}}}$$

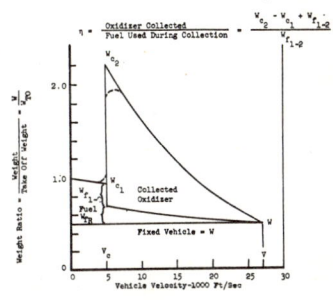

Figure 4. Oxidizer Collecting Vehicle Weight History

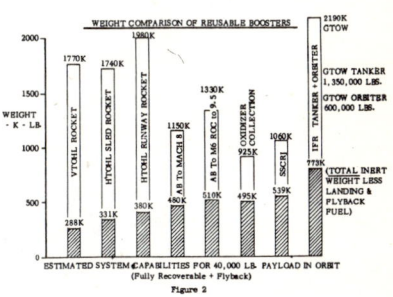

SIZE COMPARISON - FIXED WING REUSABLE BOOSTERS
40,000 LB. PAYLOAD TO 300 N.MI. ORBIT

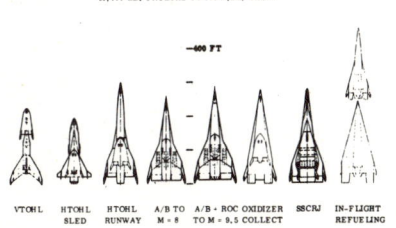

FIGURE 1

WEIGHT COMPARISON OF REUSABLE BOOSTERS

ESTIMATED SYSTEM CAPABILITIES FOR 40,000 LB. PAYLOAD IN ORBIT
(Fully Recoverable + Flyback)

Figure 2

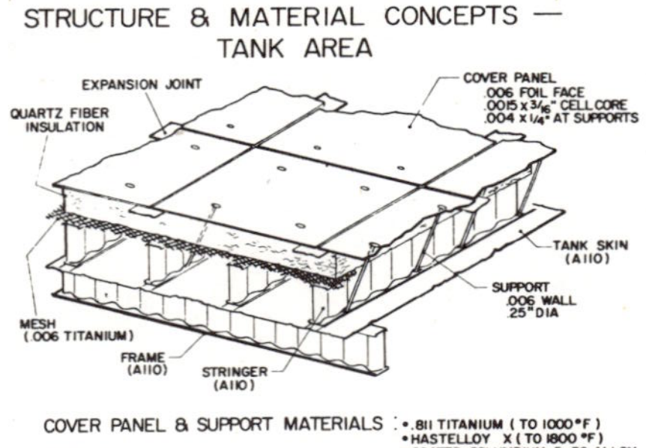

Figure 15

Table 1 - Weight Comparison ~ Fixed Wing Reusable Booster Concepts

Item	Weight ~ 1000 lb							
	VTOHL	HTOHL Sled	HTOHL Runway	Air Br.	Air Br. Rocket	ACES	SSCRJ	In-Fl. Refuel
Stage 1								
Structure	129	144	187	225	240	194	217	290
Propulsion	71	84	90	150	178	196	143	160
Equipment & Misc.	8	23	27	25	25	30	22	45
Total Inert Weight	208	251	304	400	443	420	382	495
Fly Back Fuel	44	53	60	44	44	13	38	15
Weight at Staging	252	304	364	444	487	433	420	510
Boost Propellant	1043	961	1141	231	458	322	245	160
Total Stage Weight	1295	1265	1505	675	945	755	665	1590
Stage 2								
Structure	44	44	44	44	37	44	105	157
Propulsion	23	23	23	23	19	23	40	99
Equipment & Misc.	15	15	15	15	13	15	19	22
Total Inert Weight	82	82	82	82	69	82	164	278
Payload	40	40	40	40	40	40	40	40
Weight in Orbit	122	122	122	122	109	122	204	318
Propellant	353	353	353	353	276	384	191	282
Total Stage Weight	475	475	475	475	385	506	395	600
Total Inert Weight	288	331	380	480	510	495	539	773
Gross Lift-Off Weight	1770	1740	1980	1150	1330	925	1060	2190
Relative Inert Weight	1.0	1.14	1.32	1.67	1.77	1.76	1.87	2.68
Relative GLOW	1.0	0.98	1.12	0.65	0.75	0.52	0.60	1.24

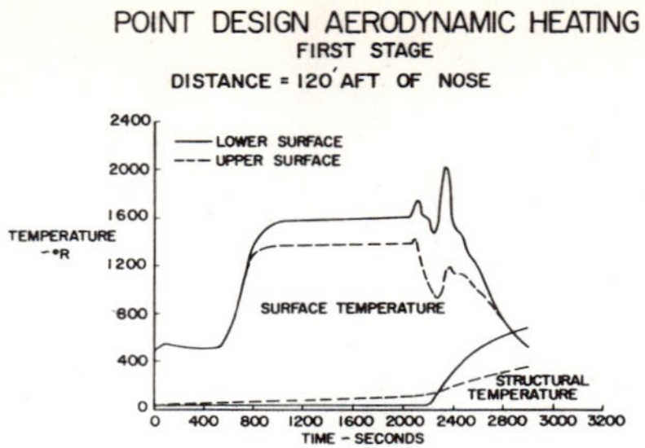

Figure 13

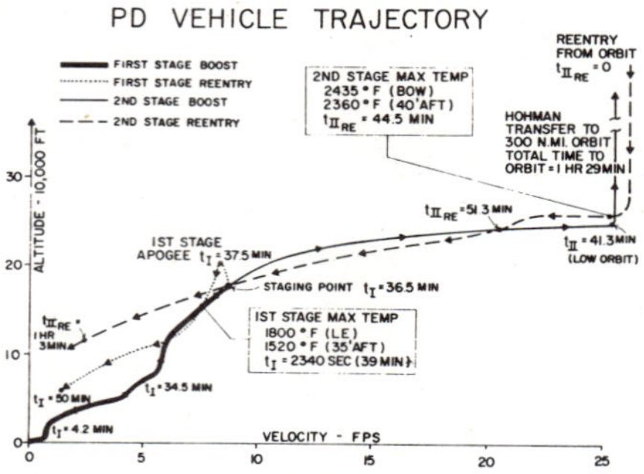

Figure 11

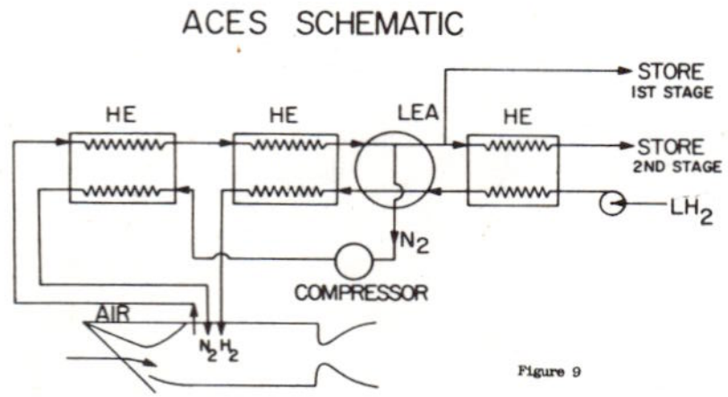

Figure 9

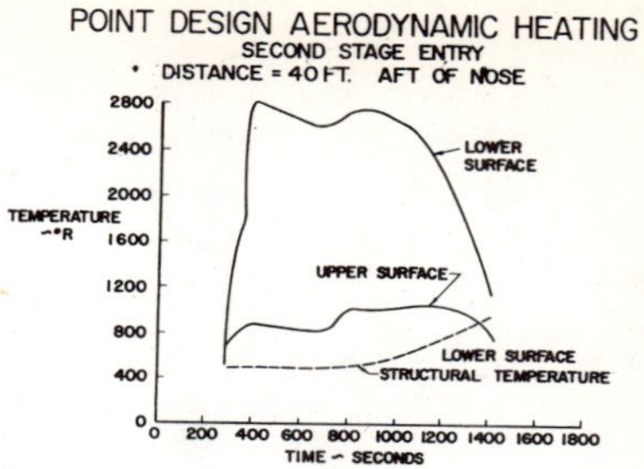

Figure 14

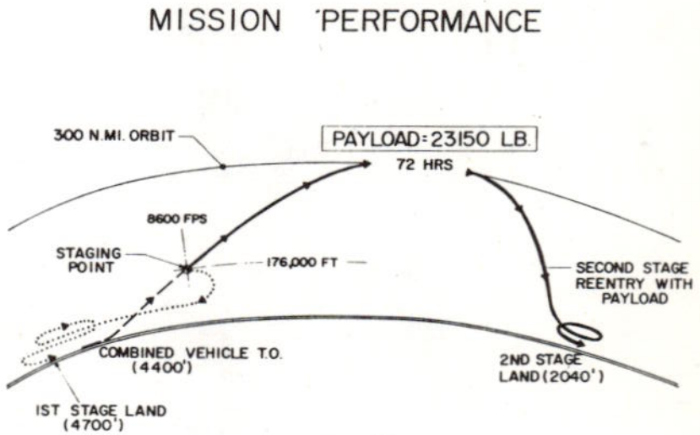

Figure 12

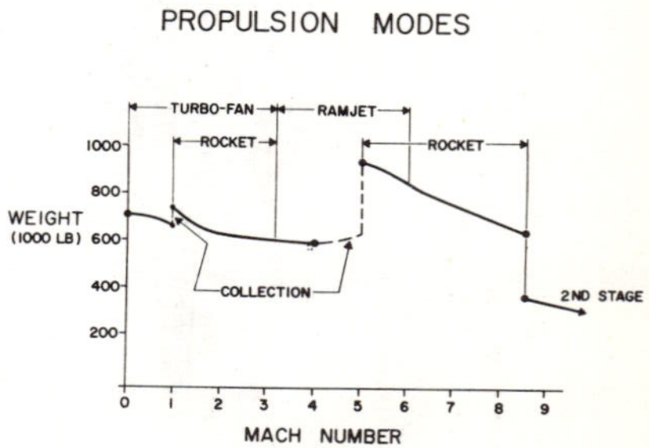

Figure 10

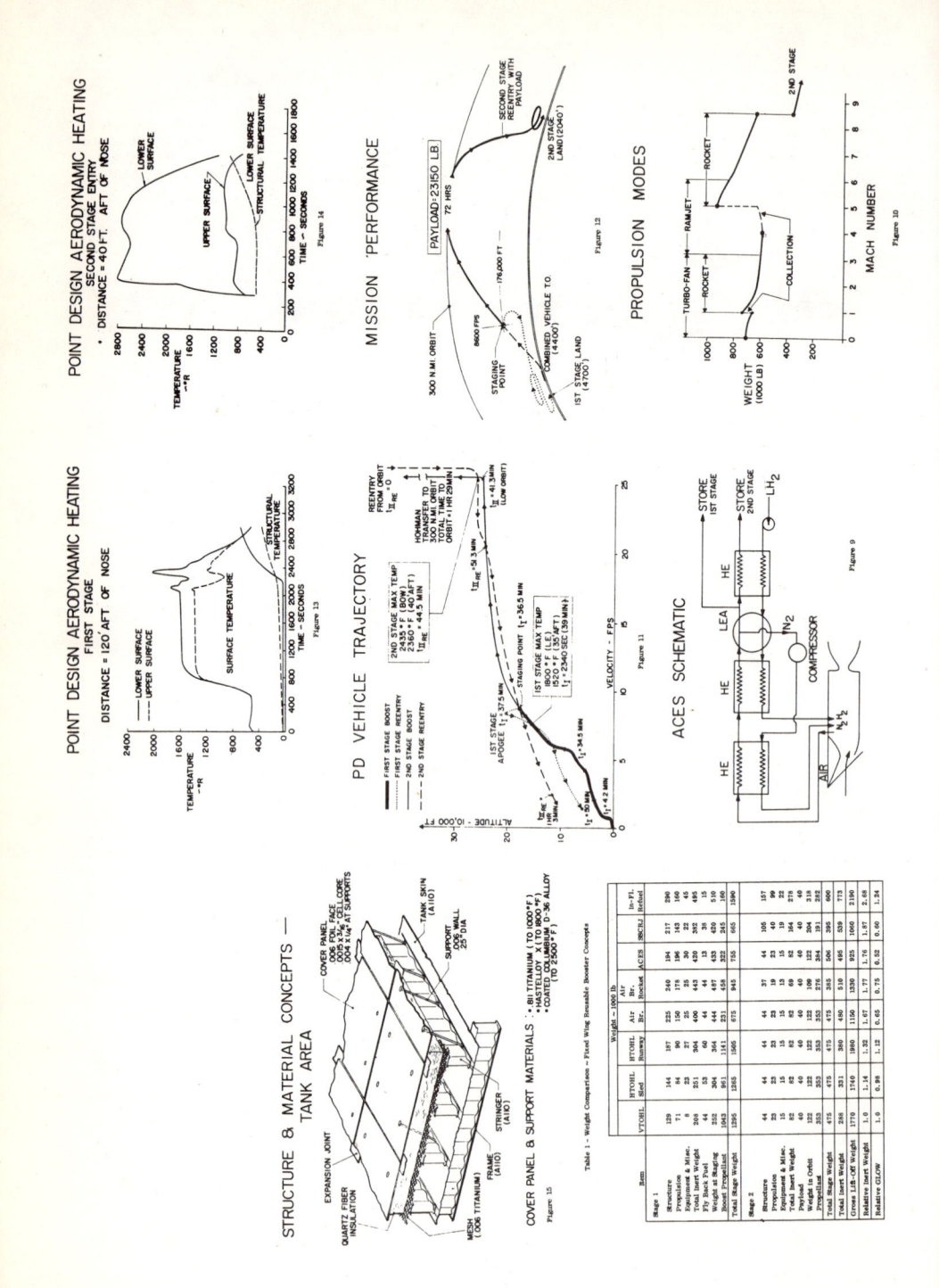

QUESTIONS/ANSWERS FOR "ENGINEERING IN SPACE FLIGHT"

Mr. Rudi Beichel*

Question: What is the advantage of mixed-mode propulsion in an SSTO?

Answer: The mixed-mode propulsion concept involves combining, in the same vehicle stage, two propulsion modes operated sequentially or in parallel. High density propellants are used for liftoff and early ascent, and low density propellants for final ascent to orbit. This produces greater stage performance than possible using either mode separately. The greater performance is achieved when the specific impulse for Mode 2 is greater than that for Mode 1, and the propellant density and density impulse for Mode 1 exceeds that for Mode 2.

The simplest trajectory for mixed-mode application involves a sequential burn, that is, startup and shutdown of Mode 1 engines, followed by operation of the engines in Mode 2. It is also possible to have some overlap of Mode 1 and Mode 2 engine operations. Overlap, approximately 25% of the total thrust, from liftoff to orbit is called "parallel burn". The difference in performance between parallel and sequential burn is a tradeoff from the slightly higher weight of a dual-fuel engine and the loss in performance due to the reduced mixed-mode effect. In general, the degree of gain in performance due to mixed-mode propulsion is affected by the structural weight (mass fraction) of the vehicle.

Question: How much more does a dual-mode rocket engine (one that burns both H_2 and hydrocarbon fuel) weigh than one that only burns a single type fuel?

Answer: There is no general simple answer. The key factor over many others such as chamber pressure, engine cycle, etc., affecting the weight of a liquid rocket engine is the density of the propellants used. For instance, a LOX/hydrocarbon pump feed engine can be designed and built with a thrust-to-weight ratio of approximately 125:1; whereas LOX/H_2 using the same level of technology will range between a ratio of 70-80:1. A dual-fuel engine, shown in Figure 1, actually is a LOX/RP-1 engine with a hydrogen pump feed system attachment kit and extendible nozzle for vacuum operation. The pump systems are operated in sequence. The engine is oxidizer-cooled and the main components - - the combustion chamber, main injector, gimbaled mount, thrust structure, control and the oxidizer fuel system - - are common to both modes of operation of the engine. Both fuel feed systems are self-contained and have independent oxidizer-rich preburners to operate the oxidizer turbopump system. The controls for the engine are, therefore, relatively simple. Since each individual pump feed system drives its own turbine, each preburner requires the same low combination temperature, approximately 1100 degrees F, thus also eliminating the need for an inner propellant seal in one of the turbopumps and, most important , the requirement for turbopump housing and injector manifold cooling. The subsystem can also be arranged and assembled to a LOX/H_2 engine as shown in Figure 1.

----
*Aerojet General Corporation, Sacramento, California

Weight advantages and operational features of the engine worth noting are the change in chamber pressure from 4000 psia during Mode 1 to 3000 psia during Mode 2, the change in expansion ratio (through use of the extendible (retractable nozzle) from 40 to 200 to obtain near optimum performance at both sea level and altitude (Mode 1 and Mode 2), and the rise of a LO_2/LH_2 mixture ratio of 7 for Mode 2 operation. The reduction in chamber pressure aids the heat transfer situation for Mode 2 and, more importantly, equalizes the propellant injection momentum for the main injector during both modes of operation.

The engine weights are shown in Figure 1. The LOX/RP-1 pump feed system weighs 1145 lb per engine. That is dead weight during Mode 2. The weight of the LOX/H_2 pump feed kit is 2565 lb and is dead weight during Mode 1 operation. The weight of the extension nozzle 200:1 is 1020 lb, and is included in the weight shown in Figure 1.

For the specific application mentioned in Question 1, the vehicle consists of eight hydrocarbon engines of 680,000 lb_f at sea level and two dual-fuel engines with the same sea level thrust and a vacuum (Mode 2) thrust of 588,400 lb_f. The total weight of the propulsion system is given in Table I.

TABLE I.

PROPULSION SYSTEM WEIGHT

	SEQUENTIAL BURN	PARALLEL BURN (same vacuum thrust)	PARALLEL BURN (same sea level thrust)
Eight O_2/HC	42,680	42,680	42,680
Two O_2/HC+ H_2	17,840	-	-
Two O_2/H_2	-	15,550	21,110
	60,520	58,230	63,790

For a parallel burn propulsion system with two O_2/H_2 engines (501,000 lb_f sea level and 588,400 lb_f vacuum) the system weight is seen to be 2290 lb_m less. In order to produce the same liftoff thrust, however, the two O_2/H_2 engines would weigh 21,110 lb_m, and have higher vacuum thrust than necessary to accomplish the same mission. The propulsion system would weigh 3270 lb_m more in this case. The question regarding engine weight, therefore, cannot be answered without consideration of the vehicle and its mission.

Question: The shuttle has a GLOW of 4.2 million pounds, what would be the GLOW of an SSTO using mixed-mode propulsion and advanced materials?

Answer: The most recent data available are for advanced earth-orbital transportation systems. A winged single-stage-to-orbit, all LOX/H_2 vehicle has been established as a baseline for comparison with a mixed-mode propulsion concept. Both vehicles in this comparison are launched vertically and return to the landing site horizontally, similar to the orbiter of the shuttle. Advanced structures and materials improvement of 25% by weight and uprated LOX/H_2 engines have been considered in the design. The same structural improvement has been applied for the mixed-mode vehicle with the exception of the engine weights. They are current technology and not yet uprated for future growth.

The two vehicles compare as follows:

Vehicle		Glow	Dry Weight	Payload
Baseline	LOX/H$_2$	2.661M lb	251,000 lb	65,000 lb
Mixed-Mode	LOX/RP-1/H$_2$	2.102M lb	182,000 lb	65,000 lb

Question: Please explain why high combustion pressure is so important to improved performance.

Answer: There are many reasons for high pressure propulsion. For instance, a single-stage winged vehicle acts as a vertical liftoff rocket during ascent to orbit and as an airplane during descent. Such a vehicle poses conflicting aerodynamic requirements on propulsion system packaging. A smaller base area is required for flyback compared with that allowed for liftoff. Since the thrust level of a rocket-propelled vehicle is a function of the gross liftoff weight as well as the vehicle base area, the smaller base area requirement for flyback must be accommodated by the propulsion system without a reduction in thrust level. Performance density optimization is required to achieve the rocket engine performance in the required vehicle base area. This means the utilization of the highest chamber pressure with today's technology giving the maximum specific impulse (maximum expansion ratio) in the allowable area. Should larger payloads be required, then the base area available for propulsion becomes less favorable because the vehicle volume (weight) increases as a cubed dimension, whereas the base area only as a squared dimension. Higher pressure engines would be able to achieve a higher ε (and thrust level) in the same base area, but lower pressure engines would not be able to meet the performance and the vehicle thrust requirements in this base area.

For comparison, nozzles at three different chamber pressures are shown in Figure 2. The sea level nozzle length is indicated by dashed lines and the lower value of ε. Each engine delivers approximately 680,000 lb thrust at sea level using sea level nozzle length. The important thing to note is the large difference in nozzle diameter and length, and to consider this fact in conjunction with the packaging of the engines in the single-stage vehicle.

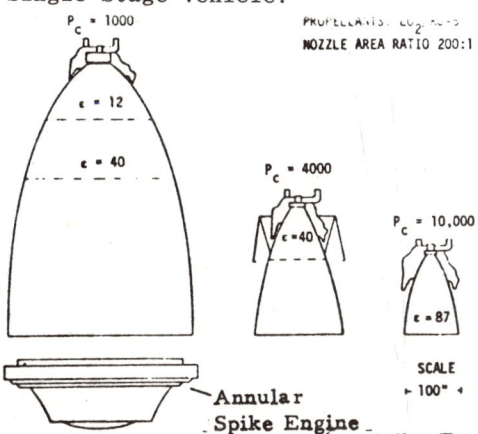

Fig. 2. Pressure Effect on Bell and Annular Spike Engine Size

Another valid argument for high chamber pressure is the increase in specific impulse. To illustrate, Figure 3 shows the effect of chamber pressure on both sea level and vacuum performance, the range of pressures are shown from 500 psia up to 20,000 psia. RP-1 has a specific impulse three seconds higher than RJ-5 but the density of RJ-5 is 33% higher than RP-1. Note that a 1000 psia engine operating at a nozzle expansion area ratio of 11:1 delivers 296.5 seconds (theoretical maximum) at sea level and 326.0 seconds in vacuum, the sea level value being 91% of the vacuum performance. A 4000 psia engine, on the other hand, produces 333.2 seconds ($\mathcal{E}$ = 29.1 optimum) at sea level and 352.8 seconds in vacuum, the sea level being 94% that of the vacuum value. The high pressure engine delivers closer to vacuum performance at low altitude and its sea level performance is 7.2 seconds higher than the vacuum performance of the low pressure engine. A low pressure engine with altitude compensating nozzle, therefore, can never compete (as implied in the literature) with a high pressure engine with any kind of reasonable nozzle design.

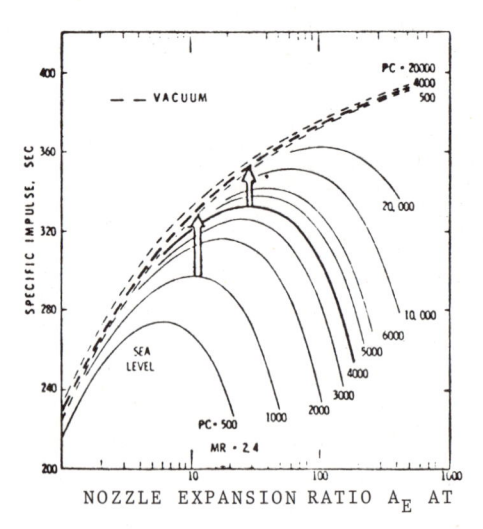

Fig. 3. LO_2/RJ-5 Propellant Performance

The trend in performance continues as chamber pressure is increased, with a 10,000 psia engine generating 96% of its theoretical vacuum performance at sea level.

Pressure has a major influence on nozzle design as shown in Figure 4 where a comparison is made between engines operating at 1,000, 4000 and 10,000 psia, using the highest performance that a given nozzle can deliver as represented by one-dimensional isentropic flow with chemical equilibrium. To avoid any confusion that could result from the use of parametric data (C_T vs P_c/P_a), I have used specific impulse and altitude (U.S. Standard Atmosphere, 1966). Note that the performance for the higher area ratio ($\mathcal{E}$ = 200:1) nozzle is less than that for the lower area ratio ($\mathcal{E}$ = 40:1) nozzle below altitudes of about 35,000 ft and 66,000 ft, respectively, for chamber pressures of 4000 and 1000 psia. Also note the points where nozzle separation occurs; that is separation of the gas flow along the nozzle wall when the exit pressure P_e is less than 0.3 (or 0.4) of the ambient (altitude) pressure P_a.

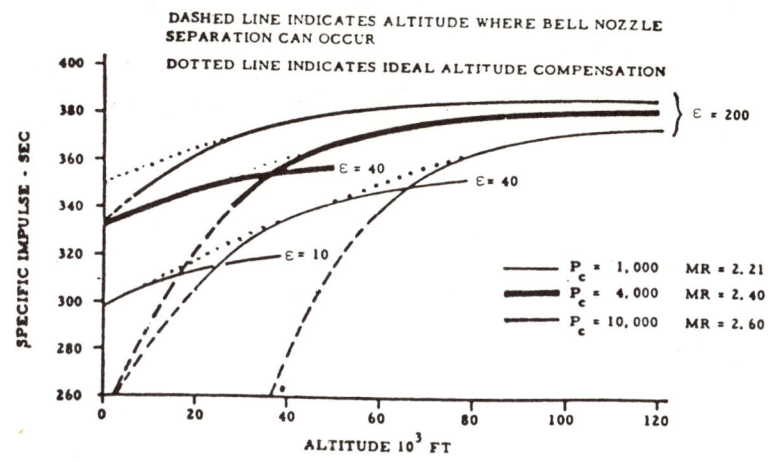

Fig. 4. Effect of Chamber Pressure on Theoretical Preformance

The performance of an ideal altitude compensating nozzle of design area ratio 200:1 is also indicated by the dotted lines in the figure. The ideal altitude compensating nozzle obeys as though it were a "rubber" nozzle able to adjust its expansion area to the optimum condition for each altitude. For a bell nozzle to approximate this performance, it must have multi-position capability. It would have a minimum of three-position, two-position and one-position for a P_c = 1000, a P_c = 4000 and a P_c = 10,000 engine, respectively, as indicated in Figure 4 and noted in Figure 2. The multi-position capability can be accomplished in a number of ways, for example, as has been demonstrated with extendible nozzle types, and with gimbaled bell nozzles on a plug.

In conclusion, it can be stated that the high chamber pressure reduces the size and weight of the propulsion system, the engine compartment of the vehicle, shrinks the propellant tanks, and correspondingly results in a smaller vehicle. It should be noted that this gain is in addition to the mixed-mode improvement.

Question: One scheme for achieving high combustion pressure is to regeneratively cool with liquid oxygen. Can such a system be made to fail safe for either a clogged cooling passage or a leaking (into or out of the combustion chamber) cooling passage?

Answer: The use of oxidizer as a coolant provides the design flexibility and, furthermore, opens the door to the use of propellant combinations (fuels) not otherwise practical for use at high pressures. In this category belongs the hydrocarbon, which is very desirable as a fuel but not practical as a coolant. The idea of using the oxidizer as a rocket motor coolant is not new. In the past a number of engines have successfully been developed to be as fail proof as any fuel-cooled engine can be. For instance, the Agena engine has a flight history of well over 1000 successful flights with no malfunctions. Delta E is oxidizer-cooled. Vanguard and the Able Star are constructed of aluminum and also are oxidizer-cooled. Not one failure in the above mentioned engines has been experienced in the test stands or during flight that is contributed to oxidizer cooling.

During the 1960's, Boelkow developed a high pressure, oxidizer (oxygen) regeneratively-cooled engine. The turbines are also powered with an oxidizer -(oxygen) rich preburner. The engine is a staged combustion cycle.

The Ares development program at Aerojet also successfully demonstrated the use of oxidizer as a coolant for high pressure engines.

All that is required to obtain a fail safe system is that the design be compatible and not intermixed with fuel and oxidizer. The risk of blocking or leaking of coolant passages in or out of the chamber is no different than in any fuel-cooled engine concept.

Question: Are there any other schemes for obtaining high P_c for a LOX/hydrocarbon engine?

Answer: There is always the possibility of a third propellant as a coolant. For instance, a 600,000 lb sea level thrust LOX/hydrocarbon engine requires approximately 20 lb of hydrogen flow rate for cooling at 4000 psi chamber pressure. This hydrogen can be injected at high pressure in the main injector together with the hydrocarbon, but also might find additional use as a turbine drive gas.

There are losses with this concept due to the reduction of mixed-mode effect and losses in specific impulse because the turbine exhaust will have to be dumped in the nozzle skirt at low pressure.

ENGINEERING IN SPACE FLIGHT
Roundtable - Questions

Subrata K. Sarkar*

I. QUESTIONS

Geosynchronous orbits will most probably be used by more and more satellites representing a variety of uses.

a) Please discuss the conflicts that may occur in r.f transmission to and from geostationary satellites as the population of such satellites increases.

b) Is it possible to control the side lobes and the bandwidth of the power transmission from a solar energy power station to avoid significant interference with other transmissions from space?

c) Excluding the solar energy power station, what kind of satellites do you predict will provide the greater demand for r.f. transmissions from space?

II. ANSWERS

a) Angular separation between space stations in geostationary orbit and distance separation between earth stations are maintained to protect receivers of space station and earth station from harmful interference. With the increase of satellite systems, constraints are imposed on these "required separations" of angles and distance in order to comply with performance objectives of the respective services.

Several means are available to relax or minimize these "separations." These are:

1) improved new technology (station keeping, satellite antenna pointing and side lobe reduction, control of spurious emmission through filtering, use of orthogonal polarization, appropriate shaped and spot beams satellite antenna, etc.)

2) monitoring and defining through measurements of the different interference levels (e.g. harmful, acceptable)

3) adherence to the current Radio Regulations. There will never be a serious conflict, as through continuous updating, all the above three areas maintain the balance between demand and supply of satellite services.

b) Phased arrays antennae, which are directly coupled to high power amplifiers, would be the choice for solar energy power transmission. The

---
*Questions see page 104, Proceedings of the Dedication Conference of the International Space Hall of Fame, Alamogordo, N.M., 5-9 Oct.1976.

amplitude and phase of each individual array can be adjusted to provide the optimum beam shape, lowest side lobe levels and beam directivity. Due to high effective radiated power, there will always be significant side lobe levels and spurious emissions (sidebands, spectrum broadening, harmonics). These will require (depending on shared or exclusive use of frequency bands) large angular separation between space stations and wider bandwidth to protect other services. How much angular separation and bandwidth are required to avoid harmful interference? These are questions to be answered during the stages of development of the relevant technology. This also depends on the point in time that the system is to be operated. Allocation and assignment service will only be possible when these matters have been thoroughly understood. The best way to understand the problem is to develop the technology and at the same time receive direct benefits. This is possible by utilizing the phased arrays antennae and high power amplifiers in similar configurations, but within a comparatively very small physical dimension, such as in the telecommunications, mobile and broadcasting satellites.

c) Frequency spectra for the Fixed Satellite Service are presently in highest demand. These are telecommunications satellites for global, regional, or national purpose and are used for telephony, video, data and telex services. There is a foreseen increased growth for Broadcasting and for Mobile (aeronautical, maritime, land) satellite services where satellite means are superior to terrestrial.

CHAPTER IV

SCIENCE IN SPACE FLIGHT ROUNDTABLE

"...The exploration of our own planet may be said to have been completed. In the exploration of space, no such end can be predicted...The exploration of space will go on forever and ever..."
    Willy Ley, 1968, (Germany-U.S.A.)

OVERVIEW OF THE VIKING PROJECT AND
MOST RECENT CONCLUSIONS

Dr. James Martin*

Thank you, Jack. It's been my pleasure to be here this morning. Viking turned out to be a very exciting mission, and what I'd like to do is show you some pictures, both orbiter and lander, and discuss a little bit the results we have found so far in some areas, particularly the question of life. Our findings are not conclusive at all, and contrary to some popular belief, we are not withholding information to promote a follow-on mission. We just simply cannot understand some of the data we are getting.

Viking is a continuation of NASA's planetary exploration program, actually started with Mariner 2, I believe, in 1962, which went to Venus, and in '64 and '67 there were three mariner flights to Mars, and then in '71 the Mariner 9 went into orbit around Mars and took a spectacular set of pictures for about a year. In fact, most of our current pre-Viking knowledge was based on Mariner 9 findings. Viking missions were started actually in 1968, following the cancellation of a program NASA had firmed up in the 60's called Voyager. Voyager was a very large, expensive Saturn-launched planetary voyage, and Viking was intended to be a smaller version of it. I think it has turned out to be that.

The two Viking missions were launched last August, 1975, from Cape Kennedy, the first one on August 20th, and the second on September 9. The first Viking arrived on Mars on June 19th and went into orbit, and the second one arrived on August 7. I'll show you now some slides, if I can have the lights and the first slide. No. 1 (NASA085066a)

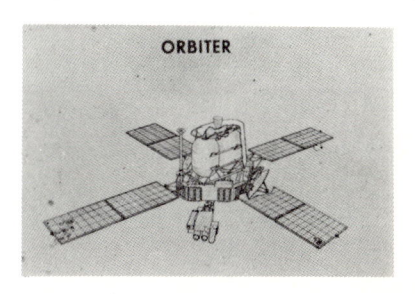

We'll have the second one please. No. 2 (NASA085064a) First a brief description of the orbiter and lander. The orbiter weighs about 5500 lbs. at launch, 3000 lbs. being the propulsion module, which is here. This propulsion module is very large because of the energy requirements for putting the spacecraft into orbit around Mars while you have the lander attached. The orbiter is powered by four solar panels with a wing span of about 30 feet. The science instruments on the orbiter, two TV cameras, and two infrared spectrometers are mounted on this device which is called a scan platform. The scan platform can move in two axes such that you can point the platform at the planet or at any point on the planet while still maintaining the spacecraft locked on the sun and celestial references. The IR spectrometers on the orbiter are to measure the water in

---

*Viking Project Manager, NASA, Langley Research Center

the atmosphere of Mars and to measure the temperature of the planet. These have been very useful in some of the findings I'll subsequently discuss.

Next slide please. No. 3 (NASA085064b) This is the lander in the descent capsule mode. The lander is unique in that it had to be very compactly folded for launch and for cruise. You see the legs are stowed and the antenna stowed here. It's packaged in an aeroshell which is used in taking out most of the energy in descending into the Martian atmosphere. There is a base cover in the back along with a parachute container.

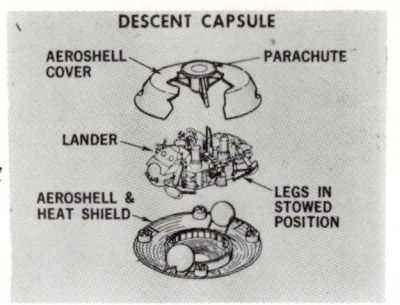

Next slide please. No. 4 (NASA P-16123BC) This is a color picture and is displayed on page 88. This is an artist's picture, but it is a good representation of what the lander looks like on the surface. Here the appendages are deployed. The landing legs, of course, are down. This is a high-gain antenna which is up and points to earth. The meteorology equipment is on this boom. This boom is deployed right after landing, and on top here are the meteorology sensors for measuring wind direction, wind speed, temperature and pressure. This is the scoop on the end of the soil sample boom. This is a retractable boom which gets a sample and then empties the sample into one of three hoppers shown back here, because the organic instrument analysis, the inorganic analysis, and biology instruments are located inside the lander body, and they receive their dirt sample through these hoppers. These two cylinder devices are two cameras. The cameras take black and white, color and infrared pictures.

Next slide please. No. 5 (NASA P-16735AC) This is a color picture and is displayed on page 88. As we approached the planet, we took pictures with the orbiter and we have some of the first color pictures taken. This one happened to be from about three days out, and that's about 400,000 miles. We were closing with the planet at this point at about 7000 miles an hour. This picture shows the biggest volcano on Mars, Olympus Mons and these are the three Tharsis Ridge volcanos, known in Mariner terminology as South Spot, Middle Spot and North Spot. Very imaginative. It is very interesting of Mariner that, if you recall, the spacecraft arrived right in the middle of a global dust storm, and as the dust storm subsided, these four volcanoes were all one could see for several days, and it was rather speculative as to what in the world we were watching.

Next slide please. No. 6 (NASA P16870 (color)) This is a color picture and is displayed on page 88. Now this is a day closer. We are about two days out from the planet and we took this picture. We took a series of color pictures over a 24-hour period such that as the planet rotated, we were able to get full coverage of the planet, but they were half-lit sections, as shown, all in color. Here we have the very large valley that was found by Mariner 9. It's a very large rift valley on Mars believed to have been caused by the contraction of the crust and splitting apart, causing this very big rift. It's about 3000 miles long, and in the deepest place about two miles deep, and something in the order of 100 miles across. It's a very large Grand Canyon. We were about 240,000 miles from the planet when we took this picture. The landing site we went for is right up about in here for the first lander. The landing site for the second lander is right on the other side of the planet; you can't see it. The north pole is just slightly over the top of the picture here, and the south pole is right down here.

No. 7 (NASA P-16848 (3A07)) Viking 1 photographed the crater Yuty, located near the spacecraft's potential landing site, from a range of 1877 kilometers (1165 miles) on June 22. Yuty, 18 kilometers (11 miles) in diameter, has a central peak and probably was made by the collision of a meteorite with the surface of Mars. The lobate flows are layers of broken rocks thrown out of the crater by the shock following impact. The leading edge of the debris flows forms a ridge similar to great avalanches on Earth. The whole area has been worn down by wind and possibly water erosion that accentuates the surface detail. The rim of Wabash crater, about 40 kilometers (25 miles) across, lies at the edge of the picture. Yuty was named for a village in Honduras, Wabash for a town in Indiana.

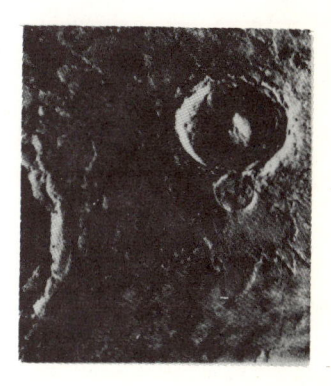

We went into orbit, as I say, on June 19th, and started to take a very spectacular set of photographs. I'm just going to show a very few. We were aimed initially at landing site selection, having been aimed at a site in the Chryse area, and initially as part of the bonus of site selection we were able to take some of these pictures that were in nearby areas. This is an impact crater typical of many on Mars. This is about 20 miles across. The interesting feature of Martian impact craters is the ejecta blanket that you can see out here as though it flowed out. If you look at ejecta patterns on craters on Mercury and the Moon, you see more of the typical rays of blocks that are thrown out. And in fact, there are several craters on Mars that have these rays of ejected material, but more typically on Mars, we find these more, rather strange-looking, flow patterns. The speculation is that something did flow, almost. One theory is the planet is covered with permafrost -- a permafrost layer, and that the impact of a very large meteorite caused local heating enough to melt the permafrost (ice), and you actually would end up with a slurry sort of a mudlike material, which did in fact flow and make these flow lines, and then froze in place. There may well be other explanations, but this is one of the unique features about mars.

Next slide please. (See next page.) Here is a photograph of that very large canyon that I mentioned, showing how the walls have slumped in. This depth from the mesa up here down to the canyon floor is something in the order of a half a mile. This on detailed examination turns out to be a very large dune field. Now you must understand one of the difficulties of landing Viking was the fact that our best resolution for identifying any object on Mars was in the order of a hundred meters, so that the smallest identifiable object we could see turned out to be the size of the Rose Bowl or a football field. And yet the lander scale is more in the order of three meters. So our dilemma was looking for an area that appeared to be a process that we could correlate to something we knew on Earth, that one could construe to be safe. We thought that the sand dune fields would be safe, and we actually looked for those for landing. Unfortunately, this canyon just isn't big enough to land in because of

No. 8 (NASA P-16941 (12A52/53/54)) A field of sand dunes (lower left) some 50 Kilometers long (30 miles) and a great avalanche can be seen on a Martian canyon floor in these pictures taken by Viking 1. The general view is of the north wall of Gangis Chasma (Ganges Canyon), one of the branch canyons of the huge canyon system along the equator. Walls of the canyon are fluted by wind erosion. Many craters caused by meteorite impact dot the plateau surface which may be covered by ancient lava flows. These three pictures are amoung 40 obtained by Viking 1 on July 1 of an area a few degrees south of the equator and near a possible landing site for Viking 2. The dune field is located at about 7.5° S. Lat., 45° W. Long. The avalanche is to the right of the dunes.

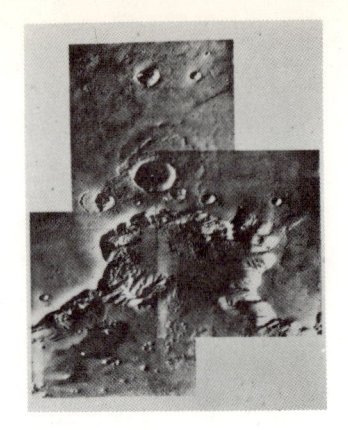

all the various error bars we have in the descent process. So, even though we thought this would be a safe area, we could not go there. Here you see another Martian impact crater with this rather strange ejecta blanket where something has flowed out, and there's a little one there. Next slide please.

No. 9 (NASA P-16952 (14A30)) This is a view taken by Viking 1 on July 3 from a range of 2000 kilometers (1240 miles) looking southward across Valles Marineris --Mariner Valley. This huge equatorial canyon of Mars, about 2 kilometers (1.2 miles) deep, is a few degrees south of the C-1 landing site for Viking 2. The area shown is 70 kilometers (43 miles) by 150 kilometers (94 miles). Aprons of debris on the canyon floor indicate how the canyon enlarges itself. The walls appear to collapse at intervals to form huge landslides that flow down and across the canyon floor. Linear striations on the landslide surface show the flow's direction. On the canyon's far wall in the picture, one landslide appears to have ridden over a previous one. Streaks on the canyon floor, aligned paralled to the length of the canyon, are evidence of wind action. Layers in the canyon wall indicate that the walls are made up of alternate layers of lava and ash or wind-blown deposits.

Here's another view of the big valley. This one is rather spectacular in that people believe that these walls have fallen in due to Mars quakes, but you get an avalanche-like condition where you can see these flow lines and this ejecta, mudslide or whatever it is, and we don't really know, has flowed out. Here's another one that came out on top of it, so it must have happened at a later time. Speculation is that this valley is certainly in the order of billions of years old. Next slide please.

No. 10 (NASA 211-4987 (Mosaic 4A50 to 4A54)) This mosaic of five Mars pictures shows the eastern part of the Chryse region near the prime Viking 1 landing site. The Viking Orbiter cameras took the pictures from a range of about 1600 kilometers (992 miles) on June 23. Braided channels record water flowing on the planet in the past. Fine grooves and hollows on the upstream side of flow obstacles also are seen. Shore of the channel is at lower right.

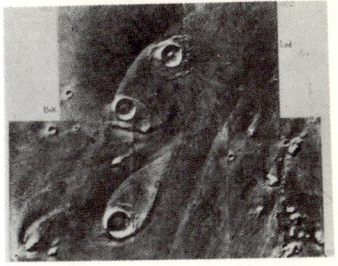

The Mariner 9 photography indicated that they saw river-like features on Mars and some evidence of water. Well, Viking has confirmed the fact that there must have been copious amounts of water on the planet at some time. And we took many pictures like this, showing river channels where this action, this stream-lines shape must have been caused due to water. No one has been able to postulate that this has been a wind erosion process. You can clearly see flow lines, channels all over, at all scales in fact. One of the interesting features of this particular picture is the fact that this ejecta blanket layer had been worn away by the water activity here, whereas this one has not, which sort of suggests that this one happened after the water, and this one was there before the water. Next slide please.

No. 11 (NASA 211-4988 (Mosaic 3A11 to 3A16)) Viking 1 took these six photos of Mars June 22 as it overflew the northeast portion of the Chryse region. Meandering intertwining channels flowing north (toward top) are vividly displayed. It is believed that these channels were cut by running water on Mars in the planet's geologic past. Each frame covers an area of about 2000 square kilometers (775 square miles).

Here is another view of the channeling, showing how apparently the water flowed and then broke through this area here and caused a new channel, and broke through here, very typical of what we see here on Earth in some of the great river systems. We believe from measurements of the length of the channel here, the difference between the water and upland is probably in the order of 100-200 meters, so maybe this gives some indication of the depth of the water.

Next slide please. (See next page) One of the questions on Mars always has been, where did the water come from, and where did it go. We find several instances that looked like this, where there is a definite channeling effect down-stream in this area, but the headwaters, if you will, appear to be a chaotic terrain with no evidence of tributaries or rivers feeding into it. But one of the things is that somehow the water came out of Mars, like a spring or melting permafrost, but suddenly there was a great gushing out of water. It traveled downhill and eroded this channel we see here. We find that most of these big river systems do flow into basins such as the Chryse basin. Chryse basin is some two and one-half kilometers below the mean "sea level"

No. 12 (NASA P-16983 (14A67/14A69))
The sinuous rille at the top of this
mosaic of eight photos probably indicates
flooding of the high plateau in the
vicinity of the potential Capri (C-1)
landing site for Viking 2. In the foreground is a valley, probably caused by
down faulting of the Mars crust. The
hummocks on the valley floor look like
chaotic terrain. It has been suggested
that the subsidence is partially caused
by melting of the subsurface ice. The
large areas of the collapsed terrain
show the regional extent of this phenomenon. These pictures were taken by
Viking 1 on July 3 from a range of 2300
kilometers (1400 miles) and cover an area
of about 300 by 300 kilometers (180 by 180 miles). South is toward
the top as seen from the spacecraft.

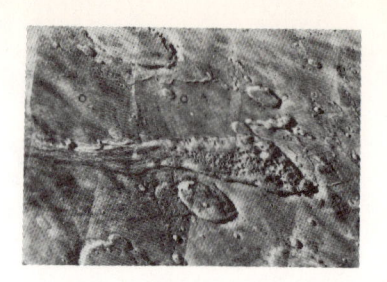

of Mars, and you can think of this as very large dry lakes or dry oceans
at this time. It certainly is a mystery as to where all this water
came from, and also where it all went.

   Next slide please. No. 13 (NASA P-18078) This is a color
picture and is displayed on page 88. Here is a color picture that
was taken from the orbiter. This was taken from fairly high up, so we
have much less resolution, but this is a color picture of this channel
area, of this rift Valles Marineris Valley, showing some of the features,
and we feel this to be a good representation of the color of Mars.
Next slide please.

No. 14 (NASA P-17041 (27A33)) Cross on
this photo taken yesterday by Viking 1
indicates the aiming point for next Tues-
days landing on the surface of Mars. In
the Mars coordinate system, the landing
site is located at 22.4^O N. Lat., 47.5^O W.
Long. The area, in the western part of
Chryse Planitia, is a smooth plain with
many small impact craters peppering the
surface. It is probable that the surface
materials are lava flows, possibly basalt
compositions. The surface may be covered
by a mantle of sediments laid down by the
streams flowing from the highlands to the
west (left). The wind may have reworked
the surface materials into smaller dune
forms. This photo was taken from a range
of 1551 kilometers (962 miles) and
measures 42 kilometers (26 miles) from
east to west and 46 kilometers (28½ miles)
from north to south. North is at top.

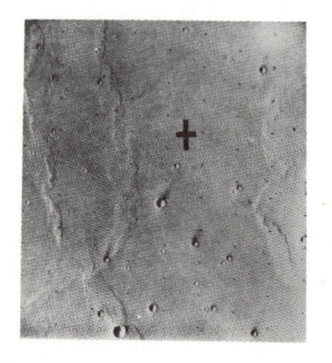

   Now to give you a feel for what we were looking at in the site
selection process. this represents our best resolution over the landing
site, and that was the selected landing site. It really looks smooth,
you see a few small impact craters, and you can see wind tails behind
these craters which we believe is caused by the fine material being blown.
We saw evidence that these wind tails have shifted in direction as much
as 30° in the same crater from the Mariner 9 photography, so there has
been some wind in that four-year interval, but this represented our
state of knowledge, and we thought it looked like a pretty good landing
site so we descended to the surface on July 20, and started taking pictures, and were somewhat surprised by the rocks we found. Next slide
please.
     P-17053 (Sol 0))   78

No. 15 (NASA P-17053 (Sol 0)) This is
the first photograph ever taken on the
surface of the planet Mars. It was obtained by Viking 1 just minutes after
the spacecraft landed successfully. The
center of the image is about 1.4 meters
(five feet) from Viking Lander camera
#2. We see both rocks and finely granulated material--sand or dust. Many of
the small foreground rocks are flat with
angular facets. Several larger rocks
exhibit irregular surfaces with pits, and
the large rock at top left shows intersecting linear cracks. Extending from
that rock toward the camera is a vertical
linear dark band which may be due to a
one-minute partial obscuration of the
landscape due to clouds or dust intervening between the sun and the
surface. Associated with several of the rocks are apparent signs
of wind transport of granular material. The large rock in the center
is about 10 centimeters (4 inches) across and shows three rough
facets. To its lower right is a rock near a smooth portion of the
Martian surface probably composed of very fine-grained material. It
is possible that the rock was moved during Viking 1 descent maneuvers,
revealing the finer-grained basement substratum; or that the fine-
grained material has accumulated adjacent to the rock. There are a
number of other furrows and depressions and places with fine-grained
material elsewhere in the picture. At right is a portion of footpad
#3. Small quantities of fine grained sand and dust are seen at the
center of the footpad near the strut and were deposited at landing.
The shadow to the left of the footpad clearly exhibits detail, due
to scattering of light either from the Martian atmosphere or from the
spacecraft, observable because the Martian sky scatters light into
shadowed areas

This is the first photograph taken by the Viking 1 lander. This is
the foot pad of the lander sitting on the surface and you can see the
many small rocks. We saw some evidence of a few of these rocks having
been moved a little bit due to the landing impact. This footpad hit at
about two meters per second velocity. Next slide please.

No. 16 (NASA P-17155) This Mars view
looks northeast from Viking 1 and completes the 360° panorama of the landing
site begun earlier with the spacecraft's
other camera. A layer of haze can be
seen in the Martian sky. Large dark
boulders dominate the scene. The largest
boulder (center) is about 3 meters (10
feet) wide and one meter (3 feet) high.
Rocks in the foreground are lighter and
appear mottled. The rocks may have been
derived from lava flows or stream deposits
which are visible on orbiter images.
These deposits may have been redistributed by impact craters. The fine
material visible between the rocks has
dune morphology and appears to have been
deposited by wind.

Now here we found a very large rock, large in comparison at least. This rock is in the order of eight meters from the lander. It's about five meters in distance this way, and about a meter and a half high. Typically, the rocks around the lander and some of these others in this general rock field are something in the order of twelve inches high or less. Not very large rocks. Those we can accommodate in landing. If we'd hit on this rock, it would have toppled the lander over and smashed it. Next slide please.

No. 17 (NASA P-17429) Those Martian weather reports, received daily from more than 200 million miles away, start right here at Viking 1's meteorology instrument. Mounted atop the extended boom, the meteorology sensors face away from the spacecraft. They stand about four feet above the surface and measure atmospheric pressure, temperature, wind velocity and wind direction. The cable parallel to the boom is connected inside the spacecraft body with the electronics for operating the sensors, reading the data and preparing it for transmission to earth. A second Mars weather station will begin operation when Viking 2 lands somewhere in the planet's northern latitude. Viking 2 arrived at Mars and went into orbit August 7, 1976.

Here's another picture, and this is another view of a different rock. There's a bigger one out here. This shows some of the sand dunes in the area. You see in the photograph some beautiful etched sand dune fields that look like they're sculptured almost. Here are some small rocks in the foreground. Next slide please

No. 18 (NASA P-17430 (Sol 14)) This spectacular picture of the Martian landscape by Viking 1 Lander shows a dune field with features remarkably similar to many seen in the deserts of Earth. The dramatic early morning lighting--7:30 a.m. local Mars time--reveals subtle details and shading. Taken August 3, 1976 by the Lander's camera #1, the picture covers 100°, looking northeast at left and southeast at right. Viking scientists have studied areas very much like the one in this view in Mexico and in California (Kelso, Death Valley, Yuma). The sharp dune crests indicate the most recent wind storms capable of moving sand over the dunes in the general direction from upper left to lower right. Small deposits downwind of rocks also indicate this wind direction. Large boulder at left is about eight meters (25 feet) from the spacecraft and measures about one by three meters (3 by 10 feet). The meteorology boom, which supports Viking's miniature weather station, cuts through the picture's center The sun rose two hours earlier and is about 30° above the horizon near the center of the picture.

Another picture showing the sand dunes. This particular boom is the meteorology boom you see here. We are looking out through it.

Next slide please. No. 19 (NASA P17164 Sol 1) This is a color picture and is displayed on page 88. This is the first color picture of Mars taken on July 21. One of the interesting features, many of the rocks on Mars appear to have been eroded in this concave kind of section. You find quite a few rocks that are almost scooped out, and this is rather unusual and would suggest that the rocks have been there quite a while, because with the low pressure on Mars, about eight millibars, the cue is such that we believe it would take a lot more wind and a longer time to do this sort of erosion you see here.

Next slide please. No. 20 (NASA P-17173) This is a color picture and is displayed on page 89. This is a picture taken a couple of days after landing early in the morning about 7 a.m. You can see the sky is darker and the land is somewhat darker. The sun is shining on this RTG wind cover that has the flag, bicentennial seal, and another emblem on it. Next slide please.

No. 21 (NASA P-17254) This image shows the trench excavated by Viking 1 surface sampler. The trench was dug by extending the surface sampler collection head in a direction from lower right toward the upper left and then withdrawing the surface sampler collector head. Lumpy piles of material at end of trench at lower right were pulled by plowing from trench by the backhoe which was used to dig trenches later in the mission. Area around trench has ripple marks produced by Martian wind. The trench which was dug early on Sol 8, is about 3 inches wide, 2 inches deep and 6 inches long. Steep dark crater walls show the grains of the Martian surface material stick together (have adhesion). The doming of surface at far end of the trench show the granular material is dense. The Martian surface material behaves somewhat like moist sand on Earth. Evidence from the trench indicate a sample was collected and delivered to the experiments after repeated tries. The biology experiment level full indicator indicates a sample was received for analysis. The X-Ray fluoresence experiment has no indication to show it received a sample. The GCMS experiment level full indicator suggests no sample was received, but this matter is being investigated.

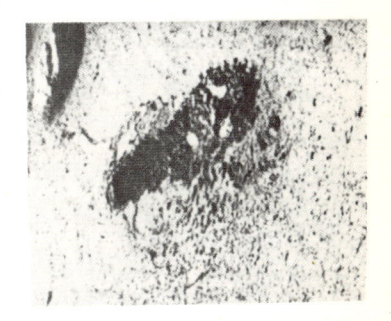

This is a picture of the first soil sample dig we made. The scoop went in at this end, pushed along, and dug up a sample. This is the initial sample for our biology test. The soil appears to be the consistency of lunar nominal soil, maybe a little finer. It handles sort of like moist sand, except it is very fine and does stick to the scoop, but it has some interesting characteristics considering how dry the planet is. In fact, there is no moisture at all on the surface that we can see.

Next slide please. No. 22 (NASA P-17614) This is a color picture and is displayed on page 89. This is a color slide of the soil sample scoop posing for its portrait. This picture was taken during one of the periods the soil sample scoop quit working, and we wanted to know where it was. We took the picture and found it. Here it shows in color two of the soil sample digs that have been taken.

Next slide please. No. 23 (NASA P-17165 Sol 4) This is a color picture and is displayed on page 89. This is another picture. I should remark that the sky is in fact pink. It is believed to be caused by particles, small particles, of this red material, which is all over Mars, suspended in the atmosphere. It is falling out. We've noticed at this site of the first landing, that the sky is less opaque now than it was by a factor of two over the two or three months we've been there. This is a picture of Mars, and this happens to be a little piece of the spacecraft here showing the terminal descent engines, and this is some of the wiring. Interesting enough, one of the problems when we got to Mars was what color Mars really was, because when you look at Mars through a telescope, you get the impact of all the Earth's atmosphere, and it really distorts colors enough that when we got there we didn't really know for sure what color it was. Just before getting to Mars, I warned some of our color scientists that whatever color they picked, they were going to have to stick with, because I didn't want Mars to be a different color every week. Besides, no one can tell whether we're right or wrong, so whatever color we came out with would be alright. They decided that it really should be red, and interestingly enough, that particular cable, if you look at a piece of it on Earth, is exactly that color. It is that precise red color of that cable, and so therefore I believe that this is a fairly good representation. The red color we believe is caused by a ferrous oxide. You can't tell quite what, but it just appears to be that because of the fact that there's a lot of iron in the soil, about twice as much in the samples on Mars as on Earth.

Next slide please. No. 24 (NASA P-18096 (Sol 35) This is a color picture and is displayed on page 89. Here's another picture of that big rock. I should tell you a little story about that rock. When we first got a picture of it a couple of days after landing, it was named Big Bertha, and it got into the press as Big Bertha. About two days later we got a telephone call from a women's rights group in San Francisco complaining quite bitterly that we had named the rock Big Bertha, so it was renamed Big Joe, and I have received no complaints about that. The rock is covered, you can see it in this photo if you look closely, with a red fine material. It appears that it has collected during a dust storm some of this very fine material that covers the planet, and you can see it on top. You can see how it fits into some of the crevices and what have you. The rock itself is of a different shade, though it's reddish, it's a much darker rock. This is a piece of the lander RTG cover in the foreground here. Next slide please.

No. 25 (NASA P-17681 (Sol 0)) Viking 2's first picture on the surface of Mars was taken within minutes after the spacecraft touched down on September 3. The scene reveals a wide variety of rocks littering a surface of fine-grained deposit. Boulders in the 10 to 20-centimeter (4 to 8 inch) size range--some vesicular (holes) and some apparently fluted by wind--are common. Many of the pebbles have tabular or platy shapes, suggesting that they may be derived from layered strata. The fluted boulder just above the Lander's footpad displays a dust-covered or scraped surface, suggesting it was overturned or altered by the foot at touchdown. Just as occurred with Viking 1's first picture on July 20, brightness variations at the beginning of the picture scan (left edge) probably are due to dust settling after landing. A substantial amount of fine-grained material kicked up by the descent

82

engines has accumulated in the concave interior of the footpad. Center of the image is about 1.4 meters (5 feet) from the camera. Field of view extends 70° from left to right and 20° from top to bottom. Viking 2 landed at a region called Utopia in the northern latitudes about 7500 kilometers (4600 miles) northeast of Viking 1's landing on the Chryse plain 45 days earlier.

Now I'll talk a little bit about the second lander and some of its photographs. Here is the first one taken by the second lander. We landed on September 3rd. One difference was the first lander landed at about 4:30 in the afternoon, Mars local time, whereas this one landed about 9:30 in the morning, Mars local time. So we got quite different sun angles on the initial photographs. We have now taken pictures of Mars at all the sun angles, because we have a tape recorder on board, so we can take pictures whenever we want to, put them on tape, and send them back to Earth. The unique feature with this second landing site was the fact that a lot of the rocks are full of what are believed to be gas holes. The vesicular nature of this would suggest that these were formed in a volcano and were thrown out. These are gas pockets. Now there are some people who believe that this is a wind erosional effect, but it's rather unique in this general area where almost all the rocks exhibit this vesicular appearance, whereas in the first landing site area, none of them seem to have this. So there is something different. We haven't been able to find any major volcano in this area at all, so if these are volcanic products, we don't know where they came from. Next slide please.

No. 26 (NASA P-17688 (Sol 1)) The rocky Martian plain surrounding Viking 2 is seen in high resolution in this 85-degree panorama sweeping from north at the left to east at right during the Martian afternoon on September 5. Large blocks litter the surface. Some are porous, sponge-like rocks like the one at the left edge (size estimate: 1½ to 2 feet); others are dense and fine-grained, such as the very bright rounded block (1 to 1½ feet across) toward lower right. Pebbled surface between the rocks is covered in places by small drifts of very fine material similar to drifts seen at the Viking 1 landing site some 4600 miles to the southwest. The fine-grained material is banked up behind some rocks, but wind tails seen by Viking 1 are not well-developed here. On the right horizon, flat-topped ridges or hills are illuminated by the afternoon sun. Slope of the horizon is due to the 8-degree tilt of the spacecraft.

This is a picture of the second landing area. One distinction here is we found no big rocks. These are all generally average in size, about 8 to 12 inches, and they seem to be widely spread in all directions of a very uniform size. We don't quite understand the processes that caused this. You can see, not too well in this pictue, but in some of the other photographs, you can see what appear to be little channels that wander around through the rocks almost like you would expect in the desert after a rainfall. It certainly suggests that water flowed here sometime.

Next slide please. (See next page) This is another view of the area around the second lander. This object is a contamination cover, which is over the end of the soil sample scoop to keep it clean. Now this soil sampling mechanism was cleaned here at White Sands, incidentally, all of

No. 27 (NASA P-17876 (Sol 5)) Shining on the Martian surface near the Viking 2 spacecraft is the aluminum shroud, or cover, which protected the collector head of the surface sampler instrument during Viking's year-long journey from Earth. On September 5, two days after Viking 2 landed, the surface sampler was rotated from its parked position atop the spacecraft and pointed downward about 40 degrees. The shroud was then ejected by a set of eight springs positioned around its base. It struck the porous rock at the bottom of the picture, bounced about 20 inches, hit the surface again and bounced another 20 inches. The scar left by the second bounce is faintly visible halfway between the shroud and the rock it struck. The shroud is 12 inches long and 4½ inches in diameter. The large rock just beyond it is about 2 feet long and about a foot thick. At lower right is the support structure of one of the spacecraft's three landing legs.

them were, because they went through a very super-cleaning process. We were searching for organics in the soil of Mars, and our instrument can measure to about one part in a billion, and we were trying to reach cleanliness levels that wouldn't contaminate our Mars sample results with Earth contamination. This is that contamination cover which is thrown off. It bounced a couple of times back here and landed up here. Next slide please.

No. 28 (NASA P-17687 (Sol 1)) High-resolution photo of the Martian surface near the Viking Lander 2 shows a few square meters (yards) at one of the possible spots for acquiring a soil sample. The sample was collected Sept. 11 by the Lander's trenching scoop and delivered to the spacecraft instruments. The rock in the right foreground is about 25 centimeters (10 inches) across. Most rocks appear to have vesicles, or small holes, in them. Such rocks on Earth can be produced by either volcanic processes or by hypervelocity impacts of meteorites. Some areas are lighter than others, suggesting the presence of two kinds of fine-grained materials, which also can be produced by both volcanic and impact processes. A nearby large impact crater, named Mie, may be the source of the rocks and fine-grained material at the landing site.

Here is just a picture of one of the areas I wanted to show you. This is a high resolution photograph, and it's blown up. These little squares are the pyxels in the picture that represent the included bits of information from which the picture is made. But what I wanted to show you is this crack that runs along through here. It suggested that this was a crusty-like material, and this crack had somehow opened up, and maybe either something had escaped from inside, or it was a habitat for organisms, so we decided to dig in this particular area as being somewhat unique and different than anything else we've seen on Mars. The next slide will show you some of the results. Next slide please.

No. 29 (NASA P-18066 (Sol 21))
Operation of the surface sampler in obtaining Martian soil for Viking 2's molecular analysis experiment September 25 was closely monitored by one of the Lander cameras because of the precision required in trenching the small area--8 by 9 inches--surrounded by rocks. Dubbed "Bonneville Salt Flats", the exposure of thin crust appeared unique in contrast with surrounding materials and became a prime target for organic analysis in spite of potential hazards. Large rock in foreground is 8 inches high. At left, the sampler scoop has touched the surface, missing the rock at upper left by a comfortable 6 inches, and the backhoe has penetrated the surface about one-half inch. The scoop was then pulled back to sample the desired point and (second photo) the backhoe furrowed the surface pulling a piece of thin crust toward the spacecraft. The initial touchdown and retraction sequence was used to avoid a collision between a rock in the shadow of the arm and a plate joining the arm and scoop. The rock was cleared by 2 to 3 inches. The third picture was taken 8 minutes after the scoop touched the surface and shows that the collector head has acquired a quantity of soil. With surface sampler withdrawn (right), the foot-long trench is seen between the rocks. The trench is three inches wide and about 1½ to 2 inches deep. The scoop reached to within 3 inches of the rock at far end of trench. Penetration appears to have left a cavernous opening roofed by the crust and only about one inch of undisturbed crust separates the deformed surface and the rock.

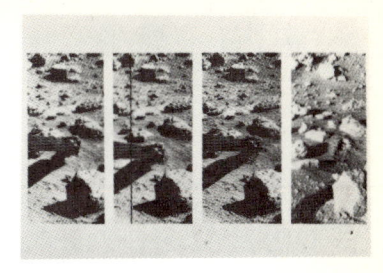

   This shows our digging in that area. This is a series of four pictures. This is the soil sampler scoop just reaching the surface and you can see the crust-like material breaking up here. Now it's pushed all the way in as you can see. Now it's withdrawn and that's the trench that was dug, and it went right across the crack. We had hoped that this was different and might show some different characteristics, but unfortunately, it showed the same confusing results in terms of biological activity, and we found no organisms in this sample or any of the others. Next slide please.

No. 30 (NASA P17689) On a clear day on Mars, you can see tens of thousands of rocks. Two high-resolution scans by one of Viking 2's cameras were mosaicked to create this scene looking northeast to the horizon some three kilometers (two miles) away. The rock in the lower right corner is 10 inches across. The largest rock near the center of the picture is about two feet long and one foot high. What appears to be a small channel winds from upper left to lower right. Slope of the horizon is due to the eight-degree tilt of landed spacecraft.

   Now we decided that maybe to look for organisms or life, we should look underneath a rock. So we decided that we wanted to push one of these rocks out of the way, so we could dig under it. So Monday night, we pushed on this particular rock right here. It seemed big enough and looked like it was sort of loose on top, and we wanted to push it out of the way, so we could dig under it.

Next slide please. No. 31 (NASA85066b) Here is a picture showing the soil sample scoop right at the rock. This is the rock we're pushing on. We pushed with a force of about 30 lbs., and that rock didn't budge! It is either the tip of the iceberg, or it is fastened to the planet pretty sturdily, because it would not move at all, and in the laboratory we have a mock-up of a lander and a pusher, and we were able to push rocks of this size quite easily. You can see where the back-hoe actually hit on the soil here, that little indentation, and we were pushing right about there. Well, we were discouraged, but not completely, because Friday of this week, tomorrow, we are going to push again on another rock, which is a little smaller, and we hope it will move. We are still anxious to dig material from underneath a rock, because one of the things that might kill life or affect the organisms on Mars is the fact you do get a high ultra-violet flux from the Sun. There is no atmosphere, no ozone layer, so that it is believed that the flux could be high enough to kill any surface life as we know it. Perhaps these rocks, if they have been there long enough, might have provided the protection from the UV, and that's why we are interested.

Next slide please. No. 32 (NASA211-5178BC) I just want to say a few words about the major findings to date of Viking. First off, in the atmosphere, we have detected Nitrogen, Argon, Krypton, Xenon and traces of Neon. These were all unknown before Viking. In addition to these, we have also measured isotopes of Nitrogen and Argon by comparing isotope ratios. Our atmospheric scientists now believe that the atmosphere on Mars was, in fact, much heavier at sometime in the past, whereas today, it is 6-8 milibars. They suspect it could have been in the hundreds of milibars, certainly enough to support liquid water. We have been measuring the winds at two different locations, and they are very light and variable. The highest wind velocity we have measured is about 20 mph, and it blows in the daytime, very much like it does most places on Earth, and quits at night. At midnight, there is almost no wind at all, and it comes up again in the morning. We have determined that the soil elements are very similar to Earth in that we've measured about the same amount of Titaniem, Aluminum, Magnesium, Calcium and Silicon. The only thing that is grossly different is where there appears to be something like twice as much iron on Mars as on the Earth. We have found no organic materials at all. In fact, it really makes us wonder. It's very difficult, of course, for our scientists to conceive of life without having organic materials from the bodies of life that has died, but we find no organics, no compounds with Carbon and Hydrogen. A surprising amount of surface material appears to be magnetic, more so than on Earth. We don't quite understand that, but it might be related to the more iron. I've shown you pictures of the extensive fluvial activity, the fact that so much water must have existed at one time on the planet is a very interesting discovery, and it raises more questions than it answers.

A recent discovery that we made just about a week ago is that the north polar cap in this season - and we're in the summer season right now, so the cap is about as small as it ever gets, so it's a permanent cap - is essentially all water, H_2O water, whereas it has been speculated originally that it was frozen CO_2. We believe as the winter comes on, another 6-8 months from now, we will see CO_2 ice form at the cap, and it will cover the water, but this water ice was determined from the orbiter instruments in that the water atmosphere detector showed that the atmosphere above the ice cap was saturated, and the temperature measurements showed the temperature to be about 100 degrees warmer than it should have been if it were frozen CO_2. So quite clearly, it is water ice. We see rather dramatic changes in water vapor around the planet. At the equator there is essentially no water vapor in the atmosphere, and as you go toward the pole, it increases. You can see all over the planet, it's latitude independent; as you go farther north there is more water in the atmosphere until you actually reach a saturated condition. It probably does actually snow on Mars, although we don't think it rains.

Then we have this question of the complex surface chemistry, where our biology instruments have gotten positive indications of life in terms of biology, but no one quite believes that this is really life, because of the fact that we cannot find organics. It is judged right now that this is more likely to be some complex chemistry going on that we don't understand. We hope that in the continuing months of Viking, we can better answer this question. The plan is to continue the prime mission until November of this year. We have a three-week period out for a Martian conjuntion, and we will pick up again, and we hope to continue to operate the lander, take pictures and measurements, and perform some long-term biological investigations for several months into next year. The plan is to continue the mission until the summer of 1978. Thank you.

No. 4

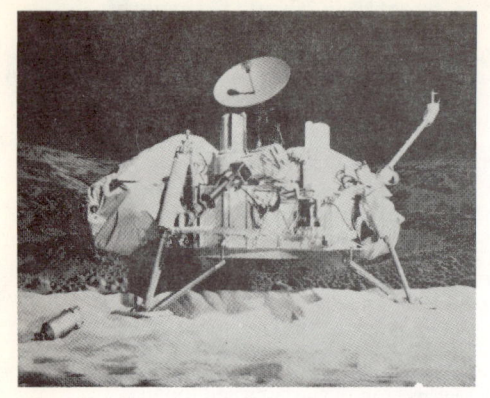

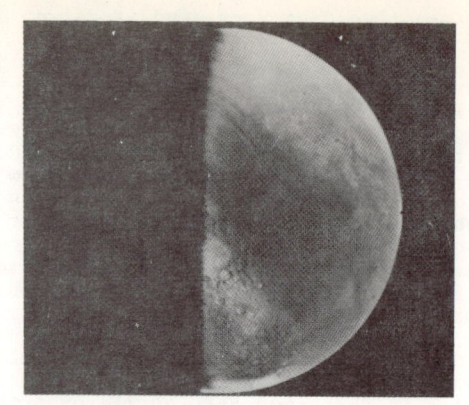

No. 6

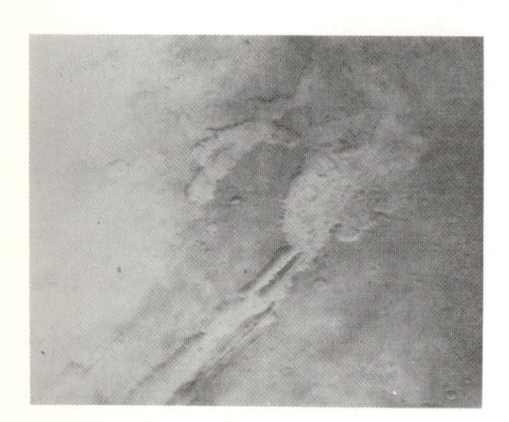

No. 5

No. 13

No. 19

88

No. 20

No. 22

No. 23

No. 24

89

ASSESSMENT OF SOLAR SYSTEM BY OPTICAL ASTRONOMY
ON SITE OBSERVATION AND CRUST SAMPLES

Professor Audouin Dollfus*

This afternoon we have a roundtable on optical astronomy. The purpose of such a roundtable is to clarify, if possible, the state of the art, the knowledge of the Universe and the solar system, and, if possible, try to foresee what will be future needs, what we need for future analysis. In order to achieve this task, we have a certain number of distinguished panel members. Present is Dr. John Evans, solar physicist, head of Sacramento Peak Observatory. We have Professor John Strong of the University of Massachusetts at Amherst, dealing in infrared optics, and pioneer in baloon astronomy. Also present is Dr. Lubos Perek, Czechoslovakian astronomer. He takes a great role in the International Astronomical Union as General Secretary, and is now head of the United Nations Department of Outer Space Affairs. We have Dr. Arthur Code, who was an astronomer at Mt. Wilson and Hale Observatories using large telescopes, then a radio astronomer, increasing his field to the entire Universe. He is head of the Department of Astronomy at the University of Wisconsin. Finally, there is Dr. Maxwell Hunter, project manager for the Large Space Telescope which is supposed to play a great role in the future. There are some absent, and we apologize for them.

Dr. Code will be the moderator of these sessions, so he will do most of the job. Also, we agreed that there are two domains, two fields, as a topic for this roundtable, first, the solar, including the sun and also the galactic and extra-galactic studies. Dr. Code will make his introduction for the galactic and inter-galactic studies.

May I say some words about our problems in our solar studies. Since a quarter of a century, if you like, a lot of data have been gathered on the solar system, and it is this gathering of data which helps the work in solving important questions like the origin and the evolution of the solar system, the creation of the Sun and planets. Among this data which were obtained either by classical telescopes or by space craft, we see simple things like a description of the solar system objects, dynamical and physical processes on them, the planets, the satellites, the asteroids, the comets and the dust. Their distribution, therefore, in space is now clarified, and so known. The size of distant objects is properly measured. We know things mostly through classical astronomy, and then through refined space technology; we know the orbital parameters for most of these objects. For the majority of all the large objects, and a certain number of small objects, we have the mass; we have their densities; we have their character, the harmonics of their gravitational field, and, in some cases, of the Moon for example, we have the mass anomalies, or mass cons. We also can infer from all this data, some results about the internal structure of these bodies. We have data on the optical properties of the surfaces of these bodies, such as the albedo,

---
*Dr. Dollfus is Director of the Observatory of Meudon, Paris, France

the polarization, the reflectance spectrum, etc. From this data, we can infer characterization of the crust at the surface of these bodies.

Another kind of results which were obtained piece by piece, are the chemical and isotropic analyses of the solar system bodies. From telescope observations we have, we made spectroscopics of the atmospheres of the planets, identifying hydrogen in the major planets as being a major constituent and CO_2 being an important constituent of the telluric planets. Also the spectrography of comets is very important because, while given a spectral signature as emission lines in the spectroscope. So this is a telescopic return for the chemical composition. For the orbital science, using orbiting satellites, new categories of data are being provided which permit the determination of the crust's chemical composition by x-ray fluorescence techniques, by gamma ray spectroscopy, and also by refractance spectroscopy of the compositions of at least the surfaces of the solar system objects, and now we have the means of in situ measurements of landing spacecraft on Venus, the basaltic composition of the surface, and of the Surveyor landing on the surface of the moon, and the Viking results about the composition of the Mars surface, which were summarized this morning. The most interesting are samples, Lunar rocks, we have avilable for laboratory measurements. With a sample in hand and with all the techniques available in terrestrial laboratories, we can make chemical, mineralogic approaches, and also dating of the samples. And we have three categories of samples for measurements. We have meteorites which fall upon the Earth, and this is very fortunate. We get three types of meteorites, the iron, the chrondites, and the achrondites, which have very different origins in the past. We have the Lunar samples returned by the Apollo mission, and all obtained analyses were extremely well coordinated by NASA. We also have the cosmic dust. Thus, we have some information about the composition of the celestial objects we like to study.

Another source of information is the atmosphere of the planets. From telescopic observations and from spacecrafts, we know the composition of the atmospheres of planets, and we have some models for vertical structures of them, the pressure, the temperature, their variations with height, and also of latitude and position with all the parameters when needed. We have in some respects, some indications for the vertical cloud structure of the atmosphere of Mars, the dust storms, the dust which rises from the surface due to strong gusts of winds and the nature of the white clouds that float and move with the wind in the atmosphere of Mars. We know in some respects the composition of the upper atmospheric clouds of Venus, which is sulfuric acid. From infrared measurements, we obtain results on the energy budget of planets, and from telescopic observations and space capsule observations, too, the circulation of the atmospheres of planets in motion, which is very strong, for the fast rotation of the upper atmosphere of Venus which was extensively studied by Mariner 10, and after this, discovery by telescopes, and the forces of the equatorial team on Jupiter's fast acceleration of clouds of upper atmospheric strata at the equator of Jupiter, and of still another category of data which was collected by the Viking mission.

We know, therefore, that most all of the planets having no or thin atmospheres, have the marks of big impact craters and very large basins, and the Earth also has impact marks at the surface. So, impacting was a major process of shaping the surfaces of celestial bodies. We see stresses, ridges, folds, due to motions of the upper layers and crusts on planets. Mercury is a compressed planet. The Moon, in contrast, undergoes periods of compression and periods of stress successively. Mars has canyons which opened and fractured the surface. And from this research, we can infer the histories, the evolutions of the interiors of these bodies. We have vulcanism also. On Mars, we have big volcanoes, huge volcanoes.

On the Moon and Mercury there are lava flows, more fluid, which cover large fractions of the surfaces, and these comprise direct indications about the interior structure of these planets. Also, on Mars we have running water, and etc., etc.

This means that we are really in a period in which a large number of data are gathered step by step. I would now like to see in what level we stand for the exploitation of all this data for an understanding of the origin and of evolution of the solar system. I shall be brief, obviously, but roughly speaking, the model of Laplace apparently is the one to retain. So a contraction of an interstellar cloud is basically what emerged to be the true model for the origin of the solar system. This event of contraction occurred from 4.6 to 5-billion years ago. It is an irreversible process because of the large scale involved. If you compress gas, the concentration of mass is enough to accelerate the process, so more and more interstellar gas is collecting together. This physics of the primeval nebula is complex because the contraction starts with a small rotation, and when the contraction increases, the rotation also increases. So spin accrued and caused substantial momentum forces, followed by flattening, etc. Many unknowns remain about these processes although the general scheme is rather clear. The situation is probably complicated because temperance occurs also. The compression produces thermodynamic heating of the whole system which involves a change of heat radiation and absorption processes in this primitive nebula, and because the high temperatures still complicate the picture, ionization proceeds to hydromagnetism, and etc.

Although all the observational results now give a somewhat clearer picture of the general processes of contraction of the Laplace model in its details, many unknowns and challenges remain to be solved in the near future. For example, to search in the galaxy, some configurations which represent equivalence of the formation of the solar system could be identified as a plate solar system configuration at the different stages of evolution. If we can find such configurations in the galaxy and the outer solar system, it would help to explain the different processes observed in primitive nebula. Also, a very major unknown, surprisingly, is at what period the Sun appeared in this process. Does the Sun ignite its nuclear heating at the beginning, or after a certain evolution of the solar nebula has taken place? This seems to be a naive question, but it is completely unknown, and several astrophysicists compete with arguments to prove that the Sun was at the beginning or that the Sun appeared as a body with its proper light later in the scheme. Also there is a problem with angular momentum conservation. The momentum of the Sun is far too small as compared to the planets. So something happened in the transfer of momentum which is absolutely not clear.

Returning to my contracting nebula, I said that the contraction would release heat. There is heating, and one of the results of this heating of the primitive nebula is vaporization of granular matter, which converts this to vapor, which later cools; this cooling, or this re-cooling of the matter is a fascinating process because refractory elements re-solidify first; then the more volatile follow in sequence. Thus, we have a cosmological sequence of reorganization of the chemistry of this nebula.

Many controversial points still remain to be solved. Does this process involve chemical equilibrium between the granular matter and vapor or not? Also, volatile components are deficient in the inner part of the solar system. Why? Is it because of the condition of escape? We need more theory and more observations to clarify all these problems. At any rate, after this heating and cooling of the material, and fractionation into the different compositions with time and with this distance to the Sun, the following process is accretion. We still deal with a nebula which is more grains of re-cooled material, and these grains have to accrue, to group themselves, to create planets and the solar system objects. This accretion is a cold process at low temperature because of the sticking of the grains- - - these two grains are nearby so they adhere, and there are more and more, so it is a very complex system which occurs and it is not very clear why the grains are so adherent. So you have a large number of bodies which grow by the adhering of grains and again, there are many unknowns to be clarified in these processes.

We have two schools: the American school argues for very rapid process, and the Soviet school is arguing for a multi-stage, lower work process which accrued larger bodies, and larger bodies accrued again by collisions, etc. As you see, the general scheme is clear, but the details are not yet clarified and more work is needed to clarify these accretional processes. Also, why the asteroids are aborted accretions, if you like. They failed to accrue into a large planet like the other parts of the solar system and the comets also failed to accrue.

Now, observing the solar system, as I said, we see a large number of impacts, the traces of bombardment by solar system objects, and apparently it corresponds to the tail of the accretion process, but we don't know if this number of celestial bodies produced by mutual collision is a continuing process of cataclysmic phases. And something else - - - a naive question which should be solved - - - is why in the present context of the solar system the interplanetary medium is so clean... why so little interplanetary dust remains in the solar system...why is it so perfectly clean in the present context?

Returning in the last few minutes, with the larger telluric planets in the solar system, we see some processes of heating, remelting, differentiation of these bodies, the Earth, the Moon, Mercury, and Mars. The creation of a crust, a core, the center, a mantle in between, and all these processes are very complex, and should be clarified, and in this light, along these lines, the space missions to those planets are indeed very helpful. This is a very rough description of the situation. If I like to look to the future, it is my expectation that if the effort for exploring the solar system by space craft mostly, and also by ground-based telescopic approach is to be continued, if the gathering of data continues to be at least at the present level, I foresee that most of these problems will be clarified, that the origin and the evolution of the solar system will be so clearly understood that it will be taught at school, and high level students will learn at school in perhaps 20 years only, what were the phases of the formation and the origin of the solar system. If such a level is reached, I think this part of research will have reached its goal.

I have just a concluding remark. While indeed the concluding remark should be that we thank all the speakers and we may go a little deeper in the field. While Art Code interested us in what he said that perhaps a new scientific evolution is needed to understand the Universe problem, I think so, too. I think really what was achieved up to now is just a preparation, that the basic knowledge perhaps is not yet acquired. It could happen that the space telescope be precisely the occasion for which the scientific revolution is needed in this field, and I hope so. For the problem of the origin of the solar system which is more limited, I don't think a scientific revolution is needed. It seems to me that we do have already all the basic data needed, and I see the future as follows: It seems that science until the middle of the century was in some respects formative. It dealt with the basic fundamental principles of physics.

The particles, the quanta, the gravitation, hydromagnetics and so on, the basics, the tools which we can plan and use to understand the universe. It was formative if you like. But in the present context, we are about to open a new aspect of science which unveiled more new and unexpected problems on the basis of all these basic laws. We now attack problems such as origin and evolution of the universe, of the solar system, of the Earth, the origin of life, and so on. This is more towards understanding our origin if you like, what was evolution, etc., and in order to achieve this new level of science, we need a more sophisticated, if you like, control of knowledge and training of brains, and a more elaborate plan of approval is needed. I foresee the future, if all goes well, as still going on this way, and when these problems of evolution and origin will be more clear to our mind, and at that time our brain will be ready to attack still more difficult problems, more dealing in fact with metaphysics, as such as for example why, not by what processes but why the universe was formed, why this solar system was created, why we have life on earth. These problems are indeed far more complex to attack and I'm sure before we reach this final goal of science, we need several revolutions, scientific and intellectual revolutions, and we are living in a period in which we already made an attempt to attack the physical problems in preparation for the metaphysical problems. This is what I hope for the future. Thank you.

Dr. Ernst A. Steinhoff -

I think we all thank Dr. Dollfus and his fine team for the presentations which they have brought us today and we are looking forward to those of Dr. Drake. In the meantime, we would like to take about a 5 minute recession in which we will make a small presentation from the scientific community to the Space Hall. I have here the chairman of the Governor's Space Hall Commission, Dwight Ohlinger, and the Executive Director, Hess, and I would like to have Dr. Kaplan - is he around? - and Dr. Draper to come here to take some administrative action of transmission of personal property to the Space Hall. One is a newly written book by Dr. Kaplan which has very much to do with space. The other items also have very much to do with space, we are supposed to have changed hands, going from the individuals to the Space Hall, to become permanent property of the Space Hall. After we are through with this then, we will start Dr. Drake's session.

ASSESSMENT OF FUTURE TASKS OF ASTRONOMY

Dr. Arthur Code[*]

The talk I have prepared, which you have before you (AAS 43, 152), concerns not the solar system, but a grander picture about the other 10^{44} of the mass of the Universe. Keep in mind, however, that a global picture, like I have described, is made up of many, many details like the things we have heard from Dr. Dollfus and Dr. Evans. A lot of hard work has been done and there are many kinds of basis, even if it may be wrong, for the generalizations that I put forward. But there is also a logical reason now for this transition from discussion of the solar system to the more distant stellar universe, and that is what characterizes modern astronomy today. It is measurements throughout the entire electromagnetic spectrum, and measurements of all kinds of objects at great distance.

The advent of space vehicles has made an important contribution to this change in how astronomers operate. For centuries, astronomers were limited to the bottom of this ocean of air that surrounds us. Probably my credentials for talking to you are that I'm a user of space vehicles for astronomy. Space vehicles provide two things for astronomy. They allow you to carry out experiments in astronomy which we couldn't do before except for analyzing a few hunks of meteorites that hit the surface of the earth. With space crafts you can go outside of the earth's atmosphere, make in-situ measurements, sample the interplanetary medium, go to a planet and make measurements, do chemical analyse of its compostion. There is this class of experiments. In addition, you can make observations, and you can make them better than ever before, and that was the final comment that Jack Evans made. He referred to making better observations of the sun, although you can imagine sending probes into the sun, getting close anyway, and measuring particles of the sun. But once you get significantly beyond the solar system, it's going to be a long time before you carry out experiments.

Astronomers are still limited with their observations. These consist of measuring radiation and particles that come to us from celestial objects, primarily radiation, although cosmic rays are important. We measure light, and what can you tell about light? You can determine the direction it comes from, the intensity, the spectral distribution, and the polarization. Build a machine that measures radiation within a 10^{-4} second of arc with a spectral resolution of 10^7, and determine the state of polarization, and you've collected all the data. If you keep doing this, you measure the direction of propagation, the intensity, spectral distribution, and polarization of objects as a function of time. That's observation astronomy. I've described it all to you, except for a few little details. Well, sitting here on the surface of the Earth, only a little bit of the light gets through, just a little tiny window in the optical region and a much bigger window in the radio region. Most of the infrared, all of the ultra-violet, x-rays and gamma rays are blocked out by the Earth's atmosphere. And many objects radiate most of their energy in one of these inaccessible regions, so for centuries, we didn't see all of the universe. We did not know anytning about the high energy universe characterized by x-rays and gamma rays. Until the advent of radio astronomy, we did not know anything about the low energy universe, like the "Big Bang", but that's the story I have described to you.

----

[*]Professor of Astronomy, University of Wisconsin

As far as observing is concerned, we've got three advantages to making observations above the Earth's atmosphere, freedom from atmospheric absorption, so you can observe at all wavelengths except you can't observe hydrogen absorption because of the hydrogen interstellar medium, the interstellar medium becomes opaque, but otherwise you are free from atmospheric absorption. The sky is darker; there isn't much scattered light so you can get better signal-to-noise ratio. You can see fainter things, and thirdly, you don't go through the turbulent atmosphere which distorts the appearance of planets when seen through a telescope. On the average, it is just a fuzzy ball, and you can't do much. And stars, they ought to be nice defration points, and they usually are fuzzy blobs that float around on the focal plane, and you try to get the light down the slit. At any rate, you can expect to get high spatial resolutoin above the Earth's atmosphere. Three things: freedom from absorption, darker sky, high spatial resolution, actually realizing the resolution of a telescope, and of course it's always clear.

This is the observational part; we get the measurements, we interpret them by theory, physical theory that works here on the surface of the Earth, and the physical theory we use today was the scientific revolution of the past. These were the theoretical ideas that weren't common sense and must be wrong. Copernicus had trouble. From about that time on, these scientific ideas really represented the great intellectual advances of mankind. It is these great ideas that are the enduring part of our heritage. If one looks back in history it is not the politicians or the great wars we shall be remembered for. Certainly the 18th century is remembered for Newton's laws, these indeed would be one of the great achievements of mankind. In the 19th century the great intellectual advance was Maxwell's law. Just think today I can know what somebody is doing, what they look like, what they are saying, exactly how they are moving, right now, by just turning on my TV set. That's the heritage of Maxwell's law, and that's the kinds of thing we will look back at from this century. I think that the great intellectual adventures and the great things this century will be known for are the quantum theory, general relativity, and the decoding of the DNA molecule to understand life. These three things are intimately interwoven in the fabric of astronomy. And it is in that context that I want to describe what modern astronomy seems to tell us about the universe we live in, and then talk a little bit about why it's probably all wrong. And then I hope the panel here can find the solution to why they are wrong.

Some things we are finding today say maybe we need another scientific revolution. We think maybe we see black holes, and there are problems with that. There are also problems connected with elementary particles physics, at any rate, that some people consider very serious with black holes. Quasars have been around now for almost two decades, and we still don't know what they are. We have an awful lot of observational data, and maybe we have to bend the physical laws a little bit. Maybe we have to build a new system. The microwave background of the universe, maybe that's telling us something new. The comments presented here by Professor Stron depart from orthodoxy and perhaps he is right, although I don't think so. In any event the new and exciting observations of modern astronomy demand some original thinking.

But I'm going to talk about what it's like in terms of general relativity and the relativistic quantum electrodynamics. That's our modern physical theory and it works pretty well. O.K. What I'm primarily concerned with is model building, building a model universe. How do we build a model universe? As I said, we use general relativity and general relativity is based on the principle of covarience and the principles of equivalence. The principle of covarience says there's nobody else in the universe that's any better off than we are, that any kind of laws of physics have to be independent of the coordinate system.

There shouldn't be anything subjective about what you learn. Namely let's write physical laws in terms of invariance, namely tensor calculus so that the form stays the same in any coordinate system. The second principle of equivalence sort of recognizes the same thing, that there isn't any special coordinate system, one way of saying it is that gravitational mass and inertial mass are the same or,... Well, I was going to drop my watch, but I'll drop this pencil instead. O.K, I could say that it's dropping in some force field and accelerating at 32 ft/sec^2. Well, it depends on what kind of coordinate system is used. If the grid spacing is changed then it's dropping uniformly in this coordinate system which is distorted. O.K. This was Einstein's idea, that you can substitute geometry for complicated physical laws. He had to construct a relationship between the geometry and the mass energy. How does mass of the sun distort space around the sun, so that the path, the normal falling or the geodesic path of the planets is the planetary orbit. This relation is called a field equation. We had to construct one. I won't go through the arguments for why it came out the way it did, but you're sort of forced to that as the simplest within the framework of requiring that all our conservation laws work. So there's a relation between geometry and mass. If I know the geometry in the universe, I know how the mass is distributed, and if you tell me how the mass and energy is distributed, I can describe the geometry from the field equation. Now the Brans-Dicke theory proceeds the same way except the field equation is a little different, instead of a tensor field, it is a tensor-scalar field. I think that there are good reasons now for not believing the Brans-Dicke theory. So it's lucky the sun came back to being spherical.

I'm going to ask a very simple question. First of all, is the universe open or closed: Is it curved positively or negatively? In principle, that's pretty easy to answer. Is the sum of the angles of a triangle 180 degrees in space, or is it greater than, or less than, 180 degrees? So, if it is greater than 180 degrees, the universe is closed, a positive curvature. Thats the appropriate geometry. Doesn't matter what the universe is like, but that's the kind of geometry you have to use if you want to describe it. We can't do things quite as simple as measuring angles, rather we predict from the models such things as redshift relations and galaxy number counts, and compare these with observations. The correct models are those that correctly predict the observed quantities. Today, we can not yet answer my simple question, but the future is encouraging and observations from space may unlock the secret.

OPTICAL ASTRONOMY

Dr. Lubos Perek*

Mr. Chairman, Mr. Moderator, and distinguished guests. Thank you very much for the privilege you are giving to me. I will just say a few remarks, you see I was just thinking of what a historian would say in the year 2001 about the development of astronomy in the 20th century, and as you remarked earlier the astronomy before the advent of space technology was actually restricted to a very narrow spectral range, through a very narrow window on the bottom of an ocean thru an almost opaque medium.

For instance, in the purely optical field, the discovery of astronometric companions of stars I think is a very important point because this gives us some basis for computing the probability of the occurrence of planets or planetary systems with other stars. In another important point -- I think that the tendency of building observatories is excellent -- at exceptionally excellent sites with good climatic conditions is also a step which deserves quite an attention. On the other hand, maybe it's a good point that not all observatories move to excellent climatic conditions, because one of the reasons, it might not be the only reasons, but one of the reasons why theory developed is that frequent cloudiness happens at the astronomic observatories. Then of course, if we consider the development of theory, then we should pay some attention to the influence of computer technology which it had on astronomy, not only in making it possible to compute, to make computations which are about the linear possibilities of the human being, but making new discoveries.

For instance, even in such classical fields as the three body problem, the discovery of the third integral which has quite different properties of the known integrals of differential equations of motions. This had very great influence on this whole area, and on stellar dynamics. Then, the historian in the year 2001 will have to talk also about radio astronomy and I'm afraid that radio astronomy will have very little to say here, so they must probably wait for the year 2001 to be really appreciated.

Then what else I here have to say will be to stress the importance of international cooperation. I think this is a very important point. International cooperation played a very important role in astronomy since its beginning many, many centuries ago, but in the recent past time it became particularly important because the communications and the transportation became easier so the cooperation developed much more easily.

One excellent example is the long baseline intraferometery, and then, just to come to my present job, I think that international cooperation developed also on other levels, on intergovernmental levels, and this is something in which or where the development of space technology was of paramount importance. If you look at the development of relations regarding law of the sea, then it took centuries for the development, but in outer space the law and intergovernmental agreements developed very quickly in the last two decades, which actually come into question.

----
*) Chief United Nations Outer Space
   Affairs Divsion.

We've already four important intergovernmental agreements which I cannot describe to you in detail because I would like to leave some time for Dr. Hunter to speak about the large space telescope. I would just like to express a personal wish to Dr. Hunter. If he could somenow smuggle into the problem of the very large space telescope, the observation of faint planetary nebulae which move with exceedingly high velocities of up to 250 kilometers per second and which are located close to the galactic center. I would very much appreciate to see what they really look like, with the very small, very high spatial resolution which his telescope will provide.

Thank you.

AMMENDED PAPER OF E. W. SILVERTOOTH AND JOHN STRONG
Published in The Eagle Has Returned Part I
American Astronautical Society, 1976

Dr. Code has provided us with an excellent summary of astrophysical observations bearing on current scientific cosmologies.

We consider some other observations that were unexpected and disturbing, in addition to Dr. Code's Mention of the low velocity (300 km/sec) of the sun relative to the 2.7K radiation background:
2) Red-shift observations indicating an apparent acceleration in the rate of expansion of the universe;
3) Anomalous red-shift observation of quasars;
4) Total lack of observation of large blue shifts.

The scientific community is acutely aware of the problems of interpretation that these items bring up: Of the apparent acceleration, Tinsley and Gunn have stated, "something is terribly wrong." Of the anomalous red shifts a theoretician is reputed to have said, "no Band-Aid will fix this one, something is wrong with the laws of physics." And on the absence of large blue shifts, Burbidge wrote, "somewhere way back, we have taken a wrong turn."

All four of these aspects of the observational material can be accommodated by the following suppositions and postulates. Although our suppositions and postulates are not in conflict with experimental facts, they clash with the second postulate of special relativity. These astrophysical observations should shock us into assessing alternatives -- which we do here. We feel it makes us heretics, since the second postulate is now dogma. The experimental facts are another matter -- not to be violated.

The alternative we elect supposes a unique inertial frame in the universe; that a clock in it runs at frequency, υ ; that a measuring rod in it has length, ℓ ; and that the sun is moving in this frame at a negligible velocity -- say less than a thousandth of c. In addition, we use two postulates:

$\upsilon' = \upsilon/\gamma$ by established time dilation;

$\ell' = \overline{\ell} \cdot \overline{e}/\gamma$ by the established Lorentz-Fitzgerald contraction.

As has been deduced elsewhere[1] the above, for a star at distance L, with radial velocity and acceleration v_r and v_r, yields the red-shift equation:

$$\upsilon_1 = \upsilon/\gamma \ (1 \pm v_r/c) \ (1 \pm Lv_r/c^2)$$

And for $v_r = 0$, (See Appendix A)

$$\upsilon_1 = \upsilon/\gamma \ (1 \pm v_r/c)$$

which contrasts with the equation that the second postulate of relativity yields:

$$\nu_2 = \upsilon \gamma \ (1 \pm v_r/c)$$

Our equation does not violate experimental facts and observations. Indeed, the observation[2] of delayed decay of μ-mesons in the atmosphere; and the Mossbauer time-dilation experiment[3], both, clearly put γ in the denominator of the red-shift equation. Furthermore, it can be shown that the experimental tests to determine the existance, or not, of a unique inertial frame -- the work of Michelson-Morley. Kennedy-Thorndyke. (See Appendix B), Turner-Hill, Cialdea, (See Appendix C) and more recently, the VLBI observations of Cole -- should all give null results regardless of the existence, or not, that they tested for. Tyapkin[4] has shown this for the first order experiments; the second order experiments gave null results when the $\upsilon \lambda = c$ relation is considered. Marinov[5] has carried out a conceptually valid test and his result is 130 km/sec. This is comfortably within the limit, 300 km/sec, that Dr. Code mentions. (An experiment still different from Marinov's, that promises high precision has been devised by one of us (EWS). It is conceptually valid also, and will be performed.) (See Appendix D)

The figure shows the dependence of υ_1 and υ_2 on v_r. With it we are prepared to re-consider the discordant aspects of the observational material:

2) A known QSO has z = 3.53. The υ_2 equation interprets v_r at 0.91c; the υ_1 equation, as 0.7c. The lower velocity for this and all large red shifts removes the argument for an acceleration of the rate of expansion of the universe. (See Appendix E)

3) Anomalous red shifts are accounted for by assuming apparent radial accelerations with respect to the observer.

4) The figure shows υ_1 reaching a maximu of 1.299υ at the value for $v_r = 0.5c$ -- thus accounting for the total absence of large blue shifts.

References:

1. Silvertooth, Applied Optics 15, 5 (1976) 1100 - 2

2. Rosser, An Introduction to the Theory of Relativity, Butterworths, (1964) 119-21

3. Champeney, Isaak, Khan, Proc. Phys. Soc., 85 (1965) 821-5

4. Tyapkin, Lett. Nuovo Cimento, 4 (1973) 821-5

5. Marinov, Czech. J. Phys. B24 (1974) 965-70

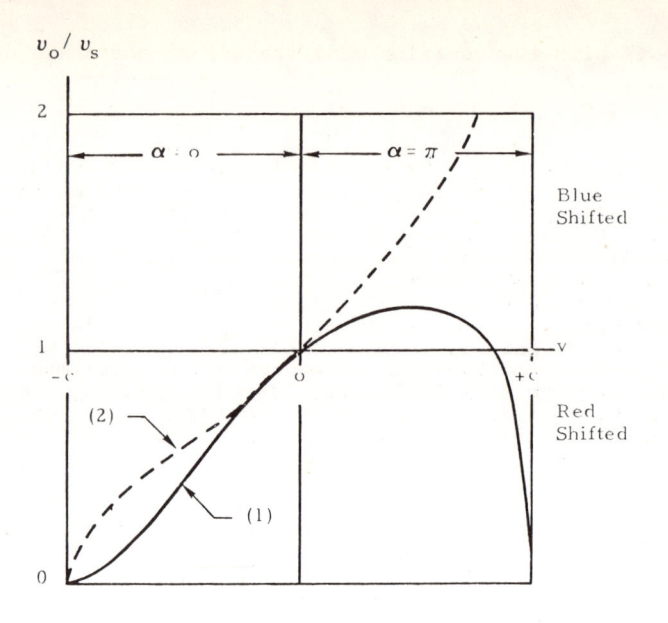

APPENDICES

A. Anomalous Red Shift Equation:
   The phase change over a length L at rest with respect to the reference frame is:
   $$\delta = \frac{L\,v}{c}$$

When the source, attached to one end of L is moving at a velocity v with respect to the reference frame (L, v colinear), the phase shift is:
$$\delta' = \frac{\left(L' = \frac{L}{\gamma}\right)\left(v' = \frac{v}{\gamma}\right)}{c \pm v_r}$$

where: $\gamma = \dfrac{1}{\sqrt{1 - \dfrac{v^2}{c^2}}}$

then: $\Delta\delta = \delta - \delta' = \pm \dfrac{L\,v\,v}{c^2}$

or in general, when L, v not colinear, $\Delta\delta = \dfrac{\vec{L}\cdot\vec{v}\,v}{c^2}$

Where: $L' = \dfrac{L}{\gamma}$ takes account of the Lorentz-Fitzgerald contraction

The anomalous red shifts are explained by the extended red shifts equation from reasonably postulated accelerations. Above we obtained the phase shift:

$$\delta - \delta' = \pm \frac{L\, v\, \upsilon}{c^2}$$

With periodic acceleration (time rate of change of phase), the frequency is modulated by the increment

$$\frac{d}{dt}(\delta - \delta') = \pm \frac{L\, \dot{v}_r\, \upsilon}{c^2}$$

Adding this increment we get the extended equation:

$$\upsilon_o = \frac{\upsilon}{\gamma}\left(1 \pm \frac{v}{c}\right)\left(1 \pm \frac{L\, \dot{v}_r}{c^2}\right)$$

To explain the anomalous red shifts we suppose the situation that the quasar is in a galaxy at a distance L, and at a distance from its center, about which it is rotating with constant angular velocity ω where

$L = 10^{27}$ cm
$a = 5 \cdot 10^4$ light years
$\omega =$ one revolution in 10^8 years

These suppositions yield a postulated acceleration that conforms to astronomy. In this case:

$$\frac{L\, \dot{v}_r}{c^2} = \frac{L\, \omega^2\, a}{c^2} = .21$$

Thus when a quasar is observed with ω^2 at a maximum with respect to the observer, one expects a 21% difference between the red shift of the quasar and that for its milleau.

B. Michelson - Morley Null:

Although there are several ways to predict a null effect, reference frame or no, for Michelson-Morley type second order experiments, here we deduce a null phase (fringe) shift for each of the two beams separately. The incremental form of the phase shift equation

$$\delta = \frac{L\, v\, \upsilon}{c^2}$$

may be written

$$d\delta = \frac{\upsilon}{c^2}\, \bar{v} \cdot \overline{ds}$$

Thus the total round trip phase shift for any path is

$$\delta = \frac{\upsilon}{c^2} \oint \bar{v} \cdot \overline{ds} = 0$$

In general, where the source and fringe pattern are not collocated, the round trip phase shift for each of the paths, quite without regard to their lengths or directions, is:

$$\delta = \frac{v}{c^2} \int_a^b \bar{v} \cdot \bar{ds}$$

That is, the phase shift is the same for each path, hence no fringe shift will be observed.

C. First order null

We owe the prediction of null effects for first order experiments to Tyapkin. We give a parody on his proof by the following typical model which relates conceptually to all the observations:

A phase shift is looked for between a source (clock) at the center of a roating table and a second source (clock) at the rim of the table as a function of the angular position of the table with respect to the direction of the velocity vector. The phase shift earlier derived along a path of length L (radius of table) varies as:

$$\delta = \frac{L\,v\,v}{c^2} \cos\theta$$

where is the angular position of the path. When the table rotates at a constant angular velocity $\dot{\theta}$, then the time rate of change of the phase, a frequency, is:

$$\dot{\delta} = \frac{L\,v\,v}{c^2} \sin\theta\,\dot{\theta}$$

Hence for one quadrant the frequency change is:

$$\dot{\delta} = \frac{L\,v\,v}{c^2} \sin\theta\,\dot{\theta} \ \Big|_o^{\pi/2} = \frac{L\,v\,v}{c^2} \dot{\theta}$$

The clock frequency at the rim of the table, to a second order approximation, has the velocity v (with respect to the reference frame) at $\theta = o$, and $v + v_1$ at $\theta = \frac{\pi}{2}$. $v_1 = L\dot{\theta} \ll v$, the tangential velocity of the source.

It is algebraically convenient here to caluculate: $v^2\big|_{\theta=o} - v^2\big|_{\theta=\frac{\pi}{2}}$

where:

$$v^2\big|_{\theta=o} = \frac{v^2}{\gamma_1^2} = \frac{v^2}{c^2}(c^2 - v^2)$$

and:

$$v^2\big|_{\theta=\frac{\pi}{2}} = \frac{v^2}{\gamma_2^2} = \frac{v^2}{c^2}(c^2 - [v+v_1]^2)$$

$$= \frac{v^2}{c^2}(c^2 - v^2 - 2vv_1 - v_1^2)$$

104

$v_1 \ll v$,

$$v_{\theta=0} - v_{\theta=\frac{\pi}{2}} = \frac{-v\, vv_1}{c^2} = \frac{-L\, v\, v}{c^2}\dot{\theta}$$

which cancels the time rate of phase change term through the second order in v.

D. Proposed new experiment

A single source is split into two beams which are later superimposed at a detector in the vicinity of the source. Each beam is sent to a returning reflector located in opposite directions, say E and W, at substantially equal distances L. Consider the radiation emitted by the single source (laser or microwave) to swing 300 MHz at a rate of 3 MHz (and alternatively at 1.5 or 6 MHz for a second part of the observation). The reflectors are at such a distance that the beams are each returned to the detector in one cycle of the modulating frequency (2L = 100 m for our example of 3 MHZ). The rationale of the experiment is clarified by recalling that in the first order interference the phase changes are proportional to frequency:

$$\Delta \delta = \frac{\overline{L} \cdot \overline{v}\, v}{c^2}$$

and these cancel if the beams are unmodulated.

For our example suppose the victor $\overline{v}$, the velocity of the experimen relative to the rest frame, is 150 Km/sec directed EW. The phase increment of the outgoing E beam is approxmately $8.3 \cdot 10^{-11}$ ($v \pm \Delta v$) and $-8.3 \cdot 10^{-11}$ ($v - \Delta v$) on the return path, while the outgoing W beam is $-8.3 \cdot 10^{-11}$ ($v + \Delta v$) and $8.3 \cdot 10^{-11}$ ($v - \Delta v$) on its return path. Thus the two beams superimpose with a phase dither (maximum) of $8.3\, 10^{-11}$ ($4\Delta v$) or approximately $36°$ of phase dither which is readi measurable.

This proposed experiment is the easiest of execution of any which have confronted the problem - confronted it correctly as in the case of Marinov's experiment, or incorrectly, conceptually, as in all other instances. The experiment is practical in that the two distances L_E and L_W may be adjusted to be equal at any time just by adjusting one of the path lengths to produce a minimum dither in the detected phase difference. The switch to 1.5 or 6.0 MHz will eliminate the dither at any desired time. Since the equipment is diurnally rotated with respect to $\overline{v}$, confidence is established in any measured dither at 3 MHz which is diurnal in its amplitude.

E. Referring to the figure in the remarks, it will be seen that a given ratio between the frequency of the observed source and the reference frequency (comparison spectrum) yields a lower radial velocity for equation:

$$v_1 = \frac{v}{\gamma}\left(1 - \frac{v_r}{c}\right)$$

than for the relativistic equation:

$$v_2 = v\gamma\left(1 - \frac{v_r}{c}\right)$$, a consequence of the second postulate,

since $\gamma \geq 1$.

Thus it is seen that the universe is expanding at a lower rate by the first equation than by the second, and the need to consider an acceleration in the expansion rate is removed.

The second postulate of special relativity is clearly negated by the observed time dilation of the muon and by the transverse dilation of the Mossbauer experiment.

It is further negated by the conceptually valid experiment of Marinov, an experiment devised to detect the motion of the earth with respect to the reference frame.

References

1. Silvertooth, Applied Optics, 15, 5, (1976) 1100-2.

2. Rosser, An Introduction to the Theory of Relativity, Butterworths, (1964) 119-21.

3. Champeney, Isaak, Khan, Proc. Phys. Soc., 85 (1965) 583-93.

4. Tyapkin, Lett. Nuovo Cimento, 4 (1973) 821-5.

5. Marinov, Czech. J. Phys., B24 (1974) 965-70.

"PAST INTERPRETATIONS OF MARS IN RETROSPECT"

Clyde W. Tombaugh*

With the new knowledge of the planet Mars revealed by the Mariner and Viking missions, it may be of interest as to how and why various erroneous interpretations of Mars prevailed in their day. A great deal of scientific knowledge is based on probabilities from limited information available at the time.

This paper may be of special interest, coming from one who has observed Mars hundreds of times with the same telescope that Percival Lowell used. For two decades, I have worked closely with three members of Lowell's staff, all of whom are now deceased. Naturally, I picked up much lore about Lowell. I read all of the books and Observatory bulletins that Lowell wrote about Mars. I have seen much of what they saw, and studied the optical and meteorological parameters which led to their deceptions. Astronomy borrows heavily from the other physical sciences for interpretations. The state of knowledge of geology and meteorology used by Lowell was primitive compared to what is known today. Indeed, within the past 25 years, several dozen investigators have published various interpretations which were partly confirmed by the spacecraft missions and partly wrong.

Let us begin with some of the early observers of Mars and briefly trace the evolution of interpretation.

Since the beginning of telescopic astronomy in the 17th century, the source of the fallacies was the assumption that the Earth was a typical planet. Slowly over the past three centuries, planetary astronomers have been forced to revise this concept many times as discoveries revealed a greater range of physical conditions than they had ever imagined.

The earliest observers of Mars saw the white polar caps and noted that they changed in size in accordance with the seasonal tilts of the axis, which happens to be very similar to that of the Earth. Also, on the disk they saw large bluish dark areas inferred to be seas of water. The other areas appeared reddish ochre and therefore, must represent continents. Temporary whitish patches veiled these areas at times and were interpreted to be clouds, implying an atmosphere. So far, Mars appeared to be Earth-like, and might well be inhabited by various forms of life, perhaps even intelligent beings. For a long time, Mars was viewed with awe.

Beer and Madler observed Mars through five consecutive oppositions (1830-1839), identifying the same dark markings and noting that their positions remained fixed in position on the planet's disk. Madler noted that the south polar cap underwent greater changes in size than the north polar cap; also that the reduced north polar cap in its summer

---

*Dr. Tombaugh is Professor Emeritus of Astronomy, New Mexico State University, Las Cruces, New Mexico 88003.

was concentric with the polar axis, while the south polar cap was de-centered from the pole by some five degrees.

In 1840, Beer and Madler published the first map of Mars, assigning latitudes and longitudes to the various permanent markings.

In 1865 Dawes observed that the ruddy color of the "continents" was most pronounced when in the central portion of the disk, which indicated their non-atmospheric nature. The pastel fading in color toward the limbs indicated substantial veiling by the atmosphere. (This fading of color toward the limb is also true for the bluish dark maria-due to the permanent dust haze.) This effect led to an overestimation of the density of the Martian atmosphere - for a long time regarded as 85 millibars of pressure.

An improved map of Mars was constructed in 1869 by Proctor from the better drawings made by Dawes.

In 1877, Green drew a few small, bright, white, detached portions of the dwindling south polar cap visible during its summer season. Green regarded these as patches of snow resting on elevated terrain after the snow had melted away on the surrounding lower ground. These were later named "The Mountains of Mitchell". They are observed every Martian year in the same place about Martian equivalent date of June 4th and disappear a few weeks later.

The memorable opposition in early September 1877, marked a new epoch in the study of Mars. At this time Mars was at perihelion, only 35 million miles from Earth and the disk subtended the largest possible angular diameter; namely, 25 seconds of arc. While executing a trigonometric survey of the planet's surface, Schiaparelli using an 8 3/4-inch Merz refractor at Milan, Italy, detected a curious network of channels, the so-called "canals" dissecting the ruddy continents and connecting the various seas. Their regular geometrical pattern was very curious. They extended for several hundred to a few thousand miles in straight lines as arcs of great circles on the globe of Mars and appeared to be of nearly uniform breadth of about sixty miles. Inspection of some drawings made by Dawes, Secchi, Holden and others previous to 1877 recorded several such markings in the same positions.

When Schiaparelli announced his finding of the canal network in 1877, he was immediately scorned and ridiculed as suffering from hallucinations. He saw the canals again in the same positions in the following opposition of 1879.

The next opposition two years later, Schiaparelli saw the canals in their places again, but about 20 of them were seen in close parallel duplicate similar to railroad tracks, the components of each pair separated by 400 miles, but for some canals as close as 200 miles. Schiaparelli was dumbfounded at such an unnatural enigma. He found that this curious property of "gemination" occurred only between the spring and autumn equinoxes of the northern hemisphere. This phenomenon was plainly apparent to Perrotin and Thollon during the opposition of 1886, using a 15-inch telescope at Nice.

Schiaparelli interpreted the bluish dark collar surrounding the summer polar caps as a temporary shallow sea drained from a thick polar cap during the season of rapid melting. This seemed to be followed by a progressing darkening which he regarded as an inundation from the melted polar snows. Also, he noted that some of the maria turned brown, resembling more the color of the permanent "continents". This he attributed to very shallow seas drying up by evaporation.

The concept of the aqueous nature of the maria began to run into trouble. The enormous amount of solar insolation required to transform such great quantities of water, especially at Mars' greater distance from the sun, was formidable. No one ever saw the brilliant subsolar image that should have been observed on the specular surface of a sea. Tests for polarized light gave negative results. Spectroscopic observations for water vapor lines were extremely weak or non-existent. The final blow came during the favorable opposition of 1894, when several observers at Lick and Flagstaff saw canal-like dark bands traversing the Mare Erythraeum.

It made more sense to explain the blue-green darkish maria and their seasonal variation in color to some hardy type vegetation -- perhaps similar to our lichens and mosses since Mars must have a tundra type of climate. The vegetation interpretation found general acceptance for some seven decades up to the time of the Mariner spacecraft. However, some proposed that the maria areas were dominated by certain minerals that would alter their coloration in response to changing conditions with the seasons. But the vegetation concept was championed by Lowell and many others, including myself.

Schiaparelli's canali particularly interested Percival Lowell, who borrowed an 18-inch refractor and set it up at Flagstaff, Arizona in 1894. Lowell was convinced that these unnatural features on Mars were built by intelligent beings who understood engineering and surveying to lay out such a "canal" network over the planet for the purpose of conducting water from the melted polar snows to lower latitudes. From his experiments on the visibility of wires at considerable distances, Lowell proposed that the long canals were irrigated strips of vegetation only a few miles wide. The narrow, bluish dark band seen surrounding the edge of the shrunken white polar cap during the spring season, he interpreted as glacial melt-water.

W. H. Pickering and A. E. Douglass (the latter of tree-ring fame) were on Lowell's observatory staff during the first few years. Pickering did <u>not</u> think well of Lowell's theory of the artificiality of the canals. Pickering suggested that the canals might owe their origin to cracks or faults from which volcanic steam might nourish natural vegetation along their courses. Fault lines can be quite straight for hundreds of miles. Pickering's explanation had greater appeal to me over the years of my study of the planet.

Lowell believed that Mars was endowed with oceans of water in its geologic past. He implied that Mars had rain and rivers in its past, which eroded all the mountains and highlands to a planet-wide peneplain, and that the present dark maria occupy the sites of its former oceans. A nearly level planet surface was essential for his canal theory. The Mariner 9 and Viking photographs reveal a number of meandering dried-up river beds. Thus Lowell seems to have been correct in this contention.

During my 17 years at the Lowell Observatory, I was intensely interested in the Mars problem. With my considerable training in geology, I became very skeptical about Lowell's concept of a level and smooth planet surface. Radioactive heat would continue to cause diastrophism, producing mountains and table lands, and peneplaination would never catch up. The melting of the polar cap could hardly be expected to yield a temporary polar sea more than a few inches in depth, yet the darker border of the North cap appeared to be continuous and regular in outline. It would be asking too much for the surface to be that level. So I concluded that the dark band could be no more than a zone of damp ground, which dried out when it disappeared during the early summer season.

Lowell set up a 24-inch refractor telescope of his own at Flagstaff in 1896. I used that same telescope hundreds of times in observing Mars. I listened to long conversations about Lowell from Lowell's long time assistants, V. M. and E. C. Slipher and C.O. Lampland. I studied the optical properties of this telescope and did some observing experiments. I think I can partly explain what led to Lowell's deception on the canals.

It is well known by observers that telescopes with apertures of 12 inches or less, yield more satisfactory views than larger telescopes most of the time because larger apertures are more sensitive to inhomogeneous disturbances in our own atmosphere. The human eye is particularly sensitive in perceiving lines. Telephone lines can be seen at great distances when the width of the wire is far below the resolving limit. The sensation signal on one cone or a dozen cones in the retina is too weak to register in the brain, but if many more bits of detail happen to be aligned in a particular direction, the combined signal strength is sufficient to produce a conscious perception as a line. This is only one of many kinds of illusions that deceive the brain.

With Earth-based telescopes, all of the features seen on the disk of Mars are albedo features, not topographic features, with the possible exception of a few who claimed to have seen a few craters. An experienced observer with normally good eyes can detect a darkish feature which is only 2 or 3 percent darker than the adjacent area, provided there is no spilling of light from haze or unsteady air. Most of the so-called canals are near this critical albedo difference. I shall consider only those canals which have been seen and drawn by several experienced planetary observers independently. That they appear as lines may well be illusions, but there are some real markings involved which are below the resolving power to be seen as individual, separated darkish patches or spots. Any approximate alignment gives the impression of a continuous line. Especially is this so if a lower magnifying power is used, causing the planetary image to be too bright and irradiation encroaches upon the dark canal from both sides, making the canal appear finer and more linear. This is the mistake that Lowell made. Lowell habitually used a power of 310 with the telescope diaphragmed to 16 inches aperture. This made the canals appear finely linear and continuous, which apparently was the way he wanted to see them so as to imply artificiality. Many times I have used the same power and aperture and gotten the same impression. Also, I experimented with higher powers and larger apertures, and studied the optical properties of the telescope.

In large refracting telescopes, the crown and flint lenses can achieve achromatism for only a small portion of the visual spectrum,unless the telescope is given an unusual long focal ratio, such as f25 or so. Very long telescopes have the disadvantage of requiring very large and very expensive domes. By comparative Hartmann tests with other large refractors the optical figure of the Lowell 24-inch refractor is excellent, but the focal ratio of aperture to focal length is 1:16. The story told is that Lowell had the telescope designed to be short enough to fit inside the dome he built for the borrowed 18-inch two years earlier. This was a most unfortunate choice, which must have contributed strongly to the history of the Mars canal furor. I found that the image began to "wash out" with magnifying powers higher than 500 diameters, even when diaphragmed down to 21 inches. The use of a yellow filter does improve the image, however.

A 24-inch refractor with a focal ratio of 1 to 30 is used at Picdu-Midi in the Pyrrenes. The long focal ratio permitted the use of magnifying powers from 700 to 900 with excellent results. They resolved the canals into individual darkish patches of detail. I had a similar experience in April 1950, using a de-centered circular diaphragm of 27 inches with a power of 660 on the 82-inch reflector at the McDonald Observatory in West Texas. Thus, it appears that Lowell missed ascertaining the nature of the Martian canals by only a factor of 2 or 3.

One of the easily visible canals, the Agatho-Cophrates, was revealed as the super-canyon by the Mariner 9 mission. Several other canals were found to be topographic relief features or patches of dark surface rock. For many of the other canals that I have seen, I was hard put to find any features that might produce a canal. Perhaps the contrast factor was too low to be recorded by the Mariner vidicon camera. When I was at the Jet Propulsion Laboratory in Pasadena on 3 September for the Viking 2 Lander, I looked at a number of mosaics prepared from photographs of the Martian surface taken by the Viking Orbiters. The surface detail was amazingly good, better by a factor of 5 or 6 over that obtained by Mariner 9. I did not have time to inspect these adequately for possible identification of canal sites. The amount of high quality detail taken by the Viking Orbiters is fantastic and will keep a number of geologists busy for a long time.

There is now abundant evidence to indicate that Mars had much more atmosphere and water in its geologic past, which supports the contention made by Lowell in the first decade of the 20th century that Mars had aged from a more favorable climate to a desperately dry and harsh desert planet. However, there appears to be more water present on Mars than formerly thought a decade ago. The canyons show landslides that appear to have been caused by the sublimation of a thick stratum of permafrost. The feeling now is that the polar cap remnants are thick deposits of H_2O ice or snow. However, I deduced this conclusion years ago. I observed that the very large caps coming out of their polar nights into the spring season shrink rapidly, implying sublimation of the solid carbon dioxide, and reach minimum size at about summer solstice time. The caps then shrink very little more during the entire summer season, implying that the substance of the cap remnant requires a higher temperature for sublimation. The north polar cap remnant is much larger than that of the south polar cap. This is not surprising since the summer season of the northern hemisphere occurs near the aphelion of the Martian orbit. Also, I noticed that after the northern summer solstice, some white areas developed in the lower latitudes, such as Nix Tanaica at 45 degrees North latitude. Several weeks later, Elysium at 20 degrees North latitude becomes white. I interpreted this time sequence as indicating that water vapor was involved which required more time for transport to the lower latitude to be deposited as frost on those areas. The above named areas do not whiten in the northern hemisphere winter season because the air is so dry that there is no moisture to freeze out.

For more detail of my views see: "A Survey of Long-Term Observational Behavior of Various Martian Features" in ICARUS Vol. 8, # 2, March 1968; and "Geology of Mars" in Advances in Space Science and Technology Vol. 10, 1970, Academic Press.

JUPITER - A LABORATORY FOR METEOROLOGISTS

Dr. Reta F. Beebe*

"The Great Red Spot is a 300 year-old hurricane." This statement, currently appearing in popular literature, implies that atmospheric circulation on Jupiter is grossly different than on earth. Just how different is a topic of current interest among meteorologists.

More similarities exist between the two atmospheres than meet the eye. Both atmospheres are composed mainly of non-condensable gases: hydrogen and helium in the case of Jupiter, and nitrogen and oxygen in the case of earth. Both atmospheres contain small amounts of condensable gases such as water, ammonia, and others which upon condensing release latent heat and could therefore play a major role in the dynamics of the atmosphere. Although the diameter of Jupiter is 10.8 times larger than the earth's, the surface gravity 2.7 times larger than earth's, and the rotation period less than half as long as earth's, these differences can be considered to be differences of degree rather than strong dissimilarities.

Major differences occur in the effect of the depth of the atmosphere and the manner in which energy is deposited in the atmosphere. While surface friction plays a major role in modifying circulation patterns in the earth's atmosphere, the atmosphere of Jupiter is extremely deep and it interfaces with the interior in a continuous manner. The rate of deposition of solar energy in the earth's atmosphere is strongly modified by the obliquity of the polar axis, causing strong seasonal dependence on a time scale of one year. On the other hand, because Jupiter's equator is tipped only 3-degrees to the plane of its orbit and it is farther from the sun, it suffers a smaller seasonal dependence on an 11-year time scale. The effect of this is further modified by the fact that infrared observations, both groundbased and from Pioneer 10 and 11, indicate that Jupiter has an internal heat source which supplies roughly as much energy to the atmosphere as is available from absorption of solar energy.

Long-term observations from groundbased observatories indicate that the major features in Jupiter's atmosphere are stable on a time scale which is extremely long compared to cloud systems in the earth's atmosphere. A basic explanation for this is that the rate of energy loss in the Jovian atmosphere due to radiation in the infrared, is much slower than for the terrestrial conditions because of the lower temperatures and molecular emitters which exist in the Jovian atmosphere.

The dominating regime in the global circulation of Jupiter's atmosphere is axisymmetric, although considerable attention has been given to the non-axisymmetric features such as the Red Spot, white ovals, and south tropical disturbances. Within the stable bright zone-dark belt configuration, small features serve as markers for tracing the latitudinal and time dependence of the velocity of the zonal winds. The position and velocity of the currents have been quite constant over

---

*Dr. Beebe is in the Department of Astronomy, New Mexico State University, Las Cruces, New Mexico 88003.

the past 15 years, for which we have detailed measurements; however, within this period the relative albedoes of the resulting belts and zones have changed considerably. An example of albedo change has occurred in the Equatorial Zone, where in 1959 a darkening began, reaching maximum darkness in early 1964. It then brightened rapidly during the first half of 1965, stayed bright throughout 1966, darkened again by 1968, and continued dark until the present time. This zone extends from +7.4-degrees to -7.0-degrees in latitude; therefore, these represent changes over a major portion of the planet.

Observations indicate that the dark belts emit more infrared radiation than the light zones. An explanation for this is that the zones are regions of high cloud cover while the radiation from the belts is emerging from a lower, warmer region of the atmosphere. A model has been proposed by Ingersoll and Cuzzi (1) which proposes that the zones are regions of high pressure and rising currents, while the belts are regions of lower pressure and descending air currents. Such a model is consistent with observational data; however, the interpretation of the observed equatorial darkening on this basis is rather difficult unless the hypothesis is accepted that some secondary disturbance has introduced material into the upper cloud deck which generates the desired albedo without altering the general circulation.

This belt zone model, combined with Coriolis forces due to the rotation of the planet, leads to a consistent pattern of circulation which predicts easterly-directed currents along the equatorward edge of the belts and currents moving toward the west along the poleward edge of the belts. This is consistent with observations, where, for example, currents at the southern and northern edges of the North Temperate Belt (24-31-degrees N latitude) are +122.40 and -12.02 m/sec relative to System III (2), respectively.

Although the above model explains the general circulation patterns in the atmosphere of Jupiter, it does not describe the basic mechanisms which drive the wind currents or explain the long-term stability of the large-scale flow patterns.

Several models have been constructed (3) assuming differing mechanisms and obtaining various degrees of success. Among these models are a baroclinic model (Gierash and Stone), where the sun provides the driving force for the atmosphere; a model based on latent heating effects to generate horizontal and vertical potential temperature differences (Ingersoll); and a convectively-driven model (Williams) which assumes the internal heat is the source of energy in the model. These models all fail to reproduce the observations to varying degrees. For example, the baroclinic model does not reproduce the observed equatorial jet, while the convective model can do so but requires an internal heat source larger than is observed. An additional model based on transfer of momentum from the stratosphere (Maxworthy) has been proposed as a mechanism for producing the equatorial jet (105 m/sec relative to System III).

The ideal model may well be a hybrid of all of these models, however, in order to understand the Jovian dynamics, observations extending over a long time span must be available because of the longevity of the atmospheric features or the large radiative time constant in the Jovian atmosphere. A combination of high-resolution "snapshots" which will yield high spatial resolution of atmospheric features with lower resolution data extending over a long time line is desirable. Pioneer 10 and 11 have given us our first close-up view, while earth-based observations, smeared by the earth's atmosphere, have been utilized to supply time-dependent information. The Mariner/Jupiter/Saturn (MJS) Mission will provide short-term data over a time interval approximately 40 times longer than for Pioneer 10 and 11 and will be capable of spatial resolution 40 times greater. Such resolution should provide detailed information on the boundary regions between belts and zones, as well as flow patterns around large atmospheric features such as the Red Spot and white ovals. This resolution should also yield information on the development and dissipation of cloud systems on a local scale.

The vertical structure of the clouds will be studied by stereoscopic projection and their relationship to the belted flow patterns and dynamical properties of the atmosphere determined. A better understanding of the high optically-thin scattering layers and the nature of the colored material in the clouds should be forthcoming.

Although the MJS mission will provide improved spatial resolution, an attractive additional source of data will be diffraction-limited, one-meter sized earth-orbiting telescopes. The resolution of such telescopes would be equivalent to the resolution of Pioneer 10 and 11 and observations could be carried out periodically over a long time line within the constraints of the proposed shuttle system. The lack of variable smearing from the earth's atmosphere would greatly enhance the possibility of establishing the identity of features from observation to observation.

Jovian orbiters carrying atmospheric probes for pressure-temperture sensing on site in the upper atmosphere would add invaluable information for modeling the Jovian atmosphere. The problem of composition as a function of depth could also be solved by such probes. The hydrogen-to helium ratio, as well as depth of water condensation and amount of latent heating could be obtained.

Within the next two decades a combination of higher spatial resolution, data from atmospheric probes, improved dynamical modeling and consolidation and modification of the current theories should greatly enhance our knowledge of atmospheric dynamics of another planet with greatly different boundary conditions than the earth. This should lead to improved understanding of atmospheric dynamics and heat balance in planetary atmospheres.

References

(1) Ingersoll, A.P. and J.N. Cuzzi (1969). Dynamics of Jupiter's Cloud Bands. J. Atmos. Sci. 26, 981-985.

(2) System III is the longitudinal system based on the observed decametric radio sources, with a rotation period of $9^h 55^m 29^s .71$.

(3) Additional information on dynamical models can be obtained from Jupiter, the Giant Planet, T. Gehrels, editor, published by the University of Arizona Press, 1976.

UNSOLVED PROBLEMS IN SOLAR PHYSICS

Dr. Jack Evans *

I've been asked to describe some of the important, unsolved problems in solar physics. When I sat down to list these, I suddenly realized that we're losing ground. The important problems are appearing faster than we can solve them. Of course they're problems only because they are difficult. All the easy things have been done. And, I think, rather more important than a list of the problems are the reasons why solar astronomers feel that these problems are important.

Beyond the fact that solar astronomers find the sun a challenging and exciting object for study, it is of some importance to other fields in astronomy and in physics. First of all, the Sun is a prototype for a very numerous class of stars in the Universe. The things we learn about the sun presumably apply to these stars and to stars that aren't too different from them. Another thing is the fact that the sun is the fount and the driving force for the solar wind and energetic particles that impinge on the Earth and perturb the ionosphere and the magnetosphere in some ways that are of practical importance to the people on the Earth. The third thing is that the Sun is a very large, large-scale hydromagnetic laboratory where all sorts of reactions between plasmas and magnetic fields take place on a scale and under physical conditions that we cannot reproduce in the laboratory. We can learn some things about the physics of plasmas from study of these things that occur on the sun. So the sun has in the past made significant contributions to the fields of stellar astronomy, plasma physics and ionospheric and magnetospheric physics. And, I think that it will be able to make many more contributions in the future as we solve some of these problems that we see before us now.

The important thing, it seems to me, in studying the sun as a star are the peculiarities of the global circulation and rotation of the material in the sun. Probably you realize that the sun is not a solid body and does not rotate as a solid body; instead, the equator rotates at a faster rate than the higher latitudes and finally the poles. The best measurements we have indicate that it takes 27 days for the equator to make a complete rotation and that this period increases up to 40 days when you get to the poles. But this is not a simple regular change in rotation. We find that we get entirely different results when we use different methods of measurement. You can, for instance, determine the period of rotation by watching a tracer, like a sunspot, move across the disk of the sun and reappear after 27 or 28 days, whatever it may be, and time this and thereby determine a rotation rate. You can make spectroscopic observations of the surface of the sun at the two limbs. The west limb where the material is going around away from you, and the east limb where it's coming toward you, measure the difference in these velocities and, knowing the diameter of the sun, you can then again calculate the rotation rate. And, at the equator, you get the same rate for these two methods, but as you go to higher latitudes, you find that the rotation rate as measured spectroscopically decreases quite a bit more rapidly than measured by the tracers. There isn't any clear rule yet, but it is beginning to look as though the lower down the objects you are using to measure the rotation are in the atmosphere, the more of this differential between the equator and the higher latitudes you find, until finally, when you get up into the corona, the sun appears to rotate very much like a solid body, and exhibits very little of this differential rotation.

----
*Director, Emeritus, Sacramento Peak Observatory, New Mexico.

Well now, all this of course must have its cause deep down in the sun. All we're seeing is the surface, but we can ask how deep into the sun do these effects go. What is it that causes the acceleration of rotation at the equator or the loss of rotation rate at the poles? We don't know, but this is a hint at least that is given to us about the internal structure of the sun, and the calculation of the material within it. When we find that, it is almost sure to have some importance in our understanding of some of the anomolies we have found, deviations from what we thought was firmly established about the internal structure of a star like the sun. I will come back to that in a moment.

Another interesting and important problem, I think, even for the subject of stellar evolution, which Dr. Code is going to talk to you about, is the mass flux of material from solar flares. The solar flare is a terrific outburst that occurs on the sun almost always in conjunction with a sun spot group where there are very strong magnetic fields, which have something to do with the flare we are quite sure, and when these occur, it usually is a matter of a big flare brightening up in say half an hour and then gradually fading away. In the next hour or two, all sorts of very high energy phenomena occur. We have a burst of x-ray radiation and bursts of radio radiation with very peculiar signatures and blasts of high energy particles, mostly protons that go out from these flares. But we find in addition to protons, a very abnormal abundance of deuterium. Deuterium, you know, is the heavy hydrogen atom which has a nucleus composed of a proton and a neutron rather than just a proton, and is a very interesting material because of its relation to the various nuclear reactions that are supposed to have occurred in the original "Big Bang" of the Universe and in various processes that have happened since then, including processes of nuclear energy production inside of stars. One of the astronomers has recently been speculating a little bit -- I guess you'd have to call it speculation, but its a good example of how the work on the sun can be of some importance in a subject like the evolution of the Universe. These flares occur on the sun with a fair frequency. But we find that there is a class of stars known as "Dwarf M. E. Stars" which have more frequent flares and flares that are something like a thousand times as energetic characteristically, as flares that occur on the sun. It's now being suggested by many astronomers that these dwarf m. e. stars are by far the most numerous in the universe, and in fact that the mass of the galaxy is composed largely of these stars. If that is so, and if, as we now know they do, they have many flares that are shooting out into space material with this high content of deuterium, this could be enough to seriously affect the calculations about the origin of the universe, i.e., the Big Bang, because we have normally assumed that the abundance of deuterium in interstellar space is what's left over from the big bang, and hasn't been too much affected since then. But here is a possible source of deuterium which some people think might be sufficient to account for 80% of the deuterium in interstellar space. If so, then our theories about the big bang have got to be modified to some extent.

Another interesting problem is the global oscillations of the sun. Some astronomers feel pretty sure now, that they have shown the sun changes diameter in a period of 45 minutes. Not a very big change, something we can just barely detect and, because this can just barely be detected, to most of us it looks as though the signal-noise ratio here is just about one. It is an observation which has been roughly confirmed by more than one observer, so even though the signal-to-noise ratio is a little low, we are taking it very seriously. This is a difficult problem and has to be pinned down because of its importance in our thinking, again, about the internal structure of the sun. It has to be something that could produce oscillations of this sort. The transport of energy. The sun is powered by a nuclear source in its center. The energy leaks out in the final stage by convection and if there's any subject which we find it difficult to deal with mathematically and accurately, it's convection. We don't really know as much about it as we need to know, and the sun is a fine place to study its characteristics. One thing about solar convection which has not been explored at all is the fact that convective elements come in two, maybe three, different sizes which are quite discrete and quite contrary to the theoretical predictions. We have little convection cells 2000 km in diameter ± 1000. Then we have medium-sized cells, what we call super granulation, where the convective cells are 30,000 km in diameter ± 10,000, but nothing in between you see. There is nothing like a 5000 and there is nothing like a 50,000, but just those in between. And then we suspect that there are what we call giant cells. And the evidence for them is still pretty indirect, so I won't say too much about them, but if they exist, and we're going to find out about this, they do have a tremendous significance for the interpretation of stellar spectra.

Another problem that's been with us for a very long time, I hope it isn't going to be with us that much longer, is the heating of the solar corona. The corona is that great globe of very tenuous gases around the sun that people are used to seeing at solar eclipses. It is there of course, whether there's an eclipse or not, and we are able to observe it from the ground and even better from satellites without the benefit of the eclipse. The peculiar thing about it is that its temperature is up around a million degrees, a million degrees and up, while the surface of the sun down below it has a temperature of only about 10,000 degrees if you're well up into the chromosphere with a minimum temperature of about 43000 degrees or 45000 degrees, How does that energy get into the corona? Certainly the flow of heat by any natural process is in the other direction. The corona's hot; the surface of the sun is cool, so the heat flows down from the corona to the surface, and it certainly does, and very fast. Something is replacing that heat in the corona, and we have only the vaguest idea of what it can be. Just by a process of elimination, we say we can't think of anything else, so it's got to be some mechanical process. And people work very hard trying to devise mechanical models that would do the job. Well, the trouble is, we don't know enough observationally about the details to be able to decide on questions like this, and this is one of the places especially, where we expect the space program to help solve these serious problems.

Another most important problem is the structure of the solar magnetic field. We used to think the sun had a magnetic field not too different from that of the earth with a smooth distribution of field strength over the surface and pretty much a simple field that you could describe by, perhaps in some cases, just two magnetic poles like a bar magnet, although usually a little more complicated than that. Now it's beginning to look as though things aren't that simple. Indirect evidence indicates that the magnetic field of the sun in undisturbed regions is granular; that is, it consists of little points on the sun of extremely strong field separated by regions of practically no field. When I say practically no field, I mean below the threshold of measurement. These little points are only about 200 km in diameter. This is indirect now, but you calculate, based on observations that can resolve areas only as small as maybe 1000 km, so this is quite an extrapolation. The direct evidence is that we can observe very large fields in very small areas. That's about as quantitative as it gets. These are visual observations made by methods that do not permit accurate quantative measurements. But this structure of the magnetic field of the sun is certainly an extremely important thing to know for sure because it has a strong influence on the migration of energy from the low levels to the upper levels, and the things that are required for it are things that we are going to achieve best from space. There are a lot of things we can do from the ground and should do, but there are some things that can be done only from space and these are the ones that we're going to look to the space program for, of course.

A very interesting problem is the obliqueness of the sun. Until recently, all the measurements that have been made of the sun suggested that it was awfully close to being a perfect sphere distorted by whatever amount one should expect of a fluid that is rotating on its axis. But it was suggested that maybe Einstein's theory of relativity should be modified and that the future of the sun could be one of the distinguishing characteristics that would decide between Einstein's version and the Brans-Dicke version. So the very careful measurements were made to determine whether the sun is round, and the first result was, no, the sun was not round. It's out of round by about five parts in a hundred thousand, and this accounts for the difference between the Brans-Dicke and the Einstein theory, the Brans-Dicke theory is correct, and so on. Some very, very careful measurements suggest that the sun is not oblique, and this is a problem that needs to be solved, of course, because of its very wide-spread implications in such fundamental aspects as the theory of relativity.

Space observations do give us in solar physics two important advantages. The first is that we can look at a part of the spectrum of the sun that is completely inaccessible from the ground. These are wavelengths in the ultra-violet and x-ray regions, that simply do not penetrate the earth's atmosphere and can be seen only from above. This part of the spectrum is very important to us because it is the place where most of the resonance lines of the solar spectrum are, and by observing these, we see a part of the solar atmosphere that normally is quite invisible to us. These radiations originate in the upper chromosphere and in the corona and they can be seen directly against the disc in spite of the very low density of these layers of the solar atmosphere. So they're telling us a great deal about these parts of the sun that we have no other way of observing.

The other information that we can hope for from the space program, and have not realized yet, is higher spatial resolution. We can see small details on the sun, like these little magnetic fields I've been talking about. That's simple because, again, we do not have to look through the earth's atmosphere, which is a very bad optical element that spoils the sharpness of any image we look at through it. As soon as you can observe from, we achieve a solar astronomer's dream of a perfectly sharp, steady image that stays there hour after hour, so we can watch and follow the evolution of sunspots and prominences and other pertubations with the kind of detail we need to describe the magnetohydrodynamic phenomena that are going on in there. Of course, we'll be able to describe them, but first of all, we must have a good picture of them, and this is one way we can achieve it. The telescopes for this kind of objective are being designed, and I hope the time will come before very long, especially with the shuttle program, when we will be able to perform this kind of observation as well as the more sophisticated observations in the X- and UV region of the spectrum than we have now. So I think the outlook for the future is very bright. All it takes is lots of dollars, unfortunately, and I am sure these will be forthcoming. Thank you.

THE LARGE SPACE TELESCOPE
AN INTRODUCTION INTO THE FUTURE OF SPACEBORNE OPTICAL ASTRONOMY

Maxwell Hunter II *

I will take just a few minutes and show you a few view-foils as to what the Large Space Telescope (LST) Program is at this point. I'm not an astronomer, I'm an engineer. During the last several years I have become acquainted with many astronomers rather well, and so I'm faced with the problem of one, providing equipment for them to use, and to periodically having to explain what it is they are up to, to people in Washington such as Congressmen or members of the Office Management and Budget and other places like that.

I did bring with me an illustration of how you can take what you've heard in the last two hours here and make it understandable to a congressman. That would take too long to go through, and rather than that I'll just give you a brief updating as to the essence of what the Large Space Telescope is, and what its program is in this point in time.

If I can have the first foil there. It is briefly stated about the largest optical telescope that you can conveniently get into a space shuttle; the telescope looks like any space telescope, it's a tube with a few solar arrays on it for power. It's put into orbit and will be serviced by the shuttle which is a very important thing. Of course this is a total system. It will communicate through the TDRS system with owner communication directly to the state and ground, there will be a Mission Operation Center and a science institute. As usual in space programs, we are dealing with a total system here. See next page, Fig. 1.

The shuttle cargo bay is about 15 ft by 60 ft long and this particular device has a 2.4 meter primary mirror. You could have stuffed about a 3 meter in there but it's a little easier and a little cheaper this way. In our design this external diameter is in the order of 14 ft and the total thing is in the order of 40 ft long, so it just about fills up the cargo bay. It takes up only about half the weight that the shuttle can carry to the orbits that we intend to put it so there's lots of ability to make use of simplified structures, use a little weight to save cost and that sort of thing.

This program didn't come about because somebody in industry conned the government into an idea. It really has its origins back to over ten years ago in some National Academy of Sciences reports. Where the realization on the part of astronomers as to what you could gain in space was very high and they laid out the requirements. They said why don't we put one up and why don't we do a very good job, make the objects defraction limited and then force the engineers to a system that was so good that they did not deteriorate the basic optic capabilities so that you would have a facility that presumably would last almost forever. Since they got to the requirements before I got to the requirements, they then took such things as the actual (aberation) disk which works out for these diameter optics with diffraction limit which is somewhere in the order of a .004 arc seconds. It was decided that, to keep people like me from lousing it up, we ought to have a pointing instability as low as only one-tenth of
----

*Lockheed Space Division, Sunnyvale
Large Space Telescope Project Director

120

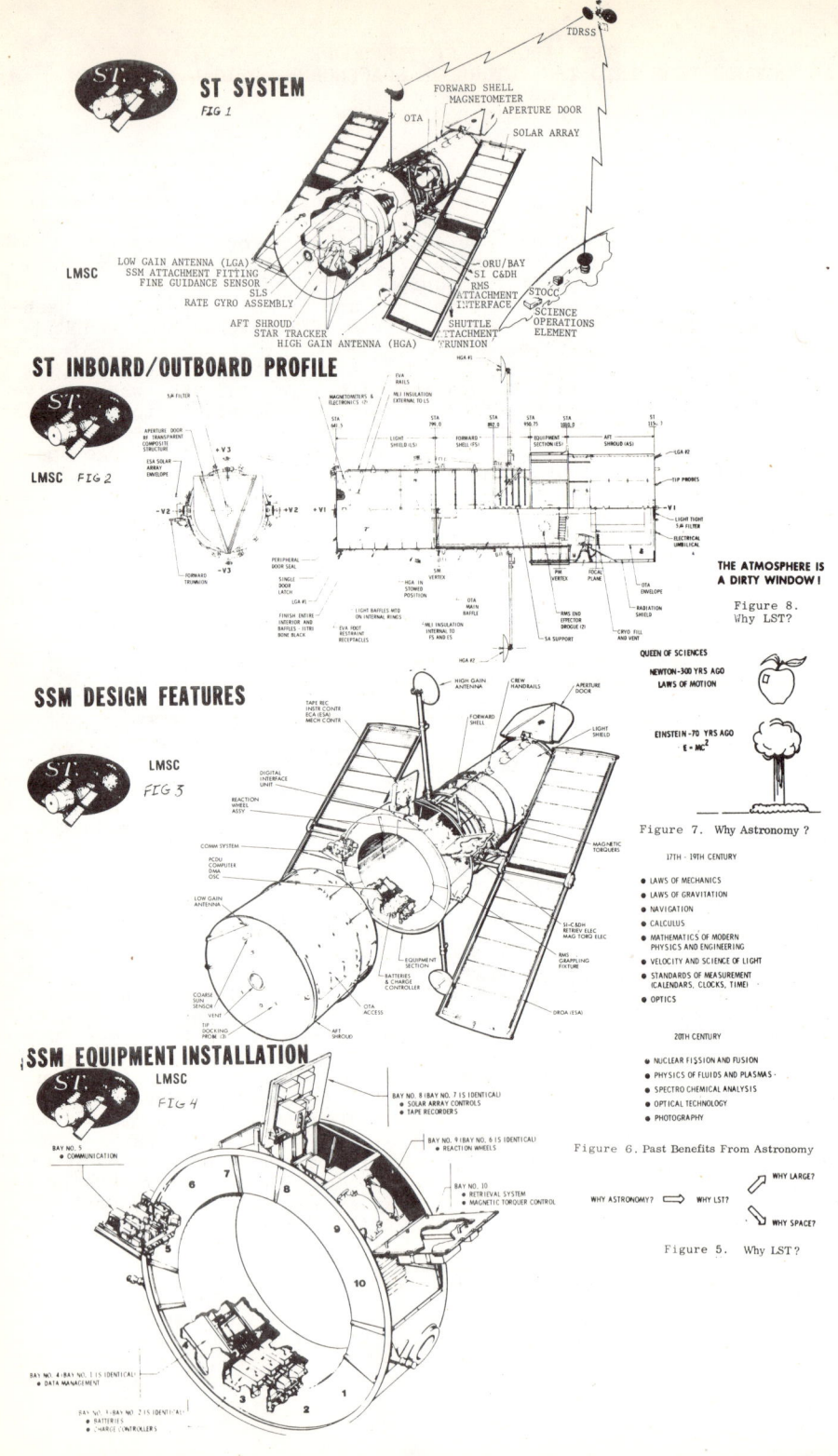

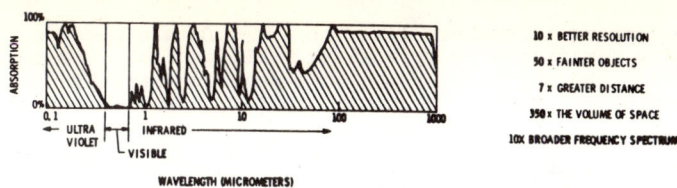

Figure 9. Atmpspheric Absorption

Figure 13. Gain-With LST

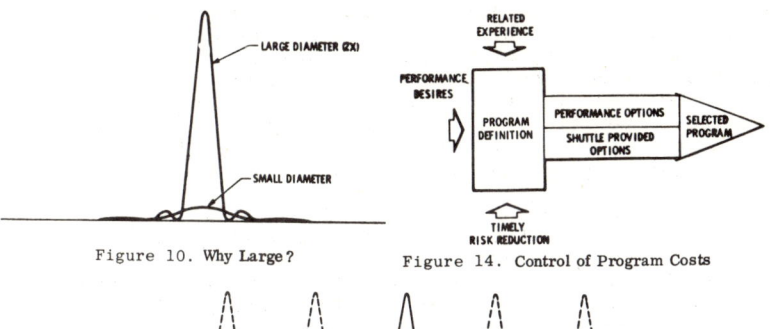

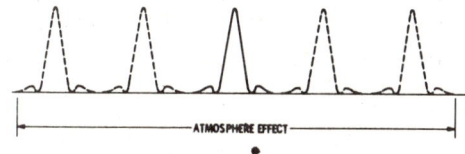

Figure 10. Why Large?

Figure 14. Control of Program Costs

Figure 11. Twinkle, Twinkle, Little Star

Although the twinkling effect does not seem like much when observing stars with the naked eye, in a large telescope the amount by which the image wanders vastly exceeds the size of the image itself. Thus, the total image blur size is set by the atmosphere, not by the basic resolving power of the telescope. This effect is so bad that on the clearest nights imaginable, the sort that occurs only once every ten years, the resolving power of the 200" earth-based telescope is only equivalent to that of a 15" space-based telescope, and on an average seeing night at Palomar, it is equivalent to only a 4" space-based telescope. Astronomers have lived with this twinkle problem from time immemorial and have become very skillful at extracting a great deal of information in spite of it. The extraction procedures must involve a large amount of time. If faint objects are involved, the blurring of the image means large amounts of additional background light must be suppressed; and it is not possible to observe faint objects anywhere nearly as effectively as they can be observed in space.

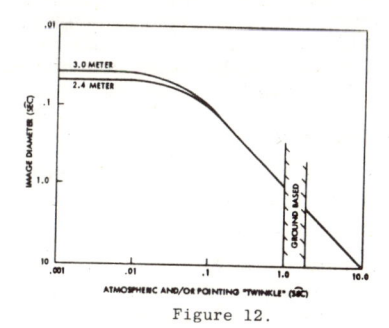

Figure 12.
Image Size Governs Performance

121

that. We're working to numbers like, no it's .04 arc seconds. It is the diffraction limit, and we're working to things like .007 for some of the allotments, .003, 5 or so arc seconds pointing stability. This is probably the greatest single technical problem in the whole program. Fortunately, we think we can do it.

If we can look at the next slide here, the next chart. This is an extremely simplified view and illustrates several points. What we have here is a primary mirror; there a secondary. The whole outer shell is the spacecraft, and for those of you who are interested in the state of this program, we had quite a flap this year since it was not a new start in this year's budget, quite a lot of discussion back and forth among Congress and Washington. It is widely expected that it will be a new start next year, that the request for proposals will come out on the 20th of January, that the competition will be held next year and the awards will be announced in the middle of July. So that's what I'm doing with my life for the next 60 to 80 months.

This feature that the shuttle takes into orbit is a very crucial thing because for the first time astronomers will be able to use this facility much as they have been using other facilities. It will be revisited. The instrument load will make it possible to change instruments throughout the years. They will be repaired and refurbished in orbit, if necessary. In other words, up there will be a basic facility just like a telescope, like those of Palomar, just like the large optical telescopes now in use. After many, many years we are making measurements that were never dreamed of by the people that built the basic facility. Because we can get to this, because we can repair it, refurbish it, reservice the changed instruments, it will probably go on and on for a long time and will no doubt be used before it is over, in matters that we can't dream of today, I would imagine.

I'm not sure about this business of tracking any very rapidly moving objects. It's a device that is inertially oriented in space. We have very weak reaction wheels on it. When we are in a hardover slew, just taking tight turns at 40,000 feet as fast as we know how, we move slightly slower than the minute hand of a watch. That has a very interesting feature about it, because one of the things that we want to never do, is never lose this telescope. Therefore, no matter what we want to shuttle, to be able to get it back. If we're embarrassed because it doesn't work, and we have to repair it, that's bad. I'll probably get fired, but that is not nearly as bad as losing the facility. The shuttle people say they can rendezvous with a device, anything as long as its maximum angular rates are less than a tenth of a degree per second. We cannot slew faster than that no matter what we do. We can inhibit the reaction wheels that way, and so we expect to have a capability that this thing is totally dead which we never expect it to be, that the shuttle will always be able to get to it. We will never have to start over again completely. Perhaps with new leaders in the program but not with new equipment. Our ability to track fast moving devices is pretty slim. This is one of the slowest, neatest maneuvering devices I've ever been near in my life. It has a lot of interesting consequences.

Would you pick the last one out of the bottom. We don't have time to go through that perfectly hypnotic other story as to how to explain astronomy to congressmen.

The basic question is, why you want an LST, as we have been asked many times to answer, and we've decided you just can't have Art Code go there and try to explain it. So we've made up simplified explanations in any other language if you will. We go through much the same story as a matter of fact only in different words. In fact we broke it down into never-get-asked-why you want a large space telescope, never answer that question. Break it into three questions, why do you want astronomy and then talk about why you want a large telescope and why you want it in space. Never get trapped into answering the three at one time. I've also decided never to ask an astronomer to tell you why astronomy is important because he will drag clear back to the primordial roots of man. He will come up with some of the greatest and most fantastic stories some of which you have heard today. It will be so overwhelming that it will sound like a "snow job". You can ever hear the music from "Aida" playing in the background when he goes through this. All we say is, you know, when they do that it's right. They really have an awesome story. The basic problem is that we have to reduce the diffration limit to a very small image, or a point. You cannot do it with the ground-based telescopes, even though their optics are good, because the atmosphere is thrashing that image back and forth on the focal plane in terrible ways. It's amazing how they have lived with it for hundreds of years, not just blurring the image a little but thrashing it back and forth. What we had to do was put it in space for all of the reasons that were said then you've got to keep people like me from lousing it up because if our pointing and control is as bad as the atmosphere, we're really right back to where we started on all this question of resolution and faint object capability. We work to be sure we are way over on the leftside here. We have a lot of reason, a lot of experimental data, to believe we can do that. The name of the game is that it must be very, very quiet. The work we do on micro-g's, microvibrations paying fiendish detail to everything that is turning and moving in this thing. We worry about the state of the bearings and reaction wheels and the gyros, things things like that. That is the basic technical question, as long as we can do that, we're going to make this kind of requirements, and the net result is something that can see 10 times deeper into space than anything based on the ground. That's a thousand times the volume of space that's never been seen in these frequencies before, and that's why everyone is so excited about this program in nice simple terms.

What I would like is those viewgraphs to explain astronomers to congressmen, and see if it works on my wife and son, too.

I sometimes answer the question why is astronomy important in terms of history. All societies in the past, that haven't devoted a significant fraction of their gross national product to science have declined and decayed, which did not devote a substantial portion of the gross national product to astromony, have declined, so we ought to try putting in a good percentage.

What I would like to do now is have a discussion but I think we will have to terminate and turn this over to the radio astronomers who have something to say too.

VERY LONG BASELINE INTERFEROMETRY (VLBI) IN MODERN RADIO ASTRONOMY

Dr. Barry Clark*

I would like to take about half a minute for a personal reminiscence namely it's about 20 years ago last spring that I was a student in Art Code's freshman astronomy class. I'm very happy to have had the demonstration today that in the interim he has lost none of his enthusiasm for the subject.

I would like to delve into considerable detail in one of the small areas of radio astronomy that make up the whole rather than trying to give an overview of something as Frank Drake has done. This is the subject of the very small objects that are found far from the earth, far from our galaxy at a distance comparable to the radius of the universe itself. In the early days of radio astronomy, such things weren't really suspected. We thought that all the radio emitting objects we saw were large things, objects comparable to the size of a galaxy, that even though they might be very distant still occupied a fairly large section of the sky. Because they are big, many tens or hundreds of thousands of light years across, we would expect them to be very stable, slowly varying objects, so it was with considerable surprise in the early '60's that we discovered that some of these radio objects that we looked at were changing their flux with time and therefore were not hundreds of thousands of light years across but were a few light years across since the light that we see them by was changing on a time scale of a year or so. If you have an object, say a light year across, and it's located perhaps a good fraction of the radius of the universe away, a day in light years away, then you know that it's going to subtend a billion angle indeed from the earth, something of the order of a thousandth of a second of arc. So that when we try to investigate what the structures of these objects are, what is really physically happening in them, we need very high resolutions indeed. In radio astronomy we are almost always defraction limited; so to find the size of instruments you need for this sort of resolution, you simply multiply the wavelength by the reciprocal of the angle that you wish to resolve expressed in radians. We're looking at objects of a thousandth of a second of arc at wavelenghts of a few tens of millimeters and when that multiplication is carried out, you find you need an instrument a few thousand or a few tens of thousands of kilometers in diameter, a very large instrument indeed. We're not contemplating filling up an instrument of that size, so to obtain samples of an instrument that size, we resort to usual techniques of interferometry where you have two elements separated by a very large distance and by making observations at varying distances and at varying position angles, one can reconstruct the brightness distribution that would be mapped by the large instrument.

To go back to elementary physical principles, it's sort of interesting to look and see what we expect from these very distant and comparatively small objects that we're looking at. The first physical principle is that radiation that we see in the radio wavelengths comes from the synchroton mechanism of the ultra-relativistic electrons spiraling around in a magnetic field. In knowing this, we can calculate what sort of energy requirements we have in the objects and we find that

----

* National Radio Astronomy Observatory, Socorro, New Meixco.
    The NRAO is operated by Associated Universities, Inc.
    under contract with the National Science Foundation.

they are very high indeed. As Frank said earlier, they're of the order of a million solar masses converted directly to energy. To get this much energy, one needs an extremely efficient process. The process that powers the sun, for instance, the burning of hydrogen to helium just doesn't release a large enough fraction of the available energy to power these objects. The only mechanism we know of that can release a large fraction of the rest-energy is gravitational collapse where an object collapses to Schwartzschild radius and in so doing, may make available a large fraction of its rest energy. There are arguments on general principles that an object collapsing to its Schwartzschild radius may break up in the process into smaller objects and as a result you have energy coming out in little bursts of a typical energy, and that is roughly of the order of the rest-energy of the smallest mass which can collapse to its Schwartzschild radius. Any masses smaller than the limiting mass, can be supported by the electron degeneracy pressure and will never get to the limiting radius in any normal manner.

The first equation below expresses this limiting mass, and the amount of energy that converting it entirely to energy can provide. This minimum mass is approximately

$$M_{lim} \approx \left(\frac{hc}{G m_p^2} \right)^{3/2} m_p$$

The expression in the parentheses is, maybe, recognizable. It's the gravitational equivalent to the fine structure constant and measures the strength of a gravitational interaction. The fine structure constant reciprocal is 137. The expression in parentheses is about 10 to the 39th. Anyway, that's the amount of matter, or energy that the collapse of the minimum mass to the Schwatzschild radius may make available.

It is a reasonable premise that, in the collapse of a larger mass, it will fragment, so that it will appear to release energy in bursts of

$$E \approx M_{lim} c^2 = \left(\frac{hc}{G m_p^2} \right)^{3/2} m_p c^2$$

$$E \approx 3 \times 10^{54} \text{ ergs}$$

If the energy is distributed in an object of size S, the resulting energy density is

$$U \approx E/S^3$$

The next elementary physics we can apply is the law of equipartition of energy between electrons and the magnetic field.

$$B^2/\mu_v \approx m_e c^2 n_e \gamma \approx U$$

where ne is the number density of electrons, and

$$\gamma = (1 - v^2/c^2)^{-\frac{1}{2}}$$

is the ratio of their energy to their rest mass.

These ultrarelativistic electrons emit radio radiation through the mechanism of synchrotron radiation. Let us set down the laws of synchrotron emission. An electron radiates most of its energy at frequencies near the critical frequency:

$$\omega_c = \frac{e}{m_e} B \gamma^2$$

The electron will radiate its energy in a typical time given by the synchrotron lifetime:

$$T_{\frac{1}{2}} = \frac{m_e^3 c}{\mu_v e^4} \gamma^{-1} B^{-2} \qquad T_{\frac{1}{2}} \approx S/c$$

This imposes a restriction on the size of the object--the energy transport is by motion of the electrons, so they may not radiate all their energy before they can cross the object. Unusual though this plasma is, it must obey the laws of thermodynamics, and its brightness cannot much exceed that of a black body whose kT product is the mean energy of the particles in the plasma. Now, the radiation, Plank's Radiation Law, is well known and if you say that this electron plasma is emitting radiation up to the electron critical frequency then at the limiting brightness, then you can calculate the power it is emitting or equivalently within the gas. The limiting energy density, which if it is written out, is that expression.

$$T_{lim} = \frac{m_e c^2 \gamma}{k} \qquad U_{lim} \approx \omega_c^3 \, k \, T_{lim} / c^3$$

$$\approx e^3 \, m_e^{-2} \, c^{-1} \, B^3 \, \gamma^7$$

We have one more bit of physics we can call on. If we try to put too much electron energy in a small volume, the magnetic field strength due to the synchrotron radiation may exceed the strength of the static magnetic field. The resulting radiation is known as inverse Compton radiation, and the condition that this does not occur or is just beyond the margin of occuring is that the magnetic field energy is equal to the limiting energy of the synchrotron radiation. Then if you throw in the condition that the energy available divided by the cube of the size is also about these two energies, then we have three equations and the three unknowns of the electron energy, the magnetic field and the size of the object. We may solve these equations with the following results:

$$\gamma = 300$$

$$B = 0.005 \text{ Gauss}$$

$$\omega_c = 6 \text{ GHz}$$

$$S = 100 \text{ light years}$$

If such an object is situated at a distance of one Hubble radius, it will subtend, from the earth, an angle of 0.001 second of arc. Since it radiates primarily at a frequency of 6 GHz, we can calculate the baseline of the interterometer necessary to resolve this object:

$$L = 10000 \text{ km}$$

Therefore, from elementary physical principles we deduce the necessity of an interferometer with a baseline about one earth diameter.

This length results directly from the particular combination of physical constants derived above. It is an interesting coincidence that the diameter of the earth is about equal to this combination of h, e, me, mp, c, and the Hubble radius.

$$L = H \left(\frac{m_e}{m_p}\right)^8 \left(\frac{hc}{Gm_p^2}\right)^{-\frac{12}{25}} \approx 7500 \text{ km}$$

THE VERY LARGE ARRAY AT MAGDALENA, N.M.

Dr. Victor Herrero*

I would like to very, very briefly describe the progress that we are making in the construction of the VLA, a very large radio telescope, one project similar to the other large project that Frank has described that very appropriately is being built here in the State of New Mexico, very close as a matter of fact to the Space Museum that we are dedicating this week.

We hope that this instrument, which in many respects will be the most powerful in existence when completed in late 1980, will help us to learn a little more about the nature of space and the universe.

I'll show the first slide now. It is the location of the instrument 50 miles west of Socorro which in turn is 75 miles south of Albuquerque. The instrument consists of a collection of 28 antennas, which can be located by means of a system of three railway tracks each approximately 13 miles long. The concept is to synthesize the image that a single dish, such as the 1000 ft. Arecibo instrument that you saw a few minutes ago, that 25 miles in diameter in round numbers would produce. By having a small collection of antennas that make measurements, one can in a computer mathematically synthesize an image.

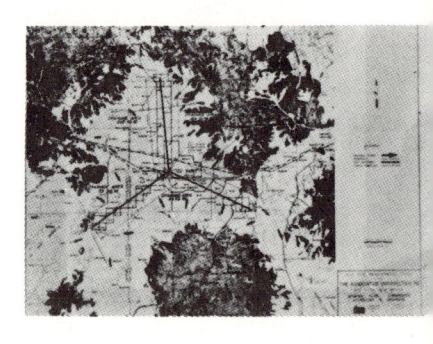

Next Slide. This is an artist's conception showing the tracks and a collection of nine antennas spread over one of the arms.

---

*National Radio Astronomy Observatory, Socorro, New Mexico. The NRAO is operated by Associated Universities, Inc. under contract with the National Science Foundation.

Next. Slide. This picture was taken a few months ago, at which time we had two antennas. We now have six, three of the antennas are operational, and are undergoing testing. There you see a fourth one being assembled in this 100 ft. high building. The antennas are taken out of the building, set on an alignment pad, and then taken to stations for further testing and eventually scientific use. Here on the left you see this red object that is a transporter that slides under an antenna, picks it up, and allows us to move the antenna, and place it anywhere on the system of 60 miles of trackage.

Next slide. This is a view from the air showing the southwest arm, the antenna assembly building and the maintenance and services building. Signals from the antennas are transmitted through a waveguide system, that I'll show you with a couple of slides in a while. The data is collected in a control building where the data processing equipment is installed.

Next. This is a close-up view of the dish. The dishes are 82 feet in diameter, and as I said there are to be 28 of them, 27 in normal operation, and one undergoing maintenance. It's a casegrain system, you see the primary mirror, the secondary mirror, and a collection of feeds. The secondary is off-axis, and by rotation it can direct the radio beam to any system of four feeds. The system operates on four frequency bands. The lowest lies between 18 and 21 centimeters wavelength. The next one is centered at 6 centimeters and the other two at 2 and 1.3 centimeters. With the array, there are basically four working configurations that effectively produce a zoom-lens effect, where you can go from a fully spread array with a capability of synthesizing images with an angular resolution of a couple of seconds of arc at the longest wavelength, and a tenth of a second of arc at the shortest wavelength to a fully compacted configuration where you have a broader field of view at lower resolution.

Next Slide. This is just to illustrate the process of assembling antennas. It's really a production line for antennas and makes you think of this as mini-cyclops in which every two months an antenna is fully assembled from raw materials. The reflector dish is assembled on the ground. Simultaneously, the yoke structure of the pedestal is assembled on a separate station, then the reflector is lifted and the pedestal is slid under the reflector, and the two units are mated. Then the panels are installed, and after their alignment, the antenna is taken out for tests of its servo system.

Next slide. This is again another view of the mating process. You can see the transporter that has picked up the yoke pedestal structure, and there, the reflector is just about in the process of being bolted together to the yoke.

In the antenna, there's a large vertex room, where the very sensitive front end amplifying equipment is installed. This is one part of the electronic equipment that you could see, there is the front end. The heart of it is a dewar, a vacuum assembly, where a set of six parametric amplifiers for two channels of operation, operate at cryogenic temperatures, at 18 degrees Kelvin. The signals, highly amplified, then are transmitted through the waveguide system to the control building.

Inside the dewar there are the parametric amplifiers. The fact that we operate at four frequency bands results in a great complexity in the waveguide system, since you have signals from a total of eight different waveguide ports coming in.

Next slide. This is an illustration of the waveguide that we use. It's a circular waveguide 60 millimeters in diameter and it operates in the 25 to 50 gigahertz range with a tremendous capacity for information transmission. We have one waveguide channel for each antenna in that 25 to 50 giga- hertz range. Each waveguide channel would have a capacity for the equiv- alent of 100 television channels. We actually will use eventually up to 4 50 megahertz bands.

Next slide. This is an illustration of the process of installation of a waveguide in a trench. There you see a very fine spiral copper wire that is used to select the T01 mode of transmission that is designed to have low propagation losses in the waveguide.

At the control building there is a complex array of computers that process the data, the samples, essentially the samples of measurements of the electromagnetic field at each antenna are combined in a Fourier Transformation process to synthesize the picture such as an optical telescope could produce.

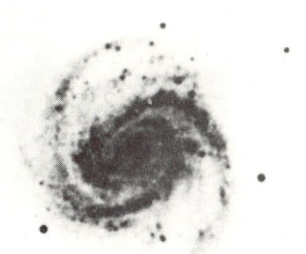

To illustrate the purpose of the instrument, this slide shows what an image of an optical telescope such as the 200 inch telescope would produce of a galaxy.

OPTICAL

One of the older generation radio telescopes such as our 300 footer in West Virginia will produce a highly blurred picture.

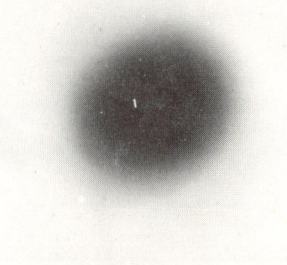

NRAO 300 FT.

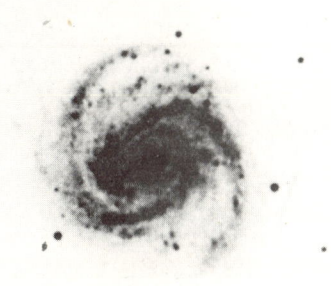

The VLA will produce, essentially an image with a comparable resolution to the resolution of a ground-based optical telescope.

VLA

Here, the fourth picture is the kind of resolution achieved with perhaps the most powerful interferometer currently in operation at Westerborg in Holland.

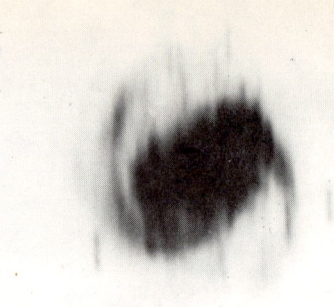

WESTERBORK

CHAPTER V

LIFE SCIENCES IN SPACE FLIGHT ROUNDTABLE

> "This is the goal: To make available for life every place where life is possible; to make inhabitable all worlds as yet uninhabitable and all life purposeful."
>   Hermann J. Oberth, 1923
>   (Rumania-Germany-U.S.A.)

Picture of the International Space Hall of Fame, taken on Dedication Day, 4 October 1977. Copy Rights NMRI

Picture of the International Space Hall of Fame, taken on Dedication Day, seen from the East. Copy Rights NMRI.

INTRODUCTION TO LIFE SCIENCES IN SPACE
ROUNDTABLE

Dr. Charles Berry*

I'm Dr. Berry and we have had some changes here necessitated by some illness that our organizer, Dr. Paul Campbell, suffered. When I was asked to do this task, I was somewhat overwhelmed because I hadn't planned to do it exactly that way, but I was delighted because this has been a very deep part of my life, and I think all of you should know before we start this session that I am noted as an optimist. I think that is what most Americans are. Someone was saying yesterday, I think Dr. Stever, said that's what we're all called, and we're not really an American if you're not. I certainly have been an optimist in this area. In our contracts with people in aerospace medicine around the world, I found that they came to be much the same way. I'd say, in general, our people from the Soviet Union have not been as optimistic about this activity as we have, but they certainly are progressing at any rate.

Now I believe that Space flight is really tied to manned space flights, and efforts that are necessary, whether they be manned or animal flight, or whatever, in order to make that possible. We've taken some lumps over the years for some of our optimism, but we've taken some gambles, and maybe had some luck as Dr. Gilruth noted last night but, in any case, some of those gambles, or the luck that we had, has allowed us to reach the state at which we are now, and have the possibility of having man in space for long periods of time. That's the question we want to address here today - exactly where we are. You have not been privileged to have the information that we have circulated to members of the panel. I'd like at this time to introduce the members of the panel, and I will just go around the table here. You will hear something from each of them later on this morning.

Let me just quickly run around the table. The first person here on my right is Dr. Paul Campbell. Dr. Campbell is now living in San Antonio. He is retired. He is the former commander of the School of Aerospace Medicine at San Antonio. I first met Paul many years ago in that position. He is well known in the IAA and has been a consultant and has even joined the Press Corps in covering space missions and all sorts of aerospace activity, and Paul is certainly a pioneer in this area.

The next person on my right is Dr. Harald von Beckh. Dr. von Beckh is currently the Director of the Medical Research at the Naval Air Development Center in Johnsville, Pennsylvania. Dr. von Beckh was previously a neighbor of yours, having lived in this city, and was at Holloman for a number of years, and he'll talk about that later in the morning.

The next person is Dr. Heinz Fuchs. Dr. Fuchs is from Germany. He is a Major-General in the German Air Force. He is a Commanding General of the German Federal Armed Forces Medical Command, all of their Armed Forces, and he is currently the President of the German Society of Aviation and Space Medicine, as those of you who were at the dinner last night may remember.
----

*Co-chairman and Moderator. President, The University of Texas Health Science Center at Houston, Texas.

The next person is Dr. Hisashi Saiki. Dr. Saiki is the Director of the Space Medicine Laboratory at the University School of Medicine in Tokyo, Japan.

The next person is Dr. Clayton White. Dr. White is the President of the Oklahoma Medical Research Foundation in Oklahoma City, Oklahoma, and was formerly a resident of this state, as you probably know, and was for many years President of the Lovelace Foundation for Medical Education Center and Research at Albuquerque, New Mexico.

The next person is Dr. Harold Sandler. Dr. Sandler is currently the Chief of the Biomedical Research Division at Ames Research Center for NASA. It's a NASA center and it's located at the Moffett Field in California.

The last person, closest to you over there, is Dr. Ulrich Luft. Dr. Luft is Head of the Department of Physiology at the Lovelace Foundation for Medical Education and Research in Albuquerque, New Mexico. Now, all of these people are close friends. We've known each other for may years and we've worked in this area for many years and have faced some of these problems together. With this, let us begin this program.

I was asked to pull together what I felt the current status of man's response to this environment was. What I have tried to do is do that in a succinct manner as possible, and you can imagine the amounts of data that have been gathered over the years. We wish we have more, but certainly a great deal has been gathered concerning man's capabilities to adapt himself to a space environment and then readapt himself to the 1g of earth following space flight. That's really what it's all about and what our concern has been for many years.

We have now completed almost, well we have completed 38,000 man hours of exposure to the space environment if you look at both the U.S. and the U.S.S.R. flight program. In addition, we have accumulated a number of hours in both zero g and 1/6g on the surface of the moon in the extravehicular situation. We've had men totally isolated for long periods of time as they pass behind the moon during the Apollo Lunar missions. Since the time that I compiled this information and sent it around to our colleagues, we have had an additional two Russians that have had a short flight and have added a few more man hours to that situation with the Soyuz 22 Flight.

Now, it would take a lot of time to totally review the painful experiences that we faced and the criticism of scientific colleagues in the development of our flight programs to date. It's well that we remember, however, that space flight thus far has indeed been an experimental program. Much of the medical information that was gained on the early flights was obtained in a very particular environment, one which was heavily engineering oriented, and that was done in order to get man safely into space and get him back. It required a great deal of faith in man's capability on the part of the space medical community in order to commit him the collection that it was possible to obtain in order to make the predictions that were necessary, and we had great difficulty with our engineering colleagues as we tried to do that because they were able to get much more information concerning their hardware than we were able to have at our command as far as man was concerned. If one compared the time frame of the development in duration of exposure of aviation and of space flight, however, I don't think you can help but be impressed with the very rapid progress that we have made in flight durations in the field of space flight in contrast to the much slower progress in the periods of years that was done in aviation. We could recount for hours some of the dire predictions that were made, the validity of only a few of these, and then the flight proof that we have been able to obtain some of these, in fact, many of them were indeed erroneous. That's not to say anything derogatory about the people who made these predictions because they were made with the best of scientific data they had available to them at the time, and based upon their own particular experience in their discipline whatever that might be.

In a very few sentences I think that I could summarize what I feel the current status of man's responses to this space environment are: First, that with long or short durations up to 84 days, while there has been some individual variability, man has always been able to perform his inflight task and to suffer no permanent physiological change postflight, only time related readaptational changes, if we do the following things: if we supply an adequate atmosphere, adequate food and hygiene facilities, adequate exercise and adequate workload planning and rest, adequate time to acclimatize or adapt, and finally adequate countermeasures where these countermeasures are warranted. Now it's important to remember these ifs, for the development of future flight programs can easily fall into the trap of taking a very cavalier attitude towards these ifs because it looks as if man has done so well, and he has. He has shown himself able to adapt over a period of days to weeks inflight to even discomforting physiologic changes, such as vestibular abnormality resulting in motion sickness. He has further shown himself able to readapt through a series of physiological changes, on the return to 1g here on the surface of the earth. Time phasing of these adaptations is of considerable interest and as one would expect, it varies with individuals. It appears that there is some sort of general adaptation or at least acclimatization that starts within the first hours of space flight and some of the body's systems have adapted or at least acclimatized within a matter of hours. The vestibular responses have all shown adaptation within a period of one week, and if we look at overall body systems it appears that the major shifts occur with a week's period and that certainly within the 60-day flight period most of the lability or the swings in the system response has leveled out. The postflight time to readapt to 1g has, in fact, seemed to decrease with increasing flight duration, the reverse of what one might expect - 24 days following a 28-day flight and 5-7 days following 59 days and 4-5 days following 84-days flights. It will be a continuing interesting exercise, as man is exposed further, to detail the individual system adaptations on a time frame and with careful checking of each of these systems.

For purposes of brevity today I'm going to outline the principle facts observed in the body system most affected by space flight and to list those systems having shown little effect or change by the measures used thus far. We will then address what we consider to be the present view concerning the need for countermeasures to these effects and the predictions concerning very long duration flight, and then try to outline some of the gaps in the data, and to conjecture as to how we might fill these gaps.

Histories taken from our astronauts and cosmonauts have revealed a general sense of enjoyment and well-being in the weightless state. To be sure there are sensations of fullness in the head and even occasional motion sickness symptomatology associated with the freedom of movement in the spacecraft and ease of activity in this very peculiar environment where up and down is related only to the placement of objects within the spacecraft.

The first system to evidence any physiological change has continued to do so throughout the duration of any flight to date, and that system is the cardiovascular system. We early noted what we call ortho-static-hypotension in the immediate postflight state, and this has been investigated throughout the 84-day flight duration. In case we are using big words that may confuse some of you, what that means is that when the individuals stand upright, they tend to pool blood in the lower portion of the body and they get an increase in heart rate to try and make up for that. They may indeed get a sensation of faintness and, indeed, some people have fainted. This was evidenced further in decreased exercise capacity on the bicycle ergometer postflight coupled with a marked decrease in cardiac output measured by an indirect method. This is understandable when related to the pooling of blood in the lower extremities which would occur on return to 1g and was shown in postflight studies on the 84-day flight were echo-cardiography coupled with lower body negative pressure demonstrated that we were observing changes due to blood volume and not due to inherent defect within the myocardium itself. There was grave concern that this was an effect of weightlessness upon heart muscle itself and not due to the shifts in blood volume. We think that the early data at least leads us to believe that these are indeed changes involved with shifts in blood volume and not with damage or change in the muscle cells itself in the heart. All of the cardiac electrical activity, measured in detail by vector-cardiography, was not significantly altered. There have been minimal episodes of cardiac arrhythmia noted with the slowing of rate in Apollo, and a single episode during exercise in Skylab.

In the hematology area we have observed a loss of red cell mass. The mean losses as observed postflight were 9.4%. There were individuals who had losses as high as 15,8.6 and 5.9 respectively for the 28, 59 and 84-day missions and you'll notice that those continue to go down, 15, 8.6, 5.9. Now, it appears that there is a suppression of red cell production but that it is governed in some way limiting this suppressive effect so that it does not tend to increase with increasing flight duration. In addition, there appears to be a replacement curve developing which tends to return the red cell mass toward the normal level. This is an interesting, and it's very facinating finding, which we think deserves considerable further evaluation because while that appears to not be a great import to us now as far as the spaceflight activity is concerned, it could have great import to us here on the ground, and be a great help to us.

In the neurological area, the most important positive finding involved the development of motion sickness early in flight by four of nine skylab crewmen. This had been observed earlier in our program in a number of Apollo astronauts, probably more than have been reported, and in a number of U.S.S.R. cosmonauts. In spite of predictions about who would develop this condition, which appeared to work just the reverse of the prediction, and of the determinations of threshold tolerance to head movements which was evaluated in the preflight condition, several Skylab astronauts did develop stomach awareness, and in some cases frank nausea and vomiting, in the early inflight period. This did respond to oral medication and adaptation did occur as shown by the capability of the crewmen within seven to ten days inflight to develop a marked increase in their tolerance to head movements and rotation compared to their preflight baseline. It became virtually impossible to create any symptoms in these individuals at the maximum number of rotations in the chair and head movements during those rotations during the adaptive period inflight. On return to 1g the crewmen had some obvious vestibular disturbance evidenced by abnormal gait and a sensation of vertigo which lasted for several days postflight. The threshold tolerance to motion sickness on head movements with rotation then, after return to 1g on the earth, also gradually decreased to return to preflight baseline.

An immediate postflight hyper-reflexia was noted in the Skylab crewmen, and this was quantitated in the postflight study.

In the musculo-skeletal area, we have seen moderate losses of calcium, phosphorous, and nitrogen throughout the flight duration. The data of these detailed balanced studies indicate that the losses are comparable to those observed in bedrest studies for equal durations, and that the trend is continuing with the increased durations that we have noted todate, in short, this has not leveled out. There has been a relatively low but measureable mineral loss also from the os calcis (the heel). The rate of loss would appear to be some 6 grams of calcium per month for roughly ½ percent of total body calcium per month. The loss in bone density of the os calcis appears to be predictable, based upon initial bone mineral content and upon some hydroxyproline levels.

In the area of fluid and electrolytes determined through the numerous biochemical observations taken before, during, and following flight, it is clear from the Skylab data that there are changes showing an increase in aldosterone secretion, increased sodium and potassium excretion, increases in osmolality and an increase in circulating cortisol levels. These changes are consistent in many ways with the hypothesis which we advanced prior to the Skylab series relating to an increase in thoracic blood volume, an increase in blood volume above the diaphramatic blood level, in the weightless state acting upon stress receptors in the walls of the great vessels and in the atria, in the upper chambers of the heart. This apparent expansion in blood volume produces a loss of water and electrolytes, particularly of sodium. A reduction in blood volume follows, and there is increased secretion of renin and aldosterone leading to establishment of new fluid and electrolyte balance which appears to be proper for the 0g situation In which the body finds itself. There are some conflicting data points, however, in that there has been some reduction in antidiuretic hormones, and in some instances increase in antidiuretic hormones. At the same time, potassium excretion levels have been elevated, as have sodium, in spite of the markedly elevated aldosterone levels which should prevent this loss. The evidence of stress on the crewmen is the increased cortisol levels. These changes in biochemical fluid and electrolyte balance have been shown to be well tolerated for the durations of flight thus far undertaken. While much has been stated concerning the physiologic changes, we have done little to state the psychological findings in our space crews. This is a repeated question from the public and one we must address in view of the public utterances of some of our astronauts. The data is quite clear that there were no dire effects upon performances, that is, problems that have been predicted of isolation, "break off," boredom, etc. were indeed not seen. The crews were originally selected with strong psyches, good self images, and both great achievement needs and accomplishments. There was some evidence of the development of body consciousness during portions of the Skylab flight but certainly this was never noted to any abnormal degree.

Most of the questions center around the Apollo astronauts who visited the moon, and then did varying activities which appeared out of context with the public's perception of these individuals prior to flight. In one instance, there was a reactive depression that was related to demands by the public for public appearances, and the individual's perception that the public had reacted only in a minor way to an accomplishment felt to be of great magnitude. In two other well known instances one individual has concentrated his life on religious pursuits, and others have followed in his footsteps. Another has devoted himself to the study of psychic phenomena and para-psychology, and it is important in both of these instances to know that these individuals had these interests preflight. I am sure everyone can appreciate the fact that there's no individual who was able to leave earth orbit and go to the moon, thus seeing our planet in an entirely different context, who could possibly remain unaffected by the experience. I think those of us who have been closely associated with the activity also have been affected. All of us develop a much deeper concern about our planet Earth, and about mankind, and about the space crew that's on our planet Earth, and what could best be done to assure its future. I think many of the activities in which the crews have engaged have been in this context.

I have tried to summarize the positive findings, and still the most remarkable thing about our long duration spaceflight data is the amount of negative data or information which would show that there has been no adverse change in the remaining body systems, at least by any measures we have today. In view of the great deal of concern about man's capability to perform in this environment, these data are probably the most valuable ever gleaned concerning man's future. We have seen that man who has been well trained for the mission has been able to enter the space environment, undergo an adaptational period, which is certainly acute for at least a week, shows marked stabilization within a 30-day period and then shows a recovery to the preflight baseline as he readapts to 1g. The recovery has taken, as we know, a progressively shorter time period with increasing flight durations. This could lead one to the conclusion that indeed a long term space flight is good for you. Obviously, however, the above findings must be considered and particularly those involving loss of calcium and nitrogen, where we have not seen a stablization as yet.

In attempting to plot the onset and duration of symptoms and findings in each of the body systems, their treatment and their prevention, we must consider countermeasures which might be used. The cardio-vascular was one of the first where this was considered.

Cardiovascular - These responses involving the heart and blood vessels do return to normal at a faster rate or in a shorter time period with increasing durations as did all of the things that we mentioned. This more rapid return to baseline may have been aided also by the amount of exercise which was done inflight that was used in some cases as a countermeasure. Still, as we look at a long duration space flight, we must consider the possibility of changing from the crutch which we and our Russian colleagues have both used, and it is just indeed that, a crutch, a positive pressure garment on the lower extremities, and that has been used on the immediate postflight period, but we must develop a true countermeasure. Lower body negative pressure inflight, at least for several days prior to return to 1g, may offer the possibility of a true countermeasure and certainly further research is needed to make this a reality.

Blood - The red blood cell mass, as we have stated, did not continue to increase in amount but appeared to be self-governed; therefore, I do not feel that there is a need for a countermeasure in this area now, but we certainly need to understand the mechanisms involved here because, as I say, they do have some import to some things on the surface of the Earth.

Vestibular - In the vestibular area, countermeasures are really unknown at the moment. Here too we've utilized a crutch in the form of medication. It's a combination of scopolamine and dexedrine that has been most widely used. It appears that selection, which has thus far been notoriously poor in this area, and perhaps training procedures, offer the only true countermeasures. Certainly, we have to do some detailed investigation into the mechanism of production of these symptoms because you cannot produce a true countermeasure if you don't understand the orgin of the symptoms. While we have a lot of feelings and thoughts about what is involved here, certainly further work needs to be done and is ongoing. One area which bears further research is the effect of shifts in blood volume which, we do know, occur. Our Russian colleagues feel this, indeed, may be the cause here. We're not ready to say that at the moment, but they feel that indeed may be the cause.

Musculoskeletal - In the musculoskeletal area many methods of trying to produce pressure upon the skeleton and varying exercise loads have been used to attempt to alleviate progressive calcium loss. These have been used in bedrest situations and none of these have been effective to date. The countermeasure which has the greatest potential from bedrest studies thus far appears to be oral intake of a combination of calcium and phosphorous. This too needs to be further studied on long-term bedrest subjects because there was some escape of the protective effect of these substances as taken orally at the end of a nine-month bedrest study.

Biochemical - In the biochemical area, there are no biochemical effects noted which did not seem to be adaptive changes of some degree. It is important to remember here, I think, that probably more than any other place that as we look at any countermeasure or as we attempt to develop one, that we must not do anything which would interfere with the body's capability to adapt and, thus, do harm rather than good. That is one of the first things that you are always taught as a physician, "If you can't do good, at least don't do any harm." This is a particularly apt admonition as we consider the biochemical area because this is one which could easily be altered by playing around with one area. You may alter that balance which could be harmful to you in the long run so there are no countermeasures recommended here at the present time.

Psychological - In the psychological area, it appears that the private contact with those on the ground is of import to every individual inflight. Careful attention to the work-rest cycles, the physical amenities of habitability and the careful following of the air-to-ground communications should be of immeasurable help as countermeasures in this area.

What about Mars mission capability then? That's the big question and a lot of data obtained to date is: Can man successfully accomplish a two-year flight of the duration and vision for a Mars mission? I would like to answer that personally in the affirmative, based upon the data that we have seen to date. If the body continues to perform as the bulk of its systems have, perhaps there is even some regulation that might control the calcium loss. If not, then we certainly must have the countermeasures available to provide the needed protection. I think we are capable of developing these and that we can support man in such an endeavor.

All of the above brave statements do not mean that there are no gaps in our data as we look at man in very long duration flight, and certainly that there is further research needed. I feel, as I've expressed on many occasions previously, that we need to procure data on man inflight for a six-month period prior to committing him to a two-year Mars mission. I also feel, however, that research should be directed toward the elucidation of mechanisms of the changes observed, particularly during the first week of space flight when so many of the changes resulting in acclimatization or adaptation are occurring. Now very short flights could even be used for this purpose if the individuals were properly instrumented and monitored. Certainly, as we look for mechanisms and for detailed information on these changes, we are in need of data which can be obtained only by invasive procedures that will require the use of animals. Animals being used in conjunction with man in the same conditions would be an extremely valuable mechanism for answering some of these open questions.

In the cardiovascular area we need to do indepth, non-invasive cardiovascular dynamic monitoring, and we need more data to support our observation that there is no direct myocardial effect at the cellular level. That data shows that at the present time there is one data point, and we ought to continue to prove that. Again, we should do invasive studies related to pressure, volume and flow, particularly in the first week of flight with animals. We still have to demonstrate the role or lack of a role for the Gauer-Henry reflex in the onset of these fluid shifts which occur on exposure to Og. We need to develop some sort of a total exercise regimen to maintain all of the muscle groups of the body. We also need to determine the role of the capacitance of veins in the deconditioning process.

In the musculoskeletal area we need to further evaluate the exact changes in loss of the bone mass and muscle mass in animals. As has been noted, further efforts to develop proper countermeasures have to be undertaken. There has been a formula developed by Vogel, a prediction formula that's related to the pre-flight bone density and to the levels of a hormone called hydroxyproline. This might be used on the basis of selection of crewmen for very long duration space flights.

In summary though, I think man has shown himself capable of adapting in specific time frames to space flight, and maintaining effective performance levels for 84 days. He has also shown himself capable of readapting himself to the 1g environment with the physiologic variables returning to the preflight baseline in the shorter time periods as we have noted. We must control the development of motion sickness in order to maintain productive crewmen during the initial five to seven days of adaptation and we should do this by a true countermeasure rather than a crutch such as medication. We also must find a true countermeasure for the calcium loss which appears to be progressive at the present time through 84 days of space flight, and if it follows the bedrest curves as it is currently doing, would continue to be progressive. We don't know that that's so though. I think man is capable of conducting a very long duration space flight if we give him the proper support as mentioned above and provide the countermeasures where they are needed. We should diligently plan to obtain the data necessary to allow us to make an early committment of man to continue this, his greatest adventure, and to open and further develop the new frontiers that are provided by space.

Each of the people here have, in one way or another, addressed some comments to these areas and I'm going to call on them individually to talk about these. That will bring out some of the questions that are raised here as concerns the long-range future. I think Dr. Campbell is uniquely qualified to speak on this subject because of his long association in this area, starting many years ago with early aviation experience. I think he can give us a philosophical view of where we are going here and what he views, and I would like him to do that.

Before he does that, there are two people in the audience I'd like to call our attention to, who were not here right when we started. One of these is on the front row here, Dr. John Paul Stapp, who has certainly worked in this area for many years, and was one of the pioneers in the field of aerospace medicine. John Paul, would you stand up please. Dr. Stapp.

Another person that we really consider the "father". We've been trying to figure out how we are related because he said, I'm his son, and he's my father, and all this sort of thing, but we really have the "father" of the discipline of space medicine right here, Dr. Hubertus Strughold. Some nine years ago (1968) he retired from the position of the Chief Scientist of the U.S. Air Force School of Aviation Medicine, a position he has held since Dr. Campbell was the first commander of the school. By the way, since his retirement Dr. Strughold is the Scientific Advisor to the Commander of the School of Aviation Medicine, and has his office near the library of the school. Dr. Strughold sitting here, and I wish you would stand up, Dr. Strughold. I'm sure he could add many comments to this, and perhaps will before the morning is over. All right, Dr. Campbell.

I am very happy to call now on Dr. Robert Gilruth, one of the first here. One of the greatest privileges of my life has been to work with him and to have found the understanding and the backing that was necessary from him to allow many of the decisions and ideas that you had, to come to fruition because he was convinced, and still it's hard to really get that kind of backing from someone with a very inquiring, scientific, and accurate engineering mind without the very large amount of data that you could give about other things. I shall be eternally grateful, and I think the nation should too, to Dr. Robert Gilruth for having done that to me. Bob, I would like you to stand up. It wouldn't have happened without you.

Now, one of the things that has been mentioned here this morning, we've heard a couple of times now, is that question, "Did exercise really alter these effects that we saw inflight?" Now we have constantly felt that if we increased the amount of exercise on each of the three flights, because we felt that we were seeing this continued loss of muscle mass, we're seeing loss of calcium and what we would like to do is to try and alter that, realizing that at the same time we were altering our results by doing that. But we felt that it was necessary, as we were exposing man longer, and because we wanted him to succeed, and to be able to perform. We have one of the real experts here in this area who has spent a good deal of his life working in this area, and who was one of the strong advisors to us as we were working with the Skylab Experiments, and that's Dr. Ulrich Luft. Dr. Luft has some comments about what is the amount of exercise and what should we really be doing, and whether exercise is as good a thing as we think it is. I think it is very provocative thought that he would like to discuss with us now. Following Dr. Luft's comments if he doesn't break a leg, we will take a few minutes and then have some coffee.

FIRST RESPONSE TO DR. BERRY'S SURVEY PAPER
Life Sciences in Space Flight-Roundtable

Dr. Hisashi Saiki*

With the prolongation of space flight missions, biological adaptations to the new environment have been noticed in the physiological functions of the crew. The investigations on the mechanism of such adaptation seems to be composing a new focus of the interests in space medicine.

This brings to mind that at the 20th IAF held in Mar Del Plata in 1969 our biomedical group had adopted adaptation and readaptation as the next objectives of our activities.

Dr. Berry's review broadly covers this problem and includes data acquired by U.S. and U.S.S.R. flights these last several years. Such data are voluminous and extremely complicated. They are not the results of a systematic study, for example, with animals, of the adaptation -- readaptation mechanisms -- this has been left for the future -- because of the nature of the primary aims of the space missions to date and the operational necessity to seek counter-measures quickly. Of course, analyzing effective countermeasures taken is also a way to study the mechanisms.

Dr. Berry has effectively summarized this valuable data in his paper. With his rich experience and understanding of space medicine and life sciences, he has deductively organized this knowledge for a systematic study in the future. It reflects the spirit of a theoretically prudent scientist who has foresight.

To start with, Dr. Berry points out that the parameters reveal a difference in the rapidity of adaptation. He also emphasizes the relationship between ease of readaptation, duration of flight and exercise loads.

With various systems, he notes the necessity to study changes of blood volume to clarify the quantity and quality of deterioration of the cardiovascular system. In the hematology area, he points to the adaptability, i.e. the mechanism of recovery, of the red cell production function. In the neurological area he notes that motion sickness develops early (7-10 days) in space flight. In the musculoskeletal area, he recognizes the peculiarity of the continuing loss of calcium and nitrogen throughout the flight duration (84 days) and emphasizes the need for countermeasures. Dr. Berry's reiteration in the summary of his views on the neurological and musculoskeletal areas reflect the degree of caution he places on these areas.

In the area of fluid and electrolytes, Dr. Berry notes increases in aldosterone secretion, in sodium and potassium excretion, in osmotality and in circulating cortison levels. As regards ADH and the role for the Gauer-Henry reflex, he indicates further study is required.
----

*Space Medicine Laboratory, the Tokyo Jikei-kai University School of Medicine, Tokyo, Japan.

As is well known, before these findings were obtained, Dr. Berry advanced a hypothesis regarding them and there is consistency between the two in the many changes noted.

These are the points in which we are the most interested, and I would like to offer here, for reference, some of our findings in simulated weightlessness research.

We learned in a 6-day water-immersion test of man that among the parameters for adaptability those that appeard early were neuro-physiological ones and those found in the enzymatic parameters. However, we also learned that the renal function as revealed by mineral excretion cannot be identified in such a short time. Seeing the need for observation over a longer period, we decided to use rats and have studied hypokinetics in a five week suspension followed by readaptation process for five weeks after return to normo-dynamics.

Through hypo-kinetic exposure with such technique, the animals suffered appetite loss, general weakness, and almost all of them suffered transient eye-nasal bleeding for a few days. They lost body weight during the initial one-two weeks. But after two-three weeks exposure, all such suffering vanished, and the rats returned to a healthy condition. After return to normal control life, the body weight of all the animals showed a good increase, even when no readaptation phenomena could be seen, in many metabolic phenomena. (Slide 1.)

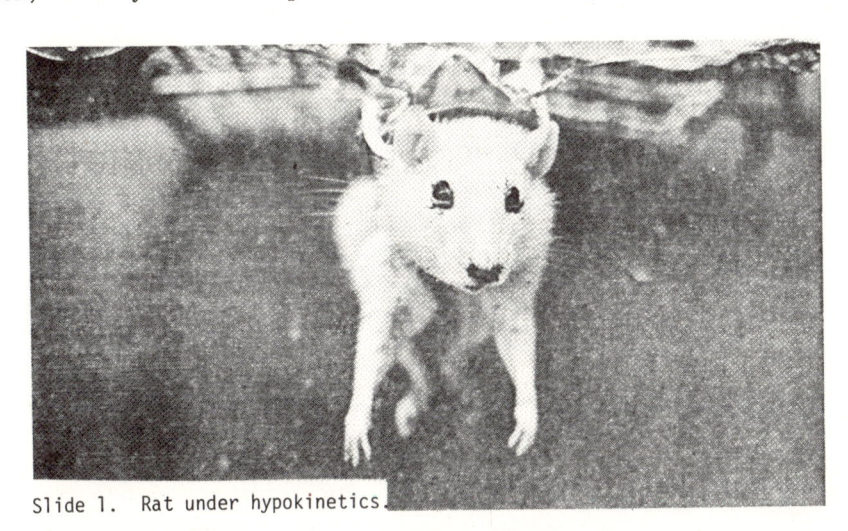

Slide 1. Rat under hypokinetics.

As the results of this experiment, on the fluctuation of rate of K, Na, Ca, (Slide 2 shows the average values for each week) we could recognize next three stages, that is, in the first week, especially in the first half of the week, the K excretion rate decreased very much. But in the second week, it returned to almost the pre-exposure normal value. And in the third week, it reached a higher plateau.

This third week seems to be the initial period of stabilized phase, and corresponding the Dr. Berry's so called adapted stage. Although, by our other experiments, this plateau formation phenomenon was found to have a tendency to be attained slower at the case of younger female rats. In such was, the first week is stress stage, the first and second weeks are preadapted stage. This stabilized stage is difficult to attain during five weeks hypokinetics on Na, that means the Na was slow to be stabilized, and in the same way, Ca was supposed slower.

Then, we placed our attention on the kidney function, especially the excretion rate of aldosterone from the adrenal glands. Effects of administration of aldosterone on the fluctuation of urinary excretion of K induced by hypokinetics during prestabilized phase of hypokinetic female rats, fully negate the anti-kaliuresis induced by the hypokinetics. (Slide 3)

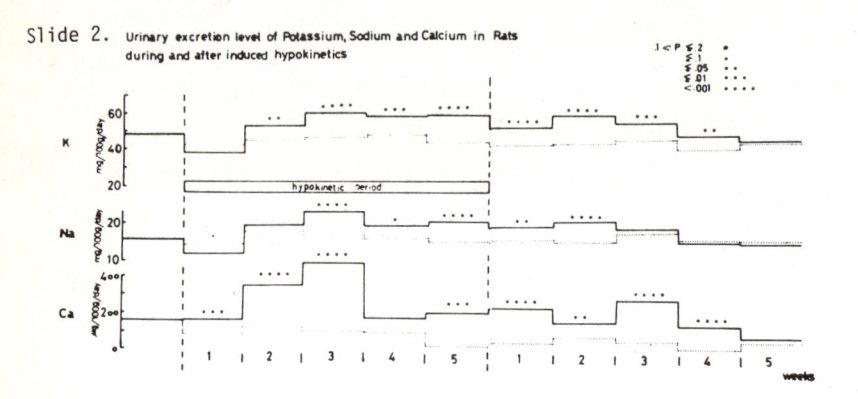

Slide 2. Urinary excretion level of Potassium, Sodium and Calcium in Rats during and after induced hypokinetics

Slide 3. EFFECTS OF ADMINISTRATION OF ALDOSTERONE ON THE FLUCTUATION OF URINARY EXCRETION OF POTASSIUM AND SODIUM INDUCED BY HYPOKINETICS (value of pre-hypokinetics period was expressed as 100%)

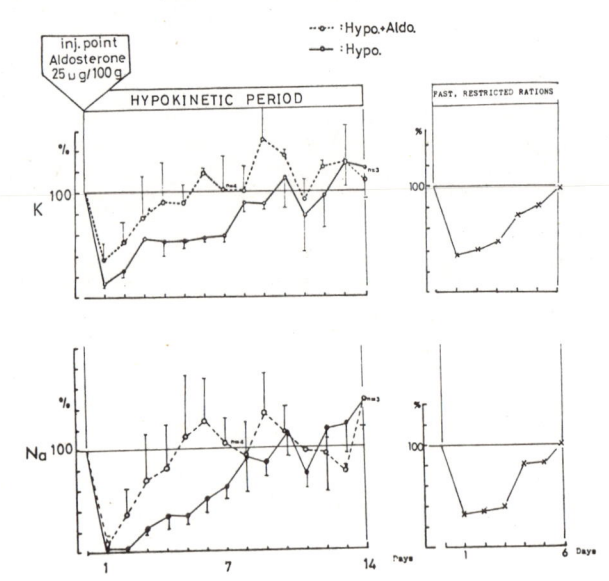

146

During stabilized period, administration of tetracycline, one of antialdosterone substances, decreased the stable kaliuresis, and it was negated by aldosterone administration. (Slide 4.)

Slide 4. DAILY URINARY EXCRETION OF K & Na INDUCED BY TETRACYCLIN AND ALDOSTERONE ADMINISTRATION DURING STABILIZED PERIOD

(average value of daily urinary excretions in preinjection stabilized period were expressed as 100%)

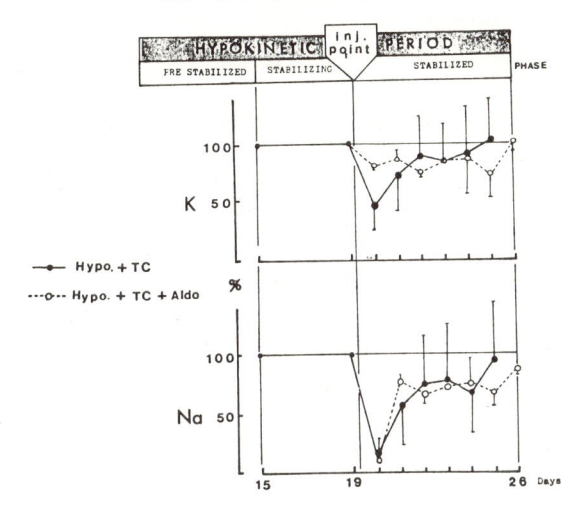

Accordingly, decreasing of aldosterone activity in the prestabilized period and the increasing of its activity in the stabilized period has seemed to be one of the main factors in renal level adaptation process to hypokinetics.

On the K contents of organ tissues, during above mentioned stablized period, only in the case of skeletal muscles, a statistically highly significant decrease was found in the K contents. Skeletal muscles seemed to be a main part of K release from the body during hypokinetics. (Slide 5.)

On the other hand, in determining enzyme activity of muscle, brain and blood, we found an increase in LDH, CPK and aldolase values in serum, and an indication of decrease in SDH value in muscle, which are evidence of muscle dystrophy. (Slide 6.)

We are interested in learning what modifications will be made to the results of these experiments in a true 0-g environment. We await the outcome of future research.

Again, turning to Dr. Berry's paper, he notes, regarding countermeasures, that Skylab results revealed that cardiovascular responses readjust faster with increasing duration of flight and that the more rapid return may have been aided by the amount of exercise done in flight.

Whether the level of adaptation (stabilization becomes shallow because of exercise, or stabilization becomes more complete because of long duration, it is difficult to conclude from Skylab conditions to date (as both factors are present). A partial explanation of this problem may possibly be expected if the stabilization mechanism includes a factor which works beneficially early in the readaptation process.

Slide 5. Potassium contents in Rat muscles on normal condition and at 2-4 weeks hypokinetics.

(M. femoris)

Groups	Symbol	n	$\bar{x} \pm \delta$	Σd^2
control	●	9	415.54 ± 19.29	3345.91
HKW2-4	○	6	367.01 ± 32.46	6332.82

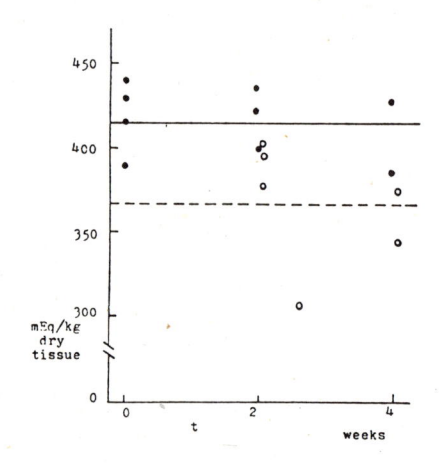

$\alpha = 0.006$

Slide 6.

LDH ACTIVITY OF SERUM OF HYPOKINETIC RATS
(Age and Sex : 3 Mon. ♀)
n=4

* p < 0.1
** p < 0.05

CPK ACTIVITY OF SERUM OF HYPOKINETIC RATS
(Age and Sex : 3 Mon. ♀)
n=4

** between 7day and 18day

I am in agreement with the affirmative answer regarding a two-year flight Mars mission. This is an opportunity for a closed ecological system to show its worth. Dr. Berry expressed the need for a six-month preliminary experiment and also research to learn of mechanisms of the changes during the first week of space flight. He also comments on the need and value of man and animal experiments under space conditions. I am in complete agreement with these views. Dr. Berry indicates that in the cardiovascular area, as regards the significance of Gauer-Henry relfex, measurement of changes in blood volume is desired, although ADH decrease has been clearly observed in Skylab flight. I would also like to stress this. We are presently planning research using microspheres and krypton-85.

I would like to express my appreciation for being given this opportunity to comment on Dr. Berry's excellent paper, which has been presented, the great attainments of space flight to date. It is my conviction that orthodox scientific research on an international scale in the various areas will supplement these attainments and thereby increase mankind's knowledge and contribute to his survival.

LIFE SCIENCES IN SPACE FLIGHT
ROUNDTABLE

Dr. Harald J. von Beckh*

I am very grateful that the invitation to this conference gave me the opportunity to return to the "Land of Enchantement" after an absence of seven years. I would like to take this opportunity to recall some major contributions of an organization to which I had the privilege to belong until its deactivation in 1970: the USAF Aeromedical Research Laboratory at Holloman Air Force Base.

The origin of this organization was intimately connected with this Nation's very first Space Efforts: shortly after World War II, a series of high altitude flights of V-2 and Aerobee rockets was launched from White Sands Missile Range and Holloman Air Force Base. This research program, originally designed for obtaining physical data on the upper atmosphere also included the exposure of biological specimens, such as Rhesus monkeys and mice.

The Life Sciences portion of this project was assigned to the USAF Aeromedical Laboratory at Wright-Patterson Air Force Base, Dayton, Ohio. Since it proved impractical to conduct such a complex project across the country, a "Field Station" was established at Holloman Air Force Base, which shortly later became independent as the "USAF Aeromedical Field Laboratory". It was later renamed the "USAF Aeromedical Research Laboratory."

These V-2 and Aerobee experiments were highly successful. They furnished for the first time telemetered data of the circulatory and respiratory system during the state of weightlessness.

When Colonel John Paul Stapp, who previously acquired fame by his rocket sled runs at Edwards Air Force Base, became chief of the Laboratory, a new main mission was added: The study of human tolerance to abrupt decelerations. His pioneering experiments on the Captive Missile Track with the sled "Sonic Wind," which made him "The fastest man on earth," are in the public domain and need not be repeated here in detail.

In 1955 the planning phase began for a series of manned stratosphere balloon flights, later known as project "Man High." In the first flight (June 1957), Captain Joseph Kittinger ascended to 96,000 feet. This was followed by the flight of Lieutenant Colonel David G. Simons on 19 - 20 August 1957 to a record altitude of 102,000 feet and 36 hours duration. The third and last flight of this series piloted by Lieutenant Clifton M. McClure reached in October 1958 an altitude of 99,900 feet. These experiments partly oriented toward cosmic radiation research, partly toward a broad range of upper atmospheric observations and investigations, showed that man was capable, physically and psychologically, to travel at space-equivalent altitudes. Also, the development and modification of life support apparatus functioning in sealed cabins under space equivalent altitudes proved to be beneficial for the later development of similar aparatus for the Mercury project.
----

*Director of Medical Research, Crew Systems Department, Naval Air Development Center, Warminster, Pennsylvania 18974

Beginning in 1967, the former weightlessness research with rockets has been complemented by human experiments flying Keplerian trajectories in the aircraft T-33, F94-C, F-100 and F104. I conducted 51 missions of more than 200 weightless trajectories investigating the problem of alternation of weightlessness and acceleration, and vice versa, because these conditions were anticipated for future orbital flights after burn-out and at the beginning of the reentry.

In these experiments it was desirable to simulate as nearly as possible the conditions of rocket ascent and reentry. Therefore, a high G load-producing pattern was adopted in which periods of from 4 to 6.5 G accelerations were obtained by flying continuous steep turns. These maneuvers were termed "diving spirals." The weightless state was obtained for periods of up to 45 seconds by flying Keplerian ballistic trajectories.

It was shown that the subjects experienced higher strain when they were exposed to a G load after the weightless state (post-weightlessness acceleration), and that their G tolerance was lowered. This "deconditioning effect" of weightlessness, was later reconfirmed by all the manned space missions, and is still now subject of interest and concern.

In 1958 the responsibility for the chimpanzee flights of Project Mercury was assigned to the Laboratory. Although chimpanzees have been used previously as sled subjects in the diodynamics program, extensive physiological baseline studies became necessary because only few biological data on this species were available then. At the same time it was necessary to establish a psychological research and training capability, because the subjects had to perform during the flight complex psycho-motor tasks, which should allow the evaluation of their performance during critical phases of the trajectory.

These efforts culminated in the ballistic flight of the chimpanzee HAM (H-A-M for Holloman Aero Med) on 31 January 1961, and the two-orbit flight of the chimpanzee Enos on 29 November 1961. HAM and Enos, thus, paved the way for the later Mercury flights of Commander Allan Shepherd and Lieutenant Colonel John Glenn.

The extensive baseline research in support of Project Mercury revealed that the chimpanzee's physiological and psychological functioning is still more similar to man than anticipated. Also his posture and his size makes this species the most valuable subject for the testing of hazardous environmental situations, when ethical considerations exclude the use of human volunteers. Still more important, however, is the fact that the chimpanzee is able to react intelligently to numerous stimuli and to perform complex psychomotor tasks, which permit detection of subtle performance changes. Thus, the use of this species warrants extrapolation of test results to man with more certainty than any other test animal.

Therefore, the Laboratory conducted a multitude of experiments with trained chimpanzees until its deactivation in 1970. Their wide range included toxicity studies of rocket fuels, the landing impact of space vehicles and other hazardous aspects of space flight.

One of the latest achievements were several series of rapid decompression experiments conducted in the Laboratory's altitude chamber. More than 400 decompression experiments have been conducted. The first three series of this effort had the purpose to simulate conditions as to be expected in the case that an astronaut's pressure suit ruptures during extravehicular activities. It was shown that unprotected but denitrogenated chimpanzees can survive near vacuum conditions for periods of up to three minutes. If retrieved and recompressed before this time period had elapsed, complete recovery, without late sequelae, has been demonstrated. A later series of experiments, had the purpose to simulate the remote but possible event of accidental cabin decompression of <u>Multi Mach/High Altitude Transport Aircraft</u>, and more specifically to explore the psychophysiological reactions of those occupants which are not able to don an oxygen mask after the decompression event and are exposed to the entire descent profile in atmospheric air.

This presentation should not be considered a eulogy for a defunct organization. Rather it should remind the scientific community that at the very beginning of the Space Age, this part of the Country, namely the town of Alamogordo and Holloman Air Force Base were the site of pioneering and significant contributions to Space Medicine.

SECOND RESPONSE TO FIRST RESPONSE OF DR. CAMPBELL
Life Sciences in Space Flight-Roundtable

Dr. Hisashi Saiki*

I was very much moved by Dr. Paul A. Campbell's response, because I entirely agree with his philosophical opinion of Dr. Berry's paper. We will be able to understand the meaning of space life science more clearly than before.

I also would like to focus our attention to the very paradox like relation between the flight time and readaptability to terrestrial environment. I think there is excellent gateway to approach the mechanism of adaptation. Dr. Campbell evaluates Dr. Berry's position on the flight to Mars, and points out the importance of logistic support.

In 1953, when the Laika of USSR orbited the earth as the first biological body in space, I pointed out that establishment of the energy requirement in space must be studied and sufficient nutritional supply technique must be provided for long duration flight. The nutritional requirement seems to be partly clarified through the direct or indirect measurement on many kinds of exercise or work under space environment. But the part of production for nutrition supply seems not to be studied sufficiently yet. I think the requirement to set up the most effective and stable closed ecological system will be one of the most important steps for such long term space missions. I think, this effort is one of the worthy ones in the space life sciences. In my opinion, success of such effort will be useful for study of the energy equilibrium and physiological and nutritional problems of our terrestrial life.

About Dr. Campbell's final comment, I would like to add that, in Japan also, space scientists went through the same experience during the past two decades.

SECOND RESPONSE TO FIRST RESPONSE OF DR. WHITE
Life Sciences in Space Flight-Roundtable

Dr. Hisashi Saiki*

I am impressed with Dr. White's all-encompassing review paper for the investigation of future space life sciences. He has proposed investigation of the adaptation mechanism through all biological systems including the effective mechanisms of artificial countermeasures, systematically.

Dr. White has set up three categories for the study of adaptation. 1.) Specific space associated adaptations; 2.) system interrelations; and 3.) Man's adaptive origins. The last one seems to be very basic and orthodox, but includes very profound problems for study.

Classification and selection of the next themes will require thinking over and much discussion. Actual data obtained may reveal to us the next problem worthy of priority study.

----

*Space Medicine Laboratory, the Tokyo Jikei-kai University School of Medicine, Tokyo, Japan.

On the Mars space mission, he emphasizes a fundamental study especially on the radio biological problems.

In summary, he emphasizes parallel study on two fronts, practical studies and fundamental and theoretical studies.

Our Japanese situation has been limited to the second studies; but now, many factors suggest to us that such parallel study is going to be possible.

SECOND RESPONSE TO FIRST RESPONSE OF DR. LUFT
Life Sciences in Space Flight--Roundtable

Dr. Hisashi Saiki*

Dr. Luft's first response is most interesting to us, because our work has been closely related to the metabolic physiology for exercises.

I have special interest in his note that of the pilots of Skylab IV the CP who had no intensive training runs was less susceptible to LBNP then the other two, who had intensive training runs before flight. They had greater increases in heart rate and in leg volume at 50 tons and also had a greater incidence of pre-syncope in flight.

In connection with those facts, I would like to add our experience. In our human experiments by water immersion technique, the athlete subjects were more sensitive on the tilt table test after eight hours immersion exposure, than non athlete subjects.

We await the results of such expertise study to evolve optional modes and amounts of physical activity suitable for use in space.

SECOND RESPONSE TO FIRST RESPONSE OF DR. GRAUL
Life Sciences in Space Flight-Roundtable

Dr. Hisashi Saiki*

Dr. Graul's paper added precious knowledge to a field in radiobiology where references are the most lacking.

Through general knowledge of radiobiology, we know the effect of cosmic radiation to be one of the most serious ones. And we are aware that heavy, high-energy particles exist in space in the course of space flight.

We have been interested in acquiring more references. In a general meeting next November, JASMPS has scheduled a special lecture on the progress of radiation medicine.

Dr. Graul's paper is a most opportune reference for the meeting and for all Japanese space scientists.

We hope to seriously study the data to gain a clear understanding of deformation effects to development ontogeny.

I trust this data will offer a fresh stimulus to all radiobiology of the world.

COMMENTS

Dr. Heinz S. Fuchs*

I'm very much afraid that I will have to call your attention back from the highly scientific level of the papers presented now to some, more or less, from this viewpoint, earthbound problems: I can only briefly and perspectively comment upon some points of operational relevance so far as my country, Germany, and Europe are concerned.

As I was already able to inform you yesterday, my country was the first creator of the European Space Agency. My country also contributes to more than 50% of the forthcoming Spacelab projects, and Spacelab is being constructed by ERNO at Bremen in Germany. Finally, one of the forthcoming Spacelab payload scientists will be a German. How much we Germans are internationally involved with today is reflected by the fact that this payload scientist, whom we claim to be a German, had an Austrian father and an Italian mother, and he married a lady coming from New Zealand!

We are now in the very happy situation of being the consumers of our American friends' wide experience gained in the Skylab missions, for which I personally could provide some logistic support to NASA. That was Dr. Eduard C. Burchard, a German Air Force doctor, who took part in the three Skylab missions, taking care, motherly care, of the crews. Dr. Burchard can now immeasurably contribute to the preparations for the forthcoming European Spacelab missions.

Obviously, no problems are put on us, so far as the selection or even the training of scientists is concerned with. No problems, too, as we've heard today by some of the papers, in the field of physical and cardiovascular conditioning or deconditioning. But since the Spacelab missions will run exactly over a 7-day period, the vestibular apparatus will post the most serious and important problems to all those concerned. I would like in this context to literally quote Dr. Berry who said in his summary "Certainly we must control development of motion sickness in order to maintain productive crewmen during the initial 5 to 7 days of adaptation, and this should be done by a true countermeasure rather than a crutch such as medication."

It might be that one of the forthcoming experiments aboard the Spacelab, the so-called Space Sled will provide new knowledge of biofeedback with regard to any countermeasures against motion sickness. It might be, however, that we have to use the crutch of drugs, perhaps with some detrimental effects as known from clinical drugs, as for instance streptomycin, in order to protect, in an operationally practical 7-day sense, the payload scientists, providing them any help to be successful in their actions during this mission. Thank you very much.

---

*Surgeon General of the Military Forces of the Federal Republic of Germany; President of the German Association of Biomedicine

COMMENTS

Dr. Harold Sandler*

Coming here in the tail end of the program, it's always a challenge to try and come up with some new concepts or ideas that other speakers have not really addressed themselves to, or were covered in great depth. The topic - I should say my topic - is going to cluster itself about: "What have we really got right now in the present, and is there really a future?"

In this regard, I think I shall be one of the optimists that Chuck talked about in his discussion, and very strongly say: "I do think that there is a great future with respect to man in space." But in that context, I feel that we are very much at a crossroads. The same crossroads were faced by civil aviation about 40 years ago. The crossroads that they faced at that time were: whether they were going to continue to carry the mail or perhaps a little bit of cargo, or whether they were going to open aviation to fly passengers and allow anyone, as we do today, to come on board. It's very obvious to us now, (some of us may have participated in the dilemma that went on at that time), about whether we were going to allow man to go aboard airplanes- and we now see that we are faced with exactly that same dilemma at this point. I feel very strongly that if we are just going to carry mail, as it were, in space, space flight will continue, but in no way to the extent or degree of which it is capable, and to reach horizons that will be available if man is permitted to get aboard in free access.

That is the tremendous challenge with which we shall be faced in regard to the question: Can anyone in this room buy a ticket and go aboard a future shuttle or space flight, and if you should, what would be the problems confronting you? This is particularly apropros if we're going to allow people in the 50-65 year age range to participate in space flight.

Chuck outlined concisely to us what we know with regard to man at this time, and obviously, we're going to have to go forward with that information because if we do not, we shall be faced with the dilemma that Toynbee described: "He who does not read history and learn what it has to tell, will be forced to repeat all the mistakes." But in the future flight that we go into, much of the data that is now available to us, might not have any pertinence; it may not even pertain. We have a whole host of questions that we must ask ourselves. Some of these are: Shall there be real dangers to people 55-65 years old in flight? Shall women be allowed to go in space, and are there dangers in having women aboard future space flights? Are there dangers with respect to flight that we do not even know about? For instance, as we follow our SKYLAB participants, will they turn up with diseases as a consequence to the exposure to the space flight environment, because we have heard today that the body has to adapt? Is it actually stretched to its limits? In that adaptation process, are there things that might be set in motion that will

----

*Director, Space Medicine Research, NASA, Moffett Field, Sunnyvale, California.

eventually wind up with problems in subsequent years? What about situations where latent disease is present, latent meaning that everything looks O. K. right now, but you are unaware that the disease is present until you stress the individual? What will happen in a situation where there is latent disease present in one who participates in space flight, a good example being dead coronary artery disease? Is this a real danger? Will space flight accelerate coronary disease processes or is there hope that it will really retard it? When one looks in a very general sense at what happens to the body in space flight, one can imagine that perhaps aging is either accelerated or it may be ameliorated. Until specific experiments are done, we really shall not know.

In this regard again, I cannot emphasize too strongly the fact that until we know the mechanism that the body uses to adapt and adjust itself to a space flight environment, we shall not actually have any of the keys or answers to our problems. There won't even be a sensible counter-measure that we can propose unless we know exactly what is the mechanism we are trying to oppose. I think there is nothing wrong with what we have learned in space flight up to this point, but if we do follow that type of course, what will happen is that we shall continue to fly, as it were, mail in space. We may be able to take instruments up, but man will not be able to really participate in a whole-hearted or functional manner. So the job that has faced those here on Earth is to try and find ways in which we can unlock some of these secrets. Every mechanism that is available to us, I think, should be utilized. Bedrest has been found to be one of those particular techniques that permits us to get to some of these problems. A great deal of information has come back, and, interestingly enough, now is the usual conclusion: we find that bedrest is not a bad analog with respect to what happens to man in space flight, and from that, I think, now we're beginning to develop methods and techniques which will allow us to really test whether we can avert some of the problems that will be prevalent in space flight, and allow man to remain there for a long time.

There are several other observations that are coming in with respect to the future that are very good. One of them is that as we look at the mission model, (that is the model and the plans that we have for participating in space flight), we find that almost 15 to one, the flights will be in near-earth orbit, as opposed to going to the Moon or perhaps going out as far as Mars. This means, very hopefully again, that man will be able to participate in near-earth flights. He may not have to stay up very long, and again, older individuals will be able to go. If you look at any technological society or structure, you always find that the people who are really directing things, who have the real insights, are always they who are somewhat older. The young fellows come along with the ideas, but it is the older guys who allow the mix and the right things to happen. I think it will be O. K. for the younger guys to go up in flight- that means the generation behind us and perhaps the one behind that - but in order to really get the maximum out of these younger people participating in space flight, some of the older individuals, the saged scientists, will have to go along. I think we, again, are going to have to provide the wherewithal for those people to get up into space.

Now the other interesting thing that has come about to me with respect to the shuttle that many individuals do not realize, is that in our shuttle design, one of the problems we're facing right now is the fact that people, or the passengers, whether it be the pilot or mission specialist, when they return will not be oriented in the g-field in the same way as he was oriented in our Apollo and Gemini missions. That means he will be seated upright so he receives an acceleration gradient on re-entry of plus-g, sub-z, (which means pulling blood from the head to the feet), rather than landing on his back. In the proposed shuttle re-entry, he will be exposed to that type of g sub-z acceleration for a period of at least 15 to 20 minutes and perhaps even a little longer. You've heard the presentation already wherein we have a problem with the cardiovascular system, so the first question is going to be: In our early shuttle flights, when we come back in a plus-g sub-z orientation, shall there really be problems, and how are we going to protect the individual? Again, the only way that I know of to do this would be through careful bedrest studies in which we are going to be able to test out various designs and possibilities, and make sure that there will be no problem.

Chuck has also stated that the vestibular problem is a very real one. We know that, because if we look at all the Russian and American space flight experiences up to now, it usually takes about four to five days for anyone, after he has gone up into space, to really orient or get himself settled with respect to the vestibular problem. If, in our early flights, we are only going to be gone about seven days, and if the first five days are spent with someone being nauseated, it seems that such will not be a very useful way in which to spend one's time in space. So, another challenge that we have to face in the immediate future will be: Is there a way either to select or train individuals so they will not be sick during that important first four days in a seven-day mission? Here again, I think some of our ground-based studies are showing some fruitful possibilities. I think Chuck already mentioned that it seems that training through bio-feedback is demonstrating great promise in individuals being able to sense when they are becoming sick and avert that sickness through various learned techniques that we are presently practicing.

Last, but not least, is the fact that calcium is indeed a problem. I think it has been mentioned now several times that bedrest and what happens during space seems to be a very good analog. It's of interest to know that if one loses calcium after a three-month bedrest period, it takes him at least six months to regain that calcium which was lost. The question will come: Is that really going to be one of the penalties or prices for space flight, and how are we going to avert that problem, particularly if that calcium may be lost from critical areas such as the spine, where again, any type of compression might produce some very harmful effects?

All in all, I should like to keep my comments brief. I feel that the challenge with respect to working in space flight has never been greater; that what we are going to do over the next several decades is going to be very critical, and not at all like what has been done in the past. The past will serve us as a very good indicator or arrow with respect to what can be done. The challenges that we face will be far different than putting a Mercury or Gemini or Apollo flight up. We shall also be challenged with putting older-aged individuals in space, opening up the whole space flight area so that everyone can participate- you and I. From what I've seen, I feel very optimistically that we shall be able to do that.

CHAPTER VI

DEVELOPMENTS IN SPACE LAW

"The exploration and use of outer space shall be carried on for the benefit and in the interests of all mankind."
United Nations resolution, 1962

INTRODUCTORY ADDRESS ON
"THE DEVELOPMENT OF THE INTERNATIONAL LAW OF OUTER SPACE"

Dr. Carl Q. Christol*

Ladies and Gentlemen. I'm Carl Q. Christol, Professor of International Law and Chairman of the Political Science Faculty, University of Southern California. I've been asked by the last speaker to be the moderator this morning. I would like to tell you how I think we should proceed. I would also like to offer a few comments at the beginning and then call upon Dr. Gorove to make his basic presentation allowing him as much time as he thinks he would like to take, and then to turn to the participants, in order, and allow them to either comment on the totality of Dr. Gorove's presentation, or to focus upon a single element of the presentation. Then I would like to proceed from one of the panelists to the next panelists. However, if there is a large interest on the part of a member of the audience in asking a question or to make a comment at any time, I think that is perfectly in order because we do have a limited number of people here today. I think we should turn this into a kind of graduate seminar, if I may use that expression, which the faculty, so to speak, over on my left and on my right can join in with the balance of the students who are also over on the right -- students who have had a great amount of experience and are certainly as well qualified to be members of this faculty operation as are those that I've previously identified as faculty.

Well, let me begin then by offering you a few remarks, and I have copies of this. We'll hand them to you. May I say that these introductory remarks are based upon my having read and analyzed all the papers which have been heretofore circulated among the participants in the panel.

All of the participants in this distinguished gathering are aware of the role of law in the ordering of the conduct of mankind. Fortunately for mankind the practical achievements of the space age have been accompanied by a developing legal order designed to deal with and to meet the interdependent needs of the now age.

The international law of the space environment, consisting of outer space, the Moon, and celestial bodies, forms a part of general international law. Just as international law is the product of many contending forces, and is influenced by a myriad of interests and values, so it must be borne in mind that international space law is affected by identical considerations. It is well to bear in mind that the larger policy issues confronting the present negotiations for a new ocean regime also have a critical bearing on the future of international space law.

As we approach the analysis of our topic I am mindful of the prestige properly accorded to and rightfully enjoyed by each of the participants. I am also accutely aware of the fact that each would either structure the overall presentation, or would place emphasis on the elements of the structure, somewhat differently. Thus, it may be helpful if I, in my role as moderator, were to offer some truncated observations with the expectation that such a brief overview would provide an orientation for the subsequent presentations.

----

*Professor of International Law and Chairman, Department of Political Science, University of Southern California, Los Angeles, California.

First, we must bear in mind the history of the space age, so that sight will not be lost of the magnificent achievements that have already been recorded.

Second, we have been cautioned not to rest upon our laurels, but rather to offer constructive suggestions for a continuing emergence of a community-oriented space law policy.

Third, we have been advised that the most relevant policy and the most realistic law can be based only on a common assessment of factual data, including but not limited to the scientific and technological learning of our times. This has been put forward in the full knowledge that true facts often may be very illusive, and that it is possible for these to be varying assessments of objective facts.

Fourth, we are aware that the international law applying to the space environment has taken several forms. These forms include the great international conventions dating from 1967, but they are not limited to such international agreements. The formal conventions are not universally subscribed to, and the clientele for one is not in quite a large number of situations as the clientele for others. Moreover, such agreements are subject to differing interpretations, and there is a need to realize that general international law disavows the use of one single criteria for treaty interpretation. I expect that some our panelists may urge a preference for one or more criteria in preference for other standard approaches.

Fifth, although we frequently asset, and undoubtedly believe it factual to state that we live in an interdependent world, nonetheless the ingredients of world space policy bear the imprint of nation-States having considerably different policy objectives. Witness, for example, the common use of such terms as advanced States, less-developed countries (LDC's) and Third World. This has lead since 1974 to the notion of a New International Economic Order, a variety of approaches, depending on economic status to the World Bank, and a preference for doing business through such international organizations as the United Nations, through the United Nations Conference on Trade and Development, or through the Organization for Economic Cooperation and Development. Despite the language of confrontation that is frequently heard in some of the foregoing institutions, or within some of their organs, yet this tactic has generally been avoided in international discussions of space law. National behavior often has spoken more loudly than words. This is evidenced in national legislation calling for international cooperative activity in the space environment, and the acceptance on the part of the non-resource States of the presence and benefits accorded through such activities as those of LANDSAT.

Sixth, present such varying outlooks, it is but natural that there should be a need to identify from the point of view of world policy what priorities should be assigned to the unfinished business of international space law. Should the higher priority be assigned to obtaining international agreement relating to sensing and direct broadcasts, either with or without prior consent on the part of the State receiving such attention from space objects? Or should a higher priority be accorded to the establishment of boundaries between airspace and outer space, particularly in the light of the expected operational capabilities of the space shuttle orbiter? Should a high priority be assigned to the identification of prohibited conduct, such as the so-called "bombs in orbit" capability, and should the space lawyer at least not wonder why an issue such as this should not attract the active attention of the Committee on the Peaceful Uses of Outer Space? Should possibly lower priorities be assigned to problems associated with the establishment of space colonies, or with the exploration and exploitation of the natural resources of the

Moon and other celestial bodies? With respect to the latter, and as one considers a well-balanced legal regime for the space environment, it should be kept in mind that the efforts to establish a regime for the ocean have faced thorny difficulties in endeavoring to provide an International Seabed Authority designed to deal with exploration and exploitation of mineral resources having economic value. Further, in terms of priorities, what attention ought to be given to the manner in which the LDC's should have access to information obtained by the resource States concerning the data acquired by the latter? In all of these situations the notion of priorities is much affected by the issue of whether a given space activity, or, in other words, the specific function being carried out, meets the overriding command of the 1967 Principle Treaty that outer space, the Moon, and celestial bodies shall be used exclusively for peaceful purposes. In order for there to be a constructive space law of the future, international space lawyers will be obliged to ask and answer the question of what priority is to be assigned to the foregoing and other importnat issues. They must also be concerned for prospects.

Seventh, international lawyers generally, and space lawyers in particular, need to clarify the meaning to be ascribed both generally and particularly to such well-used but frequently misunderstood terms as "sovereignty," "sovereign rights," and "jurisdiction." The written exchange of viewpoints that has already taken place among our participants has set the stage for a more thorough analysis of this fundamental problem during our discussion.

Eighth, in any orderly effort to arrive at policies that carry the quality of authority and legitimacy there is a natural flow from the identification of facts, the exposition of interests and values, and the drawing of conclusions. Such conclusions are reflected in the adoption of laws, and in the formation of institutions. Our panelists have indicated a lively interest in the role of institutions dealing with the laws and policies of the space environment. I expect that we shall come away from this session with clearer ideas as to the efficacy of existing institutions and possibly recommendations for new institutions. Of particular interest will be the suggestion dealing with the powers and functions of newly proposed international organizations.

Ninth, I expect that we may seize on rather special provisions of any of the international agreements relating to conduct in space. One expression caught my own eye, namely, the term "orbit" as modified by the adjective "durable." To what extent should space lawyers introduce modifiers into the terms of international agreements, and having done so, what arguments can be advanced in support of the validity of such suggestions?

These are but some of the issues that have been accumulated in the dialogue that has already taken place among our panelists. Let us see what their responses to these and other issues are as we continue our critical analysis of the international space law and policy of the present and of the future.

This concludes my effort to summarize in a quick fashion some of the ideas that caught my eye as I examined the papers which have already been printed and which I expect you have now in your possession.

Now, it is my distinct privilege to begin the introduction of our panelists. The chief panelist is the organizer of this particular roundtable; this particular discussion, and he is Professor Stephen Gorove, who is a most distinguished authority in the field of space law serving on committees of a number of critically important international organizations dealing with space law problems including the International Institute of Space Law. In his capacity as Chairman of the program of graduate studies at the law school at the University of Mississippi, and as a member of the faculty of the Law School of the University of Mississippi, he is the Editor-in-Chief of a new, but I must say exciting and continually expanding, journal entitled "The Journal of Space Law." It is, I think it is safe to say, the principal English languge publication dealing with space law and space policy and one of the very few journals in the world that concentrates on this particular subject. So, we're delighted to welcome to this podium, Dr. Stephen Gorove, an old friend, who will lead the initial discussion. The floor is yours. You may take as much time as you care. Dr. Gorove.

REVIEW OF SPACE LAW DEVELOPMENTS

Dr. Stephen Gorove*

Thank you very much Carl. First of all, I would like to welcome all of you to this conference roundtable on space law developments this morning. I'm delighted so many of you could be here with us. I'm particularly happy to welcome those people who came from a very long distance. I notice several of you out here from Europe, and it is a real delight and pleasure to greet you here. I should say that the program for this particular roundtable has not been formally affixed. We kind of fouled up, perhaps to some extent, we will play it by ear depending on how large the audience is going to be, whether we will have people here who have not been on the panel and participated in the various exchanges of thought. So, what I will try to do this morning is to give you just a little bit of background of my so-called initial survey paper which you will find in the printed book, including of course the responses to that survey paper, and again the responses to the responses. At the same time, I would also like to take up perhaps a specific topic or topics. The way we have kind of visualized that we would proceed them is to see what comments, if any, each of you care to make on particular topics which happen to be of interest to you. Perhaps before we leave a particular topic, we might also like to call on the audience to participate, to ask questions, make comments, whatever they feel like doing.

In my initial paper, I have taken the historical approach. It was supposed to be a survey paper by the rules of the roundtable, and not a position paper as some colleagues said it would be. So, I have very briefly tried to review the space law developments and I have pointed out in this the historical role of the United Nations and the various organs of the United Nations very, very briefly and pointed out some of the early, and very important resolutions for the United Nations, and looked also very briefly at the major, presently qualified, portion of international space law including the major treaties just by passing reference, such as the Outer Space Treaty, the Agreement and Rescue and Return of Astronauts, the Liability Convention, and finally the Registration Treaty, which is in the process of being ratified by a number of states. I'm pleased to say the United States has ratified the Registration Convention, I understand.

Beyond that, I looked at some of the issues which are currently outstanding, and which are being discussed by the Legal Subcommittee of the Committee of the Peaceful Uses of Outer Space. These problems have been mentioned, and several of our distinguished panelists commented on some of these problems. These include such items as sovereignty, jurisdiction, the problems of the Space Shuttle, remote sensing, direct broadcast satellites, the Draft Moon Treaty, definition of space subjects, meaning of the common heritage of mankind, the peaceful versus military utilization, geostationary orbit, possible establishment of an organization for space activity within the United Nations system, etc.

Rather than rehashing what I have presented in my initial survey paper, and also some of the responses that I have made, what I would like

----

*Chairman of the Graduate Program in Law and Professor of Law, University of Missisipi Law Center. Organizing chairman of the Roundtable on Space Law Developments, International Space Hall of Fame Dedication Conference.

to do very briefly here is take up some of the matters that Dr. Christol mentioned in his introductory observations. I realize that these cannot be very easily separated into different compartments, but I have been quite fascinated with the possibilities that some scientists say here in relation to settlement in outer space. I don't know whether we have to use the word "settlement" or whether we should use the word "colonization". I don't think it really perhaps makes too much difference as long as we know what we are talking about.

Let me very briefly mention a few thoughts on this, and tell you first of all what I understand the scientific conception of such a colony is. Many reputable scientists have suggested that a colony in free space can be concerted more economically than would be done in any other way, by establishing the colony from materials transported or hurled into outer space from the moon. Space station habitats housing 10,000 people or more may be built in free space at stable libration points from approximately 12 million tons of materials mined from the moon and then hurled into outer space. The idea would then be to set up solar power stations or build them in this free habitat by the people there and that way the space settlement eventually would pay for itself taking about 34 years or so according to scientific estimates.

Even though some of our comment speakers have remarked that this is very far fetched and very far forth in the future, I feel that we space lawyers ought to think about some of the problems that can arise if a decision is made to go ahead with the building of such a colony of satellites. The very question that came to my mind was the question of whether it is legally feasible and permissible to take about 12 million tons of lunar materials from the moon for the purposes of constructing such a habitat in outer space. I think the impending problem is first of all, the interpretation of Article II of the space treaty which prohibits national appropriation of the moon and other celestial bodies by any means including claims of sovereignty, use, occupation, etc. - any means. To make it quite clear, we are not talking here about removal of just samples of lunar materials. We are talking here about removal of substantial amounts of lunar materials. Maybe it would involve about 160 football fields, the area from which the lunar materials would have to be extracted and carried then or put in the predetermined area in free space.

Now, I realize that there has been no general agreement in the literature regarding the precise meaning and interpretation of Article II of the Space Treaty. There are some very distinguished authorities including Professor Dieterichs-Vercheor from the Netherlands, currently in England, who believes that the ban on appropriation relates to the area and not to the national resources of the moon. All of us believe that it relates only to exhaustible resourcs, and still others who distinguish spatial extension resources and flow resources, and stock resources, and determine sharability accordingly. Now maybe 12 million tons of lunar soil could be regarded insignificant or minute in relation to the total available amount. If you look at it that way, maybe it is inexhaustible. On the other hand, if you look at it exhaustible in relation to the total amount available, then the ban on appropriation seems to be applicable. I think whether the resource is exhaustible or inexhaustible is a relative question which must be determined on the basis of the then existing technology and the scope of exploitation. Now, if you interpreted Article II to mean "a ban on the appropriation of an area" and not on resources, then the extraction and subsequent utilization of such resources would not constitute an appropriation. However, in such a case it seems to me, you would have to determine whether the use of a substantial area of the moon would constitute an appropriation. The answer to this question may well depend on the duration and exclusiveness of the use and the assertion of property rights with respect to such areas.

It's interesting that the Draft Treaty relating to the moon which is just in the drafting stage and has not been finalized, makes a distinction between resources and surface and subsurface of the moon largely perhaps following the idea of distinguishing between resources and areas. With respect to national resources it says that these shall be a common heritage of all mankind, but at the same time it places a ban on any property claims relating to the surface or subsurface, so parts of the surface or subsurface may not be the object of grant, exchange, transfer, sell, lease, hard gifts, or any other transferment or transaction. This is our resulting compensation.

The stipulation doesn't seem to be around the use of the area, and not only the use, but perhaps use for a longer duration because it states that the placement of personnel, space vehicles, equipment, facility stations, installations, on or below the surface of the moon as well as the free moon land is permitted. The very fact there is a provision there requiring the station to inform the Secretary General of the location and purpose of the station and subsequently, mind you, at annual intervals to inform the Secretary General whether the station continues to be in use and whether its purposes have changed. History appears to indicate that such a station may occupy the same area for several years. There's no definite time limit. There is one limitation, the Draft Treaty also says that the state establishing the station is to use only that area which is required for the needs of that station. Whether or not the establishment of the station for the purposes of selecting lunar material for the eventual use in free space, whether that would be for the needs of the station may be subject to some doubts if strictly interpreted. Also, such stations are to be installed in such a manner that they don't need the free access by other states to all other areas of the moon. Now, what are the provisions which different objects may constitute a legal obstacle to the establishment of the space colony is the stipulation and the Draft Treaty that the natural resources of the moon are a common heritage of all mankind.

Carl mentioned, I think, the common heritage and several of our distinguished panelists in the roundtable exchanges referred to it. There has been no general consensus on the meaning of this concert, but there has been suggestion that the concept places a ban on any transfer of such a resource, to the Earth by any country, for its own exclusive economic profit, and only permits transfer under the provisions of an international regime to be established. Professor Goldie, I think, discussed to some extent this question of appropriation and common heritage and I think he sort of intimated that he didn't feel that this placed any limitation on the exploitation of natural resources. It seems to me that the concept appears to recognize, if not vast, property rights in mankind. It says that natural resources will be the common heritage of mankind. It seems to me that mankind has inherited those if it is the common heritage. Now, if you've inherited something, it seems to me you have grand property rights, but some people may look at it differently.

Throughout the negotiations of the Draft Treaty, the United States has taken the position that an expressed or implied moratorium on the exploitation of natural resources until a new national regime can be established to regulate the exploitation would be unacceptable. But many developing nations have been in favor of requiring such a moratorium. What I think, he is revealing, in the diametrically opposed positions is the fact that apparently neither of the two positions regards Article II of the Outer Space Treaty as a ban on exploitation. If either of them did, then the argument to put a moratorium on exploitation would become meaningless. Now, another important question, I think, and this is something which we have discussed a little bit in the exchanges of views, is the exercise of sovereignty, and sovereign rights, if any, and other forms of supreme authority or jurisdiction and control.

My view has been that the space treaties have not abolished sovereignty or sovereign rights in outer space in the fullest manner. The relevant article which is frequently referred to is again Article II, but Article II refers to sovereignty only in connection with the ban on national appropriations, and I think, it's sad that by prohibiting other forms of expressions of sovereign jurisdiction, sovereign rights, (initiative of using extraterrestrial resources is inhibited (The Editor)). In fact, the Outer Space Treaty provides that states shall bear international responsibility for national activities in outer space irrespective of whether such activities are carried out by governmental or non-governmental entities. I think this position presupposes the exercise of some form of sovereign rights, supreme jurisdiction or control by the state whose nationals are part of the space goal. Otherwise, it is hard to see how a state could be held in a national sense liable for activities over which it had absolutely no control. Also, we may note that the Outer Space Treaty (vests) jurisdiction and control over a space object and personnel thereof in the state in whose registry the object is carried into space. I think one interesting question there is whether or not this personnel now of the spacecraft will continue to remain its personnel after the personnel leaves this spacecraft for a space colony and for an extended period. Also, we have the additional problem of whether an object originating from the moon or another celestial body would be a space object. I'm not sure that it would.

On the surface it appears that both questions should be answered in relation to this discussion in the negative in lieu of the provision of international responsibility, and jurisdiction and control are to be exercised by the state concerned. It would be difficult, I think, to accept the principle that personnel of the space object will remain personnel of that object no matter what happens to that spacecraft or personnel subsequently in outer space. I think permanent departure from the spacecraft by the personnel of the spacecraft for a space colony, would cut off the link for various reasons, not just semantic but also for logical reasons. Aside from the question of personnel, I think objects originating in launch from outer space might not be regarded as space objects under the presently existing measure of space treaties. We speak of objects launched into outer space and their return to Earth, and of objects not launched into outer space, -- rather of objects launched from outer space -- and remaining in outer space.

Now, I think my conclusion really seems to be that the current space treaties contain no clear provision of the use of lunar resources so long as these materials are available in quantities which are not presently exhaustible. The very fact that the current negotiations in the UN center around the question of a possible moratorium on the exploitation of such resources appears to indicate clearly that the negotiating states' use of the ban on appropriation is not applicable at the present time to the national resources of the moon. So far as the other area, sovereignty and sovereign rights are concerned, I think states will have to continue to exercise supreme authority and control over their nationals and also perhaps the colonies they build in outer space. While the article on the jurisdiction, Article XIII, of the Space Treaty is not sufficiently clear on this point, I think the provision on international responsibility in Article VI makes the exercise of such jurisdiction and control imperative. I think, probably it will be the task of future negotiations to settle to some degree of precision the rules governing space colonies and those including not only the use of national resources, ban on appropriation of sovereign rights, jurisdiction, supreme control, but a whole gamut of other legal problems I think which will arise in connection with space colonies.

I think perhaps I've taken more time than I had really intended to take to start the discussions. We will have additional time this afternoon to continue our discussions at 2 o'clock. We have not had any predetermined order as I've indicated of the various subject matters that our panelists would like to bring up, therefore I would, at this time, ask you or any of you on the panel that would like to approach, talk, discuss, comment on any of the topics that he or she wishes to talk about.

Dr. Carl Christol

May I say simply to Dr. Gorove that he has made a fascinating presentation and he has given us a great deal to consider, much to agree with, a great deal to disagree with or at least to ask for further explanation along the way. It's really a professional job that commends our respect and excites us to join in with a full-fledged discussion along the way.

Let me introduce Dr. Isabella Diederiks-Verschoor, who obviously needs no introduction, to you. She comes all the way from Holland or maybe I should say from Utrecht in Holland where she is Professor of International and Space Law. She is, and has been for the past several years, President of the International Institute of Space Law. Her research is in the development of air and space law. It goes back 20 or possibly 30 years or so or maybe even longer. She is clearly one of the outstanding specialists in this entire field and as co-chairman of this panel, which is her post in this connection, I now call upon Dr. Diederiks-Verschoor.

COMMENTS

Dr. Isabella Diederiks-Verschoor*

Thank you very much Carl for your kind words. My first comments are on the introductory paper of Prof. Gorove which dealt with historical facts and present law, but in viewing the comments of the panelist I came to some other ideas, and I would like to present them here.

As Eilene Galloway stressed in her paper, problems involving outer space are always multi-disciplinary, and law is only one element among inter-related factors such as science, technology, economics, politics, culture, organization and management. I agree with her that problems in this field should not be thoroughly analyzed in legal terms, but on the other hand, I do think the development of the Space Shuttle, and in the future the development of other spacecraft with flying capabilities has brought the importance of finding a demarcation line between air space and outer space once more to the fore, and I think there will be a good possibility to a legal approach on this matter as well.

It seems to me really difficult to deal with two legal regimes for flights that will occupy different areas at given moments. Dr. Herczeg has stressed in his paper the possible unification of the two legal regimes. I will suggest the following possibilities.

> A unification of the above mentioned two legal regimes which results in a general regime applicable to all spacecraft with flying capabilities and also phenomena which deals with this phase from the legal point of view. In this respect, we can make a comparison with a legal approach toward a holocaust and the legal implications in this field. The disadvantage of this system is that you cannot figure out all of the possibilities in the future regarding these phenomena now. The second possibility is making a legal regime with a unified character only for the Space Shuttle, because of the fact that we know exactly which legal implications are involved with this particular spacecraft. The first advantage of this system is that we can see what is going to happen in the future; and the second advantage of this system is to have the possibility to determine whether a certain spacecraft should fall into this regime and which will not. I think we should formulate a regime with a specific idea of the well-known Space Shuttle in mind, because of the second mentioned reason of knowing the characteristics of the particular spacecraft.

I agree fully with Eilene Galloway that sovereignty should not be confused with jurisdiction and states cannot claim sovereignty, and cannot acquire sovereignty through the use of outer space. I refer to page three of Eilene Galloway's statement - that international cooperation is the important issue regarding the use of the farsighted purpose of Article II of the Space Treaty. Let us hope, the cooperation will help us to solve all sorts of problems of the delineation between air and outer space from the legal point of view.

---

*Professor of International and Space Law, Utrecht, Holland
President of the International Institute of Space Law

Dr. Diederiks-Verschoor

I have only one question for Professor Gorove: If I understood you well, it brings up a question. If the personnel of a spacecraft is to remain under this jurisdiction, if it leaves the spacecraft, one has left the spacecraft, and I ask myself- what is the alternative? Do you think about an international agreement about the status of the personnel of a spacecraft? Or what did you have i mind as an alternative? This is my only question!

Thank you very much.

Dr. Carl Christol - Moderator, Space Law Session Roundtable

Thank you ever so much, Prof. Diederiks-Verschoor, for your presentation. You have been so brief that you will be entitled to have a lot of time later on during the debate. Your comments were extremely concise. They referred to a single regime, a dual regime, a situation involving a boundary between air space and outer space, and the acquisition of sovereignty by way of use. They are all of critical importance and, of course, relate very clearly to the basic presentation of Prof. Gorove. We thank you very, very much for that presentation.

I should call attention to the fact that on our program, Judge Manfred Lachs, of the International Court of Justice, is listed as a co-chairman, and not only is Judge Lachs a distinguished jurist in many general senses, but he is also a distinguished specialist in the field of international space law and served as the first chairman of the ad hoc committee on space at the United Nations and later carried on in the United Nations as representative of Poland in that area prior to his being elected a judge of the world court. We regret that illness in his family prevented him from coming on from New York, although he was on his way here for this meeting, so we now turn to the next participant, and I'm going to call upon these in the sequence in which their names appear on the program.

The next person on the program is General Menter who you have just had occasion to see and to become acquainted with. I should say to you that General Menter, following a distinguished career as a law officer in the United States Air Force, during which time he wrote one of the first and the most critically important analyses of the developing space law field back in the early 1960's or late 1950's, has now left his official status and is now in retirement, but continues to be extremely active in the American Bar Association, in the Federal Bar Association, in the International Law Association, in the American Society of International Law, in the aerospace field and has testified on a number of occasions before the Congress of the United States on these subjects and has followed the development carefully. Further, he is a practicing attorney with offices in Washington, D. C. So, I now call upon my long-time friend and close associate, General Martin Menter.

COMMENTS

General Martin Menter*

Thank you very much. I would like to first make just a very brief comment on the general statement of Dr. Gorove - not commenting pro or con - but just a little bit as I view the development of the law.

Whenever a bill is introduced in a legislative body, when you're trying to construe the purposes of the bill, in this case now, the interpretation of Article II of the 1967 Space Treaty, you look to see what was the problem that the legislative body was trying to resolve, and if you recall what the law throughout the ages of obtaining sovereignty over newly discovered land masses on Earth has been - not just the planting of the flag, the first person there claiming it for their particular sovereign - king or queen or democratic republic, but it was the "effective occupation" which meant that over a period of time you actually exercised the dominion over that particular territory.

Now, with that in mind, you're trying to avoid the importance of discovery, that you're sovereign and therefore it's your territory. You're trying to avoid the claim to the territory based upon your effective occupation. Look to the wording now of Article II, and it says you can't have appropriation of the property by claim of sovereignty by use or any other means. I think what we're trying to do in Article II was to negate the application of the old principle of the country taking... being able to claim that the particular property was subject to its jurisdiction as its own property, and like the song goes, "The Moon Belongs to Everyone", and that's what I think was intended to be accomplished by Article II.

Now, for my comment on the one point as a participant that I would like to discuss with you, and that has been suggested already this morning and suggested in the papers that have been circulated before our meeting, that with the advent of the Space Shuttle, we perhaps also should give priority now to the resolution of the problem of delineation between air space and outer space, that is , having a line of demarcation where sovereignty in the air space and outer space begins. I think in this, too, we should look to history. Way back, let's start with Grotius, who wrote that all air was free because it was something common to all, and that no nation over air or the sea, being common to all, should be able to claim dominion over it. That was about 1604. Later in 1625 when Grotius recognized that a nation could launch a cannonball one sea league or approximately three nautical miles from the coastal shoreline out into the ocean, and therefore the coastal state could exercise dominion over that three nautical miles that they should have sovereignty over that or dominion over that, and consequently we evolved the concept of the territorial sea.

We come along now skipping way down to the turn of this century where we had the airship which went over boundaries between states and then around 1909 when Louis Blériot flew from France to England in an airplane, it was no longer an academic question of sovereignty in air space. It became a real issue, and we had conferences starting with the

---

*Member of American Society of International Law and American Bar Association.

Paris convention of 1910, Havana Convention in 1928, I think, and then, of course, our tremendous convention which now governs international civil aviation, the so-called Chicago Convention. In Article I, of course, it says that the subjacent state has exclusive sovereignty in the air space above its territory. So we see that in looking back on history, a nation claims sovereignty, and is recognized that it has the right too over its territorial sea -- whatever that distance is now -- and over the air space above it. Now, in considering aviation law, the writers there didn't need to define airspace, because airplanes only flew in the airspace. They didn't have anything above that, that we utilized at that time, and it wasn't of course, until we came into the present space age of Sputnik I that we had the problem of sovereignty above the airspace.

    I wonder whether it is important really to resolve that question. Right after the UN resolution in 1963, in the year 1963...which as you recall, sets forth the basic principles which became the body of the 1967 Outer Space Treaty, there was the same recital that's now in the treaty of outer space, being free for all nations, and free for use by all nations. At that time when that treaty became effective, I wrote that I thought that that should set to rest the problem for some time because I thought that it wasn't a matter that we should be really deeply concerned about inasmuch as the treaty provided another protection. And, after all, in the development of the territorial sea and the claim of sovereignty in the air space above it, the whole purpose of those two was the protection of the sovereign -- the costal state or the subjacent state.

    In the '67 treaty we had the provision of, you remember, no bombs in orbit so to speak or otherwise stationed in outer space, and it didn't make any difference if a bomb should drop from 45 miles up or 65 miles up or 100 miles up, it would still come down by force of gravity and do the same amount of damage. I felt that as long as we had provisions like that in the treaty, we gave ourselves greater protection to the subjacent state than we could have by any arbitrary line.

    Now, of course, we've come into that point with the advent of the Shuttle. People are starting to question whether or not the Shuttle shouldn't change our thinking and that we should demonstrate a line. To my mind, there's no difference. The Shuttle profile on landing is almost identical to the profile of the Apollo on landing. The Apollo drog chutes open at 24,000 feet. The Shuttle coming on down doesn't rely entirely on aerodynamic lift until it reaches about 50,000 feet. The man at the controls in the Shuttle when he makes his deorbit burn, does so knowing exactly what runway he's coming into, because once he makes that burn and starts coming down he's on pretty much a computerized course. Let me quote from a letter I received just about a week ago from the head of the Space Shuttle Program of NASA. However, "The Space Shuttle is operationally constrained to land at a preplanned landing field. Once the deorbit maneuver is performed, the Space Shuttle is committed to land at a specific landing strip."

    At that time of the unveiling of the Space Shuttle a couple of weeks ago, I remember in the newspapers the statements that this was pretty much of a dead-stick landing, a straight in approach, and that the Shuttle couldn't go around again. That's exactly it. When that Space Shuttle makes its deorbit burn, it comes right on down and it's computerized into a particular runway. As I see it, the Space Shuttle is still a space object. It's not really a true airplane as we understand it, so I don't see that the introduction of the Space Shuttle really changes the rationale requiring an establishment of a demarcation in space. The protection of sovereign based on functional limitations "no bombs in orbit" is still to me the guide. Back at the time the Treaty was being considered, or even before that rather, the Legal Advisor to the Department of State in 1959, made this statement, (that was Mr. Loftus E. Becker): "I'm opposed as of this moment to any attempt to have an international agreement that says our sovereignty extends so far up and no farther, because I say we don't have enough facts to know that it is in the national interest of the United States." We don't know whether that's in the national interest of the United States. I could say the same thing on the behalf of every subjacent state. I think every state had this same concern, that is protection of the subjacent states. With the Space Shuttle, I think the same statement can be made. I don't think we have enough facts to know whether that's in the interest of any of the subjacent states. We've gone along for centuries developing the law to govern and we have evolved these concepts for protection of the coastal state and the subjacent state. I can't see that just because we have a Space Shuttle which comes in at a dead-stick landing, that that provides any real reason for changing the basic concept, and I think that those who are recommending that we do have a delineation, I think they should have the burden of showing us the cogent reasons why that should be adopted.

Looking to the future beyond the Space Shuttle, of course, we're going to have other craft. We're going to have aerospace planes that will take off, go into outer space and come on down and land. There will be two different tugs as I see it. One would be a terrestrial transport. About 10 years ago, within the Douglas Aircraft Company, some of its people came out with actual drawings for NASA of a suggested spacecraft that would blast off, go on up to outer space, and come on down. I remember one writer from Douglas Aircraft Company, indicating that they could have a spacecraft that would take 170 passengers and 18 tons of cargo to any point on Earth within 45 minutes at the average speed of 17,000 miles per hour. You could go, say, from New York to Sydney, Australia, or to Japan - just blast off and down, landing. But it would be very little different than being in an airplane coming through. The Air Traffic Control picks up the plane coming from Europe long before it gets to the shoreline, long before it gets to the demarcation of the territorial sea. They call in for instructions at the altitude they should come in, in the azimuth they come in, etc. It is the same thing on a plane that goes on up and comes down through outer space such as this one that's visualized. They would contact Air Traffic Control.

The other type of plane might be called the true aerospace plane - it would take off, go on up into outer space to a space station - perhaps go to one or more space stations, stay there a while and make a return trip. But that particular vehicle too, long before it came down within the airspace would have to contact Air Traffic Control so that Air Traffic Control could provide protection to that craft, of airplanes and other things that may be around in the airspace, and provide protection, of course, to the airplane too. So, in closing and in summary, it seems to me that it's very important that we consider this question of drawing a demarcation line in space, that those who propose it should carry the burden of why we should have it, and that in my view, that inasmuch as the Space Shuttle profile is almost identical to the profile of the Apollo, I see no reason for a change, that the protection of the sovereign is still served, best served, by the functional approach that's in the '67 treaty. Thank you very much.

Dr. Carl Christol - Moderator, Space Law Session Roundtable

General Menter, we thank you very, very much for that skillful analysis. It strikes me that you're doing one of the things that we should all be doing, and we are I think, namely getting our facts straight before we start to determine the policies that should be ultimately arrived at .and so your reference to these practical considerations is, to use an old expression, right on target. We thank you very, very much for that thoughtful presentation.

I like to turn now to the next name on the program, that of Mrs. Eilene Galloway. Words fail me when I endeavor to find those most appropriate to introduce you. But let me say that you are certainly one of the most respected, if not the most respected person in the United States in terms of law science, space science and policy, and your experience has been an extended one and a thorough one, and your experience in this field has just simply been phenomenal. For those of you who don't know Mrs. Galloway, I should say that for many years she was a member of the staff of the Library of Congress, and, in particular, the legislative reference service. Recently, she has decided that she should like to be a consultant to the space field, and has therefore retired from the official activities at the Library of Congress, but no sooner had she announced her retirement from that, than she was seized by her friends in the United States Senate, and more particular the committee on aeronautics and space sciences, and so she's now back there giving them whatever time they need. Mrs. Galloway is currently the President of the American Branch of the International Institute of Space Law. She has written extensively. She has traveled extensively. She has represented the United States at meetings of the Committee on the Peaceful Uses of Outer Space. I think I can say nothing more than call upon Mrs. Eilene Galloway for her presentation.

ISSUES PENDING BEFORE THE UNITED NATIONS LEGAL SUBCOMMITTEE

Mrs. Eilene Galloway*

Dr. Christol and Dr. Gorove. I also wish to join in welcoming our visitors especially those who have come from so far. I think that I should like to talk about what we can do between now and next year that would be very practical to help to solve some of the issues that are pending before the United Nations Legal Subcommittee.

There are four items on the agenda. The first one is the Moon Treaty which has not yet been concluded. The second one is the Direct Broadcast Satellites. The third is the Treaty on Remote Sensing of the Earth by Satellites. All three of those subjects have priority - a high priority, so I think they are all equal in priority and then they have added the demarcation or delimination between air space and outer space which has no priority. They talked so much about the first three subjects that they never got around to talking about that very much.

The Legal Subcommittee is to meet in New York at the United Nations Headquarters in early spring of next year. The first item on the agenda will be the Draft Moon Treaty, and they have agreed on a number of points to go in the treaty, but there are three points on which there's no consensus. I should like to explain to you who have not been there that all of the work that is done by the Committee on Peaceful Uses of Outer Space and its two subcommittees, the scientific and technical and the legal, all of this work is done by consensus. That means that every member has to agree to what has occurred before any agreement can be reached, and at first I thought that we never would be able to get a treaty under that provision. I mean, they were not going to vote. They were not going to come to any decision by voting. Everyone has to agree. I expressed my doubt to Dr. Wyzner, who is chairman of the legal subcommittee, and he said, "I didn't think either that we would be able to proceed on this basis." But this is one of the really marvelous things that has been done by this Committee on Peaceful Uses, and the Legal Subcommittee, because we now have four space treaties that are in force-the original 1967 treaty, the treaty on assistance to astronauts and return, the Liability Convention, and the most recent one on the Registration of Space Objects.

The method is such that it could well be emulated by other parts of the United Nations and by other international groups because when the committee can't agree, and the subcommittee can't, and this is a formal procedure, they turn themselves into a working group, and if they can't agree, they turn themselves into a mini-working group - it might be a group such as this - and then they would try to hammer out these words and reduce the area of disagreement. Now, some people think that as a result of a certain session maybe nothing has been agreed upon, but when you look at the text you will find that everything that is in agreement is down there, and everything in disagreement is in square brackets. So, at least you have identified what has not been agreed upon and you can work on it in between this session and the next session. In the case of the Moon Treaty, no consensus has been reached on three

---

*President of the American Branch of the American Association of International Institute of Space Law.

points. The first point is whether the treaty should apply only to the Moon or to the Moon and other celestial bodies. This is an issue between the United States and the Soviet Union. The Soviet Union was the first to propose the Moon Treaty and later the United States went along with this, but wanted it to apply to other celestial bodies in the same way as the 1967 Treaty on Guiding Principles. The second one was they have not decided on the nature and the timing of the kind of information that should be given to the Secretary General of the United Nations. You will see if you study the trend of negotiations which began in 1958 that there has been a trend toward giving the Secretary General more of a role to play in outer space activities, and in this case there is quite an issue on what kind of information we should give to the Secretary General, and when you could give the information.

Now, the most critical issue, and the one that will be discussed next spring the most, is whether or not the moon's natural resources constitute the common heritage of all mankind. Now, I think that these words were put in the original treaty as a sort of a "flourish" you know like in the Bible you read, "Peace on Earth, goodwill toward men." I don't think they were analyzed very well as to what they should mean. But when you speak of the province of all mankind, in English, you know, you might say "this is the province of all mankind" and you were just ending everything with a happy expression; but when that's translated in Spanish it becomes a province and a province is a territory over which there is sovereignty, and you immediately have a great deal of discussion about this. So, this common heritage of all mankind is one of the most acute issues. Now, one way in which they are trying to resolve it is to say that within a certain time the United Nations should convene a conference in the future. That is, we could get the Moon Treaty and then provide -- well we can't decide now what kind of a regime we are going to have to govern and exploit the moon's resources but perhaps the Secretary General, on the basis of information, could decide that in five or ten years we would call a conference in order to settle that matter, and that way we could get the Moon Treaty and have some future.

Now, it seems to me that anyone who is a member of the International Institute of Space Law, or anyone who is interested in this, has ideas that would be helpful to the Legal Subcommittee and especially with regard to the wording. That is, if you have noticed that the wording of some of the phrases is not exactly right, or might cause some trouble in the future in explaining it, then I think we could make quite a contribution if we would send our ideas to the Legal Subcommittee.

The second issue before the Legal Subcommittee is a matter of Direct Broadcast Television Satellites. Here, for the first time, both with Direct Broadcast and Remote Sensing, we come to two issues that have been presented in such irreconcilable terms that you don't have enough margin to get a consensus. Hitherto, by working at it over a long period of time and in the case of a liability for damage from space accidents, it was 8 years before we got the convention and that was very tricky and it was very complicated. We did get a result. But in the case of Direct Broadcast Television Satellites, the issue is presented in such irreconcilable terms that you wonder whether you are ever going to be able to eliminate the square brackets. The way this started out was, there was a fear that a nation would put up a satellite that could broadcast directly into a home set without any intervening device - any land stations or anything like that - and that this would be an invasion of the culture and of the education of this particular country and the country would object. This fear was much greater when the issue first arose, I think, than it is now, but it was especially expressed by Canada and Sweden, and the first proposal of the Soviet Union indicated almost as if they were going to shoot a satellite down if the satellite attempted to broadcast directly into a home receiver. It happened that the Soviet Union incidentally changed their position on that and said that they would take all legal means to stop such broadcasts.

Well, it happened that we didn't have the technology yet. We had the technology for part of it for a community broadcast such as we had in India and a program between the United States and India where the United States furnishes the technology and India has control over the program content and over the way in which it is to be broadcast to their about 5,000 villages. But it was going to be sometime in the future before we ever got the Direct Broadcast into an unaugmented home set. Nevertheless, this fear was expressed, even if you said that a nation didn't have to import sets or didn't have to manufacture sets and had control over it. It still didn't get away from this fear which was very real and this is what has caused all the discussion in the United Nations. Nevertheless, after about five years, we have got consensus on nine principles and they concern the purposes and the objects and the fact that international law is applicable to this activity that states have certain rights and benefits and the international cooperation is especially stressed. States are responsible for consulting with each other, if they have any kind of problem and the peaceful settlement of disputes is provided for.

The role of the Secretary General is broadened so that he can receive information and notify everybody about what's going on and copyright is protected. However, the three irreconcilable issues -- and someone at UNESCO told me he thought it would be ten years before we even got close to this -- was that a number of states have been asking that they have prior consent before a program is broadcast, and that they participate in this activity. The second issue is one of program content and there's a great deal of difference between what the various nations want. Canada, for example, would be satisfied if there was consent to putting up an operational system right at the beginning, but the Soviet Union would like to have what other nations seem to regard as censorship over the actual content of the program -- over each individual program. Other nations might be contented with sort of a general agreement over programs. Then there are some nations that say that some types of programs that have violence or commercial advertising would be unlawful and inadmissible. There is conflict on this between the United States and Great Britain on the one hand, the Soviet Union on the other hand accompanied by Canada, Sweden, Mexico, and Argentina to mention just a few nations, with the United States' position being the free flow of information and ideas. The other side of the picture is one in which the Soviet Union doesn't want another nation to broadcast directly to its people without its consent. So, if you have any ideas on how to merge these two positions and get a consensus, you would be doing a great favor to the Legal Subcommittee.

The next one is the one on Remote Sensing of the Earth by satellite. This issue is rather curious. You see, in the case of Direct Broadcast, we didn't have all this technology. It's to be developed and some of it in the future. In the case of the Remote Sensing, we already had the Earth Satellites and the LANDSAT satellites that were so -- you know they were so good in the experimental phase -- that they are practically operational. They are used in operational ways by many countries. There you have also an issue of some nations asking for prior consent before pictures are taken of their country, or (they haven't done that so far) because the pictures have been taken by the LANDSAT satellite and all of the information was for sale. It's for sale in exactly the same way as such pictures that have been taken in the past by airplanes or by ground crews or any other way. United States' laws on this kind of thing go back 100 years for distribution of maps and geological maps and geologic survey and things of these kinds. So, it was not a new policy. It was simply a case of using new technology to go on with the policy that we had. However, there are some nations that would like to get the information first about their territory before it was given to anyone else. One reason this was such a critical issue was because of the words "natural resources" which seems to many people to imply that the satellites were going to take pictures of where you could mine for copper or various minerals and that these could be exploited. It was almost like 19th Century British Imperialism. That was kind of the symbol of it. All of the other uses were neglected. The people who talked about the natural resources did not seem to take into consideration that this type of data could be used to solve certain global problems such as population problems, the food problems, the kind of problems that Kraft Ehricke was talking about an hour before I came here. Now as time goes on, some of the nations have become used to this and yet they do have a problem of acquisition and dissemination of information. Who is going to pay for it? What kind of institution has control of it? How is it going to be managed? How are you going to train people to analyze the data? So, I would say that of all these problems, this is the one that offers the greatest opportunity for us to make a contribution to it in terms of solutions, you know, suggesting different kinds of solutions, so they would have options.

 The fourth item on the agenda is this one of the definition and/or delimitation of outer space and outer space activities. Now, this is very curious the way it is on the agenda. There were a great many articles, it was surprising, written on this subject before a satellite was ever orbited. The reason was people were interested in comparing it to freedom of the seas and sovereignty so they thought about outer space and sovereignty. And then it was thought of as a line and when you read the early articles, you almost feel as if the Earth is a perfect sphere and the line is also going to be a perfect halo around the earth. So, it was something of a shock to find out the Earth was not a perfect sphere and that this line varied and that there was something in between airspace and outer space and no one knew how large it was or where it was or anything like that. Originally, we just simply disregarded it and got a treaty which combines some area approaches in so far as it mentions outer space as an environment -- as an area. It also has a functional approach of the state being responsible for its object no matter where it is. It can be in orbit or it can be a space probe like the one to Mars. It doesn't have to be in a durable orbit. If it just goes around once, it's in orbit. If it's a space probe, it goes out and the state is responsible for it no matter where it is. This is the way in which the legal profession really recognize, without saying so in so many words, the engineering concept because an engineer doesn't care where air space ends and outer space begins as long as he's been given the mission to get to the moon. You know, he wants to get to the moon, he wants to get back from the moon. However, in the way in which this has been worded on the agenda, there's a kind of sleeper here -- the definition and/or delimitation of outer space. You know, even delimitation beyond outer space into something that is outer-outer space, and outer space activities because some people are trying to use this in order to delimit outer space activities from which data is received on the Earth because they are trying to use it in order to control the Direct Broadcast Satellite and the Remote Sensing, so it's something more, you know, than just a line.

 In conclusion, I would say that I think it's the function of the law to help and to facilitate the solution of problems, and not to hinder or hamper them in ways and that we have a responsibility to consider these issues and see if we can make a contribution to them. There's a mutual responsibility of the scientist and engineers and of the lawyers

to each other to keep themselves informed because there are very few
people who know how to relate these two fields and they're apt to work
in separate pools and not get together. If they're working in separate
pools, they're thinking vertically and we've got to have some people who
think horizontally between the scientific and the technical and the
legal so that we can do what we want to do and not have any adverse
affects that we would not like.

Dr. Carl Christol - Moderator, Space Law Session Roundtable

Thank you very much Mrs. Galloway. I'm sure you all now realize
why Mrs. Galloway enjoys the enormous respect she does in this field
and why her prestige is so great. I'm very much interested in a
number of points that you made. You did address yourself to the
issue of priorities, which I think is important for us. You suggest
that we as humble citizens can make our feelings and our knowledge known
so that the governments may have the benefit of our thoughts and services,
and the invitation which you extended I think we all ought to take into
account and act upon.

There are a number of points which were made which I'm sure we'll
wish to consider in greater detail when we come to our discussions.
Professor Gorove informs me that we are expected to conclude this meeting at 11:30 because of other engagements that our panelists and participants have. So, I think that we should now adjourn until the hour of
2 o'clock at which time we will turn directly to Dr. Fasan's presentation,
and having completed that we will then go to our dialogue and general
discussions. I wonder if there are any general announcements or comments
that might take the next minute or so before we do break up - of an administrative nature? I have a small one. I have additional copies of
my prepared remarks that I distributed at the outset for those of you
who came in late if you'd like to have copies of them. I shall say later
on, but let me say again, that I think we should commend Dr. Steinhoff
for his brilliance in assembling such an enormously competent and interesting panel, and we are delighted to have at the same time such a distinguished audience, those who are not panelists, but those who have come
along to share our trials and tribulations as we look to the law and
policy of the space environment. Let's resume at 2 o'clock. Thank you
very much for being with us this morning.

Ladies and Gentlemen. Welcome to the continued portion of the
space law development roundtable. May I say that I'm going to be
obliged to leave at 4:00 p.m. in order to get a plane. Therefore
at this point, if we're still in session, I shall call upon Dr.
Diederiks-Verschoor to act as the moderator, and she will preside in
that fashion if required and then she will turn the meeting back to our
organizer, Dr. Gorove, for final comments.

I'm also obliged to say that Prof. Lay, Mr. Finch and Prof. Herczeg
have been unable to join us for a variety of reasons. Dr. Herczeg
writes that his health is not adequate to making the long trip, and I
understand that both Prof. Lay and Mr. Finch will be with us in Anaheim
for the International Institute of Space Law meeting there. Let me then
introduce to you with enormous pleasure our final panelist, a person
known to all of us for many of his outstanding achievements. You will
remember that when the International Institute of Space Law met in
Vienna that Dr. Fasan was the organizing chairman for that particular
meeting. You will know him for his writings which I, myself, consider
to be extremely innovative and creative, and you also know him as Secretary and board member of the International Institute of Space Law.
We're delighted to welcome him to the Far West. He tells me he's been
in the United States before, but never out this far. Later we should
be even equally delighted or possibly more delighted to welcome him to
California, so he can really see what the rest of the world is truly like.
An old friend, a lawyer from Austria, Dr. **Ernst Fasan**.

THE UTILITY OF MORPHOLOGY TO SPACE LAW

Dr. Ernst Fasan*

Thank you very much Mr. Chairman. Ladies and Gentlemen, in view of the fact that our community - our space community - is rather small, today, I believe that you will permit me to mix up (I hope nothing too bad in the sense of the word) my statement which I've prepared for to- day and some remarks which were paneled in my mind by the very learned and interesting papers of this morning.

Mainly, I believe, that a kind of morphological approach to space law is not only in order of the legate feranda but that a tendency of creat- ng a kind of morphological order (and I'm honoring Prof. Zwicky on this account of course) and that a tendency of showing a kind of morphological order has been inherent in the space law development even from the begin- ning. To be particular, we have at first a kind of basic legal motions for outer space, as for instance, were formulated in the United Nations Resolution 1721. This created a kind of first layer in a kind of hier- archy of norms of space law.

The first layer was then enlarged on the scientific non-binding basis by resolutions of the United Nations' General Assembly 1884 and 1962 and then came on an internationally legally binding level, the Outer Space Treaty of 1967 as has been mentioned this morning several times quite correctly. This was the first level now building or creat- ing a kind of constitution for outer space, then formed the Return and Rescue Agreement, the Liability Convention, the Registration Convention. All of them regulating in more detail some of the general motions of the Space Treaty of 1967 and in this level we can expect, as Mrs. Galloway told us in very great, enlightening detail, the Moon Treaty or Remote Sensing Convention and Direct Broadcasting Convention. But, there is on the legal.. generally legal basis, a third level. There are very specific treaties or conventions. For instance, the 1972 agreement between USA and USSR regarding Apollo-Soyuz mission. For instance, the Intelsat or the Intersputnik Treaties. For instance the ELDO, ESRO, and ESA Conventions, etc. These three layers of general legal regulations seemed to be followed in our hierarchy order by a fourth level, namely the finding of decisions according to those procedures which have been set up by the various conventions and treaties, and finally a level, let me call it a fifth level, for actual execution of those agreements which had been made beforehand.

When I prepared this paper, Professor Gorove correctly pointed out that not always the same parties are subscribers of the, better to say, have ratified all of those legal treaties, meaning that there may well be a nation which did agree upon the Space Treaty of 1967 but did not agree upon another more specific treaty. Also, this is, of course correct, we have to realize that a kind of pattern of morphology did form, or did seem to form, itself. And I would now like to ask you, not for the mom- ent but to think about it, to put some of those questions which have been raised this morning into this morphological box. For instance, General Menter discussed questions of Article II of Outer Space Treaty. There seems to be kind of an ambivalence between known appropriation according

---
*Secretary and Board Member of the International Institute of Space Law.

to Article II and use Article I, between known appropriation, Article II and ownership of durable stations on celestial bodies, Article VIII of the Space Treaty, between known appropriation and no sovereignty principle of Article II and this ownership and jurisdiction over stations according to Article VIII, between even jurisdiction and use, between free access to all areas Article I and ownership Article VIII, between free access to all areas Article I and safety provisions Article XII liability provisions Article VII.

In our morphological box, we furthermore would have to put in, although this problem is in the legal subcommittee not so urgent, the question of delimitation or demarcation between air space and outer space. There are problems which can otherwise not be solved, for instance, as Dr. Campbell provided me this morning with a kind of appendix, of space shuttle. If Austria starts a space shuttle, it could go up through Austrian air space; that's quite correct, but if this space shuttle would land in Austria, it must go according to the landing orbit. It must go through the air space of another country because Austria would be too small for such a landing. Now, if we accept, as we do, that air space belongs to the territory of a nation and if we accept that each nation has got sovereignty over its air space, than we will have either to say that the space shuttle is going through the sovereign territory of another state or that it does not because the sovereign territory does not extend so high up. That's only one problem.

The second problem is a problem of neutral states. A neutral state has to defend its own territory by all possible means. Now, when we say one warfaring state would send high altitude rockets beneath an orbit level over a neutral state, say over Austria and this is not an invention of mine, but an invention of Professor Corovan, late Professor Corovan from the Soviet Union, if such a rocket would go very high but below orbit level through, let me say over a neutral state, would it then pass through the sovereign territory of this neutral state, and would therefore the neutral state then be obliged to defend itself or not. This will be very important for the neutral state because if the neutral state would be unable or unwilling to defend its own territory, then the attack state would by way of substitute intervention, offend the territory of the neutral state and this would mean that the attack state would explode the warheads over neutral Austria, I mean legally, correctly, could do that. So, the neutral state would have to know where are the upper limits of the sovereign territory in order to argue "You are not allowed over my territory to explode a nuclear warhead which is going over my territory, not through my territory." These are some of the questions which I believe could be solved if we morphologically computerize all motions upon which agreement has been reached in the various space law treaties, and if we create an interdepending morphological system which then would permit each question either to answer or to know that for this question there is still an open hole which should be answered. Therefore, I hope that this idea of morphological approach to open space law questions would show us patterns for future work. Thank you very much.

Dr. Carl Christol - Moderator, Space Law Session Roundtable

Thank you ever so much Dr. Fasan for this very stimulating presentation and for the orderly structure that you contemplate as a scheme of analysis for us, as we pursue some of the subcommittee's problems and difficulties which we know exist and which you've called to our attention. Now, I think we are at that happy state when we can turn to each of us in turn and get comments by way of responses to those points which have already been presented, and so as I look around this room and this distinguished audience and participants and wonder who would like to make the first comments? I think Dr. Gorove would like to make his comments first. Dr. Gorove.

Dr. Carl Christol - Moderator, Space Law Session Roundtable

I think we've probably covered the orbital issue. If it goes around once it is in orbit, which is an exact quotation from Mrs. Galloway. We are concerned with "what is orbit now?", and whether or not it has to go around once or whether, as I have suggested, that it need not have gone around once but at least to have the capacity to go around once as Mrs. Galloway has indicated.

To be in orbit, you said.

All right, that's the other side of the coin, and if we can accept that conclusion, all well and good, and Steve will come back to it. I did think that Dr. Fasan made a good point. Do you remember he was citing -- now I'm back to the subject of treaty interpretation -- he cited a number of the articles of the 1967 Convention. I would say that if he were here, one would have to compare the different articles, both as to their context and as to their major purpose which again are elements that the lawyers would look to under the general heading of criteria for treaty interpretation. I would like to say, with respect to sovereignty that one should concern oneself with possibly a degree of control. That I think, is a political issue; the degreee of control which is open to a nation state. If one can say that one wants to have a very substantial degree of control, then one could argue sovereignty. If one wants to say that there is a need to exercise influence and even dominion with respect to lesser events, then one could say it's a matter of jurisdiction, and somewhere in between one might find the utility of the term "sovereign rights".

Mrs. Galloway made a point I would like to come back to a little bit. It's a pretty sophisticated point and I'm not entirely sure that I understood it. It had to do with the remote sensing situation. You suggested this subject offers a great opportunity for those who would have suggestions to make. It is agreed in terms of history that the Landsat type of satellite had become operational prior to the existence of any objections, as to the kind of sensing that they were doing. You said, I believe, that some requests had come in for prior consent prior to engaging in remote sensing. I think I understood you to say that this series of requests has been somewhat thwarted by the easy distribution of the data which the United States had been providing. I think I also understood you to say that some states want exclusive delivery of data and I wondered whether or not there was a trade off here between the claim for exclusive delivery of data and the willingness to forego the demands for "prior consent." Could you clarify that please.

I think that we ought to ask Dr. Diederiks-Verschoor if she has any comments that she would like to make, then I will call upon General Menter, then Dr. Gorove and Dr. Fasan before we get the other other participants into it.

COMMENTS

Dr. Stephen Gorove*

I would like to make a couple of brief comments on the discussion that we have had so far. First of all, let me say that Dr. Christol has an extremely sharp eye. Few people would have picked up the word "durable" out of my paper and he is quite right. That word has gotten in there in the context in which I was talking about international customary law. I didn't refer to it as my own views. I mentioned it to indicate that some people believe that an international customary law has emerged which regards satellites in durable orbit as being in outer space. This problem was very relevant and is relevant today in relation to determination of the possible demarcation line between air space and outer space. I believe it was Professor Myres Smith McDougal who used this phrase at the time when he made the observation. I don't think the statement is in any way erroneous in the sense that it applies to satellites in durable orbit. I think the very interesting question that Dr. Christol has raised relates to the problem of non-durable satellites in non-durable orbit. In other words, would we want to say that now the customary law again did not apply to satellites in non-durable orbit. I'm not sure that we can give perhaps an absolute answer to it. We have no occasion or we haven't had any objections as far as I know about any state saying that a satellite in a non-durable orbit is not in outer space. So, I think Dr. Christol's observation may very well be right that satellites which are in orbit, durable or not, are in outer space. Perhaps we have reached that point that we could even assert that. Perhaps even Dr. McDougal made the remark, this was not so clear at the time. Of course, it is also possible that, if and when, a demarcation line is going to be found, that that will be below the point of orbit and therefore then, this whole question would become kind of mute.

I would also like to come back again to this bothersome problem,—has been bothersome. I think it was indicated in our discussions, exchanges of views of the roundtable, the problem of sovereignty. When I listened this morning to General Menter's very fine remarks and how th has come about, I'm quite cognizant of the factors and intentions that the parties had at the time when the Outer Space Treaty was drafted, a also prior to it in the United Nations resolutions which in a way form the basis for the subsequent treaty. However, I think maybe one of ou problems in relation to perhaps the appearance of, or semblance of disag ment on this point, may very well be that we have not clarified sufficiently the meaning of this term. I have tried to give a definition and I realize that definitions are always subject to criticism, and al they are not fool proof because we may find that the definition is all right for a particular purpose or context but then in another context if used for some other purpose, it may not work. But if one really goe back just for a moment or looks back at history, I think you will find that John Boudon, who forced to our knowledge, mentioned the term sovereignty in his famous book on the republic which was entitled in French "Siz Livres De La Republique", the "Six Books on the Republic". He really was talking about the sovereign monarch and the supreme authori that this monarch or sovereign exercised. Of course, in that sense perhaps sovereignty does not exist as it existed at the time of John Boudon when we had so many sovereign rulers that he could take as an example. I think sovereignty is a supreme authority exercised by the state over people, resources, and institutions. If we look at it in that way then here under resources I include land, of course. If you look at it that way, I think you will see that sovereignty has really many manifestations and that's why I regard this abbreviated and perhap somewhat misleading notion today because it's not characteristic of our relationships in the world today anymore in the same as it may have been characteristic at John Boudon's time.

So, if we look at it this way, we find that the supreme authority of the state is exercised in relation to people, in relation to resources, and in relation to institutions. Certainly some elements, some manifestations of this supreme authority and control have been done away with in relation to outer space. What I have maintained is that not all aspects of sovereignty have been abolished. I think to say that the opposite would be really not to take into account the actual situation. As I indicated before, I don't believe the Space Treaty wanted to accomplish that or if the drafters had that in mind, well certainly they didn't accomplish it through the Space Treaty. They very much wanted to have the state retain control over people, over institutions in outer space. They wanted the states to retain jurisdictional control, etc. and ownership and so the state does exercise supreme authority in relation to those things. We could not send up our rockets into outer space, or spacecraft, manned spacecraft, if we had no control over the crew.

In fact more than just Article VI and Article VIII, also Article XII of the Space Treaty I think indicates that the state exercises a degree of control in relation to space stations, because it says that the access that they we're talking about is contingent upon reciprocity, so it is unlike, for instance, the Antartic Treaty where the reciprocity is not mentioned as a condition of free access. It has to be convenient for the state whose facilities are being visited and advance notice has to be given so it's not like I just walk into the room whether the person likes it or not. I have to ask for permission. There must be reciprocity. It must not be dangerous to the establishment, so from a point of view of safety, it has to be safe, etc. This very clearly indicates even in those situations, the state exercises certain sovereign prerogatives and rights. Even more so, of course, if the spacecraft is in outer space and certainly no other country would dare to enter our spacecraft or we would not dare to enter another nation's spacecraft without our or their permission. I think again you mentioned or referred to the US/Soviet Space Agreement. The linkup was based on the agreement and it was because of the agreement and the consent that we could enter the Russian spacecraft and they could enter ours. So the state very much exercises supreme authority in relation to this. Maybe you may wish to recall that in relation to the law of the sea, nations have accepted the idea that the state has certain sovereign rights outside of the territorial waters in relation to the seabed, and the exploitation of the resources of the seabed. Now, that doesn't mean complete sovereignty, but certain manifestations of sovereignty, and I think this is the same way the situation has been developing so far.

Now, I'm not saying that it may not be perhaps the legate feranda desirable to do away with certain other forms of sovereignty or sovereign rights or manifestations of sovereignty, but I have some doubts even in relation to that that would be desirable today. But I think all these provisions in the Space Treaty, particularly the ones I mentioned in a national responsibility, make it clear that the state is obligated to exercise supreme control and authority. Otherwise we couldn't say the state is internationally responsible for the activities of its nationals, or private, or public organizations.

I would like to also perhaps say a few words here on the very interesting presentation of Dr. Fasan on morphological approach. I wonder whether he includes in this hierarchy also the drafts, like the Draft Moon Treaty at the present time or would he leave that out for the time being. But, I think we have a number of really vital issues and one of them which I would like to see somehow resolved, is really the question of the exploitation of natural resources of the moon. I think what is holding up the problem, is the Draft Moon Treaty at the present time. I think if we could find some solution, in the way that Mrs. Galloway so keenly reminded us of, that we ought to try to do so; it might be helpful. I think the Draft Moon Treaty might very well be then the next complete and accepted treaty in the field of outer space law.

I might say just perhaps in conclusion that I'd be glad to come back if we have some further discussions later on. We have held a very interesting visit this afternoon and tried to learn a great deal about gyroscopes and the inertial guidance systems. Mrs. Galloway last night also thought that this was extremely interesting and fascinating, and I have been kind of amazed at the scientific details that go into this end of thinking. I asked her kind of facetiously, whether she thought there are some legal problems or can we think of any legal problems that could be somehow associated with gyroscopes or inertial guidance systems, etc. After a long moment of reflection, I said to myself, that maybe what our lesson really ought to be from seeing how the scientists are working, that we ought to be, we lawyers or legal technicians, ought to be much more precise and much more circumspect in approaching our problems. If we worked perhaps with the degree of precision that the scientists are working with, how they check and double check every single system, how precise this has to be, and how complex the operation is, if we could do that, maybe very few legal problems would remain to be solved.

I thank you.

Dr. Carl Christol - Moderator, Space Law Session Roundtable

Thank you very much Dr. Gorove. You've been hitting at some of the critical issues that are before us. I wonder if I might be enabled now to make a few remarks that have to do with this problem of sovereignty versus jurisdiction, with the term of sovereign rights also thrown in as another possible semantic stubling block. We are dealing with a concept that needs to be identified and having been identified to be used in a way that meets the needs of mankind.

COMMENTS

Dr. Carl Christol

In Dr. Gorove's written paper I understood him to say, and I may not have gotten my notes quite correctly, that the treaties have not abolished sovereignty, and now I hear him say something a little different, namely that treaties have not abolished all sovereignty.

At least let's start by taking a look at what the principle's treaty actually says. It's been quoted before, but let me quote it again, Article II reads: "Outer space, including the moon and celestial bodies, are not subject to national appropriation by claim of sovereignty, by means of use or occupation, or by any other means." There are a number of critical ideas that are reflected in that one short sentence, and I don't want to talk about the clause "by means of use or occupation", but rather to the more specific provision that suggests to me that what I have seen fit to refer to as a space environment, namely outer space, the moon and celestial bodies, is that area, that thing, that locus, that situs, that is not subject to national appropriation by claim of sovereignty. Much of what I'm going to say, I think, depends on that distinction because in the illustrations which Dr. Gorove has given to us, he frequently has referred to peoples and objects as illustrative of the points that he's making under the heading of sovereignty. My concern is that possibly his illustrations are not particularly suited to sovereignty, but might be more appropriately suited to jurisdiction, a considerably different legal concept.

I do make the distinction between sovereignty and jurisdiction, and sovereign rights. I would agree with Dr. Gorove that sovereignty is a characteristic of the nation state, and I would recall with him that Bodin when he made his observations was talking about a supreme monarch, but was also indicating that that supreme monarch was subject to some five limitations of the exercise of the so-called ultimate or sovereign world, and one of these limitations was the law of nations. That is somewhat an aside, but the thrust of my remarks, I believe, can be supported by the illustration that Dr. Fasan provided in his own reference to the notion of sovereignty. He gave the illustration, as you will recall, of the state that was going to be attacked by a hostile state, and the problem confronted by that potentially attacked state by the presence of a neutral state between the two, the attacker and the attacked. He was wondering whether or not under international law, and I take it international space law, the potentially attacked state might act in order to protect itself either in the air space or the outer space lying above the neutral state.

This, I think then, is that kind of situation where the claim of sovereignty is relevant in the sense that the attacked state would have the opportunity to use any area, no matter where situated or belonging to whom, in the event it would be necessary to protect itself against a hostile state. In short, there are some circumstances in which an attacked state may interpose itself via coercion in the area of another state, and the factor of sovereignty should not be considered as a bar to that course of conduct in that particular area.

Basically, I think sovereignty should be seen as a problem of degree, and exclusivity of control. Or to put it in somewhat different words, it must be related to the notion of interdependence, and nowadays we see that in a very practical sense, the notion of sovereignty has been considerably diminished by reason of the notion of interdependence. There is not that opportunity or occasion to claim the exercise of ultimate will, in the sense that it had been claimed at an earlier time. The whole political structure of the world, the whole notion of interdependence in the world today, has caused inroads upon the doctrine of sovereignty.

Well, what does this suggest to you, and to me? It suggests to me that we can effect a distinction between jurisdiction on the one hand, and sovereignty on the other even though sovereignty has a diminished role at the present time. It's well know that a given nation state - let's call it state A - is not restricted by the presence or absence of sovereignty in dealing with conditions which are potentially adverse to it. By the same token, a foreign state by reason of its sovereignty, is not able to withstand some influences of another state within it. Let me give you some illustrations of what I have in mind. Take the interesting question of whether or not a subpoena issued by a Federal District Court in Chicago has any legal meaning in France. One might suppose that the sovereign barriers of France would mean that the application of U.S. law in France would be impermissible. But the U.S. Supreme Court has held, and France has not protested, that a U.S. subpoena does have legal force within France. To take another well-known illustration: a taxpayer, a U.S. national in Mexico, who didn't want to pay his U.S. income taxes. Is there any way in which the long arm of the U.S. jurisdiction can extend out to apply to the behavior, or non-behavior of this particular individual? The answer is that the jurisdiction of the United States can extend out to American nationals who are beyond the boundaries of the United States. At the same time, the jurisdiction of the United States can extend out on to the high seas and to take into account some events which are transpiring there. The exercise of dominion. or control over an American national aboard a U.S. vessel can be put down under the heading of exercise of jurisdiction, as well as, the notion of sovereignty.

I would think that one could argue then that jurisdiction is a way to deal with events beyond one's boundaries as well as sovereignty. In direct response to Dr. Gorove's point, the United States can control U.S. registered space objects, or U.S. nationals aboard those space objects in the space environment under the heading of jurisdiction without having to raise the problem of sovereignty. The question of sovereign rights also poses some difficulties. I suppose that it has been borne home most particularly to the United States as result of the Truman Proclamation of 1945, which suggested that the United States should exercise exclusive jurisdiction and control over the seabed area that has now become known as the continental shelf. There was a clear effort on the part of the United States in selecting those terms to avoid the use of sovereignty. It was a claim only for exclusive jurisdiction and control. This was conditioned obviously by the fact that it was thought that the continental shelf area was below the high seas. The high seas were free and open to all, and there was no interest on the part of the United States in seeking to extend this larger claim of sovereignty into the seabed area. Well, ultimately this got before the United Nations, as we all know, and it came down in the form of the Continental Shelf Convention of 1958, at which time the term "sovereign rights" was in fact used, but it was a term that was used in order to avoid the use of the term "sovereignty", and was used in order to allow for this continued historical distinction between the high seas on the one hand, which were free seas, and special rights on the seabed and ocean floor. Even today when we talk about the concept of the exclusive economic zone, the EEZ, that's coming out of the current law of the sea negotiations, there is an effort to effect a delineation between sovereignty on the one hand, sovereign rights on the other, and so there is no claim that the coastal state will have sovereignty going out 200 miles on the surface. States are claiming that this will be treated as an area where the coastal state, or any state can exercise jurisdiction over its surface vessels and over its aircraft flying above the surface, but certainly neither sovereign rights nor sovereignty. So I think basically what Dr. Gorove is telling us, is appealing, but whether one should carry it to the logical extent that he suggests needs to be questioned. Language of Article II must continue to apply. At the same time the affected space resource state must be allowed to exercise jurisdiction and to control the behavior of its individuals, and its space objects in the non-sovereign area.

While I'm here, let me make a few other comments in terms of the notes which I have prepared as we have been listening to the presentation. I was very much intrigued with the first point that Dr. Gorove made, having to do with the problem of the natural resources of the moon and other celestial bodies. Mrs. Galloway commented on that at some length as did other commentators. May I say, Mrs. Galloway, that I shall try to resurrect the paper that I prepared for presentation at the Lisbon meeting of the IISL, which did try to look at the common heritage of mankind concept as it relates to the resources of the space environment. I will try to suggest that this may have some use for the U.N. committee. The common heritage of mankind idea is enormously complex.

I think I am influenced considerably in my own thinking on the subject by having had occasion to talk at some length with Arvid Pardo, who is the former Maltese Diplomat, who has engendered the basic discussions that are going on now in the United Nations respecting the oceans. He has stated rather categorically, that he does not consider the common heritage of mankind idea, despite its possible relation, as Dr. Gorove indicated, to the term of inheritance, in fact to be a property concept. Dr. Pardo says that it's not a property concept and says in fact rather categorically it's quite the opposite of a property concept, and it should not be construed to be a property concept. But rather that it should be spoken of in terms of uses and benefits, and that the uses and benefits should, as has been indicated in the law of the sea discussions, be reserved to the needs of the developing nations of the world, and particularly for those who are land-locked and most disadvantaged. I think we can learn something from the law of the sea negotiations. I have spent an enormous amount of time this past year in looking at the deep seabed problems of the polymetallic nodules that are out there, to try to understand just what the law is with respect to their economic utilization, and by the same token trying to understand the law, and the problems that are related to the utilization of those mineral resources, which have a potential and actual value. I see the situation very parallel, for tangible space resources, namely, that these are resources that man with his science and technology will try to capture, and, therefore, will seek to use them as effectively as possible. So the problem is essentially one of total property ownership, possibly that kind of totality that is reflected in the notion of sovereignty. But are there some lesser legal concepts that are available, some lesser terms that are available, such as the right to use, the right to harvest, the right to gather, the right to exploit, and if so, once the harvest, the gathering, the exploitation has taken place, what distribution should be made of the values of these resources once they have been brought in? Here one has to think, I suppose, of the distinction between res nullius and a res communis situation. It seems to me that just as the fish that are swimming freely in the oceans are open to those who can catch them, so also the maganese nodules on the seabed are open to those who can mine them. This, I think, is existing international law. I suggest the same is true for tangible space resources.

I would also suggest to you that the 1967 Principles Treaty does say that international law should be applied to the space environment. I don't think under the 1967 treaty that a given nation state can say it is going to exercise sovereignty on the moon. I don't think one can say we're going to exercise sovereignty on Mars. I did want to say that I think a distinction can be drawn between the exploitation and use of the resources on the moon and celestial bodies and a totality of proprietary involvement in them.

Well, let me see if I can't piece together a few other comments that have been identified here. I was pleased to hear Dr. Gorove make reference ....what he said was in effect a customary rule of international law developing. I think that fits in very nicely with the norms which Dr. Fasan mentioned earlier, and surely these different sources of international law that are well known, and which Dr. Fasan referred to, are really the building blocks that all of us must use as we try to move our subject ahead in a constructive way.

When Mrs. Galloway was referring to the need to get further understanding of the common heritage of mankind concept, I was reminded of the story about how legislation is to be compared with the manufacture of sausage. The statement is something to this effect: If you can see how legislation is made, like sausage is made, you wouldn't like it. We shall have our problems in the processing or working out of the ultimate meaning of the common heritage of mankind. It involves many pulls, tugs, and compromises, and ultimately we may not have a concept that is any sharper in its elemental sense, than the notion of sovereignty as we're trying to explain it today. Maybe that's one point that I can make that I would be willing to argue rather vigorously for, and that is that whatever these concepts are - sovereignty, common heritage, etc., they should be used in such a way as to advance the needs of the people of the world, and this involves all of the political bargaining that I referred to in my beginning remarks under the heading of the new international economic order, the selection of suitable international institutions, the working out of a way in which those seabed nodules can be distributed around the world. Basically, on the matter of resources, I do see this-- I see a basic contest between different economic structures. The free enterprise system of the West on one side, and the social-economic orientation of the socialist states on the other. I see the moratorium effort as an effort really to prevent the capabilities of the advanced states having science and technology from involving themselves in these exploiting activities. It's hard for me to believe that man's ingenuity is going to be prevented, and forestalled when there is a practical need to go out to try and find these resources. But again, the point is not just the gathering of them, but the distribution of the value, and we may learn something from the law of the sea negotiations. The United States has proposed that the mining firms that know the locations of these resources shall identify two potential mining sites. One of those mining sites shall be allocated to the seabed authority, the Enterprise, which would have the site for its use, but the other will be reserved to the country and the nationals of the country, the corporations of the country, that have the capacity to engage in that mining activity. This may be a compromise between the advanced countries and the LDC's with respect to the resources on the moon and celestial bodies. I was glad to hear Mrs. Galloway and others make reference to the need to achieve procedures for peaceful settlement of disputes. I certainly applaud that particular observation, and would again point out that in the law of the sea negotiations, there have been some very, very extended negotiations on that with the prospect that there will be two different tribunals, along with alternative procedures to deal with that overall situation. So that is a contribution that I certainly was pleased to hear and welcome.

On the matter of treaty interpretation, Professor Gorove did pick up the point that I suggested, with respect to the problem of orbiting and the term "durable orbit." I don't have any particular criticism of the use of the term "durable orbit." I am interested in what it means and whether or not there are suitable values and interests which would allow that term to be interjected into Article IV of the 1967 convention. Article IV says as we all know: "The states' parties to the treaty undertake, not to place, in orbit around the earth, etc." <u>Shall not place in orbit</u>. There are a variety of criteria to be used in interpretation of any treaty, and one of the criteria is a literal reading of the plain meaning of the words. If one takes that approach, one would not interject the term "durable orbit", but at the same time, the need would still be to identify what is really meant by "orbit". I'd like to ask on my own account whether or not orbit encompasses only successful launchings that have gone around at least once to create that orbit or is it a situation where there is an attempted orbit, which attempt is not totally fulfilled, although the prospect when initiated would indicate that it would be fulfilled; in other words, something that does not achieve orbit, although it has the capacity to achieve orbit at the time of the launching. Does that still come within the notion "not to place in orbit?" Well, you see I'm getting away from the literal or the plain meaning of the words, and I think that Dr. Gorove has gotten away from the literal and plain meaning of the words. In other instances he has rather stressed the literal or plain meaning of the words. I don't object to that, Steve, mind you, but I am concerned as to why other potential criteria are not also considered, and as they are being considered, just what are the great values and great interests that are involved.

Well, this is wonderful and I see Dr. Sarkar would like to come in, but may I comment while he's walking up here to the microphone. This inter-exchange between the historians, the lawyers, and the scientists is absolutely phenomenal. We're glad to have it. Dr. Sarkar is from Switzerland.

REMOTE SENSING AND DIRECT BROADCAST
WITHIN THE UNITED NATIONS' SYSTEM

Subrata K. Sarkar*

I'm a communications engineer, and perhaps I can define it another way...these orbit problems are that we have a communication breakdown, us and the spacecraft you know. So long as it is in space, we have very good communication, but when it's reentering then we have due to the plasma, that is, you do not hear from the astronauts or anything for a very short time. This is exactly where it happens when we are also 80 miles or 120 kilometers away. It is of that region, so the other way to say that when the communication between the astronauts and the people on the ground is just continuing from the space data break and then afterwards when they have completely reentered, they establish the communication. This is another way to describe where the limit is between the air space and outer space. If you allow me, I would like to say something different.

I would like to present tomorrow a subject which Dr. Steinhoff asked me "The Prospective in Space Communications", and in Anaheim I will be there on another subject with the space law, that is on earth resources, but the subject which I would like to discuss here in as short a time as possible, something completely different which may reflect what -- what's that qualitate -- Mrs. Galloway said about the development of remote sensing and direct broadcast within the United Nations' system. Now, I think it is appropriate to say at this time, the development of space law is also associated with the development of radio regulations, because we are quite clear that radio regulations is not just frequency allocations or frequency assignments and also the interference problem, but it's quite clear that radio allocations has something to do with orbital position of the spacecraft. That means the satellite, whether it is in the lower orbit or in the geostationary orbit, the satellite has a position which is directed to earth whether it is fixed or it is moving. Now, the radio regulation is developing, and I wanted to say in a few words how the radio regulation sees this total prospect of frequency spectrum and this orbital position. Now, as you perhaps know in 1947, just after the war, there were a lot of radio broadcasting stations, and there were a lot of aircraft. There was confusion about the frequency system - allocation problem and assignment problem - so in the Atlantic City Conference, the first time this International Telesat Commission Union met, they established this radio regulation. Now in 1959 this radio regulation was taken not only to broadcasting, or radio broadcasting, but to many other areas -- in radio systems, in other terrestrial communications -- they include many other areas. That means any area where there is a radio spectrum concerned, and wherever there is an international problem concerned, the radio regulation comes into being. Now, in 1963 when the space communication began to get established very clearly, then people felt that it was time to have some allocation of frequencies for space communications. This was the first time that we realized first of all that the sequence spectrum is quite unlimited and the problem is this - that it has been allocated already. Now nobody wanted to give away that which they had already received, and that means they had already invested money in it, and so they do not want to have exclusively to give it to satellite or space communications. So what happened with the technical people, we tried to understand the problem of sharing. How could we share the same frequency with different services? This was the first step attempted in 1963, so we have, like most of you people know, that International Intercommunication Telesat Union with Comsat worked on 4 and 6 Gigahertz.

Dr. Sarkar is with the General Directorate of the Swiss PTT, CH-3000 Berne, Switzerland. Earlier during the Dedication Conference he presented a paper in Part I p.87.

The same frequency is also allocated to terrestrial links to regulate with, and we are sharing that frequency. So we developed at '63 this type of criteria -- How to share between these two services. We don't mention interference. In '66, especially in the United States, people realized that because in '65 it was completely established that geostationary orbit is the right way to go for many, many communication systems. Up to 1963 we are not so sure what kind of role geostationary would play, because the problem was this, that the communication delay is that there is always a delay when you send radio signals as high as 36,000 kilometers in distance, and the dealing was of a 400 milliseconds, so many people were afraid that this 400 milliseconds of delay I mean return part, would intervene in the conversations. So people will take more time to adjust. So we were not so sure in '62 or '61 whether the people would accept telecommunication with such delay, but later on both things happened. First of all we were able to put a satellite in the radio geostationary position, that was one of the achievements, and second we saw that people were able to accept such a delay. They very quickly adjusted to it. So the problem started in 1966, how could we go now to include the geostationary orbit position and the frequency spectrum together. This is a new development and since '66, we are from 10 years, and I think we come to a very interesting situation. The situation is this, that you know the frequency spectrum has depolarization. It has also its direction that we can share in time. So the problems we understand about the frequency spectrum: "Now, how do you use the geostationary orbit?" What we did is place the spot beam on a different spot, using the same frequency over and over again So our position is this that also in '66 we were working in the so-called "power-limited" regions. Power limited means that solar power was very little of it at that time, and frequency spectrum we thought we had much. In order to increase general capacity within the band, we felt that we could increase more and more power until we got the maximum. Now, we came to a situation where power is not a problem. The problem is the frequency band, the spectrum available or effective spectrum, not just frequency alone, but how to combine this orbit spectrum with propagation with a spot beam with effectivness total. How to increase it. So now we are in a situation where this combination between the spectrum and orbit combined and how we can make this efficiency higher is that we would like to share this orbit spectrum with many services.

The situation in 1971, there was a space conference in Geneva, or Administrative Conference. This was the first time that the broadcasting satellites got a definition, and in the radio regulation you must have all the service provided as clear-cut definition -- must have -- otherwise we do not know how to work so the first definition was given to direct broadcast satellites. Broadcast satellites as such, at that time people felt that what do we do with the frequency. Now we got a definition, but what do we with the frequency? Now broadcasting as communication, community antenna or broadcasting as the point to point is not a broadcasting. Broadcasting means always a direct -- means direct to people because you must talk to people out in space. In broadcasting it is a point to point communication. So there was some confusion. The people say, well we mean the broadcasting that is broadcasting for the general public. Now here we came to the problem with the legal questions. Now we defined a direct broadcast satellite, which has two characteristics. One would be a community antenna and one would be an individual reception antenna, but the difference between these two is just the smallness of the antenna and the quality that you require, otherwise it is direct broadcast you see Now, later on the problem involved -- in 1963 when we were working, talking in terms of comparing terrestrial systems with the space systems. That means that terrestrial system direct broadcast is always on the AM, just use your side band AM receiver. Now wide band amplitude modulation means that one actually has to have e.g. a 6 Me-

megahertz band, and a lot of power. In space broadcasting we could not do these things because what happens is we could not provide such a high power, or if we have too much higher power then it will interfere with so many other services. So the idea of amplitude modification from the satellite was discarded completely because it simply is not only not possible, it is not desirable, and this was understood in 1971. So we cannot do any kind of direct broadcast from the satellite with amplitude modulation. We can go to digital modulation, we can go to frequency modulation, that means it's all wide band so we make trade off between the wide band and the power. The question now is coming up that when we hear that there is a front end or an antenna which has to convert a frequency mobilization to the end system, and for that we need another device to attach to our home television, so it was quite clear that we have to have whatever broadcasting you need from space. Now, what I wanted to say is that the frequency and orbit began to evolve in a very similar fashion, and it is quite clear that the way the frequency is allocated, the way the frequency has been used, and the way a frequency has been assigned to a station was very clear to the administration concerned. But there was a problem about orbital position of the satellite. Now what happened that perhaps you know is that we have in the radio regulation an article called 9A, and this article 9A provides us with how to go about coordinating when you want to implement any systems which regard radio communications. Now Article 9A especially for radio astronomy and space communication, broadcasting was excluded because broadcasting has to be shared with several different services. At that time the frequency was allocated to these 12 gigahertz band. Now, in the 12 gigahertz band what we said was that it has no limitation in power because we wanted to keep this thing for direct reception for individual time so there is no limitation of power. But at the same time, this band has to be shared with many different frequencies - different services so the problem we have to solve in the planning conference, which is going to be held in January or February of next year is how to plan this 12 gigahertz band use to share direct broadcast satellite services within the region such that in the Americas, space communication and also terrestrial broadcasting, and also terrestrial fixed time in this planning conference of 1977. Now another conference, which is the most important one, and which I think should be one of the biggest conferences for the next 10 to 15 years, will be in September 1979, and will deal with the revision of the entire radio-frequency regulation. This is where we feel, or I personally myself feel, "how to bring this problem of allocation up," or for frequency regulations to bring up the problem of allocation of orbit positions. Now, here is a very interesting case because the administrations are allocating the positions of the orbit, and they now do not need separately coordinate with other nations. If they can coordinate frequency allocation because they would give all the data for the orbital position in advance, they will need special allocation. But in my paper I give...."end of tape!"

There's two definitions that we have to understand. One is the allocation of an orbital position, of anywhere as you see, and which then the assignment of that orbital position you plan to use. The way it goes is that in that orbit frequency, I can say to you, in the World Administrative Conference meaning all the Administrators, these get together and allocate frequencies to different services, and they give the silver bands and this has been agreed on. And when this is done, each administration will assign to a particular station which will be reported to International Frequency Registration Board where it will be recorded. But, there's no such arrangement for the allocation of the orbital position for space services. What I tried last year in the World Telecommunication Forum in Geneva is to say that before we do assign an orbital position, we would have to find a procedure - how to allocate the orbital position before say a satellite has been launched. Now this is a very delicate question because it is a question of, as you said, sovereignty question. Because now what should happen if it is going through then, like in this country or any other country, that administration cannot reserve any orbital position for themselves. They have to leave completely open the international situation. Now in an international situation you have to understand that it's an international way of coordinating. It's a coordination within the administration concerned so here is a very important and very delicate problem, and I talked to several people within our engineering people and our regulator people, and they find it is an area which is political, legal, and also technical. With this statement I would like to conclude my remarks.

Dr. Carl Christol - Moderator, Space Law Session Roundtable

Thank you again Dr. Sarkar for bringing these technical problems, and some of us would be very much interested to pursue this further with you particularly when you use the term "we". I'm not quite sure you mean the International Telecommunication Union or "we" the people of the world community or "we" those who work through the Committee on the Peaceful Uses of Outer Space or just who the "we" is, but certainly we could agree that it's a legal, and a technical, and a political problem. Let me complete my remarks very quickly and then I'm sure that some of the things I have said would engender some responses from those that have been with us.

COMMENTS

Mrs. Eileen Galloway

We have had three satellites, ERTS I and ERTS II, and they changed the name of it. We now have a LANDSAT satellite. All of these satellites are experimental and are running on experimental frequencies. We can't set up an operational system without, you know, changing the frequencies. What I intended to convey is that the experimental data is so successful that there are some countries that are using it as if its operational. What I mean by that is that they...Brazil for example, in studying blight on the coffee crop, coffee tree bushes, trees, whatever they are, and the people in Japan have used the data just the same as if it were operational, but it is an experimental system. Now the distribution, the acquisition of the data, and the distribution of the data is all done by facilities in the United States, that is the Goddard Space Flight Center at Sioux Falls, etc., and the data is sold and there is no difficulty about that. That's going on as usual. The questions that have been raised are raised entirely in the United Nations in the legal Subcommittee by some nations that would like to apply the doctrine of prior consent with regard to the acquisition, and particularly the dissimination of data. So while it's an issue in terms of its being argued in the Legal Subcommittee, actually the fact of the matter, the pragmatic situation is that the data is being acquired, its being assimilated, its being used and its highly successful. Now if were going to have an operational system, Congress would have to vote money for it and you would have to decide in the future, that you were going to have a follow up program.

We have six ground stations and the most terrific pattern of international cooperation, I mean highly successful, among a great many nations. The time is coming when a decision will have to be made as to whether we are going to have money to put into another LANDSAT, and whether it's going to be operational and how it's going to work out.

It was my impression that anything that goes into orbit once around the earth is in outer space. If it doesn't make a complete orbit, then you run into a different problem. For example, the Soviet Union has a fractional orbital bombardment system and when this was used, the vehicle did not make a complete orbit, and at that time the Secretary of Defense announced that it was not in violation of the 1967 Treaty, because it didn't make a complete orbit. Then that caused some difficulty in the Senate because that problem hadn't come up in the hearings, you see, and they felt this had something to do with the demilitarization of space. I remember Senator Cooper was especially distressed because it hadn't been analyzed. The other thing is that often in the treaty, and other official documents, it refers to objects in orbit or beyond, so I don't think legally we should confine ourselves to things that are in orbit, because a space probe goes beyond. In fact, that's the insignia of the American Institute of Aeronautics and Astronautics. You know it has an orbit, then it has a thing going out that way beyond, and that is also in outer space.

Well, if it's in orbit, its in outer space.

COMMENTS

Dr. Charles Stark Draper

A very simple remark, as I listened I thought that perhaps if you substituted the word "stable" for the word "durable" that you would come closer to falling in with something that would be understood by everybody.

Well, if anything or system is in stable motion, it means that it reappears, or follows a systematic path. If it's in unstable motion, it means that it does not follow a stable path but departs in a significant way from the same path, so if you have a body in an unstable orbit, that means it might go up and go around, but it would not stay there, it would come on down. Whereas, if it's in the stable orbit, it at least would do pretty much the same thing but a number of times.

Well, the use of the word orbit implies that the path around the Earth would be a closed path, and if you say "beyond orbit" that means that the object has enough speed so that it would depart, would leave a path that would bring it ever back to the earth again. Of course, these things are ellipses, you see, and if you get going faster and faster, you finally reach a point where you get it away from the earth and it will never come back, in which case it is gone -- it's in orbit. If you like, it's open, and so you wouldn't use that term. The word "orbit" it seems to me implies that it's going to be a closed path.

This is a good distinction all right. It seems to me that the point of issue here is if you have something in a circular orbit, your far side of the ellipse has come together and so you have a perfect circle. That means that you're traveling at all times at the same distance from the sun or the Earth. Your lead ball would have a maximum mass and energy in a minimum drag, whereas if you had a structure that's going to have a lot of resistance to the air and it will pick up enough drag by coming into the lighter atmosphere, so it will do just what you say. This would be kind of a hard criteria to apply unless you arbitrarily took... that's why I'm saying. It's an artificial distinction. This is like saying that we're going with everything between this number of miles and this number of miles, and I'm not sure that is what you're searching for. The distinction between whether you're in an orbit or in space or not in space, I would like to feel it depended on where you were with respect to the atmosphere. If you're in the atmosphere, the air will supply you with something to give you lift for the wings and also give you resistance. If you're outside of this, then somebody will continue to go around by virtue of Newton's laws.

Well, its amount of atmosphere, yes.

Well, but you could pretty well tell how far you had to go up in order to get the atmosphere to the point that it doesn't drop you anymore. The people that come into the Earth from the trips to the Moon know where the atmosphere begins to grab a hold of their vehicle. That's an arbitrary number too, but that's definite, if you want too.

But you see they are coming in from way out yonder and that's going pretty fast and so when they begin to get into the outer fringes of the atmosphere, the atmosphere builds up drag on them and that pushes them back, and they feel it all right.

SPACE LAW DEVELOPMENTS
NEW COMMENTS

Dr. István Herczeg*

The intention of my first response was not a bit more than to wake the interest in the matter of space shuttles and raise some fundamental legal questions which are in close connection with the prospective advent of these aerospace vehicles.

As far as I can see the result of my short paper is encouraging. Many participants of the Roundtable Conference - Prof. De Saussure, Mr. Finch, Prof. Gorove, Prof. Lay and Mr. Menter - recognized the importance of the above-mentioned problems. There is only one participant, Mrs. Eilene Galloway who absolutely denies even the existence of any problem. At first sight her argumentation is disarming. She says: "Spacecraft... went up and down through airspace, without causing legal difficulties. Now that economies can be made by reusing the shuttle, what difference does it make if it returns through airspace and can be used again?"

My answer is self-evident: the difference is very significant. In the moment when space shuttles adopt the <u>functions</u> of aircrafts/transportation, commerce etc./ at once the question arises, which legal regime can be valid for them: that of the outer space elaborated originally for scientific exploration or that of the air-space worked out originally for transportation? And if both of them can be valid where is the delineation between the two legal regimes or whether they can be applied with some unifications?

As we can see the situation after the effective appearance of space shuttles will not be simple at all. In this respect there exists some resemblance with the advent of remote sensing satellites. In the beginning the basic principle of the free use of outer space seemed to solve every related problem. Thereafter one ought to discover that the Space Treaty of 1967 cannot be applied in itself but only in strict accordance with international law including the Charter of the U.N.O.; consequently the sovereignty of States and their exclusive rights over their natural resources cannot be neglected. Now after five years of debates it became clear that remote sensing is no more a technical but chiefly a legal problem.

We can spare at least five years if we recognize in good time that space shuttles will also cause difficult legal problems.

---

*Managing Chairman of the Hungarian Space Law Committee, Director of the International Institute of Space Law, member of the Space Law Committee of the International Law Association.

GREETINGS FROM THE AMERICAN BAR ASSOCIATION

Edward R. Finch, Jr.*

I am reminded of an occasion at the Waldorf Astoria Hotel in the City of New York when at the end of some very long remarks a speaker stood up and he said "My address is 36 West 44th Street, Thank you." Well, I can't be quite that brief because I wouldn't be discharging my official duties tonight to bring you greetings from the international lawyers of the United States, namely the American Bar Association.

We are believers in my committee that it is fully time that the law took up the quality of pioneering which is being done by the scientists in outer space and we are working with some very distinguished members in my committee towards the end that science and law shall begin to march together in the pioneering efforts of outer space.

Forgive three very brief personal words. I feel at home here at Alamogordo, New Mexico because I had the honor to serve my country in Los Alamos, New Mexico for almost two years. Secondly, I refer to the previous speaker's remarks about the V-1 and V-2. I had the honor of being on the receiving end of those particular outer space items. Third, recently my wife and I had the great pleasure of traveling in Africa, to a number of our telemetry stations with a lady by the name of Mrs. Goddard, a very fine lady whose husband has been mentioned at this podium this evening.

If I may close, I would like to leave with you the big four E's that my committee in the American Bar Association are so much concerned with in the world of outer space. I do not need to remind you that since 1967 by a treaty signed and ratified by ninety-three nations of this world, we have the fixed doctrine of the freedom of outer space. But the big four E's that we are working on now are: Energy, Economics, the bang for the buck from outer space, the Environment and last but not least the Ethics of outer space. These are things that I hope our committee in the American Bar Association can begin to accomplish, in the coordination of law, international law, and the progress of international law in the world of outer space for the benefit, the peaceful benefit, of all Mankind.

And, finally, I will close by speaking off the record if I may. We are pursuing the extension of the agreement which will permit us this coming week in California, to meet with ten of the top lawyers from Moscow, concerned with outer space. The agreement I am talking about was made in 1972 and expires on May 24, 1977, and our committee is seeking to extend this agreement which permits the Soviet attorneys and scientists concerned with outer space to work peacefully with their opposite numbers in our United States of America.

Distinguished guests, friends, devotees, and those who love the outer space, thank you for asking the American Bar Association and our committee to be present with you on this important occasion of the dedication of this new space museum. Thank you.

---

*Chairman, Aero Space Law Committee of the International Law Section of the American Bar Association.

COMMENTS

General Martin Menter*

I think we've had quite a very interesting discussion on the meaning of the term "sovereignty" as used in the treaties. I think we're all agreed that jurisdiction clearly is retained in the state that's conducting the space activity. Article VI, of the '67 treaty, talks of the non-governmental entities of the state must have the approval and continuing supervision of the state. I just think it might be interesting for you to know what the United States is trying to do. Our laws are not extra-territorial unless we make them so. So we don't have any provisions of law governing spacecraft activities at this time, except for those individuals who are in the military services. But, we have a bill right now that's being introduced in each of the Congresses and it has the same number. The Congress that's just terminating now, the 94th Congress, it's S1, and it was reported out favorably by the subcommittee which considered it on October 21, 1975. I think it was rather interesting that they defined "special jurisdiction of the United States" to be "an offense is committed within the special jurisdiction of the United States, and it is committed within the special territorial jurisdiction, the special maritime jurisdiction, or the special aircraft jurisdiction of the United States as set forth herein." Then the one part here that's germaine is the special aircraft jurisdiction and it's interesting that this bill defines the term "aircraft" as "any craft designed for navigation in air or space." In other words, any time they talk of aircraft, they're talking about spacecraft too. So it does have provisions governing aircraft and governing what they call -- what we call in the bill -- extra territorial jurisdiction of the United States, and I quote just this first part: "except as otherwise expressly provided by statute or by treaty or other international agreement, an offense is committed within the extra-territorial jurisdiction of the United States. If it is committed outside the general jurisdiction of the United States, and a) the offense is the crime of violence and the victim or intended victim is a U.S. official or a federal public servant outside the United States for the purpose of performing his official duties," and there are quite a few other listings.

The main point I want to bring up in these few minutes that I have, is one I think Dr. Fasan raised. It seems to me that we talked in the '67 treaty about the exploration and use of outer space being for the province of all mankind. We have the language there that states you'll facilitate and encourage international cooperation in scientific investigation of outer space. Then we have in the treaty, this first time in human history, where our people - our astronauts, cosmonauts - are envoys of mankind. This is something different from sailors and airmen. They are just representatives of the particular state but here now astronauts are envoys of mankind. Then in the next treaty, on return of astronauts, we speak of assisting astronauts when by accident or inadvertence, they land in the high seas or in the territory - in the territories of a state other than the launching state. It is the responsibility of the -- of any state that can reach them -- to assist them and return them to representatives of the launching state.

It seems to me that this total concept of a special relationship of the astronaut indicates an intent of the world family of nations, that
----

*Member of American Society of International Law and American Bar Association.

we should forego perhaps by waiver some of the prerogatives that are normally incident to sovereignty of the subjacent state. In construing the law to apply, we always look to analogous fields and in the sea law, we had the doctrine of innocent passage, namely for the land-locked state that's on a river to have its ship get out to the high seas, and vice-versa as long as it wasn't doing anything that was deleterious to any other state, they could have freedom of passage -- innocent passage. And you look to the reasons there and you think about space law, and it just seems to me like Dr. Fasan talks about a small country when the profile of the spacecraft coming back must by its nature transit the airspace of a neighbor state. Well it seems to me that by this concept, in this treaty of '67, and the subsequent treaty on return of astronauts, that we are cooperating with all astronautic activities. We're talking about -- again in the '67 treaty -- the state shall facilitate and encourage international cooperation, and a scientific investigation of outer space. So that when there is a scientific expedition, certainly not harmful to its neighbor, it should waive its jurisdiction, shall I say, and I think not shall waive - I think it has waived. I think there has been an implied waiver by our acceptance of these treaties. Let me quote from a very fine, well respected, space law writer, who concludes: "In summary a good case can be made for the existence at the present time of the customary rule of international space law permitting the innocent passage of space vehicles through national airspace." That was Dr. Carl Christol, who wrote that in an article in September-October, 1965, "An Innocent Passage in the International Law of Outer Space."

SUMMARY AND CONCLUSION
Dr. Stephen Gorove

Let me try to come back just for a moment or two to the question of sovereignty first of all, and then to some of the other questions. I still feel that the semblance of a disagreement here may very well be due to the differences in the definition of terms. I do include in sovereignty just as classical writers did - not only control but also jurisdiction. I think jurisdiction is an essential part of sovereignty. It's not the only part, but control is an equally important aspect. I think you have to have both formal authority and effective control in order to exercise legitimately sovereignty. If you didn't have firm authority, I think the exercise of effective control would be naked power, so you must have formal authority or jurisdiction in order to exercise sovereignty. I have never equated sovereignty and I don't intend to equate sovereignty only with jurisdiction. I think that would be an error. The courts, for instance, do have jurisdiction, but this doesn't necessarily mean that the courts would exercise control directly in a given situation. We may have controls in various contexts, and we may have no formal authority or jurisdiction, in which case, as I've indicated, that would be kind of a naked power. I don't think that I have changed my presentation of what I have said beforehand in relation to sovereignty. Let me perhaps restate that here of what I have said in my written presentation.

I said one aspect that had to be discussed here included the scope of prohibition of national appropriation and any discrepancy which may arise in this respect between the Space Treaty and the proposed Moon Treaty. Another one relates to the notion of sovereignty or more precisely I said, "the extent of sovereign rights in outer space." Then I said, "While there is general belief, particularly on the part of lay audiences, that sovereignty has been completely abolished in relation to outer space, it should be stressed that the Outer Space Treaty prohibits only national appropriation by claim of sovereignty, and not the exercise of all forms of sovereignty, sovereign rights or jurisdiction."

Now, as the subsequent response, I have said this on sovereignty: "I believe we should discuss the question of applicability of sovereignty and sovereign rights in outer space. Personally, I don't find the term "Sovereignty", so characteristic of a previous age, very useful in describing, identifying, or revealing many of the complex relationships of authority and power in the world today. However, the fact remains that the term is still with us and the extent of its applicability in outer space, whether in the form of sovereignty or sovereign rights, or sovereign rights or other jurisdictional aspects, as emphasized in General Menter's fine observation, and Professor Goldie's hypothetical, which ought to be discussed at the roundtable."

And, finally I have said in my last comments, that "apart from the doctrine of sovereign equality, one of the fundamental aspects of sovereignty involves the exercise by the state of supreme authority over people, resources and institutions." Just as I stated, the authority includes both formal authority and effective control. Formal authority or jurisdiction is an essential part of sovereignty, and as I stated, without it the exercise of effective control would be naked power. Then I quoted Article II of the Outer Space Treaty, which I don't want to repeat here, and said that the ban relates to national appropriation by claim of sovereignty, use of occupation or by any other means and not to the exercise of sovereignty by any other means. In other words, a state is not permitted to subject outer space, including the moon and other celestial bodies, to national appropriation by any means including claim of sovereignty, use of occupation. However, Article II - or for that matter any other article of the space treaty, says nothing about banning other manifestations of sovereign prerogatives such as, for instance, the exercise of supreme authority over nationals and national institutions. And then I came and explained the role of Article VIII in this, as well as referred to Article VI which makes the state parties internationally responsible for national activities in outer space, whether such activities are carried on by governmental agencies or on-governmental entities. Also, for assuring that national activities are carried out in conformity with the provision of the treaty. So by what reason or logic could a state be made internationally reliable for the activities in outer space of its own governmental agencies or non-governmental entities if it didn't have the right to exercise supreme authority, including jurisdiction and control over them. It could not. So this is what I stated in relation to sovereignty.

Dr. Christol in his remark also stressed the fact that exclusivity of control was essential, and he also mentioned a concomitant inter-dependence, I think this is very right. The reason really that I am concerned with the use of the term is because it is not very descriptive today. We have these relationships of authority at various levels within the state and also internationally. If you say a state is sovereign, it doesn't reveal too much about the relationships of author-ity and power. For instance, you may talk about countries like Albania or Romania, Czechoslovakia and other countries; if you say this country is just as sovereign as Austria is or Switzerland is, this doesn't reveal anything about what type of power relationship exists within that country, and how are they influenced in the international field. So this is really basically my quarrel with the continued use of the term "sover-eignty". I think we ought to talk about degrees of authority, exercised by whom, in what context, for what particular purpose, then we get a much better and clearer understanding of what really sovereignty is.

Let me just say a word here about the question that Dr. Diederiks has raised in relation to personnel of a spacecraft and the jurisdiction, where it would be. Frankly, at the present time, I don't know whether there is any answer.

Let me just very briefly sum up my final remarks here. I believe that insofar as the question, what would happen to the personnel of the spacecraft or other people in a spacecraft, who have settled in outer space. This is a question which has not been answered by the Outer Space Treaty. I think some regulation may be necessary, of course. Article VIII indicates that the control would not extend the jurisdiction to them if they can no longer be regarded as personnel of the spacecraft. It is true that Article VIII speaks about responsibility of the state, and their continued supervision. So there is some area there which may suggest that the states may wish to exercise or continue to exercise such supervision. I have also listened with great interest to General Menter's remark about the legislation. I would only like to perhaps observe that even though we have no legislation, implementing legisla-tion that is at this time, in relation to jurisdiction and control and supervision, I think the Outer Space Treaty is law of the land at the present time just like any other federal statute, and certainly that enables us to exercise such a jurisdiction. No question that the implementing legislation will be needed if we want to enforce particu-larly criminal type of jurisdiction.

I think with respect to the question of durable orbit, I'd just like to say that my statement really related to indicating what people have said was the customary law in this respect. I have never really referred to it in the context of Article IV that Dr. Christol was referring to, which, of course, is the treaty Text, and there I think Eileen Galloway - Mrs. Galloway - is quite right. You have to have a full orbit, I think, for that particular provision to apply. And anything like a fractional orbit or less than an orbit would not be regarded as the applicable situation.

Let me finally take the opportunity to thank you all for coming here and participating in this roundtable. I think you have given magnificent support with your participation to the success of this meeting. Let me also say that we are most grateful to Dr. Steinhoff's work for organizing this very important conference, and we would like to take this opportunity to express to him in his absence our gratitude for this wonderful meeting.

Thank you very much. With this I would like to declare this meeting adjourned.

CHAPTER VII

IAA HISTORY SYMPOSIUM

ISHF

"There will certainly be no lack of
pioneers when we have mastered the
art of flight...Let us create vessels
and sails adjusted to the heavenly
ether, and there will be plenty of
people unafraid of the empty wastes."
    Johannes Kepler, 1610 (Poland)

SELECT BIOGRAPHIES OF OUTSTANDING SPACE PIONEERS

Symposium

International Academy of Astronautics

Organizing Co-Chairmen: E. M. Emme (U.S.A.), and
V. N. Sokolsky (U.S.S.R.)*

INTRODUCTION

Three sessions of the History Symposium (biographies) of Pioneers of Space Flight are sponsored by the History Committee of the International Academy of Astronautics. For the Proceedings, these biographies are arranged in rough chronological sequence in the course of history.

It is important to give recognition to persons who made space history, from the first individual to formulate a concept, to those who expand its scientific meaning, develop rocket and other technology, and translate it into practical application. There should have been a session at the International Space Hall of Fame Dedication, if one may pardon the prejudice of historians, devoted to the great amount of space history made in New Mexico--Robert Goddard, Trinity Site, White Sands Missile Range, Holloman AFB, the Lovelace Clinic, and all the rest.

Founded in 1961, the History Committee of the International Academy of Astronautics was intended to fulfill a function envisioned at the birth of the Academy itself--to preserve and promote the history of rocketry and astronautics on an international basis. The challenge of attainment of man's exploration of outer space had deep historical roots of an obvious international character. Science, engineering, and achievement in the evolutionary story of mankind, the accumulation of all knowledge itself by the passengers of Spaceship Earth, require that the history of rocketry and astronautics be nurtured on an international basis. What was difficult to explain then is now widely appreciated.

The history of science and technology is not a well-attended vineyard among the intellectual disciplines. In a sense, it is an interdisciplinary methodology of all that has transpired in the past. There was little agreed upon in the way of assumptions for undertaking the work of the History Committee on the I.A.A., a few historians, engineers, and scientists participating. It was decided that rather than establish an institute, annual symposia on the available contributions to the history of rocketry and astronautics on an international perspective was a way to get serious attention. The First Symposia of the I.A.A. History Committee on "Contributions to the History of Rocketry and Astronautics" was held in conjunction with the I.A.F. Congress in Belgrade in 1967. They have been held annually since then, and the early symposia have or are being published in both English and Russian languages. The tenth annual I.A.A. History Symposium on "Contributions to the History of Rocketry and Astronautics" will be held in conjunction with the 27th Congress of the International Astronautical Federation meeting in Anaheim, California next week.

When the International Academy of Astronautics was consulted by the organizers of the International Space Hall of Fame in Alamogordo, New
----

*Co-Chairmen, History Committee, International Academy of Astronautics.

Mexico, it was entirely appropriate that the History Committee of the I.A.A. was requested to contribute a special symposium consisting of biographies of outstanding deceased space pioneers at the Dedication Conference. It was the Governor's Commission of the ISHF, however, which selected the nominees and not the I.A.A.

How were the pioneers selected, the biographies of some of which constitute this special symposium of the I.A.A. History Committee? The process was initiated over a year ago when the Committee attempted to prepare a comprehensive list of space pioneers from all nations, and for the entire scope of the history of rocketry and astronautics. It was wisely decided that only deceased pioneers should be considered in this initial process. A master preliminary list was agreed upon as a starter. It contained the names of obvious outstanding innovators, engineers, and scientists, as well as "dreamers" and "activists," who had made historic contributions.

It proved difficult to find many students of history who know the full history of pioneers in their own nation, and fewer still are well grounded in the history of space exploration internationally considered. Participants are best informed about their own experience within finite parameters of time and place. From the I.A.A. list, it was possible to ascertain that it was perhaps incomplete as certain pioneers are better known for their outstanding achievements. Americans now know about Tsiolkovsky just as Soviet engineers have learned about Goddard and Oberth and others. But other pioneers are not known. So it was possible to select from the I.A.A. list the names of pioneers for whom a student of the full history might prepare a solid biography for presentation at the International Space Hall Conference.

A major problem in seeking to give recognition to outstanding pioneers in the history of astronautics was a central fact: many of the most important in the more recent period of gigantic astronautical accomplishments are not deceased. When one regards all that has been accomplished in the recent period since Sputnik began the "space age," the early work of previous decades since Galileo by a few has blossomed into the life work of hundreds of thousands. The History Committee of the I.A.A. has, however, sustained the assumption made at the outset that only those deceased would be included in its "Select Biographies."

Biographies of pioneers of astronautics should indeed help contribute to better understanding of how mankind got underway in seeking its destiny beyond planet Earth. Most important, it may be appreciated that this evolution had an international genesis, and was derived from the contributions of outstanding individuals in many nations. And, as will also be evidenced, in this international consideration of space work, that several of the outstanding pioneers worked in New Mexico.

THE CREATIVITY OF ROBERT H. GODDARD
1882-1945

F.C. Durant, III*

Robert Hutchings Goddard is America's rocket pioneer. In our national bicentennial year we commemorate also the fiftieth anniversary of the launching of the world's first liquid propellant rocket. It was the first flight of several dozen rocket-powered test vehicles that Goddard would conceive, design, build and test in his lifetime. The rocket had a thrust of less than 10 lb., but on that cold afternoon in a field near Auburn, Massachusetts, Goddard's first liquid propellant rocket lifted out of its launch stand on March 16, 1926. It rose to a height of 41 ft., curving over to land 184 ft. away. The propellants were gasoline and liquid oxygen. Forty-three years later the multi-engine, three-stage Saturn 5 rocket launched three men to the Moon. The basic principles involved -- the use of metered liquid bi-propellants forced into a combustion chamber to burn and exit at supersonic speed through a tapered exhaust nozzle, and the concept of multiple, or staged, rockets -- were detailed in two U.S. Patents issued to Robert Goddard in July 1914. In 1926 these principles were demonstrated in actual flight.

These two patents, incidentally, were the first of 48 to be issued to him during his lifetime. Thirty-five others which he had applied for were issued after his death. An additional 131, based upon his notes, sketches and photographs, were applied for by his widow and issued as late as 1956. In 1960 the U.S. Government paid the sum of $1,000,000 to acquire the rights to use these 214 patents.

Robert Goddard devoted his entire professional career to developing the technology of rocket propulsion for the achievement of high altitude flight. He never considered any one rocket was a complete solution to design. He wrote in 1936:

> The inquiry is ... sometimes made as to when the research will be completed. The answer is that it will probably be the same year the automobile and the airplane are completed. It has not the finality of building an individual machine, but is a new method of transportation, which I feel certain will have many more applications than the sending of recording instruments into the high atmosphere, important as this phase of the subject is.

Between 1926 and 1941 he conducted dozens of static tests of rocket motors and thirty-five rockets of increasing sophistication were launched successfully. During those years he refined his designs of sounding rockets, developing turbopump systems; gyro-stabilization; aerodynamic and jet-deflector flight controls; automatic, sequencing launch systems; flight trajectory tracking and recording devices; gimbal-mount clustered

----

*Assistant Director, National Air and Space Museum, Smithsonian Institution, Washington, D.C. A past president of the American Rocket Society and International Astronautical Federation, Mr. Durant has authored numerous articles on the history of rocketry and space flight. This paper prepared for presentation at the Dedication of the International Space Hall of Fame, Alamogordo, New Mexico, October 1976.

2. April 19, 1932. First rocket to be gyro-stabilized with flight corrections by both aerodynamic vanes and rocket exhaust jet deflectors. R. H. Goddard is second from right.

1. Robert H. Goddard stands beside rocket in launch frame shortly before the first flight of a liquid propellant rocket at Auburn, Mass., March 16, 1926.

3. Gyro stabilization device located in structural frame of April 19, 1932 rocket. Storage batteries provided electric power for solenoid-operated, pneumatic servo units connected to flight controls.

4. Goddard, center, flanked by friends and supporters, Harry Guggenheim and Charles A. Lindbergh, September 15, 1935.

rocket motors; parachute recovery -- and many other techniques later to be refined by others. But he did it first!

Born 5 October, 1882, and educated in Worcester, Massachusetts, Goddard received his B.Sc. degree from Worcester Polytechnic Institute in 1908 and his M.Sc. and Ph.D. degrees from Clark University in 1910 and 1911. He conducted research at Princeton University (1912-1913) on a post-doctoral fellowship, but for reasons of health returned to Worcester. Majoring in physics throughout his educational training, Goddard embraced the academic life and taught physics, first as an instructor and soon after as Professor, when he was not engaged in research on rockets.

From 1917 to 1929 Goddard's research on rocket development was supported by the Smithsonian Institution. From 1929-1941 increased funding was provided by Daniel Guggenheim and the Daniel and Florence Guggenheim Foundation for the Promotion of Aeronautics. Initiators of this support were Charles A. Lindbergh and Harry F. Guggenheim. These funds made possible a move in 1930 to Roswell, New Mexico, where his most advanced development and test work was conducted.

Goddard clearly saw that rocket power had potential for both peaceful uses and for weapons. In both World Wars he offered his services and creative capabilities to the armed forces of his country. During World War I he developed several rocket propelled devices demonstrated before U.S. Army observers a few days before the signing of the Armistice. One of these weapons was to become, after further development by one of his colleagues, C.N. Hickman, the well-known World War II "bazooka" antitank rocket. His prescience and ability to visualize applications of rocket power were many. In 1933, considering possible weapons, he postulated an antiaircraft, rocket-propelled missile with infrared homing guidance. From December 1941 until his death, August 10, 1945, Goddard developed Jet-Assist-Take-Off (JATO) and variable thrust rocket engines for the U.S. Navy.

Webster defines "creativity" as "To bring into being; to cause to exist". As a physicist, Goddard understood fundamental Newtonian mechanics and thought deeply on applications of these principles for flight to the Moon and the planets -- even to the stars. Because such matters were not to become "respectable" for some decades, Goddard did not publicize such thinking. Instead he set down his thoughts and ideas in memoranda and reports to be preserved for the world to read later.

Charles Lindbergh wrote:

> Robert Goddard's dual approach to astronautics is a fascinating study, now cautious and realistic, now adventurous and fictional. On one day he would be a conservative scientist, confining the objective of his "high-altitude research project" to measurements relating to the earth's upper atmosphere. On another, he would let his mind run freely through fictional accounts of interplanetary warfare, or put down in writing new ideas for human emigration to distant areas of space.

In 1959 Wernher von Braun stated:

> Dr. Goddard's contributions have provided a solid basis for the progressive development of rocketry as the means by which to achieve his shining ambition -- the exploration of the space regions where the silent planets, stars, and galaxies await the adventurers who follow in his giant footsteps.

5. Goddard beside rocket sent to Smithsonian Institution in November, 1935. The rocket is on exhibit at the Smithsonian's National Air and Space Museum, Washington, D.C.

6. This August 16, 1937 flight was the twenty-seventh successful launching of a liquid propellant rocket by Robert H. Goddard.

7. View of largest and most sophisticated rocket design built by Robert H. Goddard. This rocket, 18 in. diam., 22 ft. long, is on exhibit at the Smithsonian's National Air and Space Museum, Washington, D.C.

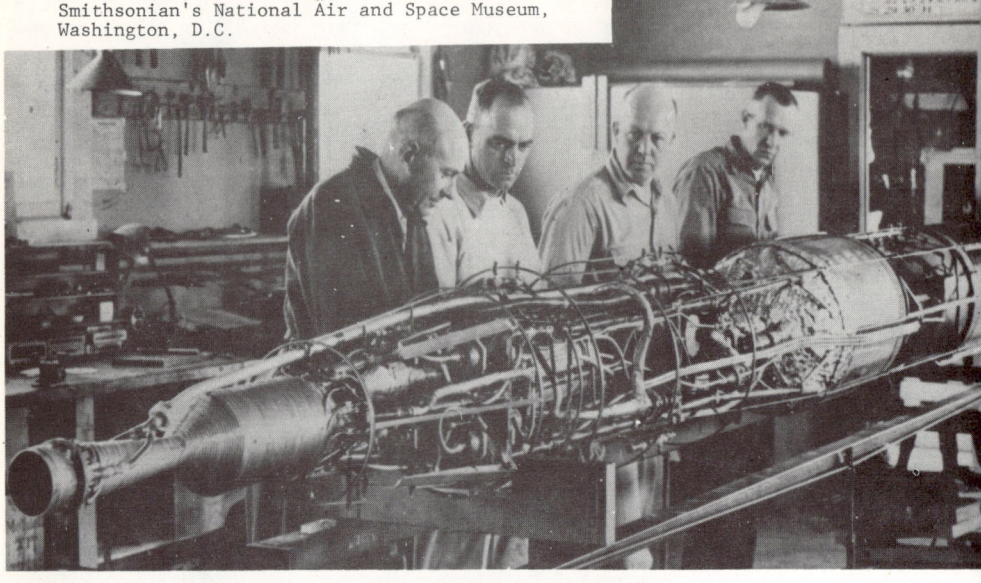

8. Robert H. Goddard, American rocket pioneer.

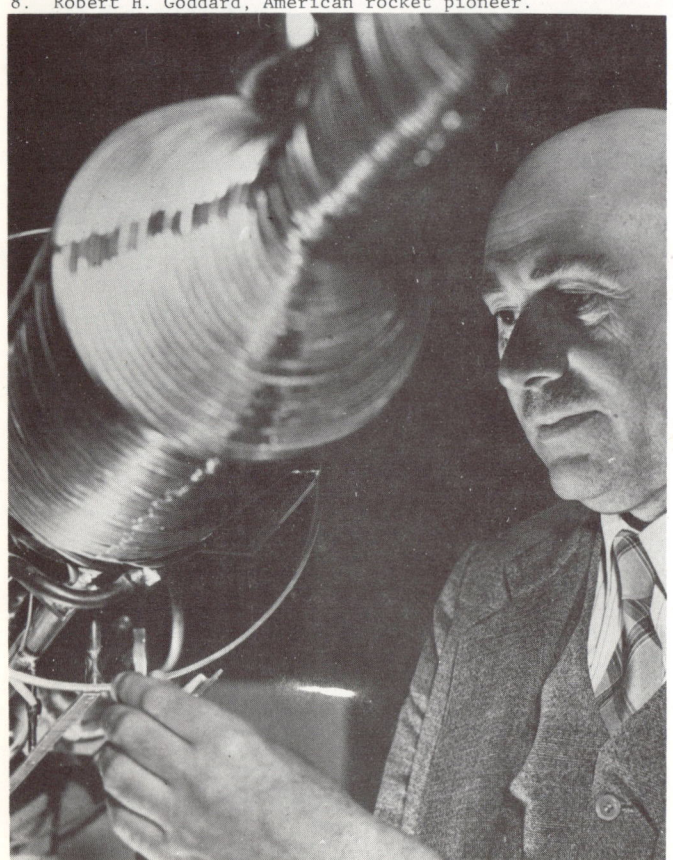

Robert Goddard was a trail blazer in the development of rocket technology for the achievement of space flight. He did not live to see man walk on the surface of the Moon, spend months in earth orbit, or the exploration of the surface of Mars. His writings and his lifelong pioneering work, however, anticipated the day when such exploration would be achieved.

REFERENCES

1. Durant, F.C., III, <u>Robert H. Goddard and the Roswell Years</u>, presented at Seventh History of Astronautics Symposium, Baku, U.S.S.R, October 1968, to be published.

2. Durant, F.C., III, <u>Robert H. Goddard and the Smithsonian Institution</u>, presented at Second History of Astronautics Symposium, New York, October 1968, published in <u>First Steps Toward Space</u>, US GPO #4705-00011.

3. Goddard, Esther C. and Pendray, G. Edward, eds., <u>The Papers of Robert H. Goddard</u>, 3 vols., McGraw-Hill, New York, 1970. This work is outstanding; a rich source of information and reference.

4. Lehman, Milton, <u>This High Man</u>, Farrar, Straus and Co., New York, 1963; a popular biography of R.H. Goddard.

WILLIAM RANDOLPH LOVELACE II
(1908-1965)

Jacqueline Cochran*

Those of us who knew Dr. W. Randolph Lovelace II, the late President of the Lovelace Foundation for Medical Research in Albuquerque, feel it highly appropriate that he be inducted into the International Space Hall of Fame in Alamogordo, New Mexico. Dr. Lovelace, who, with his wife Mary, died in a plane accident in 1965, had made many contributions to the United States aerospace program and to aviation and aerospace medicine. Pilot, physician, surgeon, research scientist, humanitarian, . . . Randy Lovelace was undoubtedly one of the great pioneers in our nation's aviation and aerospace history.

Randy Lovelace's life spanned fifty-eight years--years of great change and great accomplishments. Born in Springfield, Missouri, Randy Lovelace was only six months old when his parents moved to Sunnyside, New Mexico, which is now known as Fort Sumner. It was in Sunnyside that Randy's beloved "Uncle Doc" (for whom he was named) established his first medical practice and became one of New Mexico's medical pioneers. Randy's parents followed Uncle Doc to Albuquerque a year after Doctor Lovelace I transferred his practice there in 1913. Some five years later, his parents were divorced and Randy, then a boy of 11, went to live with his Uncle Doc and his grandmother in a magnificent home just off Central on 12th Street.

It was probably this early exposure to medicine that captured young Randy's attention for he often accompanied his uncle on house calls. In 1922, Dr. William R. Lovelace I, and another uncle Dr. Edgar Lassetter, joined practice to form the Lovelace Clinic. Three years later, Randy graduated from high school and went on to St. Louis where he enrolled in his uncle's alma mater, Washington University. That was the beginning of 11 years of liberal arts and medical education. In 1933, two important events happened; Randy married Mary Moulton, herself from an old New Mexico family, who had just graduated from Northwestern University in Evanston. The second event was Randy's transfer from Washington University to Harvard Medical School where he received his medical degree in 1934. A surgical residency at Bellevue Hospital in New York followed. The famed Mayo Clinic in Rochester, Minnesota, offered Randy a Fellowship in surgery, and in 1936, Randy and Mary moved to Rochester. The idea for the Lovelace Clinic came in large part from Dr. William Lovelace's longstanding friendship with Drs. William and Charles Mayo. This friendship began back in 1915 when Dr. Lovelace I was made a life member of the Mayo Surgeons Club. Randy's enthusiasm about going to the Mayo Clinic was well understood.

During his six years in Rochester, all the elements and patterns of Randy's future began to coalesce. He received his M.S. in Surgery from
----

*American aviatrix who attained more records than any other woman in the history of flight, and was a sponsor of laboratories and foundations, including Lovelace Clinic in Albuquerque, NM. Paper read by Brig. General Charles Yeager, USAF Ret., first pilot to attain supersonic flight.

the Mayo Foundation and the University of Minnesota in 1939. In that same year he was awarded a Mayo Foundation scholarship for the study of surgery abroad. This sent him to the various centers of surgery throughout Europe and aviation medicine centers, particularly in Germany. With Europe moving ever more rapidly toward an active war, he and Mary came home late in June prior to the blitzkrieg against France and the Lowlands. In 1942, Randy became Staff Surgeon and head of a surgical section at the Mayo Clinic, a position he held throughout the war, even while he was on active duty with the Army Air Corps.

Randy Lovelace's interest in aviation also began early in his life. At Mayo, he came under the influence of Dr. Walter M. Boothby, a long-time specialist in military medicine and, by then, a pioneer in aviation medicine. In 1937, shortly after joining Mayo, Randy had attained the rank of Flight Surgeon and First Lieutenant in the Army Medical Corps reserve. Later that year, at the National Air Races in Cleveland, he first met Jaqueline Cochran and her financier husband, Floyd Odlum, both of whom were to play an important part in the establishment, ten years later, of the Lovelace Foundation for Medical Education and Research in Albuquerque. In 1937, Jacqueline Cochran was already a well established aviatrix, having won the International Harmon Trophy as the outstanding woman aviator in 1937. She was to win the International Harman Trophy twelve more times after that In 1938, she won the Bendix Race against a field of nine men. It was in 1938 that Randy met the equally famous aviator, Jimmy Doolittle, who discussed with the young physician the problems which pilots were encountering as they flew newer planes to ever increasing altitudes. A short time later, the Aeromedical Laboratory at Wright Field requested Dr. Boothby, Dr. Lovelace, and Dr. Arthur H. Bulbulian to develop a high altitude oxygen mask. Randy then became a dollar-a-year consultant to the Army Air Corps. It was Henry H. Timken, Jr., of Dayton, Ohio, and the Timken Ball Bearing Company, who used his single engine plane to test the mask developed by both Lovelace and Bulbulian at 15,000 feet. Mr. Timken later became a major financial contributor to the Lovelace Foundation. Indeed, the mask was later named BLB, after the initials of its developers, and is now a part of aviation history.

When Randy Lovelace returned to the Mayo Clinic after his trip to Europe, he and others expanded on their work on pressure chambers and cabin pressurization. In February, 1942, Dr. Lovelace went on leave from Mayo and into uniform as a full time Major in the Office of Air Surgeons Headquarters, Army Air Corps. When he retired from active duty, he was a full Colonel, and was made a Major General posthumously in 1966. In the spring of 1942, Dr. Lovelace visited the Pacific, Alaskan, and Mediterranean Theatres of war to learn first-hand the flyers' needs, which came within the scope of aviation medicine.

A year later, on June 24, 1943, Lt. Col. W. Randolph Lovelace II dropped through the bombay doors of a B-17, 40,200 feet above the wheatlands of central Washington state to test bailout equipment and procecures. This was Randy's one and only parachute jump--he had never jumped before and he never jumped again. Randy Lovelace had made the experimental jump to personally experience and determine what was happening that caused injury and death to pilots forced to abandon their planes at high altitudes. There are some indications that the Army Air Forces had not been informed of the intention of the jump, and in fact, the War Department Bureau of Public Relations only became involved with after-the-fact handouts, because they had not been notified about the planned experiment in advance. The jump made national headlines, and Life and Look magazines, the two major photo-journalism publications of that day, carried pictures and a text with the result that Dr. Lovelace attained a hero's status.

What did actually happen when Randy went through the bombay doors? First, a cord attached to the plane pulled the chute open immediately. It was thought that there would be less shock if the chute was opened in the thin, high air, rather than if the airman dropped down to thicker air before opening it. But the opposite proved true. Randy suffered 40 or more G's of deceleration when the chute opened and was knocked unconscious. The violent pendulum swings of his body below the chute nauseated him, and his hand was deeply frostbitten after his gloves were torn off by the initial shock. Fighting his way back to slight consciousness at 8,000 feet, he eventually landed hard in a field of wheat stubble.

In September of 1943, Lovelace was advanced to Chief of the Aeromedical Laboratory at Wright Field, and in that capacity supervised a staff of 225 who were engaged in research in the various aspects of aviation medicine. He supervised studies aimed at eliminating effects which would adversely affect and limit capabilities of flying personnel. He supervised development and testing to many items of life support, including oxygen equipment, and airplane ambulance equipment. During his work at Rochester and his wartime work at Wright Field, Dr. Lovelace explored all known stresses affecting flyers under then-current conditions and helped arrive at solutions, thus establishing a strong base of knowledge from which to look into the future to anticipate the stresses on flying personnel and the equipment needed for high speed, high altitude aircraft. From that point on, it was not a difficult projection to assess and visualize the requirements on men orbiting in space capsules.

After the war, Randy again returned to the Mayo Clinic. By that time, his family included a daughter and two sons, and it was in 1946 that tragedy struck twice. Little Randy III was five and one half years old when he contracted polio in Rochester and died on July 7. A short time later, the youngest boy, Chuck, came down with the same dread disease, was paralyzed, and died on August 13. The family returned to Albuquerque and Randy Lovelace, overcome with grief, decided to not return to the Mayo Clinic, and two other daughters were later born in Albuquerque.

It was then that plans for a Lovelace Clinic patterned after the Mayo Clinic emerged and Randy, his Uncle Doc, and Dr. Lassetter formed a partnership. In the fall of 1947 the Lovelace Foundation for Medical Education and Research was born. Its first Chairman was Floyd Odlum, husband of Jaqueline Cochran. Dr. William Randolph Lovelace I was its first President. Randy served as Vice President of the Foundation until he moved up to the Presidency in 1963. The first Foundation building, often referred to as the Lovelace Clinic Building, was dedicated on November 5, 1950, on the present Gibson Boulevard site of Lovelace Bataan Medical Center. Following the dedication, those attending moved east to an adjoining ten-acre site which had been donated by Uncle Doc Lovelace to the Methodist Church where ground was broken for Bataan Memorial Methodist Hospital which opened two years later in 1952.

As the Lovelace Clinic and the Lovelace Foundation prospered and personnel were added, Randy's interest in aviation medicine remained constant. In 1947, he became Chief Medical Officer for Trans-World Airlines, and later a consultant to other airlines and aircraft manufacturers. In 1951, with the Korean War fully underway, Randy accepted an appointment as Chairman of the Armed Forces Medical Policy Council to the Secretary of Defense. In company of the younger Dr. Charles Mayo, Randy traveled 30,000 miles across the continent inspecting medical installations of U.S. Armed Forces and of our allies. For several years, Randy literally commuted between Washington, D.C., and Albuquerque.

In 1952, Randy helped found the Aerospace Medical Panel of AGARD, an advisory group for aeronautics research and development, of the North Atlantic Treaty Organization (NATO). When Sputnik orbited the earth in 1957, the United States hurried efforts to answer with the Explorer. Suddenly, the accent was on rockets, satellites and space, and Dr. Lovelace's interest by then centered on space medicine. In April of 1958, Dr. Lovelace became Chairman of the Working Group on Human Factors in Training of the Special Committee on Space Technology of the National Advisory Committee on Aeronautics. Serving with him on the eight-man committee were Dr. Ulrich Luft of the Lovelace Foundation and Dr. Wright Langham of Los Alamos. This group performed a study exploring the working environment of placing man into space, and later in that year, Randy became Chairman of the Special Advisory Committee on Life Sciences for Project Mercury of the National Aeronautics and Space Administration.

Dr. Lovelace and his associates helped design, develop, and program the extensive series of tests to be given to the 32 military test pilots from whom the seven Mercury astronauts were selected. The tests were conducted on the pilots at the Lovelace Clinic and Lovelace Foundation in Albuquerque. It was in the Lovelace livingroom in Albuquerque that a then-unknown Lt. Col. John H. Glenn, Jr., said, "I didn't know the human body had so many openings to explore, probe, poke, jab, squeeze and scrutinize." From the years of the testing of the original astronauts to the day of his death, December 12, 1965, Dr. Lovelace continued his medical and surgical research activities at a pace that can only be described as truly remarkable. He continued to travel widely and consult on various aspects of aviation and space medicine, and at one time, in addition, even served as President of the Air Force Association. He was only 57 when he died in the crash of a private light plane which was planned to carry him and his wife Mary from Aspen, Colorado, back to Albuquerque. His achievements, his honors, and his contributions to aviation, aviation medicine, and to the newer science of aerospace medicine are too numerous to recount at this time. To this day, the Lovelace-Bataan Medical Center reserves a special row of showcases which display many of his awards, citations, and cherished photographs, and these continue to attract the attention of new and old patients.

In closing, may I quote from a tribute to Dr. W. Randolph Lovelace II written by Brig. General Ernest A. Pinson, Commander, Office of Aerospace Research, United States Air Force. Said General Pinson, "While Randy possessed many commendable talents and characteristics, there were two of these for which I had special admiration. The first of these was courage. He had the mental courage to believe firmly in his convictions and the physical courage to act upon them. He never let the effort required by, or the hazard involved in, any necessary action, deter him from fulfilling the absolute full measure of his responsibilities. The second of these was foresight. He had a rare talent for visualizing the significant actions required to be done now - for future needs. I believe this characteristic, more than any other, was responsible for his greatest contributions to science, to his country, and to mankind."

ROBERT ESNAULT-PELTERIE
1881-1957

Edmond A. Brun*

It is difficult to say what stands out in the personality of Robert Esnault-Pelterie: His originality as a physicist? His skill as an experimenter? His knowledge as an engineer? His ability as a builder? Esnault-Pelterie was certainly an inventor, but his approach to invention was always scientific. He made this point clear in the forword to his classic book, l'Astronoutique (1930):

> In the early days of aviation, ignorant but daring inventors sometimes took the lead over more methodical investigators... True, at this time, the science of aerodynamics did not exist... The beginnings of astronautics took place in an altogether different fashion; science provided the means of studying nearly all of the questions connected with the problem beforehand; empiricism could do nothing here... It is our duty to warn inventors by bringing out the many questions at stake and the only method for tackling them: the scientific method.

And before any other development in the book, he presents the theory of the rocket, and studies its movement in a vacuum and in the air.

Robert Esnault-Pelterie was born in Paris on November 8, 1881, in a family of industrialists. After initial studies at the Lycée Janson de Sailly, then at the Sorbonne, where he obtained a Bachelor's degree in Physical Science, he turned to the problems of aerodynamics which were then being raised by hopes for the conquest of space.

In 1900, he built a kite with a surface area of 18 m^2 which enabled, him to study the aerodynamic zone of wind at variable angles of incidence. In 1904, when the Wright brothers had just flown their first plane, he built a biplane, based on the model recommended by the French-American Octave Chanute. He was soon to replace warping of the aerofoils by mobile wing-tips.

In 1905, he resumed testing in a more logical manner by installing aerofoils on a test car which could reach a speed of 100 kilometers per hour, which was quite high for the time. His experiments enabled him to determine the proper form for tail rudders, demonstrate the advantage of profiling tubes and stayings, etc.

He then began to study and build his first "aeroplane" which was finished in 1907 and which he piloted himself. In order to do this:

- he devised a single-lever control to ensure that the plane could be handled by the wing-tips and diving rudder, since known as a "broomstick"; the arrangement

---
*Vice President of the International Academy of Astronautics and Director of the Laboratory of Fluid Dynamics, Paris, France. Paper read by Michelle Pige, Associate Secretary of the International Academy of Astronautics, Paris, France.

met with worldwide success because of the simplification it made in piloting;

- he developed the oleopneumatic brake, which eased the shock on landing and kept the plane from rebounding;

- in 1906 he studied and built the first light aero-engine, a radial engine with an uneven number of cylinders and a single cam, established the theory of stability for it.

As soon as he was sure that his planes could fly, Esnault-Pelterie had a veritable factory built in Billancourt, probably the first one in the world devoted to aviation. It is from this factory that various types of engines and aircraft used by the French and foreign armies during the war of 1914-1918 came.

He personally undertook flight testing for the planes; one of the flights in Toussus-le Noble, in June, 1918, would have had a tragic ending if he had not taken the precaution of fastening himself to the seat with an elastic belt!

Pioneer of aviation, Esnault-Pelterie was one of the four great forerunners of astronautics, along with Goddard, Oberth, and Tsiolkovski. He had long been struck by the mistake of Jules Verne, and his thoughts had led to the conviction that "the only solution for interplanetary travel would be rockets." The first trace of his ideas is found in a book (1908) by the French Air Force Captain Ferdinand Ferber, "From peak to peak, town to town, continent to continent," in which the author developed the idea of a jet engine, linking Archedaion, Esnault-Pelterie, and H.G. Wells to his reflections.

In this era, however, space travel was considered a utopian dream. In a paper presented to the French Physics Society on November 15, 1912, Esnault-Pelterie was forced to conceal his real aim under a harmless and curious title: "Considerations on the Results of Indefinite Lightening of Engines." Further difficulties arose when the paper was to be inserted in the Journal de Physique which served as a bulletin for the Society, and only an incomplete, abridged form was printed.

Following the 1912 paper, the Belgian Andre Bing contacted Esnault-Pelterie to inform him of a patent he had taken out on June 10, 1911, for "a device designed to make possible exploration of the upper reaches of the atmosphere, however rarefied." In spite of the lack of thorough calculations, Bing's patent allowed for the possibility of reaching an almost indefinite altitude with a system of successive rockets, each of which is abandoned as soon as it is consumed; this is precisely the principle which was later developed by Goddard.

It was learned that the latter's work had started in 1912 at the University of Princeton, and went on at Clark University in 1915-1916, where he experimented with ejecting rockets in both open air and a vacuum. In 1919, Goddard sent Esnault-Pelterie his complete works, published by the Smithsonian Institution.

Esnault-Pelterie stated that he had not known of any other work when he gave a lecture on "Rocket Exploration of the Upper Atmosphere and the Possibility of Interplanetary Travel" to the Astronomical Society of France on June 8, 1927. It covered results from as early as 1920, and showed that attainable ejection speeds made it possible to make the ratio of mass almost acceptable. The paper also included the theory of gas expansion in a diffuser nozzle.

Following the lecture, he was informed of the German work of Hermann Oberth (1923-1929) and Hohmann (1925). In 1928 the first international prize, created by Esnault-Pelterie and his friend Andre-Louis Hirsh for the best work pertaining to astronautics was awarded to Hermann Oberth. In 1939, the last prize was given to a young American, Frank Malina, who later founded Aerojet Engineering Corporation with Theodore Von Karman, and others.

In 1930, Esnault-Pelterie published his masterpiece, l'Astronautique in which the conditions for effecting a trip around the Moon, making it possible to photograph the Moon's hidden side, are already given. Without attempting to list the contents, it should be pointed out that the following items were covered in the book and its 1935 complement:

- thermochemical studies and the first idea of using free radicals for propulsion (anticipating the use of atomic hydrogen);

- considerations on nuclear propulsion, even though fission had barely been discovered in 1930;

- aerodynamic studies for movement in the lower and upper atmosphere;

- calculations for orbital routes and what are now called transfer orbits;

- means of piloting and navigation, with a comparative study of various designs (based on viscosity, gyroscopics, pendulum);

- the idea of photonic propulsion, applying the law of relativity;

- living conditions aboard a spaceship, with regard to temperature and apparent weight.

After this veritable treatise on astronautics, conditions were ready to carry things out.

At the time, solid propellants were less powerful than liquid mixtures and required higher combustion pressures. A liquid propellant rocket, which was furthermore easier to handle than a solid propellant rocket, would therefore be used.

Esnault-Pelterie quickly realized that practically nothing could be done without studying new fuels. In his search for the best combination of fuel and oxidant, he tested tetranitromethane, a powerful oxidant. One of the tests went wrong; he lost four fingers, but was not discouraged. He turned to a mixture of liquid oxygen and petroleum ether.

Everything still remained to be done in the actual construction of the engine: jets, mixers and pumps had to be studied. In order to find the right dose of fuel and oxidant, he was thus led to study a three-phase cam pump, driven by a turbine located in a suitable part of the thrust nozzle, and, in order to prevent cavitation during the induction phase of relative liquids like liquid oxygen, he thought of a precompression centrifugal pump.

In the early part of 1932, a well-equipped static testing center was set up in Satory. At the outset, the propellants which were tested developed a thrust of the order of 100 kg, and firing testing was more and more encouraging.

The silent partners, however, grew impatient at not seeing any rockets take off; with his concern for perfection, Esnault-Pelterie refused to launch a rocket which was not equipped with a gyropilot. In addition, he became bitter about the lack of interest in his astronautical work, in the official world as well as among men of science. The war of 1939 soon put a definitive and infinitely regrettable end to the astronautical activity of this great engineer, a member of the Academy of Science and the Institute of France since 1936.

Robert Esnault-Pelterie died in late 1957, two months after the launching of the first man-made satellite. He had the satisfaction of seeing his views borne out and the dream of his life-time achieved.

Robert Esnault-Pelterie

1881-1957

SIR WILLIAM CONGREVE
(1772-1828)

Frank H. Winter*

BACKGROUND

The name of Sir William Congreve is inseparable from the thousand-year history of the rocket. This fact was internationally recognized in 1970 when the International Astronomical Union designated a crater on the far-side of the Moon in his honor. Congreve Crater, situated at 0°, 168° West, pays tribute to the man who virtually reintroduced the rocket into Europe during the Napoleonic period after centuries of neglect. Congreve may also be said to have established the first modern solid-propellant rocket technology in that he initiated the mass production of rockets, significantly altered and improved their design and performance, greatly improved black powder propellant formulations and processing, and established an entire "rocket system" and launching means, by both land and sea, for calibers of from 3 lbs (1.36 kgs) to 1,000 lbs (453.59 kgs).** Congreve was also one of the first to adapt the newly invented parachute (claimed by Jean Pierre Blanchard in 1785) to the rocket, though a kind of wind sock arrangement had been tried at least once, much earlier.

Through his influence Congreve also precipitated some of the first mathematical studies of rocket flight and the possibilities of manned flight by rocket. "Congreve's influence," wrote Willy Ley, the famed rocket historian, "was enormous." From forty to fifty books were written on Congreve rockets during the 19th century and thirty or more countries adopted his rockets--and rocket technology. Judging from dictionaries of the period, Congreve's name also entered several languages. In short, Congreve became synonymous with the rocket and vice versa.

Prior to Congreve's first studies of the rocket which began in 1804, this projectile was nothing more than an object of amusement in firework displays and a night signal upon the battlefield and over the sea. Rockets are generally believed to have originated in China during the Sung Dynasty, ca. 960-1279 A.D., and appear to have spread into Europe by the 14th century, probably via trading ships entering northern Italian ports. It's academic whether gunpowder and guns preceded the rocket or not. What is important is that almost as soon as it appeared the rocket was cast aside and its technological development neglected. In contrast, the gun was immediately and continually lavished attention by technologists so that in effect the rocket almost died a still birth. This was understandable considering that the earliest rockets, such as described in the anonymously-written Liber Ignium (Fire Book) of ca. the 12th century, were crude affairs made of rolls of parchment paper closed at one end and filled with coarsely ground gunpowder. Nonetheless, the perfor-

----

*Research Historian, Department of Astronautics, National Air and Space Museum, Smithsonian Institution, Washington, D. C.

**To early pyrotechnists, calibers of rockets were generally determined by the weight of a lead ball that would fit into the interior of the choke or orifice. Congreve chose the weight of the rocket itself as the designation. Larger calibers were designated by rocket diameter.

mance of these first rockets must have been mystifying to the medieval mind in their inexplicable ability to move by themselves. They seemed to possess their own souls. They were also very unpredictable and very feeble in firepower compared to the first guns and thus seemed to have no potential. The rocket therefore languished in its technological evolution in both East and West for 600 years.(1)

Only the rocket's unique ability to propel itself saved it from complete extinction. This innate power, which some pyrotechnists literally attributed to a "soul" which they called the conical cavity from which the exhaust gases issued, was adapted readily by pyrotechnists for firework displays. Skyrockets could be made to burst into brilliant colors in the air or to climb into beautiful spirals and explode. Without their long stabilizing guidesticks, their pushing power could be put to work driving or moving other amusing pyrotechnical pieces such as Catherine wheels, artificial fish, dragons, and birds. With very little modification, military and naval men adopted the same skyrockets as signals. Yet outside of a very short-lived period as weapons of war, their application and potential for growth was severely restricted. Even up to Congreve's day, except in India, the construction of rockets differed little from those described in the Liber Ignium. They were still made by hand, but with several sheets of pasteboard and glue replacing the parchment and received a more carefully proportioned and prepared mixture of gunpowder.(2)

INTRODUCTION OF CONGREVE ROCKETS

The threat of an invasion of England by Napoleon's fleet at Boulogne-sur-Mer across the English Channel in 1804 led Congreve, the ingenious mechanically-minded son of the Comptroller of the Royal Arsenal, to attempt to improve the rocket as a "new" weapon of war. At first, out of his own pocket, he purchased the largest skyrockets he could find in London, probably 6-pounders (2.72 kgs). The first investigations were made about September, 1804. "None of them," he found, were "capable of more than a five or six hundred yard (457.2 or 548.64 meter) range according to the usual modes of construction. Various plans suggested themselves to my mind, and were successively tried." These trials, unfortunately the details of which are not known, were first conducted privately and then afterwards, with proper approval, with the aid of the facilities of the Royal Laboratory at the Woolwich Military Works. The later test flights were conducted on the firing ranges of Woolwich. This initial work proceeded, at Congreve's own cost and expending "several hundred pounds in different experiments," until he succeeded in increasing the range of ordinary rockets from 600 yards to 1,500 yards (540 to 1,350 meters). The fundamental means of improving upon the performance of rockets, Congreve found, was not only in strengthening the propellant composition and generally improving the construction of the case, but also determining the best firing angles. In time, Congreve worked out a complete set of range tables for his rockets. By April 29, 1805, with the threat of invasion still looming, Congreve was given permission to convert an old shed at the back of an old proof-butt at Woolwich as a temporary workshop for driving the rockets and he was also afforded the assistance of some of the workmen of the Royal Brass Foundry and Engineering Department of the Arsenal to make the rockets he had designed.(3)

While it readily consented to the wishes and plans of the energetic Congreve, the Board of Ordnance became increasingly overwhelmed over the growing magnitude of the rocket project. The inventor's requests grew in great number and his experiments branched out in all directions. The Board now admitted that the scope of Congreve's work was "much greater than they had first imagined." The work progressed at such a

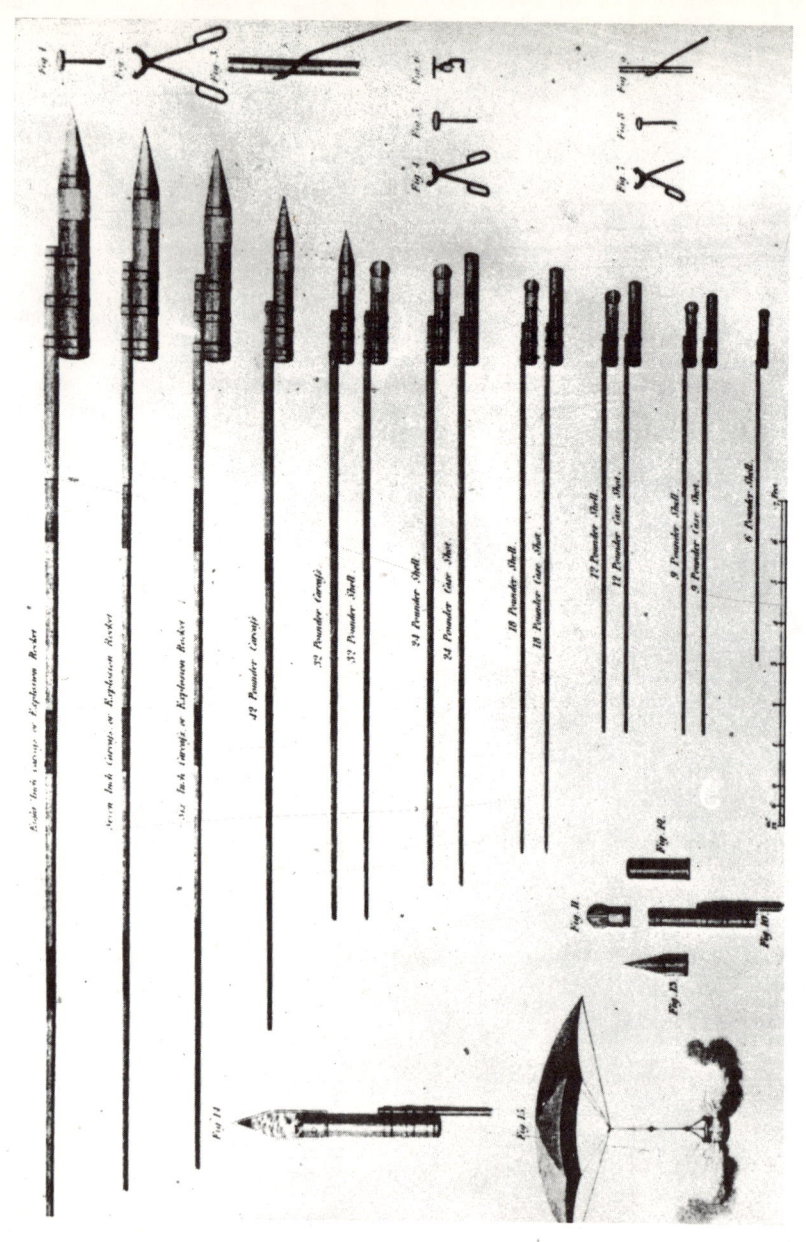

From William Congreve, Treatise on the Rocket System (London, 1827)

rate, however, that by September, 1805, he was ready to promote the use of the rockets against French shipping at Boulogne. He managed to gain an audience with one of the country's leading advocates of scien-- tific and military matters, the Prince of Wales (afterwards, King George IV) at the Royal Pavillion at Brighton. It was a most fortuitous meeting. The inventor won both the immediate and highest placed support for his project and a friendship and patronage of lifelong standing. Congreve was to face considerable opposition enough from severe critics of his weapon system both from the military and the politicians. In a sense the story of this early phase of Congreve's rockets may be likened to the acceptance or disapproval of a major new weapon or space project of today. The scrutinizing questions were the same: Was the system efficacious? Was there a national need for it? What were the expenditures? For the same reasons, Congreve arrived at the first cost effectiveness studies on the large-scale manufacture of his complete system. Very likely, the earliest cost effective realizations of the approach were made before the Prince of Wales at Brighton. (4)

From Brighton, the scene shifted to Walmer Castle, at Deal, Kent. Here Congreve had been sent in the Prince's own cutter "that no time might be lost" to speak with the Prime Minister, William Pitt the younger. The Prince of Wales had also written a letter "in his own hand to Mr. Pitt," Congreve later recalled, "recommending his immediate attention to its (the rocket's) adoption." "The result," concluded Congreve, "was that preparations were immediately commenced for the attack, which took place in the autumn of that year." These preparations and the attack itself were neither as swift nor as smooth as Congreve's glossed remark might convey. For one thing, friction developed between two inventors of alternate means of attack, Colonel (afterwards, General) Henry Shrapnel who was beginning to lay before the Government his fragmentary shell, and the American Robert Fulton who was proposing his submarine torpedo mines against the Boulogne fleet. There was also opposition from very high naval circles, including Lord Nelson and senior members of the Admiralty. The attack itself was almost farcical. Both Fulton's torpedoes and Congreve's rockets received their baptism of fire together. On the night of November 21, 1805, ten launches fitted with 48 rockets each of 8 pounds (3.63 kgs) caliber, including three pounds (1.36 kgs) of "unquenchable" incendiary "carcass" composition, set out from the main British attacking squadron under Sir Sydney Smith towards the Boulogne Basin. Almost at the moment of attack, the wind suddenly shifted and soon became a gale. Five of the launches became swamped and the handful of rockets that were fired either burst in the boats or otherwise failed to go in the intended direction.(5)

Living up to his family motto, "Persevere," Congreve tenaciously overcame both the technical, political, and military difficulties that confronted him, and by the following year personally helped direct a successful combat trial of 200 improved rockets at Boulogne. On October 8, 1806, the greatly improved and enlarged rounds, now 24-pounders (10.89 kgs), had conclusively proved themselves to the Admiralty, and prepared the way for future adoption. The path was slow, even though the Royal Navy and later the Royal Artillery employed the weapon on a frequent, but for several years, expediency basis. It was not until 1814 that two "Rocket Troops" were officially organized in the Army. In the meantime, Congreve's rockets had achieved an impressive record. At the bombardment of Copenhagen in 1807, about 300 of them (contrary to published reports of 20,000 or 40,000) wrought such destruction that the Danes were led to initiate their own rocketry experiments and afterwards created their own rocket troops. Congreve himself surveyed the damage of the city after the bombardment dressed in disguise. In the Peninsular War and at the famous "Battle of Nations" at Leipzig, in 1813, which led to Napoleon's abdication, the "Rocket Brigade" was the solitary British unit present, and gave such a good account of itself that several of the men were personally presented the highest Russian, Prussian, and Swedish decorations from the sovereigns of those nations. Congreve himself was conferred the knighthood Order of St. Anne of Russia by Tsar Alexander and the Sword of Sweden by Bernadotte, the newly elected King of that country. Congreve rockets were also in evidence in America, during the War of 1812. The Baltimore lawyer, Francis Scott Key, saw them as "The Rockets' red glare" over Fort McHenry when he wrote "The Star Spangled Banner." In that memorable engagement it was the Erebus, the second rocket ship designed by Congreve, that launched those projectiles. The Erebus had formerly been an 18 to 20 gun sloop but was entirely refitted under Congreve's supervision to fire either 24- or 30 pounder (10.89 kgs or 14.51 kgs) rockets from twenty specially made "scuttles" fit into the sides. (Congreve's first rocket ship was the Galago, which rendered good service off Holland, in the Scheldt operations, 1809-1810, though from the Boulogne days there had already been numerous smaller rocket boats.) At Napoleon's downfall at Waterloo Congreve's rockets were also in action. While the conclusion of the Napoleonic campaigns would seem to have marked the death-knell of the Congreve rocket, in fact it turned out to be its initiation.(6)

INFLUENCE OF CONGREVE

Congreve rockets and their successors, the Boxer and Hale models, were literally used around the globe until 1899. Hales, stickless or rotary rockets, the final stage of evolution of the 19th century gunpowder-propelled war rocket, remained on the official Royal Army weapons lists until as late as 1919, when it was finally declared obsolete. In bearing these weapons around the world, the Royal Navy and Army was not only forging the British Empire and maintaining Pax Britannia, but was also internationally demonstrating their nation's rocket technology. The end result was considerably more beneficial than would first appear. As will shortly be evident, Congreve war rockets were not confined to taking lives but also to saving and enriching them. They led directly to the life-saving rockets, some of the first concepts of rocket-propelled flying machines, and other diversified applications.(7)

Amongst some of the innumerable post-Napoleonic actions in which Congreve or Hale war rockets were: the bombardments of Algiers in 1816 and 1824 to suppress piracy from that quarter; the Anglo-Burmese War of 1824-1826; in Canada, during the rebellion there in 1837; during the First Afghan War, 1838-1842; the Opium War in China, 1839-1842; in New Zealand during the first Maori Wars, 1845-1846; the intervention against the dictator Manuel Rosas of Argentina, in 1846, at Saint Lorenzo; throughout the Crimean War, in Russia, 1854-1856; the Sepoy Mutiny in India, 1856-1858; the later Maori wars in New Zealand, 1860-1863; in Africa, during the Abyssinian War, 1868, in which British Hale rockets replaced Congreves; the Zulu War, Africa, in 1879; the First Boer War, Transvaal, 1880-1881, upon the same continent; and in countless minor engagements ranging from military assistance to Loyalist Spaniards during the Carlist War of 1836-1839 to punitive expeditions against Maylay marauders in 1873-1875 and rebels in British Honduras, 1867.(8).

Simultaneously with this wide usage of Congreve and later war rockets by the British was an equally intensive activity on the part of rocket troops of other nations. All of this activity in turn emmanated from the success Congreve met. The Danes, as indicated above, were the first to follow the British example. Following the bombardment of Copenhagen in 1807, Second-Lieutenant of the Engineers Andreas Anton Frederick Schumacher obtained permission from King Frederick VI to first duplicate and then improve upon the English system. Starting from an "unburned" rocket he had found, Schumacher feverishly developed his own system by 1811 in a specially created secret laboratory on the island of Hjelm in the

Kattegat Sea. In the same year a Danish rocket factory was established
at Frederickvaerk and a flottilla of lugger boats that guarded the Belt
and coasts of Zeeland were already armed with rockets. On February 16,
1813, Schumacher, now a Captain, was placed in charge of an officially-
formed "Raket-Compagniet" (Rocket Company) and two rocket batteries were
also assigned at Fynn. In effect, the Danes preceded the British themselves in approving the first officially or regularly-armed rocket units
in both their navy and army. Ironically, some of the Danish rockets saw
action against the British when some men-of-war passed through Langeland
in 1814. The Compagniet was afterwards elevated to a Raket Corpset
(Corps) and lasted until 1851. It notably saw action against the Prussians in the Schleslwig-Holstein War of 1848-1850.(9)

   The Austrians followed suit after Denmark. At the battle of Leipzig in 1813 Artillery Major Vincenz von Augustin personally witnessed
the success of the British "Rocket Brigade" in that crucial engagement,
and like Schumacher and Congreve before him, devoted the remainder of
his professional career to rocketry. Augustin, who afterwards became
a General and Director of the Austrian Ordnance, seems to have also been
afforded a first hand look at Congreve rockets at Woolwich where he also
consulted with Congreve himself. In addition, by secret diplomatic arrangement worked out by Prince Clemens Metternich, von Augustin was also
able to visit Schumacher on Hjelm. Consequently, von Augustin established a "Kriegs-Raketen-Anstalt" (War Rocket Establishment) at Weiner-
Neustadt, the Woolwich of Austria. Weiner-Neustadt thereafter became
known as "Raketendorf," literally "Rocket Town." In 1815 von Augustin
was able to report a complete rocket battery was in the field ready for
action and that it was equipped with 2,400 rockets of different calibers.
The "Rakenbatterie" was to expand enormously until it became a Raketen-
Regiment of some 20 individual batteries. As such, the Austrian rocket
establishment was the largest during the Congreve era, far exceeding
that founded by Congreve. Its rockets were also the most sophisticated
and in turn were extensively copied. The Austrians purchased Hale rockets in 1860 and likewise greatly modified them. Austrian rockets, whether Congreve's or Hale's, were also amongst the most widely used.
Throughout their lifetime, from 1815 to 1866, von Augustin's "Raketen-
batterien" fought at Huningue, France; Naples, Italy, 1820; Montenegro,
1838; Beirut, Syria (now Lebanon), 1840; throughout Italy and Hungary
and parts of what is now Rumania during the Risorgiemento and Sardinian
War of 1859; and near Königratz (now Hradec Kravlove, Czechoslovakia)
in 1866.(10)

   Captain (later General) Josef Bem of Poland was introduced to Congreve rockets at the siege of Danzig in 1813. Rocket experiments were
afterwards pursued in Warsaw and culminated in a classic bi-lingual
(French and German) book on the subject and the formation of the Polish
"Rakietkorpusu" (Rocket Corps) which lasted from 1823 to 1831. The
Russians formed their first rocket troops in 1827, and appear to have not
disbanded them until as late as 1885. Their rocket factories (situated
first at St. Petersburg and later at Nikolayev) continued to produce
thousands of life-saving and signal rockets based upon Congreve-rocket
technology well into the 20th century. Their two most prominent rocket
pioneers during this epoch were the artillerymen Alexander Dimitievitch
Zasyadko and Konstantin Ivanovitch Konstantinov. The former appears to
have first seen British Congreve rockets in battle in Germany and France
in 1813-1815, and also benefitted from the researches of Bem and his associates in Warsaw. The Tsar greatly encouraged Zasyadko's own studies
as well as those of Konstantinov. Konstantinov's work was truly pioneering in that he produced the world's first rocket ballistic pendulum,
devised his own hydraulic rocket presses, utilized electricity in the
igniting and testing of rockets, designed dual chamber life-saving rockets, and examined the possibilities of adapting Congreve-type rockets
for the propulsion of airships.

The history of French rocketry during the 19th century is a complex one and began with chemical and engineering analyses of British captured rockets during the early days of the Napoleonic wars. Napoleon himself promoted them but it was not until 1827 that an Englishman who had been employed by Congreve, Robert Bedford, signed a contract with the French Government to impart his knowledge and help from the first "Fuseens" (Rocketeers). Bedford stayed at the Pyrotechnic School at Metz until 1844. Shortly thereafter, Louis Auguste Victor Vincent Susane assumed leadership of French rocketry, and under his direction tens of thousands of rockets were made, some of them (12 cm caliber) ranging up to 7,500 yards (6,750 meters). Susane also wrote one of the classic histories of rocketry. Like the Russians, the French abandoned their Congreve-type rockets late (1872), but continued to apply the technology to peaceful applications.(11)

The Swedes, Americans, Swiss, Prussians and lesser German states (Saxony, Wurttemberg, and Bavaria), Dutch, Belgians, Greeks, Portugese, Spanish, Italian states (Kingdoms of Sardinia and the Two Sicilies), Hungarians, Turks, Egyptians, Persians, Brazilians, Chileans, Peruvians, Argentinians, Cubans, Mexicans, Paraguayans, and one or two petty principalities also followed in the wake of the larger powers in adopting or experimenting with Congreve rockets, and in many cases forming their own rocket troops. Congreve rockets also made their way to China and India, the birthplaces of their ancestors. Some of these nations directly purchased their rockets and rocket manufacturing equipment from England. The American rocket pioneer, Robert H. Goddard, was therefore quite justified in placing the name of Congreve high in the priority of names in his outline of the history of rocketry. Unfortunately, however, we do not have the full benefit of Goddard's own thoughts on Congreve as he did not live to complete his book. Goddard began his own researches with large Coston signs or lifesaving rockets which were derived directly from Congreve rocket technology. The same transition applied to the solid propellant powered projects of the Austrian rocketry and astronautical pioneer Max Valier. The first JATO's, notably those tried by the Junker's Company in 1928 and the first rocket-propelled airplanes powered with Sander lifesaving rockets were very much spin-offs of Congreve rocket technology. A less tangible but far profounder legacy of Congreve is that he established the rocket as a viable instrument of tremendous potential for the new age of rockets. "He paved the way," write rocket historians Williams and Epstein, "for the men who would come after him--men whose ingenuity would supply it with the power it needed to play that role."(12)

FALLOUTS FROM CONGREVE ROCKET TECHNOLOGY

Despite the military side of Congreve rockets and their fulfillment of immediate national needs, there was also a harvest of fallout technology for peaceful purposes. This inevitable development repeated and greatly magnified itself in the 20th century when the infamous V-2 rocket became a tool for peaceful high altitude research, and eventually led to the satellite launcher, Moon rockets, and vehicles for interplanetary probes. In Congreve's day, the technological fallout translated into life-saving rockets, improved and higher payload-carrying signal rockets, the idea of powerful rockets for reaction-propelled flying machines, and the first studies of rocket dynamics. In addition, there were proposed or tried, the first mail-carrying rockets, camera-carrying rockets, weather-modification rockets, illumination (flare) rockets, scientific research rockets (for geographical surveying, rating chronometers, and the like), telephone wire-laying rockets, rockets for rescuing people from burning buildings, and rockets for hoisting scaling ladders up steep cliffs. All of these applications accrued directly from the basic technology of the Congreve rocket. Either the war rockets themselves were directly modified or the transference was made by example. In all cases the feeble power and flimsy construction of the purely firework family of rockets was not sufficient enough to be adapted to these tasks.(13)

The earliest of these fallouts was the purely scientific study of rocket dynamics. Prior to the age of Congreve, empirical pyrotechnists paid scant attention to why a rocket ascends. Speculations that were made were both faulty and nonmathematical. The leading theory held that air was the essential medium needed for the rocket to push against. The "soul" mystique of the rocket described below was also commonly accepted. With the arrival of the Congreve rocket these archaic notions became outmoded. As a result of the mass bombardment of Copenhagen in 1807, the rocket suddenly came into prominence both in scientific and military worlds. In 1807, Congreve himself speculated upon the "principles of the flight of rockets, with a view to determine the precise effects of the stick, or any other mode (wings or fins) that may be imagined for guiding them. . ." His was a nonmathematical explanation of the increase of stability by attaching fins or centering the guidestick, the latter mode he eventually adopted for all his rockets by 1819. The first mathematical investigations, however, were undertaken directly through his influence. In 1810 the Kongelige Danske Videnskabernes Selskabs (the Royal Danish Academy of Science and Letters) offered a series of prize questions on the physics of Congreve rockets. None of the papers submitted received the award because something was lacking though the authors "showed unmistakable proof of great knowledge and analytical proficiency." Nonetheless, William Moore, a gifted English mathematics instructor at the Royal Military Academy at Woolwich, was induced to take up the challenge independently, and by 1813 produced his Treatise on the Motion of Rockets, the world's first mathematical dissertation on rocket dynamics. Applying differential calculus, Newton's classic laws on motion, and the ballistic findings of Benjamin Robins, inventor of the ballistic pendulum, Moore arrived at formidable propositions for the motions of rockets both with air resistance and in a vacuum. He was also the first to suggest the rocket ballistic pendulum for the more accurate determination of performance parameters, as well as thrust, rocket velocity, and the significant value of specific impulse. Even more remarkable was Moore's hint of a space rocket. Set only as a mathematical demonstration of the ultimate speed of a rocket if unchecked by air, he calculated that "it would never return, but continue to move forever, or fly off to an infinite distance."(14)

Though taught at the Military Academy at Woolwich and probably under the eye of Congreve himself, these exercises were wholly philosophical and scientific in scope. Similar investigations appeared elsewhere and were likewise precipitated by the Congreve rocket. In the 1820's the eminent mathematician Dr. Paulo Ruffini of the Italian Dutchy of Modena based his mathematical theories of rocket flight upon known principles of hydraulics as well as ballistics. The results of his findings were published posthumously as a lengthy paper by the Royal Academy of Sciences, Letters, and Arts of Modena in 1833 under the title "Osservvazioni intorno al moto dei Razzi alla Congreve." A fellow Italian, the chemist, mineralogist, and soldier Colonel Giuseppe Mori produced his Sul Rinculo delle Armi da Fuoco di Movimento de' Razzi in Naples, in 1839. Armand Quillet's theories on rocket thrust were presented at the meeting of March 3, 1851, before the French Academy of Sciences in Paris under the title "Sur les Fusees de Guerre: Recherche de la Loi de Variation de la Force Impulsive." While nine years later, in 1859, the height of ascent of rockets was formulated by the German mathematician and editor of the journal Zeitschrift für Mathematik und Physik of Leipzig, Gustav Emil Kahl. His article appeared as "Die Berechung der Steighöhe der Raketen." In 1861 one of the earliest essentially speculative, and certainly the most interesting experiments were undertaken by Louis Dufour, Professor of Physics at the Academy of Lausanne, Switzerland. Climbing the Swiss

Alps to the summit of Chenalletes, a height of 9,700 feet (2,956.56 meters) above sea level, Professor Dufour fired a series of 64 rockets at different altitudes to ascertain the effects of decreased air pressure upon their combustion.(15)

In the realm of strictly military-sponsored rocket ballistic research are many more names of contributors to the new science. The French Artillery Captain Dominique-Nicolas Munier, for example, wrote his Theorie du Mouvement et du Tir des Fusées in 1830 as a result of the earliest known bonafide rocket ballistic experimentation and was carried out at the Pyrotechnic School at Metz apparently with the aid of dynamometers. Rocket ballistics and the influence of Congreve also penetrated the technologically-backward Ottoman Empire. General Emin Mehemet Paşa devoted his book Memoir sur un Nouveau System de Confection des Fusées de Guerre (Paris, 1840) to seeking mathematical solutions to instability of rocket flight. One answer was the rotation of the projectile. Emin Paşa had thus anticipated the first successful spin-stabilized rockets of William Hale by four years. In 1848 the Swiss-born man of letters, Lieutenant-Colonel Adolphe Pictet of the Artillery of the Swiss Confederation detailed in his Essai sur les Proprietes et la Tactique des Fusées de Guerre the results of his trials of rockets outside Turin to reduce lateral dispersions and longitudinal deviations and how they could be predicted, with and without the influence of wind. These same problems were also taken up by the French artillerist Lieutenant-Colonel Charles Victor Thiroux in his 1858 report, "Quelques Observations sur la Theorie Actuelle des Deviations des Projectiles," as well as in his Observations et Vues Nouvelles sur les Fusées de Guerre (Paris, 1849). The Belgian Director of the arsenal at Antwerp, Major-General Casimir-Erasme Coquilhat went several steps further and, like Moore, considered rocket flight in a vacuum. In 1872 appeared his own 33-page treatise, Trajectories des Fusées Volantes dans le Vide, later printed in Memoires de la Société de Liège in 1873.(16)

By far, the most advanced program of rocket ballistic research during the Congreve era was accomplished in Russia by Konstantin Ivanovitch Konstantinov, commander of the St. Petersburg Rocket Establishment from 1849 until it moved to Nikolayev in 1867, and head of the Nikolayev plant until his death in 1871. In 1846 he constructed his first rocket ballistic pendulum which was followed in 1849 by a much larger machine set on a granite base and meant for rockets of higher impulse. Through this apparatus which was partially electrically driven, Konstantinov was able to static test his rockets and plot their performance curves on graphs. In this way he was able to arrive at a basic law of the relationship of exhaust velocity to the momentum of the rocket. An instrument was also devised for measuring rocket altitudes as well as an optical range finder. Lectures were delivered on his findings, several books written, both in Russian and French, and numerous papers published in the country's leading military and scientific journals. These inquiries into the nature of the rocket, initiated under the impetus of Congreve, were carried forward into our own century. Whether these findings in themselves contributed to the foundations of modern rocket ballistics is problematical. What they do represent for certain, to reiterate an earlier theme, is the continuation of the technology of the 19th century powder rocket to its ultimate limits as a basis of departure towards the liquid fuel rocket. More importantly, they are proof of a continuity of the inquiring spirit of rocket scientists of both ages.(17)

Even if the Congreve war rocket all but disappeared by the turn of the century, its technology was not only extablished but viable. This was exemplified by the life-saving rocket. Congreve himself had sought to improve upon this invention from 1818 t0 1827. In 1825 he had also prepared a set of instructions, Sir William Congreve's Method of Saving the Crews of Shipwrecked Vessels without Assistance from the Shore. In actuality, however, the first successful lifesaving rocket is credited to Henry Trengrouse of Helston, Cornwall, England, whom Sharpe suggests was strongly influenced by Congreve's rockets fired upon Copenhagen in 1807. The first rocket bombardment in modern history thus paradoxically heralded inadvertent benefits to mankind. Sharpe's prize-winning 1968 Goddard essay of the development of the lifesaving rocket fully chronicles the subsequent story of this most humane invention. Sharpe also makes the cogent and paradoxical observation that more lives were saved by the rocket in the 19th century than were taken by it. "Recordings of battles in which rockets were used from Copenhagen in 1807 to the V-2 bombardment of London in 1944," he says, "are sparse in account of wholesale slaughter by rocket. On the other hand, statistics reveal that at least 15,000 lives were saved in the British Isles alone between 1871 and 1962." The lifesaving rocket not only provided Goddard, Valier and others their first testbeds of rocket power with which to pursue their goals towards space vehicles, but it has continued down to today to rescue lives around the world. Again, according to Sharpe, the lifesaving rocket seems assured a permanent place in apparatus used for saving lives of the shipwrecked. The International Conference of Safety at Sea of 1948 and subsequent conferences resulted in the requirement for ships of all member nations to carry line-throwing rockets. "The fact that lives are saved each year in basically the same way as they were in the days of . . . (Congreve) . . . also seems to assure the rocket a continuing role in this humanitarian application of a device developed originally as a weapon."(18)

Congreve war rockets also generated some of the earliest concepts of reaction-propelled flying machines. Congreve himself dreamt of flight, both above the earth and beyond it. In a letter written to his father in 1785, when he was away at school, and two years after the Montgolfiers made their epochal ascent in a lighter than air balloon, the 13 year old Congreve sketched his own machine and boldly declared himself "fully bent on going to the Moon in an aerial balloon." The dream of flight never quitted him even though it was unfulfilled. Two months before his death, in March, 1828, he submitted an illustrated plan for a wind and muscle powered "Aerial Carriage" to The Mechanics Magazine. Congreve undoubtedly died too soon. Who can say that he would not have eventually applied the packaged energy of his rockets to this machine? For very shortly after, in 1831, there materialized an anonymous project for a "reaction aerostat," a balloon fitted with a cluster of large rockets for giving it direction. The plan was printed in the eight page booklet, Scoperta della Direzione del Globo Aerostatico, printed "at the Bridge of Moses" in Venice. The Englishman James Nye specifically recommended Congreve rockets when he proposed his own "rocket balloon" in 1852 in his Thoughts on Aerial Travelling. The Nye apparatus was a 337 foot (102.72 meter) long airship buoyed up by hot air in lieu of the volatile hydrogen then preferred. Its diameter was 67.4 feet (20.54 meters) and its lifting capacity 8,550 pounds (3,878.21 kgs). As Nye estimated that Congreve's smaller rockets sped to a distance to two miles (3.22 km) in seven seconds, the balloon could carry five to six thousand of them, thereby providing sufficient propulsion for a voyage of 200 miles (321.86 km) at 15 mph (24.14 km/hr). From the results of trials of 32-pounder (14.5 kgs) Congreve rockets fired at the Artillery Practice Ground at Dum Dum, India, on June 22, 1816, more accurate performance values are available. Fired at an elevation of 55°, the rockets attained ranges of from 2,036 yards (1,832.4 meters) to 2,874 yards (2,586.6 meters) in times of flight of from 20 to 25 seconds. Average flight velocities therefore varied from 243-303 feet/sec (74.07-92.35 meters/sec). Smaller 12-pounder (5.44 kg) rockets faired higher with velocities of 438-456 feet/sec (133.5-138.99 meters/sec). Insofar as thrusts or exhaust velocities of Congreve rockets are concerned, these are not known with any precision. Knowing the approximate specific impulse of gunpowder (80-100 seconds); the weight of propellant of 32-pounder (14.5 kgs) rockets (10 lbs, or

4.54 kgs); and the firing time (3 seconds), it is possible to make a reasonable estimate of thrust. The thrust of the 32-pounder is found to have been approximately 250 pounds (113.4 kgs) for three seconds. This performance may be compared to the actual thrust ("pressure") exerted on a piston by a Hale 24-pounder (10.89 kgs) rocket in a test at Woolwich in 1879. Burning 9.5 pounds (4.31 kgs) of propellant for approximately the same duration as the earlier projectile, the Hale produced 300 pounds (136.08 kgs). Earlier, the French gauged thrusts ("impulsions") on dynometers and succeeded in boosting performances from 507 pounds (230 kgs) to 2,469 pounds (1,120 kgs) for their largest (apparently 12-cm) calibers. The hydraulic press, first tried by Congreve and afterwards developed by Hale, Coston and others, considerably increased the density of propellants driven by the muscle and weight "monkey press" (a kind of pile driver) accounting for much of the improved performance and stability. Densities increased from 1.3 to 1.8. At any rate, thrusts of Congreve rockets for propelling and directing James Nye's lighter-than-air machine would have been quite adequate, though enormously expanded and the system vastly more efficient with more contemporary rockets. In short, Nye's rocket system itself was probably workable if more thrust and longer burning durations were available (i.e., a higher propellant). The weights of the rockets and other parts of the balloon were also excessive. The Congreve rockets were ignited successively by means of a steam-driven wheel which automatically moved a rocket into place, ignited it, and ejected it (Nye is silent upon the disposal of the spent rounds). Rocket wheels at either end enabled forward and reverse movements. The Russian disciple of Congreve, Konstantin Konstantinov, also thought of war rockets for human flight. After deliberately testing some on his ballistic pendulum for this purpose, however, he concluded quite rightly that they were still far too weak to be considered.(19)

Rocket balloons and planes still abounded. Most of these designs were really reaction-propelled, steam, compressed air, besides pure rocket and gunpowder-cartridge fed machines. In the steam category were a curious series of steam rocket drawings dating from 1828 and appearing up until the 1860's showing an almost humorous caricature of a man stradling a rocket, head towards the nosecone and feet planted on rear movable rudders. These drawings, generally referred to as Charles Golightly "steam horses" after one of the British designs of 1841, may actually be attributed to the steam rocket patent of the American James Perkins (British patent No. 4592 of May 15, 1824). Very probably Perkins' idea itself originated from the Congreve rocket as the two men personally knew one another from their respective work beginning in 1819 on the development of printing machines for the Bank of England. The Congreve rocket may therefore be said to have led directly not only to some early rocket propelled flying machines but also to the first steam rocket. This was recognized and expressed somewhat obliquely in The London Illustrated News for April 1, 1843, which remarked in a general discussion of flying machines that: "The rocket of Colonel Congreve, and its application to the conveyance of passengers by Mr. Perkins come next on the list, and though a steam-conveyance rocket is a very ingenious invention, yet there are few persons who would like to demonstrate its practical utility for making the first trip."(20)

It was thus at the turn of the 19th century that William Congreve founded a rocket technology that was further matured by the Industrial Revolution. The Congreve era was clearly a demarcation and necessary step between the first rockets and those of our own time. Created for war, Congreve rockets produced their own beneficial and lasting technological fallouts. They disseminated the technology throughout the world, stimulated the first scientific inquiries into the dynamics of the rocket, gave rise to their exploitation as vehicles for manned flight, and provided a basic hardware upon which to start afresh and construct the space age.

REFERENCES

1. Frank H. Winter, "Sir William Congreve: A Bi-Centennial Memorial," Spaceflight, Vol. 14, No. 9, September 1972, pp. 333-334; J.R. Partington, A History of Greek Fire and Gunpowder (Cambridge, 1960), pp. 42-81; Willy Ley, Rockets, Missiles, and Space Travel (New York, 1959), p. 71.

2. Frank H. Winter, "A History of Italian Rocketry During the 19th Century," Armi Antiche, Bollettino dell'Accademia di S. Marciano, (Torino, 1965), pp. 181-182, 195.

3. William Congreve, A Concise Account of the Origin and Progress of the Rocket System (London, 1810), p. 1; Brigadier O.F.G. Hogg, The Royal Arsenal (London, 1963), Vol. I, pp. 518-519.

4. Ibid.

5. Congreve, Op. Cit.; Mary Ann Clarke, The Rival Princes (London, 1810), Vol. 2, pp. 40-41; Christopher Lloyd and Hardin Craig, Jr., (eds.), "Congreve's Rockets 1805-1806," Pt. VIII, The Naval Miscellany, Vol. IV, The Naval Records Society Publications, Vol. 92 (London, 1952), Passim.; The Dispatches and Letters of Vice Admiral Lord Viscount Nelson (London, 1846), Vol. VII, pp. 61-62, 121-122.

6. Congreve, Op. Cit.; Capt. H.B. Latham, "The Rocket Service and the Award of the Swedish Decorations for Leipzig," The Journal of the Royal Artillery. Vol. LVI, January 1930, p. 421, 451-452; Frank H. Winter and Mitchell R. Sharpe, "William Moore: A Pioneer in the Theory of Rocket Dynamics," Spaceflight, Vol 18, No. 5, May 1976, p. 179; Frank H. Winter, "A Case Study in Challenge and Response--Danish Rocketry in the 19th Century," Aerospace Historian, Vol. XIII, Summer 1966, p. 81.

7. Frank H. Winter, "Congreve, Sir William," Dictionary of Scientific Biography (New York, 1971), Vol. III, pp. 388-389; Hogg, Loc. Cit., Vol. II, p. 1379.

8. The Annual Register, Passim.; The London Times, Passim.; Publications of the Naval Records Society, Passim.; United Service Journal, Passim.; Journal of the Royal Artillery, Passim.; Major Francis Duncan, History of the Royal Regiment of Artillery (London, 1879), Vols. I and II, Passim.; J.W. Fortescue, A History of the British Army (London, 1920-1930), Vols. IX, XI, XIII, XIV, Passim.; G.A. Henty, et. al., Battles of the Nineteenth Century (London, 1897), Vols. I and II, Passim.; Maj. Gen. Sir John Headlam, The History of the Royal Artillery (Woolwich, 1940), Vol. III, Passim.

9. Winter, "A Cast Study," Op. Cit.

10. Frank H. Winter, "Baron Vincenz von Augustin and His Raketenbatterien--The History of Austrian Rocketry in the 19th Century," paper presented at the 22nd IAF Congress, Brussels, 1971, Passim.

11. Josef Bem, Erfahrungen über die Congrev'schen Brand-Raketen, bis zum Jahre 1819 in der Königl. Polnischen Artillerie (Weimar, 1820), Passim.; Bronislaw Gembarzewski, Wojsko Polskie Królestwo Polskie 1815-1830, (Warsaw, 1903), pp. 121-145; General M. (K.I.) Konstantinoff, Lectures sur les Fusées de Guerre (Paris 1861), Passim.; V.N. Sokol'skii, Russian Solid-Fuel Rockets (Jerusalem, 1967), Passim.; A. Pralon, Une Page de l'Histoire de l'Artillerie: Les Fusées de Guerre en France (Paris, 1883), Passim.; M. (L.A.V.V.) Susane, Les Fusées de Guerre (Metz, 1893), Passim.; "Rockets Lift Plane from German River," New York Times, August 10, 1929,

p. 14, col. 4; "Rockets to 'Boost' Heavy Planes," Popular Science Monthly, Vol. 115, November 1929, p. 41; Ley, Loc. Cit., Passim.

12. Ley, Op. Cit.; Winter, "A History of Italian Rocketry," Op. Cit.; United Service Journal, Passim.; M. (Merigón) de Montgery, Traité des Fusées de Guerre, Nommées Autrefois Fochettes et Maintenant Fusées a la Congreve, (Paris, 1825), Passim.; De Spectator Militaire, Passim.; Major General D.J.G. von Hoyer, System der Brandraketen nach Congreve und Andern (Leipzig, 1827), Passim.; Adolphe Pictet, Essai sur les Propriétes et la Tactique des Fusées (Turin, 1848), Passim.; Bulletin des Sciences Militaires, Passim.; Allgemeine Militar-Zeitung, Passim.; Spectateur Militaire, Passim.; United Service Journal, Passim.; Army and Navy Journal, Passim.; Journal des Armes Spéciales, Passim.; Nautical Magazine, Passim.; Naval and Military Magazine, Passim.; The Broad Arrow, Passim.; Archiv für die Offiziere de Koniglich Preussischen Artillerie, Passim.; Morskoi Sbornik, Passim.; Revue d'Artillerie, Passim.; et. al.

13. Ibid.; Mitchell R. Sharpe, "Nonmilitary Applications of the Rocket Between the 17th and 20th Centuries," paper presented at the 21st IAF Congress, Constance, Federal Republic of Germany, 1970, Passim.

14. Frank H. Winter and Mitchell R. Sharpe, "William Moore," Op. Cit.; William Congreve, A Concise Account of the Origin and Progress of the Rocket System (London, 1807), pp. 1-8 of addendum.

15. Frank H. Winter, "A History of Italian Rocketry," Loc. Cit., p. 190; Paolo Ruffini, "Osservazioni intorno al moto dei Razzi alla Congreve." Memoire della Reale Accademia di Scienze, Lettere e d'Arti di Modena, Tomo. I, 1833, Pt. II, pp. 56-79; Arm. Quillet, "Sur les Fusées de guerre; recherche de la loi de variation de la force impulsive," Compte Rendu des Seances de l'Academie des Sciences, Vol. XXXII, 1st Semester, No. 9, March 3, 1851, pp. 339-340; (Gustav) Emil Kahl, "Ueber die Berechnung der Steighöhe der Raketen," Zeitschrift für Mathematik und Physik, 4. Jahrgang, 1859, pp. 279-284; M.L. Dufour, "De l'Influence de la pression atmosphérique sur la durée de combustion des fusées," Zeitschrift für die Schweizerische Artillerie, Jahrgang. 1865, No. 4, pp. 1-9.

16. Dominique-Nicolas Munier, Theorie du mouvement et du tir des fusées (Paris, 1830), Passim.; also found in Mémoires de l'Académie Royale de Metz, XIe Année, 1829-1830, pp. 114-129; Emin-Pascha (Paşa), Mémoire sur un Nouveau Systeme de Confection des Fusées de Guerre (Paris, 1840), Passim.; Pictet, Op. Cit.; Charles Thiroux, Quelques observations sur la theorie actuelle des deviations des projectiles (Paris, 1858), Passim.; Charles Thiroux, Observations et Vues sur les Fusées de Guerre (Paris, 1849), Passim.; Major-General Casimir-Erasme Coquilhat, Trajectories des Fusées Volantes dans le Vide (Brussels, 1872), Passim.; also found in Mémoires de la Société (Royale) des Sciences de Liège, T.V., 2 Séries, November, 1873, pp. 1-33.

17. Sokol'skii, Loc. Cit., pp. 60-69; Konstantinoff, Passim.

18. Mitchell R. Sharpe, Development of the Lifesaving Rocket--A study in 19th Century Technological Fallout (Huntsville, Ala., 1969), Marshall Space Flight Center Historical Note 4, pp. 1,3-4, 7, 61.

19. Winter, "Sir William Congreve," Loc. Cit., p. 334; Frank H. Winter, "Man, Rockets and Space Travel," Mankind, Vol. II, No. 4, December 1969, pp. 13-14; James Nye, Thoughts on Aerial Travelling and on the Best Means of Propelling Balloons (London, 1852), Passim.; Great Britain, War Office, Treatise on Ammunition (London, 1881), p. 301; Goddard, Loc. Cit., Vol. I, p. 172; "Substitutions de la Presse Hydraulique au Mouton dans le Chargement des Fusées de Guerre," Journal des Armes Spéciales, Series 1, T. 1, No. 10, October 1834, pp. 457-461.

20. Ley, Loc. Cit., pp. 61-62, 86; The London Illustrated News, Vol. II, April 1, 1843, p. 233; Elizabeth M. Harris, Sir William Congreve and His Compound-Plate Printing (Washington, D.C., 1967), United States National Museum Bulletin 252--Contributions from the Museum of History and Technology: Paper 71, pp. 74, 76, 82; Jacob Perkins, British Patent No. 4952, "Discharging Projectiles by the Force of Steam," May 15, 1824.

Sir William Congreve

MAX VALIER: SPACE PIONEER
(1895-1930)

Mitchell Sharpe*

Max Valier, one of the early pioneers in the field of astronautical engineering, had a relatively brief career and died tragically in pursuing his search for a rocket engine capable of propelling a spaceship. His early life and subsequent contributions to astronautics are discussed.

Where and how does one begin to consider Max Valier and his place in the history of astronautics? Is it as an intense and dedicated young science student in his native Austrian Tyrol? Or is it as the writer of popular books and articles on astronomy and spaceflight? Perhaps it should properly begin with his role as a serious investigator in the development of liquid propulsion rocket engines.

His story began in the village of Bozen, in the southern Tyrol, where he was born on February 9, 1895. His family was descended from French Hugenot settlers in the region. In 1913, he graduated from the Pater Franziskaner Gymnasium, where he had studied astronomy with a passion but found time to write science fiction and learned to play the piano. During the time, he filled his days with serious application to his studies and had little time for the prevailing social life.

Following graduation from the gymnasium, he entered the university at Innsbruck where he studied mathematics, astronomy, physics and meteorology, again with the dedication he had shown in the gymnasium. During the period, he sold several articles on astronomy and developed a star chart with revolving scales which was sold by a publisher. In the spring of 1914, he also undertook an experiment that provoked the ire of the police. From the top of the Hotel Maribrunn, he launched a rocket propelled model airplane that caused some consternation among the populace below.(1)

With the advent of World War I, Valier entered the Austrian army as a meteorologist. However, he soon became an aerial observer in a tethered balloon and progressed from that duty to an aerial observer in aircraft. With his taste for flying whetted, he later bacame a test pilot. It was during this time that the thought occurred to him that the airplane was forever limited in the altitude and speed it could obtain because of its means of propulsion. As he later wrote, ". . . I firmly believe that airships and aeroplanes driven by propellers will never be able to rise higher than about 16 miles above the earth's surface or attain a speed much over 400 miles per hour."(2) Perhaps recalling his earlier escapade in Innsbruck, he concluded that a rocket-propelled airplane was the way of the future.

As the war ended, he was preparing a study on the use of rockets to investigate the stratosphere. It apparently was never published and a copy was not found among his papers.
----

*Historian, Alabama Space and Rocket Center, Huntsville, Alabama.

Following World War I, he began lecturing and writing articles on astronomy in Munich. In doing so, he showed a decided gift for both means of communication. He was especially skilled in presenting highly technical concepts in terms that could be easily understood by the general public, who were willing to pay to hear such exciting lectures or read their written equivalents.

However, it was at this time that his enthusiasm for astronomy led him into accepting the theory put forth by Hans Hoerbiger, a designer of coal mining machinery, that, briefly, explained the cosmos as a result of comets or large chunks of ice falling into the Sun with an ensuing transformation of thermal energy into kinetic energy.(3)

As Valier did with all his causes, he accepted Hoerbiger's complicated and absurd thesis with passion and conviction. Indeed, he became something of an evangelist for it. His biographer, I. Essers, feels that Valier may have turned to rocketry as a means of ultimately proving Hoerbiger's theory. Whether he did or not is of little importance to his later contributions to rocketry; however, his evangelism for it cost him credulence among the academic scientists of the day, especially astronomers.

As a footnote to his writings on astronomy, it is interesting to note that, in 1919, the Deutsche Buchdruckerei, in Innsbruck, published his "Spiridion Illuxt," a piece of science fiction in which Valier presented the idea of splitting the atom for purposes of destruction.

Interestingly enough, while Hoerbiger was happy to have such a talented publicist for his theory, he deplored Valier's equal enthusiasm for rocketry and spaceflight.

In 1921, Valier matriculated at the university in Munich to continue his studies in astronomy. However, circumstances beyond his control made his attendance at lectures erratic. The great economic depression in Europe was beginning, and he had married and fathered a child. Thus, much of his time was spent in lecturing and writing in order to make ends meet. He also had to turn to the more popular magazines for his articles since they paid more than did the more scholarly and scientific ones.

Three years later, Valier published his book <u>Der Sterne Bahn und Wesen</u> (<u>Orbit and Nature of the Stars</u>), which like several books over the preceeding years was heavily enlarded with Hoerbiger's theory. More important to the history of astronautics, however, is the fact that in this year he read Hermann Oberth's <u>Die Rakete zu den Planetenraeumen</u> (<u>The Rocket into Interplanetary Space</u>) and began an active correspondence with him. After reading the book, Valier realized that it had great merit from a scientific viewpoint but a limited appeal to the lay public. He suggested that Oberth convince his publisher that Valier should rewrite the book and bring out a popular version of it. Oberth agreed, but the project was never to be. Oberth simply balked at a sentence in a letter from Valier on February 10: "Your hopes are founded on the fact that you take certain (in my opinion out-dated) astronomical ideas as a basis whereas I am one of the pioneers of Hoerbiger's glacial cosmogony."

To further push his point and further alienate Oberth, albeit unintentionally, Valier continued, "You understand, professor, what for us, the followers of the theory of glacial cosmogony, the possibility of going in person to the moon, means. This means simply everything! If it is found that the moon is not composed of ice, we are completely beaten, then there is no salvation; but if the whole surface of the moon is a crust of ice, everything is won and all astronomy, astrophysics, meteorology, and geology up till now is beaten and we are triumphant!!"

In a letter to Oberth on July 16, 1924, Valier set forth what he considered to be the most logical plan for realizing manned spaceflight. He was opposed to Oberth's idea of building and launching small research rockets.

He wrote, ". . . we will build such a rocket motor on the ground, on a platform, at our leisure and clamp it down to see how the combustion gases flow through the nozzle . . . . and (then) we shall put the thing onto a railroad car that is on a straight stretch to track and ignite it . . . . I then propose that one of these days we put such a rocket motor in a quite ordinary, all-metal airplane. Initially, we take off with the usual engine (then). . . we switch off this motor and let the rocket function . . . . The next step is to make an airplane which no longer has the usual engine, but a rocket motor only . . . . then we will pass progressively from airplane to spaceship."(4)

While Oberth, the theorist, wanted to begin space exploration directly with rocket flights, Valier, the empiricist, preferred to take a more leisurely course that in ways repeated the historical development of transportation!

Fundamentally, what Valier suggested as the proper course for rocketry to follow, in some ways, later occurred. The first crude rocket engines were fired from static test stands. He himself instigated the phase of using rockets to propel surface vehicles. Several attempts were made to propel aircraft by such power. Later, of course, rocket-propelled planes such as the X-1 and X-15, in a more serious view, actually flew before the first manned rockets.

With the rather puzzling exception of his insistence on using rockets to drive railway cars and automobiles, as well as sleds, Valier seems to have foreseen the general path of development in astronautics that led to the first manned spaceships.

By 1926, Valier's circumstances had increased somewhat. He was receiving more offers to give lectures and to write articles than he could handle, but the prosperity was short-lived. The depths of the economic depression in Europe were yet to come.

In March 25, of that year, Valier received a letter from an individual generally conceded a footnote in the history of rocketry. Hermann Ganswindt wrote him a frenetic letter demanding that Valier in all future books and subsequent editions of current ones cite Ganswindt as "the first and only inventor of rocket vehicles . . . in accordance with the custom of scientific circles." Valier dutifully investigated Ganswindt's claims, and acknowledged him--not as <u>only inventor</u>--but as a contributor to rocket theory and design.

In seeking support for his plan for astronautical research, Valier also turned to other men who are today acknowledged pioneers in the field. Among these were Dr. Franz von Hoefft, of Vienna, an early advocate of the postal rocket and later founder of the Austrian Rocket Society, and Walter Hohmann, the city engineer of Essen, whose book <u>Die Erreichbarkeit der Himmelskoerper</u> (<u>The Attainability of Celestial Bodies</u>), in 1925, was an important contribution to the literature of astronautical theory. He invited their close cooperation in helping him realize his plan. In particular, he wanted their assistance in mathematically analyzing the feasibility of rockets with airfoils "like the Junker's airplanes," i.e., hollow and in which fuel could be stored. However, no such cooperation appeared among them.

By 1927, Valier's delicate relationship with Oberth reached the point of rupture. Oberth suggested that Valier lend him money and his lantern slides for his various lectures. Valier had to refuse both requests.

On June 5 of that year, an event of great importance in the history of astronautics took place in the Golden Scepter, a small restaurant in Breslau. The famous Verein fuer Raumschiffahrt (VfR), Society for Space Travel, was founded by Valier, Johannes Winkler and seven others.(5)

As the year ended, Valier began a collaboration with two men to pursue the second phase of his plan for rocket research: the application of rocket power to surface vehicles. The men were Friedrich Sander, a fireworks manufacturer of Wesermuende, and the famous Fritz von Opel, the automobole magnate. Sander had earlier provided solid propellant rockets that Valier statically tested in accordance with phase one of his plan. These tests continued through the early months of 1938 at Sander's plant, where Valier had devised a special spring balance for determining thrust from which he derived thrust-time curves. The safety precautions he took in doing so were detailed and followed to the letter, in contrast to his later experiments with liquid propellant rockets.(6)

The performance data he so gathered helped to convince von Opel, an engineer as well as a keen businessman with an eye for advertising, that a rocket-propelled car was practical. Even before approaching him, Valier secretly outfitted an Opel car with 5 and 9-centimeter Sander rockets and had driven it at Bremerhaven. Von Opel needed little additional persuasion. While he was engineer enough to know that rocket propulsion for automobiles was nonsensical and impractical, he foresaw the value accruing to him through the attendant publicity.

On March 12, 1928, a small Opel with a 5 and 9-centimeter Sander rocket mounted on the rear circled part of the test track at the company proving ground at Ruesselsheim. Valier wanted to be at the wheel, but von Opel refused to let him drive. The task went to Kurt Volkhart, a professional race car driver. It was a success, but there was no publicity concerning it. On April 11, Volkhart again drove a car in secrecy.

While these tests were in progress, Valier on March 13 visited Alexander Lippisch and Friedrich Stamer, well-known airplane designer and pilot, to discuss the feasibility of mounting solid rockets on a glider. His research schedule was moving apace.

Opel Rak 1, driven by Volkhart on April 11, made newspaper headlines around the world. Von Opel was ecstatic. On May 23, he himself drove Opel Rak 2, having again decided against Valier's driving it. The test took place on the Avus speedway in Berlin before a crowd of 300, including many members of the press. Within two minutes, the rockets had burned out and von Opel had reached a top speed of 118 miles per hour.(7)

Meanwhile, the rocket-driven glider project was progressing.

It was in part funded by von Opel, too. A specially designed delta wing plane with lightweight rockets made an unmanned flight, but the design proved unsuitable to rocket propulsion. Next a craft with canard controls and rocket propulsion was attempted. Again the results were discouraging. However, on June 1, 1928, Stamer took to the air in such a plane called "The Duck." It was equipped with two Sander rockets producing between 7 and 9 pounds of thrust each. The plane was launched by an elastic cord, but one of the rockets exploded upon ignition. However, the pilot escaped with his life.(8)

Between June and August, several attempts were made to propel an Opel on railroad tracks. All attempts were unmanned, and the vehicle had a negative airfoil at the front to help keep it on the tracks. One such run on June 23 resulted in a speed of 169 miles per hour. After a vehicle exploded on August 4, local railway authorities forbad further trials.

At this time, Valier severed his relations with von Opel and Sander. He affiliated himself with J.F. Eisfeld, in Silberhuette-Anhalt. There he designed several unmanned rocket cars for use on rails. The solid propellant rockets, manufactured by the Eisfeld Company, were mounted at a small angle toward the ground so that the thrust vector would help to keep the vehicle on the tracks. Several models were constructed and tested on a railway track within the company's facility. Larger models were tested at speeds up to 108 miles per hour on a railroad track nearby. Before the ban on such experiments, Valier himself finally got to drive one of his rocket-propelled vehicles.(9)

In October, 1928, Valier parted company with the Eisfeld firm, and began investigating the possibilities of a rocket-propelled sled.

Experiments with such vehicles began in January 1929, first in secret at the airport at Schleissheim, in which speeds of 65 miles per hour were realized, and a month later on the frozen Eibsee. During one of the latter tests, Valier's wife piloted the sled becoming the first woman to travel thusly.(10)

As the year progressed, it found Valier anxious to move on toward his goal. The next step was to demonstrate an automobile propelled by a high-pressure gas stream, which he did in the fall of the year. Using cylinders of liquid carbon dioxide and a throttable nozzle, he had a jet-propelled vehicle which he assumed would give him the handling characteristics of the vehicle he intended to build next: the liquid-propellant rocket car.

At this same time, Valier made what must have been one of the most difficult decisions of his life. Germany, he was living in Munich at the time, was almost at the bottom of its greatest depression. Valier was considerably in debt. Industrialists were not inclined to finance his research. Propituously, he was offered a well-paying job as a teacher of science and mathematics by a headmaster who was impressed by the ease with which his students assimilated the subjects after reading Valier's books and articles.

After considering the tempting offer, Valier wrote to his wife, ". . . who knows if the others really will make it? If they experience a setback, won't they perhaps lose the desire to reach for the stars? Sander will return to his ships' rescue rockets, Oberth to his school-boys, and Opel to another sport. And should I be the first to forsake the stars? No, nothing doing!"(11)

By this time, too, Valier had fallen into disfavor with the organization he had founded a few years earlier. VfR disapproved of his lecturing, writing popular articles, and building rocket-driven automobiles, airplanes, and sleds. Its members felt that such exploits cheapened what should be a more dignified approach to research.(12)

However, the pettishness that cropped up in the "Kritsche Ecke," "Critics Corner," of Die Rakete, official publication of the VfR, had little effect on him. He stuck to his plan for achieving space.

His immediate concern was the development of a liquid propellant rocket engine. By the end of 1929 he had found such a sponsor in the form of Dr. Paul Heylandt, a producer of gaseous and liquid oxygen for industrial uses in Berlin-Britz. Valier was to achieve financial support for research only; no money was available for his living expenses. The facilities of the Heylandt plant were at his disposal also. Included in the assistance offered by Heylandt were the services of Walter H. J. Riedel, a research engineer with the company.

In January, 1930, developmental work on a combustion chamber for liquid propellants began in earnest. A working model was ready within two months. It consisted of a steel chamber with a nozzle at one end and an enclosed "shower head" plate for introducing the oxidizer (liquid oxygen) into the chamber. The fuel was introduced through a port in the side of the chamber so that it impinged on the oxygen flow. The fuel was alcohol. The propellants were stored in pressure vessels some distance away from the engine, which was mounted on a commercial scale. With the nozzle pointing downward on the scale, the thrust was measured directly. (13)

Tests of the motor found Valier at the valves controlling the flow of the propellants to the combustion chamber while Reidel pressurized the propellant tanks and another assistant Arthur Rudolph ingited the engine with a plumber's blowtorch--all this done with the personnel in close proximity to the test stand. Indeed, Valier's face was often a fiery red after a test run. Thrusts of up to 18 pounds were recorded.

During this period, Valier took time out to seek additional funding from the Shell Corporation, in Switzerland. He hoped that they would be persuaded to finance the fuel side of his research as Heylandt was the oxidizer side. The company showed some interest.

On March 22, the little rocket engine was mounted on a modified car, and Valier drove it around the Heylandt plant for some 22 minutes, until the propellants were exhausted. However, gaseous oxygen was used for the test.

Much to the displeasure of the VfR, Valier again made the headlines on April 17 when he drove the Valier-Heylandt Rak 7 around the factory grounds for the press. Two days later, he repeàted the performance at Berlin's Tempelhof airport, where one of the younger members of the Society for Space Travel looked on with enthusiasm; he was Wernher von Braun. Thus occurred the first runs made by a liquid-propellant rocket power automobile, hardly a noteworthy event in the history of astronautics from a "breakthrough" point of view.(14)

In order to convince Shell that his engine would run as well on kerosene, Valier adapted it to use that fuel. One problem arose, however. Alcohol and water are miscible. Thus, a mixture can be obtained that will produce a desired temperature, one that will not melt the walls of a combustion chamber. Kerosene and water will not mix, but they can be emulsified to produce the same effect. Valier, Riedel, and Rudolph developed an emulsifier that was placed in the flow stream before the mixture entered the combustion chamber, and it worked well enough. The idea of regenerative cooling or even water cooling of the engine, though known, was not undertaken by Valier, probably for the sake of expediency. He feared that Oberth would launch a liquid-propellant rocket before he could.

Tests using the new fuel and emulsifier were encouraging. On May 17, 1930, Valier decided to push on with still more, despite the pleas of Riedel and Rudolph to call it a day. What happened is best described by Riedel: "At nine in the evening everything was ready. The combustion chamber was ignited and the pressure in it raised to seven atmospheres gage pressure by the old, well-known procedure, by regulating the hand valves for the propellants and water. Just as the combustion chamber had attained this pressure, there was a violent explosion. I immediately closed all the propellant valves and sprang over to Valier, who collapsed. I was just able to catch him as he fell, and laid him on the floor. While a machinist and Arthur Rudolph looked after him, I went off in search of a car. When I came back ten minutes later, Max Valier had already died. A tiny splinter had struck him in the pulmonary artery."(15)

What then was Valier's contribution to the advance ot astronautics?

Certainly his book Der Vorstoss in den Weltenraum (Advance Into Space), first published in 1924 and revised for several years afterwards, was in some ways a pivotal as the works of Oberth. Valier's great talent lay in explaining highly technical things in ways that could be easily understood by the man in the street. As Arthur Rudolph later pointed out, "In those days, nobody but nobody had formal training in rocketry. So the ones who did do something--write, experiment, were real pioneers; and I think Valier was foremost among them since he was the first one in Germany to experiment with useful hardware . . . . He concentrated on hardware experimentation and was not bogged down in theory, but he knew the theory . . . . In those days, everything in rocketry was a step into the unknown . . . Valier was very determined; he knew exactly what he was doing; he knew the limitations and what had to be done . . . therefore, Valier's contributions are great."(16)

While one can today wonder why such a talented man would adopt causes such as furthering the theories of Hoerbiger, one can only do so with the hindsight of four decades of advances in physics and astronomy, the latter discipline which has been advanced by a quantum jump through astronautics. One can also wonder why Valier should insist on such a pedestrian approach to spaceflight as he did, using rockets to propel first surface vehicles, then airplanes, and finally space vehicles.

However, in summary, Max Valier was one of those few men that appear on the scene at the appropriate moment in the development of any technology. He was dedicated and sincere above all. If his dedication and enthusiasm sometimes led him down paths that we know today were dead ends, it still does not detract from him and his place in the history of astronautics. Max Valier was a doer when things needed to be done in a time when there were more thinkers than doers.

REFERENCES

1. I Essers, Max Valier, Ein Vorkaempfer der Weltramfahrt, Duessldorf, VDI Verlag, 1968.

2. Max Valier, "Flying Beyond the Earth." Airways, Nov. 1929, pp.57-59.

3. Essers, op. cit.

4. Essers, op. cit.

5. "Verein fuer Raumschiffahrt E.V.," Die Rakete, Juli 1927, S. 82.

6. H. Gartmann, The Men Behind the Space Rockets, Weidenfeld and Nicolson, London, 1955, pp. 74-78.

7. F. von Opel, The Historical Development of Rockets and the Purpose and Limits of Technology., NASA TTF-13,436. National Aeronautics and Space Administration, Washington, D.C., 1970, pp. 31-33.

9. Gartmann, op. cit.

10. Max Valier, "Meine Versuche mit dem Raketenschlitten," Die Rakete, Heft 2, Febr. 1929, Ss. 24-26.

11. Essers, op. cit.

12. B. Williams and S. Epstein, The Rocket Pioneers on the Road to Space,

13. W. H. J. Riedel, "Ein Kapitel Raketengeschichte der neueren Zeit," Weltraumfahrt, H3, Juli, 1953.

14. W. von Braun and F. I. Ordway, The Rockets' Red Glare, New York, Doubleday, 1976, p. 132.

15. Riedel, op. cit.

16. Letter to author from Arthur Rudolph, August 6, 1976.

MAX VALIER (1895-1930)

G. ARTURO CROCCO
(1877-1969)

Luigi Crocco*

It is rather unusual that the son of an outstanding man be invited to present his father's biography. No doubt, this choice involves the danger of not complete objectivity. However it is obvious that no one better than a son, working in a close field of activity, can be aware of the details of the life and of the accomplishments of his father.

For instance, who knows today, even in Italy, that in 1912 an hydroplane, designed by my father as an application of his theoretical stuies on airplane stability, made its first (and, unfortunately, last!) flight over the lake of Bracciano under the yet unskilled hand of Commander Ginocchio (who came out uninjured)? A regrettable consequence of the accident was the order to my father, from his superior military authorities, to stay away thereafter from the heavier than air vehicles, and concentrate on those being lighter than air, being closer to operational use. As a result, my father did not find again an opportunity to materialize an aircraft of his own conception until 1929, when he designed the CC20 together with the Breda Industry.

Similarly, no one seems to remember today, writing the history of airships, that my father was the first one to conceive the idea of placing engines in more than one separate propulsive nacelle. This happened in 1912, when he designed a 40.000 cm rigid airship which went through complicated vicissitudes, mainly for the reason that, in Italy, the technology of aluminum alloys was not yet sufficiently mastered. For this reason the airship manufacture was planned to take place in England, with the collaboration of the British Government, except for the nacelles of which a prototype was built in Germany in 1913. Shortly thereafter, engine nacelles of this type made their first appearance on the Zeppelin airships. I remember quite vividly the shining prototype lying somewhere in the Instituto Centrale Aeronautico, where my father used to take me as a child to satisfy my enthusiastic curiosity. The outbreak of World War I brought an end to the project in 1914.

And who would know today that during the First World War, my father invented and successfully tested an original ultrasensitive fuse designed to explode against the light airship envelope, although unaffected by strong air forces? Or that he conceived and tested a 24-barrel gun for defense against aircraft? Or that he was the first one to dare placing the then-famous French 75 mm gun on the nacelle of an airship, much against the categorical objections of the French designer? The gun was used many times to blind the enemy search-projectors during the First World War.

Again, very few people are aware today that in 1924 my father designed in great detail a "laboratory of the extremes" for the study of physical and chemical environment at very high or low pressures and temperatures, or that in 1943 he conducted a detailed study (and built a model) of an airplane catapult based on the use of two water jets functioning as rails and at the same time producing forward propulsive effect?

---

*Formerly Professor at Rome University (1939-1948) and at Princeton University (1949-1969). Presently Professor at the ECOLE CENTRALE in Paris, France. Paper read by E. M. Emme.

These examples, and even more the better known highlights of his career as an inventor and a scientist, give a measure of the fertility of my father's mind. Evidently, better than anyone else, his children have been aware and are witnesses of these qualities illuminating not only his scientific career, but also down to the little events of every day's family life and keeping their young minds under a continuous state of fascination.

But let's follow now a more conventional presentation of his life and works. Gaetano Arturo Crocco was born in Naples on the 26th of October, 1877. His father, an esteemed civil engineer, was Neapolitan, from the old Basile family, Parmelitan. His childhood and most of his schooling took place in Palermo, including the first two years of his University education. He chose the field of Engineering, much to the great disappointment of his high school teachers, who had predicted a brilliant career in the literary field. Indeed, he already had a noticeable literary production, including a never-published philosophic and almost science-fiction novel. But the technologic promises of the twentieth century stimulated his mind, and proved to be a stronger attraction. In 1897, he left Sicily for the Military Academy and the Scuola de Applicazione de Artiglieria e Genio of Torino. In 1900, he graduated as a Lieutenant of the Technical Corps of Genio Militare, and was immediately assigned to the Specialists Brigade, in charge of study and application of advanced technologies, where he became responsible for the study of telephotographic cameras for use from ships and from airships. In 1902 he was transferred to Belgium to study at the then-famous Montefiore Institute of Electrotechnics at Liege.

At Liege, he graduated first of his course and was immediately offered a brilliant position at Westinghouse of England, an offer which, had he accepted, would have completely changed his career. But his heart was already taken by a Palermitan girl, Bice Licastri Patti, whom he wanted to marry and whose father would never have allowed a marriage leading to expatriation. So he declined the offer and went back to the Specialist Brigade. Here he conducted, in collaboration with the Navy, a photographic campaign from the sea with a pendular camera of his invention, thus having an opportunity to study the equations of the oscillations of ships. The transition to the study of oscillation and stability of flying machines was almost inevitable, as shown by a paper on airplane stability presented in 1903 at the French Academie des Sciences. The new fascinating field was on the verge of reaching, with the historic flight of the Wright brothers, the age of explosive growth. It is certainly no coincidence that of the 11 papers Crocco published in 1904 and 1905, only one, related to naval photography, does not concern aeronautics. These papers gave great relevance to the study of propellers, both for propulsion and sustentation, and of the stability of flying machines, bringing to light for the first time the fundamental importance of the dynamic derivatives. This discovery was recognized by the Academie des Sciences with a beautiful gold medal in my possession today.

These and the following years of intensive work, directed at theory, experiments and realizations; and one wonders how such a busy person could find the spare time to marry Miss Patti in 1905, and how he could manage to be an always attentive father for the seven children they had between 1906 and 1925 (I being the third) constantly taking a very active and stimulating part in their education until they became of age.

From those first years are the realization of a rather primitive wind tunnel alimented by the discharge of a gasometer, the first rational study of aircraft propellers, the first suggestion of the utilization of propeller autorotation, culminating in a 1906 patent. Also from those first years is the realization, determined by the need of testing full-sized propellers, of two hydrofoils with aerial propulsion. The first, a monopropeller, was operated on the Tiber, first, and next on the lake of Bracciano, where the second, a bipropeller, reached the speed of 100 km/h. The propellers were well in advance of the times, being all-metallic and variable-pitch. It is right to observe that hydrofoils had already been experimented elsewhere. However, the V-shaped lifting surfaces used on modern hydrofoils made their appearance for the first time, and were actually the object of a patent which had unfortunately fallen into the public domain when, over fifty years later, they were widely applied.

In those years Crocco began also to take a serious interest in the realization of an airship. The task was undertaken very systematically making use of known experimental techniques and inventing new ones. A rigorous theoretical analysis of the stability of the airship showed that the existing theories were strongly in error because they ignored the dynamic derivatives. A conventional Froude water tank was built to measure the static aerodynamic coefficients and to check on models the predictions of the stability theory. Another unconventionally circular water tank was conceived and used to measure the dynamic derivatives. The airship was to be of the semi-rigid type. The shape to be assumed by the hydrogen-filled envelope was studied both theoretically and experimentally using new modeling techniques and a simple type of analog computer. The spacial limits of flammability of the jets of hydrogen when discharged through the valves were carefully determined, again on models. All these details I mention to indicate how, in those pioneering years of intuitive design, my father inaugurated in aeronautics the method of systematic prediction based on theory and experiments, the only one valid today.

The construction of the airship took place in 1907-1908 and its first flight, in the sky of Rome, on October 31, 1908, under the inexpert pilotage of my father himself. Born just 3 months later, I was aged only a few weeks when my father took me and my elder brother on a flight, making us perhaps the first children to fly.

After the first airship, many more followed, with the required changes and improvements, and had their fire-baptism in the Lybian war. Altogether, in about 10 years, 30 semi-rigid airships of four different models and sizes were built, and widely used during the First World War for exploration, bombardment and coast-guarding missions. Even a 40.000 cm rigid airship was designed, as I have mentioned before, but the outbreak of the war interrupted the construction.

It is important to single out here an important publication of the pre-war period. It was in 1911 indeed that Crocco published his most complete paper on airplane stability, the first study in the world to present a rigorous and complete treatment of the problem. In 1914 Crocco founded the Istituto Centrale Aeronautico to concentrate and coordinate the aeronautical research and testing. The experimental plants already functioning at that time (such as the two water-tanks and a 1912 wind-tunnel) were housed there. New plants and structural as well as chemical laboratories were added. Among them the 1914 double-return wind-tunnel which was still in use in 1935 when the Aeronautical research was transferred to the new center of Guidonia, designed itself under Crocco's inspiration and supervision.

For the next six years Crocco directed the Institute, which constituted during the war a center of intensive research, with the precious collaboration of many university professors, including the great mathematician Vito Volterra. The Institute rapidly became a center of attraction and an object of visit for scientists of the whole world. Crocco, with his co-workers, conceived there many interesting devices. For instance, in addition to the few previously mentioned, a "tele-bomb" and a "tele-torpedo" were designed and successfully tested, to be dropped 20 km away from the target, with a gyroscopic guidance system, the first realization of what can rightly be considered a precursor of remote range missiles. Another example is, among other instruments, a course indicator, later acquired and employed by Alcock and Brown for their 1919 airplane crossing of the Ocean. In view of his intense activity and outstanding achievements, Crocco advanced in grade very rapidly. He was a Colonel when, in 1920, he left the Army to try his way in private industry. The decision was partly due to the demobilization of the spirits after the war, but was certainly precipitated by the pressures of his father-in-law asking him to bring back his home from Rome to Palermo. However the attempts in the industrial field did not lead to durable results. The only one worth mentioning here is the design, under the sponsorship of the industrial Usuelli, of a large semi-rigid airship of 120.000 cm for passenger transportation. The sudden death of the sponsor prevented the realization of this very interesting airship.

In the meantime Fascism had come to power. In 1923 Mussolini, in what may be considered his first and last opening toward the opposition, called the renowned antifascist physicist Corbino to be minister of the National Economy. In turn Corbino, well aware of the many facets of Crocco's personality, asked him to assume the Direction of Industry. Naturally Corbino's ideas were very soon in contrast with those of Mussolini, and the experiment ended in a divorce. But the two years it lasted were enough for Crocco to become deeply involved in the basic problems of modern industry and write three important papers, one on the role of scientific research in industrial independence, another on the economic revolution due to the great progress of chemical synthesis, and a third on the juridical aspects of scientific property.

From those years are also the conception of the already-mentioned "laboratory of the extremes," of a Scientific Institute for Industrial Synthesis, and of a "machine" (that can be considered as a forerunner of modern atom-smashing machines) destined to the transmutation of the elements.

Also from 1923 is a paper in which he suggested the use of radioactive energy (the only form of atomic energy available then) for space propulsion, discussing ways of realization. However these new involvements did not in the least diminish Crocco's ardor for aviation, as witnessed by the 17 publications he wrote on aeronautical subjects between 1923 and 1925, most of them concerning commercial aviation, and the others on the helicopter problem, on hydroplanes and a few other subjects. In fact, right after the School of Aeronautical Engineering was created in 1926 in Rome, he was invited to teach a course on the theory and construction of airships. Of 1926 are also three rather essential publications on the future of rocketry as a more powerful tool than gunnery to reach high speeds and ranges, as well as a patent application for a jet airplane (granted in 1930). A very fundamental publication of the same year was on "super aviation," establishing for the first time the importance of associating high speed and altitude. In these papers we can see Crocco's mind always pointing toward the new possibilities offered by technical progress. His publications on rockets, as well as the one previously mentioned on the applications of atomic energy to extraterrestrial propulsion opened the way to his future work on astronautics.

The following year 1927 saw Crocco's nomination to the permanent position of Professor in the School of Aeronautical Engineering. In the same year the General Staff, impressed by his publications on rocketry, asked him to start a practical study of rocket motors and rockets and provided some funds. In 1928 Crocco was recalled as a reserve General to be the Director of Aeronautical Constructions in the Ministry of Aeronautics of recent creation.

At this point starts, to last until 1933, what Crocco himself defined as the most laborious period of his professional activity. Indeed only his rigorous discipline of work, his power of synthesis and his rapidity of decision can explain how he could manage to fulfill very conscientiously and with great efficiency his ministerial tasks while at the same time he did not miss teaching a class, and directed with substantial personal involvement the theoretical and experimental research of rockets for the General Staff.

It must be observed that his ministerial position was soon supplemented by additional official tasks, such as his nomination to President of the Technical Committee of the Italian Aeronautical Register (1928-1960) in charge of the procedures for acceptance of commercial airplanes, to member of the Superior Council for Public Education (1928-1945), to President of the Aeronautical Section of the National Research Council (1929-1943). But, independently of these additional charges, it is easy to realize how intensive his activity as Director of the Aeronautical Constructions must have been if one considers that those were the years of the great Anglo-Italian competition which was the Schneider Trophy and of the famous Balbo's Atlantic cruises, and that the laborious preparation to those enterprises took place under the direct responsibility of Crocco's service.

His professional position was also distinguished by intensive developments. In 1929 he switched to a more actual and appealing course on "General Aeronautics," a broad title under which the content of the course could be continuously adjusted to cover more advanced questions of airplane aerodynamics and design. Part of his ideas on these subjects were published first in a 1930 limited edition as "Elements of Aviation," the first Italian didactic book in this field, and in 1931 offered with revisions and additions to the more general public.

In the same period 1927-1933, 18 papers on several problems of great actuality were also published, among them two on airplane structural problems (connected to the design of the CC20, to be mentioned later), 4 on blind flight, 2 on flight mechanics, 2 on ramjets and lifting ramjets and one quite fundamental on the very high speeds of flight for which he again demonstrated the importance of high altitudes and jet propulsion. This prophetic paper, published in Great Britain in 1931, was received with ironic comments by the more conservative aeronautical engineers.

The research on rockets had started just one year before he became involved with his Ministerial position. Considering that this Symposium is particularly devoted to astronautics, it seems appropriate to deal a little longer with this side of Crocco's activity. During that year he personally designed and conducted, with my enthusiastic help (I had entered the University in 1926), a series of experiments on deflagration of solid double-base propellants at the relatively low pressures appropriate to rockets, measuring the burning rates, checking the ignition and charge holding procedures of constant surface grains, in view of safety and repeatibility, and finally performing a first series of launchings of fin-stabilized rockets. This part of rocket research was conducted very swiftly under the firm hand of my father at the laboratories and testgrounds of the Bombrini-Parodi-Delfino Company.

However, the new heavy responsibilities of Crocco starting in 1928 made it necessary for him to abandon the actual execution of the experiments, leaving it to me. But despite his exceedingly busy time, he always kept supervising the research through long and detailed discussions we carried on at home during the evenings and the week-ends. The supervision and the financing were made easier by the transfer of the experiments to the Stabilimento di Costruzioni Aeronautiche, a branch of his service. Here the previous experiments were carefully reviewed and repeated in 1928 and 1929 and resulted in the choice of a better propellant. But no additional launchings were performed, maybe because my own mind, contrary to that of my father, was more addicted to speculation than to pragmatism; but maybe also because in the meantime the attention of my father had become attracted by the wider possibilities of liquid propellant rockets.

It was in 1929, indeed, that he decided to test a rocket motor of this type. After careful survey the choice fell on the storable bi-propellant combination petrol-nitrogen tetroxide, and we designed and built a 5 kg thrust testing chamber with regenerative cooling. It was tested with the assistance of a chemist, Dr. Landi, and operated successfully for a 10-minute run in 1930. But again, having established the principle, very little further work on the bipropellant rocket motor was programmed. The tests underwent an interruption of one year in 1931 due to my military service. In the meantime, Dr. Landi died and when the research was resumed in 1932, it was in a new location, my father's Institute of General Aeronautics at the University, with a new chemical help, that of Dr. Corelli, and in the brand new direction of liquid monopropellants.

Indeed the nocturnal exchanges of views between my father and myself had brought forward the apparently simpler possibility of using a single liquid, or a liquid slurry, in place of the more elaborate bipropellant systems. The first idea was to use a double base powder slurry in a suitable supporting liquid. It was soon replaced by nitroglycerine, attenuated and desensitized by the addition of 30% methyl alcohol. During 1933 this still rather dangerous monopropellant was tested in a combustion chamber, producing the expected chemical and thermal results. It was followed by mononitromethane, a powerful monopropellant we had discovered in the meantime, which was thought to be less dangerous. Experiments with nitromethane were successful, so that research on its applications went on for a certain time despite a bad accident in 1934. But it does not seem appropriate to give further details on these applications since rocketry was not involved, and moreover, my father's role finally had reduced just to general supervision until it vanished entirely.

We see what an extraordinary load of activities Crocco was able to carry in those crucial five years 1928-1933. As if this sum of activities were not sufficient, he became also involved in 1928 in the structural design of a large Breda airplane, the CC20, for which he realized an original and interesting type of structure. The airplane successfully flew in 1929.

In 1932 as a result of divergences with Balbo born during the preparation of the Atlantic cruises, Crocco moved to the Direction of Aeronautical Experiments. He spent in the new position a little over a year, after which his relations with Balbo became so strained that the only way out was the retirement from the official post. In the meantime, the creation of the new Experimental Center of Guidonia, including the construction of the supersonic wind tunnel, had been decided, and its detailed layout developed under Crocco's careful and imaginative direction.

The 1933 resignation from the official post closes the period of Crocco's multifaceted activity. From there on his central position becomes that of professor, and soon Chairman (1935-1945) of the Scuola di Ingegneria Aeronautica at Rome University, and most of his efforts are devoted to science and technological progress. However there was no production slow-down, as proved, among others, by the world-famous 1935 Volta Congress held under the auspices of the Academy of Italy of which Crocco had become a member in 1932.

The Volta Congress on high velocities in aviation came to represent in the following decades one of those rare gatherings where the bases of a future technology are presented and scientifically discussed in advance. In addition to the participation of the engineers responsible for the hydroplanes of the Schneider Trophy and for their engines, it assembled for the first time under Crocco's presidency the major aerodynamicists of the whole world (such as Prandtl, Von Karman, G.I. Taylor, Ackeret, Busemann, Burgers) for an extremely stimulating presentation of the problems and techniques of supersonic flight.

The continuous outflow of scientific and technical publications on many aspects of flight mechanics, aerodynamics and aeronautical technology (20 publications between 1934 and 1938) was slowed down in the war years (only 7 between 1939 and 1942). In those war years Crocco became deeply involved in developing a prolongation of his 1918 telebomb, an airplane-borne self-guided anti-naval bomb designed and tested in those years. The end of the war brought all activities to an end, and was followed by a period of silence. In those years of turmoil consecutive to the armistice, report and details about the revolutionary war weapons kept reaching the scientific public. Certainly what impressed by father most were the V2 and the atom bomb reports. It appeared clear to him that with these technical achievements the old dream of extraterrestrial navigation was coming close to realization, and that no doubt humanity would soon see an explosive development of astronautics. What happened next was that this aging man, close to retirement, became suddenly animated by an enthusiastic activity aimed at analyzing the new possibilities and convincing the scientific public as well as the Italian Authorities that the astronautical era was coming. Twenty-one of his 26 pre-Sputnik, and 16 of 18 post-Sputnik publications were devoted to astronautics. Quite a few contain the scientific exploration of realizable space-missions. or the scientific projections in as as yet unrealizable but exciting future. The remaining publications were written to popularize and propagandize the advent of astronautics either in technical magazine and encyclopedia articles, or in always-crowded public lectures with the help of illustrative models and simple experiments.

In these publications many new conceptions were presented and analyzed. Well before the first intercontinental ballistic missiles, Crocco studied the ballistics as well as the reentry and stability problems of what he named "geodetic missiles," spanning the Earth for military or commercial applications. The problems of staging, the possibility of using piloted, recoverable boosters of the ballistic or winged type, the solution of parallel-mounted boosting moduli (to avoid gigantism) and that of recoverable cisterns for booster refilling in space, were presented and analyzed in detail. One of the practical purposes of these studies was the cost reduction to bring the financial load into reasonable limits and allow the exploitation by private companies.

A very interesting 1956 paper presents the detailed study of an economical one-year exploration Earth-Mars-Venus-Earth, based on the new idea of using the swing-by maneuver around Venus to correct without additional fuel expenditure the perturbation due to the Mars swing-by. This round-trip mission was calculated in great detail for the true orbital data of the planets, and, under the name of "Crocco Mission" was adopted in 1957 by NASA in a request for study by industrial concerns. The exploitation of swing-by maneuvers, introduced by Crocco, has more recently greatly enhanced particularly expensive missions such as the close solar probing or the remote planets exploration.

Many additional points of interest for the future of astronautics were raised and discussed in Crocco's publications, such as the future use of atomic energy to reach the closest star; but also questions of a non-technical nature, like the foundations of space law.

Crocco's activity in the astronautical field was not limited to lectures and publications. His was the creation and organization in 1951 of an Informative Course of Advanced Ballistics at the Scuola d'Ingegneria Aeronautica of which, although retired from the teaching of General Aeronautics, he had again been called to be Chairman between 1948 and 1952. In that year of 1951, he was the founder and first president of the Italian Rocket Association which was the instrument, in the same and following years, for the official participation of Italy to Congresses of Astronautics and to the creation of the International Astronautical Federation of which Crocco was co-founder. In 1956 under his intensive direction the Italian Rocket Association organized the first large-scale congress of the I.A.F. in Rome. Crocco, with his reputation and authority, was able to raise the necessary funds for this VII International Astronautical Congress which played a greatly stimulating role for the future of astronautics and preceeded the first Sputnik by about one year. He very actively contributed to the Congress not only through his role of President, but also with an exciting inauguration speech and with the already-mentioned paper on the Earth-Mars-Venus-Earth mission.

Crocco's outstanding achievements as a pioneer and a scientist in the fields of aeronautics and astronautics were recognized during his long career not only by the Italian Authorities, but also by national and international scientific and academic organizations. He was a member of Italian and foreign academies and scientific societies, including, since its foundation, the Institute of Aeronautical Sciences (I give this detail because his Fellowship has mysteriously disappeared from the Membership Roster, but the original certificate is in my possession). He was the recipient of decorations, prizes and medals from Italy and other countries. Crocco's publications, spanning over a 58-year period which has seen the explosion of modern advanced techniques, represent an essential document for the history of aeronautics and astronautics, certainly the most relevant as far as Italy is concerned. It was therefore a great disappointment for all those concerned that the publication of his complete works, a joint enterpris of the Accademia dei Lincei and of the Ministry of Defense decided soon after his death, had to be retrograded for the lack of money to the status of selected works containing only 60% of the total material. This would not have happened were it not for the profound crisis Italy is going through.

The examination of his career, of his 138 publications and of his more than 30 patents show with evidence that Crocco's mind possessed a quite unusual combination of analytical skill, inventive imagination, and realization power with an exceptional foresight concerning future technical events. His mind had to an extreme degree the power of simplification. In every problem he was able to set aside the accessory and aim straight to the essential. Only in a second time the details and the intricacies were taken into account. All his studies, even the most analytical, were directed at a conclusion, a practical criterion, an order of magnitude estimation, a quantitative rule.

Without danger of being partial, one cannot but recognize the outstanding importance of Crocco's contribution to aeronautics and astronautics, and, concerning the latter, one can wonder about the additional, more effective directions his contribution could have taken had he been active in a country with wider possibilities. To conclude, I would like to thank the International Academy of Astronautics for offering me the possibility of bringing a contribution to the perpetuation of the memory of a man who occupies a very special position among the pioneers and the scientists of the aeronautical and astronautical era.

G. ARTURO CROCCO (1877-1969)

JOHANNES WINKLER: EARLY INVESTIGATOR IN LIQUID PROPELLANTS (1897-1947)

Mitchell Sharpe*

Born in Karlsruhe on May 29, 1897, Johannes Winkler attended the Realschule in Oppeln and the Gymnasium Johanneum in Liegnitz. For two semesters, he studied machine design at the university in Danzig before World War I broke out. During the war, he was employed in the German submarine shipyard in Danzig. Following the end of the conflict, he resumed his education, spending an additional eight semesters at the universities in Breslau and Leipzig and took his examinations in 1922. He subsequently spent a few years in an administrative position with a church.(1)

A few years later, largely at the urging of Max Valier, the well-known lecturer and writer on astronomy and spaceflight, Winkler undertook to form an organization of space-flight enthusiasts in Germany.

On June 5, 1927, a group of nine men gathered in the Golden Scepter Restaurant in Breslau. Winkler and Valier with the others proceeded to organize the Verein fuer Raumschiffahrt (Society for Space Travel). While Valier lent his whole-hearted support, he told the group that he simply could not accept the proffered position of president. The mantle fell to the shoulders of Johannes Winkler; Valier had a pressing reason for not accepting the honor. He could not devote the time to the job because he needed every minute to earn a precarious living.(2)

Winkler became editor of <u>Die Rakete</u>, the society's journal and remained so until June, 1929, when Willy Ley took over the task of editing a briefer version, really more of a newsletter.

The relatively short-lived association got off to an inauspicious start. In seeking to gain legal status, the group ran into what at first seemed a monumental bureaucratic obstacle. The authorities balked at recognizing an organization the name of which contained a word not existing in German dictionaries: Raumschiffahrt (space travel). A change of name was suggested, but a compromise was not acceptable to Winkler. A final solution to the problem was found as early member Willy Ley later recalled, "However, the court finally relented and, since new inventions require new words, it accepted the registration of the society under the condition that the document of registration itself define the name in an unmistakable manner."(3)

The charter so complied and the registration was granted.

Within a year, the membership of the VfR, as it is generally known today, grew to some 500 members. It included some of the foremost names in the fledgling science of astronautics. They included Hermann Oberth, then of Romania, and Walter Hohmann, Germany; Count Guido von Pirquet and Dr. Franz von Hoefft, in Vienna; Robert Esnault-Pelterie, of France; and Professor Nikolai Rynin, of the USSR. By September, 1929, the number of members had risen to 870, including a 16-year old lad who had just graduated from high school, Wernher von Braun.

----

*Alabama Space and Rocket Center, Huntsville, Alabama.

The organization introduced itself to the German public with a certain elan, which failed to materialize as things turned out.

On April 30, 1930, it sponsored a public lecture on its aims and space travel in general at the auditorium of the German Post Office, in Berlin. The post office had evinced some interest in a postal rocket, if such a thing could be built.

Among the invited guests were a number of industrialists and businessmen who might be persuaded by what they saw and heard to donate money to the VfR. However, the members were concentrating their efforts on the postal authorities and officials from the German weather bureau who were present; it was hoped that the latter group would see the practicality of a meteorological sounding rocket and invest in it accordingly.

The room was decorated to attract the attention of the audience, officials or lay public. There was a model of one of Oberth's rockets with its parachute deployed hanging from the ceiling as well as his "kegelduese" rocket that would be statically fired some three months later at the Chemische-Technische Reichsanstalt (comparable to the US Bureau of Standards).

Winkler gave the major lecture; but despite his serious and professional attempt to convince everyone that space travel was almost around the corner, few Reichsmarks were forthcoming from the audience.(4)

Among those present that night was founding member Max Valier. By this time, he was held in something less than high esteem by the general membership. Over the past three years, he had been making headlines and newsreels with rocket-propelled automobiles, gliders, and sleds. In doing so, he incurred the displeasure of members who felt that such unrealistic uses of the rocket were derogatory if not disgraceful and unprofessional. Winkler was especially upset with Valier: "At times, I was strongly tempted to join in with this rocket-mania, though it aimed more at vulgar popularity than scientific progress. It is always so much easier to show off second-rate rockets than to strive to obtain really worthwhile results from machines designed for high performance, as these 'thoroughbred' rockets are always liable to misbehave. Nevertheless, I have always maintained the principle of devoting myself only to serious research, the premature application of which can bring nothing but harm." (5)

Indeed, while he had no use for the popular press, he contributed little to the literature of rocketry, other than a few highly technical articles for Die Rakete. On occasion, he did make similarly serious lectures before interested groups in Breslau. However, in 1928, when the VfR sponsored a book Die Moeglichkeit der Weltraumfahrt (The Possibility of Spaceflight), with contributions by members and edited by Willy Ley, Winkler did not provide a chapter.

While Winkler was an engineer and primarily was interested in rocket propulsion, as were most other rocket experimenters in Europe at that time, he did consider the human factors of spaceflight. On July 10, 1928, he performed what must have been one of the earliest experiments in spaceflight medical research, crude as it was. A travelling carnival visited Breslau, and it featured a carrousel driven by an electric motor. The new machine immediately attracted his attention, not for diversion, but for research into the effect of acceleration of space vehicles upon astronauts.

Winkler rented the machine and hired one of the carnival performers. A man known only as Wittkuhn was to become one of the first human guinea pigs in space medical research. Clad in a clown's suit, this early pioneer entered the Cyklon Rad, and its operator turned on the motor. The machine reached a speed of 24 revolutions per minute, timed by Winkler with a stop-watch. Since the centrifuge had a radius of 3.2 meters, he was mathematically able to determine that his test subject had undergone 4.3 G's of acceleration. More importantly, Winkler noted the ability of the test subject to move his hands, arms, and legs under such accelerative loads.(6)

Thus, by more than a year, Winkler performed an experiment that would be done with less scientific elegance by fellow member of the VfR, Hermann Oberth in 1928. Oberth had been hired by movie director Fritz Lang as a technical advisor on one of the last silent films to be made in Germany, and a science fiction film at that. It was, of course, the famous Frau im Mond (The Woman in the Moon). Upon visiting the movie set one day, Lang was amazed to find Oberth supine upon the floor with stage hands stacking lead weights, usually employed to hold down "flats", upon his chest and stomach. Lang naturally inquired what was going on, and Oberth replied that he wondered what it felt like when a rocket ship took off from Earth.(7)

Unhappy with the pace at which research was progressing within the VfR, Winkler decided to follow his own course. As early as 1928, he had theoretically looked into the energies available with solid propellant rockets and decided that there was no hope for them in space travel. The future, he knew, was in liquid propellants.

By the end of 1929, he had proposed a rocket engine that employed methane and liquid oxygen. The design was tubular with a ceramic lining of the nozzle. Development of the engine apparently began in the summer of 1930.(8)

His research was partially financed by the industrialist Hugo Hueckel, and by the end of the year tests showed that the motor could produce a thrust about twice its own weight. A flight-test version of the rocket consisted of three aluminum tubes 24 inches long that supported the motor. The tubes contained liquid oxygen, liquid methane, and compressed nitrogen. The motor itself was made of seamless steel and was approximately 18 inches long. The three tubes were joined by flat plates, each of which bore one of the three words Winkler, Hueckel, Astris (the last being the ultimate destination of the rocket or its progeny).

The first attempt to launch the HW-1, as it was numbered, occurred on February 21; but it was a failure. The second attempt, on the parade ground at Gross-Kuehnau, near Dessau, was a success in the technical sense of the word. The rocket took off in a curving trajectory, reaching a maximum ordinate of some 180 feet and travelling a horizontal distance of some 570 feet.(9) However, it was the first liquid-propellant rocket launched in Europe.

The news was received with dismay and disappointment by the VfR. Winkler had earlier had differences with the organization and resigned as president. Hermann Oberth had been appointed to succeed him. Under the new direction, the VfR felt confident that a product of their labors would be the first such rocket launched in Europe.(10) However, any resentment soon disappeared; and the VfR went to work with renewed energy and the encouragement that a liquid propellant rocket could fly.

In October, 1931, Winkler, by then employed by Junkers Airplane Co., and his new assistant Rolf Engel appeared at the VfR's Raketenlugplats, Berlin (Berlin Rocketport), in the suburb of Reinickendorf. Winkler proposed to rent an outlying building from the group for a few months. He did not say why he needed it, but the roar of a rocket engine a few weeks later did not surprise the VfR.

Winkler and Engel were at work on HW-2, the successor to HW-1. The rocket was bigger and more ambitious from an engineering viewpoint than its predecessor. Engel later described it in the following terms: "This rocket (1.50 meters long and a lift-off weight of 50 kilograms)-- with spherical propellant tanks arranged one above the other, a parachute in its nose, engine in the aft part, stabilizer fins, and a streamlined hull of very thin Electron sheet metal--had a mass ratio of 4.8, which was superior to the mass ratio of the later V-2."(11) But the rocket, despite its engineering sophistication was put together hastily. The engine had been through only one static firing, and the nozzle had burned through on that occasion. Outwardly, the rocket resembled very much the one designed by Oberth during the production of the movie "Frau im Mond".

However, Winkler took pains to assure the press that he was an independent investigator and did not necessarily follow the design philosophy of his contemporaries. To a reporter he said, "To obtain the highest performance, such as is required for space flight, those rocket experts, Oberth and Goddard, suggested that two or three rockets increased in geometrical progression should be employed. The drawback to this procedure consists in the fact that it is difficult and expensive to test such large rockets in the laboratory. In addition, in view of the discrepancy between weight and recoil (i.e., thrust), it is difficult to take off with such large rockets."

"I therefore took another line," he continued, "starting from individual rockets of a given size, whose properties are known. For the highest performances, I have evolved a formula which makes it possible to work out the figure required for any given performance, in respect of every kind of rocket tested at this laboratory....The advantage of this formula consists in the fact that it shows the way along which rocket investigation must proceed."(12)

After four successful static tests, the HW-2 was ready for flight by mid-summer; the launch was to take place from Griefswalder Oie, a small island in the Baltic Sea just off the coast of another island named Peenemuende, where four years later would be established the famous rocket research center that produced the V-2. The Oie had earlier been proposed as a launch site for Oberth's rocket, but authorities forbade such activities as a danger to a lighthouse upon the island. Winkler and Engel lost several months arguing with the officials and having no more success than Oberth.

While Winkler was waiting to hear from the authorities about using the island for launching the HW-2, Engel and his fellow workers had printed up a mock advertisement, which they presented to Winkler on his birthday, May 29.

> HIGH-POWER ROCKETS AND JET MOTORS FOR SCIENTIFIC AND INDUSTRIAL USES
>
> Delivery of complete propulsion assemblies for rocket-like aircraft of any size and performance, especially for flight above 100 kilometers altitude or more than 200 kilometers range, respectively.
>
> USES
>
> Exploration of the atmosphere to its limits, transportation of urgent matter over long

distances or inaccessible territory, or ship to shore. Take-off assistance for float planes and flying boats; rescue work; influencing the weather, etc., etc. Rocket propulsion is used wherever a powerful push of short duration, coupled with low weight of the mechanism, is required. Only under special circumstances is it applicable for long duration. Rocket propulsion is used in highly attenuated air or empty space.

INQUIRIES INVITED

Precise information about the purpose, the altitude desired, range, weight of the aircraft, and sometimes permissible acceleration, etc., will save much time-consuming correspondence. For flights up to about 100 kilometers, solid fuels will prove simpler and cheaper; beyond that liquid fuels are preferable.

ROCKET RESEARCH INSTITUTE AND CONSTRUCTION BUREAU

Johannes Winkler. Berlin-Reinickendorf-West 3

The alternate launching site turned out to be the Frische Nehrung, a desolate spit of sand off the coast of East Prussia, not far from the city of Pillau.

The first attempt to launch the HW-2 took place on September 29, 1932. However, Engel recalled, ". . . the propellant valves froze, blocking the flow of propellants into the combustion chamber."(13)

The next attempt was made on October 6. To lessen the possibility of an explosion occurring from leaks of the oxygen or methane, Winkler had built into the rocket a special feature: a stream of nitrogen through the "hull" to flush out such gases prior to ignition of the motor. Apparently the system did not work or worked imperfectly. The HW-2 detonated and was completely destroyed. Only one member of the press had been permitted to witness the test, and he was sworn not to write anything about it if it were a failure. However, the news soon leaked out, and Winkler was quoted as saying, "I learned a lot today. I guess I'll have to try again."(14)

However, he never tried again.

Even the VfR was beginning to disperse, despite the record by May, 1932, set at its Raketenflugplatz--220 static tests and 87 launches of liquid propellant rockets. Only one week before the ill-fated attempt to launch HW-2, Wernher von Braun had left the VfR for the employ of the German Army Ordnance Office.

Winkler returned to the Junkers Company where he became involved in the design of small, liquid-propellant rockets for assisting heavily laden aircraft to take off. He also reportedly worked upon a sounding rocket, but neither is believed to have reached fruition.(15)

There was later some talk of his joining a group that would include Engel and the German Army organization with von Braun and others of the former VfR, which was to have been formed by the SS and was to have taken over the operation of the Reinickendorf facility. However, Hitler upset that plan by making the Germany Army Ordnance Office the sole center for rocket weapon research.

As more of the old VfR group joined von Braun at Reinickendorf, Winkler had to look elsewhere to continue his career in rocketry. In 1936, Eugen Saenger was hired away from the technical university in Vienna to form a rocket research center for the German Air Force. Located in a desolate area of the Luneberg Heath, between Hamburg and Braunschweig, the establishment was a division of the Institute for Gas Dymanics, which was in turn, a component of the Hermann Goering Aeronautical Research Institute. Winkler found a place with the Institute for Gas Dynamics, as a technical assistant in its transonic and supersonic wind tunnels.(16)

Again he was frustrated in not being able to work directly in rocket research.(17)

Winkler never once got to visit Saenger's nearby operation, nor did he get to accompany Dr. A. Busemann, director of the Institute for Gas Dynamics, on his semi-annual visits to Peenemuende, where he was a consultant on the developing V-2.(18)

For the man who had launched Europe's first liquid propellant rocket, the war years and the great strides made during them in the technology of which he was a pioneer were bitter ones. Why he was not placed in a more productive position considering his obvious experience and skills is not known. Undoubtedly the political element entered into the causes. Like his fellow pioneer Hermann Oberth, he was destined to contribute little to rocket research or technology in its most formative years.

Johannes Winkler died, in Braunschweig, on December 27, 1947.

REFERENCES

1. "Johannes Winkler", Die Rakete, H. 7, July 1928, S. 110.

2. "Verein fuer Raumschiffahrt E.V.", Die Rakete, 15 Juli 1927, S. 82-84.

3. Willy Ley, Rockets, Missiles, and Men In Space, New York, Viking 1968, pp. 108-109.

4. Ibid., pp. 124-125.

5. "Versuche ueber den zulaessigen Andruck", Die Rakete, No. 7 15 Juli 1928, S. 100-101.

6. Anecdote remembered by Lang in conversation with author during a visit to Marshall Space Flight Center, c. 1965.

7. I. Saenger-Bredt and R. Engel, "The Development of Regeneratively Cooled Liquid Rocket Engines in Austria and Germany, 1926-1942", in F. C. Durant III and G. S. James, eds., First Steps Toward Space (Smithsonian Annals of Flight, No. 10). Washington, D.C., Smithsonian Institution Press, 1974, pp. 217-246.

8. "Die erste europaeische Flueissigkeits-rakete", Mitteilungen der DGRR astronautische Berichte, Nr. 66, Febr. 1967, S. 6. (Reproduction of Winkler's letter to Engel detailing the launch of HW-1).

9. Ley, op. cit., p. 121.

10. Saenger-Bredt and Engel, op. cit., pp. 222-223.

11. A. Hessenstein, "Rocket-Ship Trips from Planet to Planet Within the Next Two Generations Are Predicted by Johannes Winkler, Noted German Scientist", Washington Post, Dec. 11, 1932.

12. Saenger-Bredt and Engel, loc. cit.

13. "Moon Rocket Blows Up, Nearly Killing Scientist", Washington Herald, Oct. 7, 1932.

14. Ley, op. cit., p 203.

15. H. Gartmann, The Men Behind The Space Rockets. London, Weidenfeld and Nicolson, 1955, pp. 96-97.

16. Letter from Dr. A. Busemann, November 7, 1974 to Fred A. Durant III.

17. W. R. Dornberger, "European Rocketry After World War I", Journal of the British Interplanetary Society, Vol 13, Sept. 1954, p. 248.

18. Letter from Dr. A. Busemann, October 1, 1976, to author.

JOHANNES WINKLER (1897-1947)

THEODORE VON KÁRMÁN
(1881-1963)

R. Cargill Hall*

"Theodore von Kármán," William Pickering has observed, "was many great men; a scientist and engineer, a teacher whose students may still be found among the leaders of astronautics in many nations, a gracious gentleman and persuasive advocate, and a builder of institutions such as the International Academy of Astronautics."(1) Pickering could have added that von Kármán was also a consultant to governments and industry, a director of aeronautical institutes, and the president of a corporation. In each of these roles von Kármán served with distinction during a lifetime that spanned aerostatics and astronautics. In the 1930's when rocketry and astronautics were generally considered an indulgence for amateurs, he encouraged some of his best students to investigate seriously liquid and solid-propellant jet propulsion systems. He and his former student, Frank J. Malina, in 1944 founded the Jet Propulsion Laboratory.

But the contributions that will likely assure von Kármán a place among the great names in aeronautical and astronautical sciences in the twentieth century were his exposition of fundamental theory, particularly in the field of fluid mechanics, (2) and his singular ability as a teacher to unite theory and applied mathematics with engineering, and communicate that important relationship to students.

IN EUROPE

Theodore von Kármán was the third of five children born to Maurice and Helen von Kármán on May 11, 1881, in Budapest, Hungary. The family possessed a strong intellectual and cultural heritage; perhaps the best known of his ancestors is Judah Low ben Bezalel, the 16th century Rabbi of Prague and friend of Tycho Brahe, who is credited with inventing a mechanical robot called the Golem of Prague.(3) As a child, von Kármán displayed a pronounced aptitude for mathematics, attending a specialized secondary school following a curriculum established by Maurice von Kármán His father, a highly regarded professor of education at the Pazmany Peter University of Budapest who had reorganized the Hungarian secondary schools, exerted a profound influence on his son; he counseled a career in engineering--counsel the boy embraced. Maurice was also strongly internationalist in outlook, a trait that his son would adopt and reflect throughout his life.

As a youth, von Kármán was also impressed profoundly by the book Science and Hypothesis. The author, Henri Poincaré, the great French mathematician and philosopher, stressed the limitations of science-- that accepted scientific truth was not absolute. Von Kármán carried a tolerance for new ideas and a love of fundamental theory with him in 1898 to the Royal Joseph University of Polytechnics and Economics in
----

*Jet Propulsion Laboratory, California Institute of Technology. Presented at the International Space Hall of Fame-International Academy of Astronautics Symposium: SELECT BIOGRAPHIES OF OUTSTANDING SPACE PIONEERS, Alamogordo, New Mexico, October 9, 1976.

Budapest, where he "soaked up the examples the principles--the march of reasoning, as Poincaré once described it "(4)

In 1902 von Kármán graduated with honors from the Royal Joseph University of Polytechnics and Economics with a degree in mechanical engineering. After a brief interlude in the Austro-Hungarian army, he returned to his alma mater to continue advanced studies. His interest in a study of structures culminated in publication of a theory on the phenomenon of the nonelastic buckling of columns in 1906, and earned the young engineer a two year fellowship at "the prince of German universities," Göttingen. There von Kármán studied under Ludwig Prandtl, "the father of modern aerodynamics," completing his PhD dissertation on nonelastic buckling of columns in 1908. That summer he visited Paris and witnessed the record-breaking flight of Henri Farman over a two-kilometer course in a heavier-than-air machine. The flight aroused his interest in aeronautics; it never waned.

Returning to Gottingen as an Assistant Professor, von Kármán busied himself with constructing one of Europe's first wind tunnels for use in the design of Zeppelins, and he began studying physics related to the phenomena of aerodynamic drag. During this period he was greatly influenced by the famous mathematicians Felix Klein and David Helbert. Von Kármán became a devoted exponent of an approach to study and teaching that united theory and applied mathematics with engineering. This approach moved from fundamental theoretical principles to concrete applications--and emphasized creative thinking. Perfected at Gottingen, he would soon spread it around the world.

In 1911 von Kármán published a scientific explanation of one important source of aerodynamic drag, known as "vortex shedding." This phenomenon occurs when the airstream fails to stick to the shape of a body, but breaks off on either side behind it in parallel wakes with alternating vortices. The theory, which involves what became known as the Kármán Vortex Street, has had important applications in the engineering of aircraft, ships, and suspension bridges. With substantial theoretical contributions in structural and fluid mechanics already to his credit at the age of 31, early in 1913 von Kármán accepted an offer of the chair of aeronautics at the Technische Hochschule in Aachen, Germany. Except for World War I, when he served as an aircraft designer and aeronautical specialist for the Austro-Hungarian Luftarsenal, he would remain at Aachen until 1929, later as the Director of the Aerodynamics Institute.

During his long association with the Technische Hochscule, von Kármán became a world figure in the aeronautical sciences and engineering. He built new wind tunnels and broke with the traditional formal atmosphere of German higher education by introducing weekly teas for students at home. His mother and sister who came to live with him in 1923 after his father's death, hosted these social gatherings. The family's mastery of languages and knowledge of poetry, literature, and history, ensured lively discussions. Von Kármán spoke fluently Hungarian, German, French, Italian, Spanish, and "what he always described as the international language, 'bad English.'"(5) The new aeronautics facilities coupled with von Kármán's personality and expertise attracted an international student body to the university. Helping to reunite scientists and re-establish the exchange of technical information among these members of the former Allied and Central Powers after World War I, he played a major role in founding the International Congress of Applied Mechanics. For the growing European aircraft industry, he served as consultant to the Junkers, Zeppelin, and Fokker firms.

At the invitation of Robert A. Millikan, Chairman of the Executive Committee of the California Institute of Technology, von Kármán visited the United States in 1926 to assist in planning the curriculum and facilities for the Guggenheim Aeronautics Laboratory (GALCIT). He continued on to Japan to oversee the establishment of yet another aeronautical laboratory, and to lecture in India. In 1929, when Nazi ideology began to infect the German academic community, von Kármán accepted an offer from Robert Millikan to become the Director of GALCIT.(6) With his mother and sister he settled in Pasadena in 1930, and a few years later became a citizen of the United States. (Figure 1) Germany, in turn, lost an eminent Jewish professor; he was among the first of numerous distinguished German intellectuals who would journey across the Atlantic and settle in the United States during the 1930's.

IN AMERICA

Under von Kármán's guidance, GALCIT developed into a second Aachen and one of the world's leading schools of aeronautics; at the same time, Southern California became a leading center for the aircraft industry in the United States. The era began with his publication of a theory on mechanical similarity in turbulence in 1930,* and concluded with his retirement in 1949. GALCIT aeronautics hosted an international student body in which Chinese, Japanese, Turkish, Romance, and Indo-European languages were all represented. Weekly social gatherings supplemented weekly seminars, and the von Kármán home became a mecca where students and visiting dignitaries mingled.

The news media soon acknowledged von Kármán to be "a great American scientist," an accolade confirmed in 1938 on his election to the National Academy of Sciences. In 1932 he had helped found the Institute of Aeronautical Sciences (an organization that would merge with the American Rocket Society in the early 1960's to become the American Institute of Aeronautics and Astronautics). All the while von Kármán participated in international symposia and lectured at universities and technical institutes in Europe and Asia. His book with M. A. Biot, <u>Mathematical Methods in Engineering</u>, written during this time, would be translated into seven languages.(7)

In 1936, at 55 years of age, von Kármán gave assistance and guidance to graduate students who wished to pursue studies in rocketry. True to his ideal to encourage research and experimentation in any area he believed held merit no matter how far-fetched it might appear to his peers, he allowed GALCIT facilities to be used by.the rocketeers. The aeronautics buiding soon rumbled to the noise of rocket engine test firings as Caltech became the first American university to actively sponsor rocket research for students. Coming under government sponsorship in 1936, the GALCIT Rocket Research Project was reorganized and renamed the Jet Propulsion Laboratory, California Institute of Technology, in 1944.

Von Kármán continued to consult for industry and governments. Using the theory of fluid mechanics, he helped the General Electric Company redesign the blades of steam turbines for greater efficiency, and performed a similar service for the Los Angeles Metropolitan Water District on centrifugal water pumps. In 1941 he found the cause of the Tacoma Narrows Bridge collapse for the State of Washington. During World War II he acted as a consultant to the Army Air Forces in aeronautics and rocketry, and helped found and became the first President of the Aerojet Engineering Corporation (now the Aerojet General Corporation) which produced rocket motors for the armed services. In 1944 at the request of General Henry H. Arnold, Chief of the Army Air Forces, he set up the Air Force Scientific Advisory Group (new the Air Force Scientific Advisory Board)

----

*In Germany, Ludwig Prandtl, who had also sought a unified solution to this problem, was moved to observe that von Kármán possessed an uncanny talent for "skimming the cream off the milk."

Fig. 1   Theodore von Kármán in 1930

composed of a number of the nation's leading scientists, to chart a future course for the development of strategic aircraft and missiles. On leaves of absence from Caltech between 1944 and 1949, von Kármán served the Air Force nearly full time. In the interim, during trips to Europe after the war, he helped organize and became Honorary President of the International Union of Theoretical and Applied Mechanics (IUTAM).(8)

IN THE WORLD

Upon his retirement from Caltech in 1949 at the age of 68, von Kármán embarked on the third and final chapter in his career. For the next fourteen years he moved even more rapidly around the world, riding the wind in airplanes his work had helped create in search of ways to unite and promote international scientific endeavors. A trail of technical papers and mislaid berets and mufflers marked his passage. "When you ask von Kármán's whereabouts," one colleague remarked, "you never find out where he is, but always where he has been or where he is going."(9)

Von Kármán succeeded admirably in galvanizing international scientific ventures and associations. In 1950 one of his proposals led to the initiation of a long-term Arid Zone Research Program by the United Nations Educational, Scientific, and Cultural Organization (UNESCO); another, for cooperative activity among scientists of the member states of the North Atlantic Treaty Organization, resulted in the Advisory Group for Aeronautical Research and Development (AGARD) in 1952. (Figure 2) His efforts to foster cooperation among scientists engaged in aeronautics brought about the formation in 1956 of what is now the von Kármán Institute for Fluid Mechanics in Rhode-St. Genese, Belgium, and in 1958 of the International Council of the Aeronautical Sciences (ICAS), supported by national scientific societies of 27 nations. In 1960 he chaired a founding committee that established the International Academy of Astronautics of the International Astronautical Federation. Although little absorbed now in original research, he also published numerous review articles and another book,(10) participated in scientific meetings, while continuing to host social gatherings for visitors from every walk of life.

For his contributions to scientific theory, the teaching of engineering, national defense, and international cooperation, von Kármán received honors and awards enough for two lifetimes. Among them can be found the Gold Medal of the Franklin Institute (1948), Lord Kelvin Gold Medal (1950), Wright Brothers Memorial Trophy (1954), and the Robert Goddard Memorial Gold Medal (1960).(11) The Gas Dynamics Facility at the Air Force Arnold Engineering Development Center in Tennessee was named in his honor.(12) In early 1963 von Kármán was called to Washington, D.C., to receive the first National Medal of Science from President Kennedy. (Figure 3) Afterwards, he journeyed to Aix-la-Chapelle to rest, preparing for a schedule of seven international meetings between June and September. But these meetings were not attended. Theodore von Kármán died in Aix-la-Chapelle on the 7th of May, 1963, "five days before completing his 82nd voyage around the sun."(13)

Always approachable and gregarious, von Kármán was renowned for his quick wit and saucy humor. After dinner talks, he advised, "must be like a lady's dress--short, interesting, and covering the essential points." It is said that the famous British aviatrix Amy Johnson, at a meeting of the Royal Aeronautical Society, asked von Kármán, "can you tell me in a few words what causes (aircraft to) spin?" He replied: "Young lady, a spin is like a love affair. You don't notice how you get into it, and

Fig. 2   Theodore von Kármán at Home, 1955.

Fig 3 Theodore von Kármán Receiving the First National
Medal of Science From President Kennedy,
February 18, 1963.

it is very hard to get out of." Hugh Dryden recalled that he defined a consultant as a man who is willing to assume everything except the responsibility, an expert as any engineer who lives 300 miles away from the home office, and a Hungarian as someone who goes into a revolving door behind you and comes out ahead of you.

Von Kármán possessed an extraordinary talent for bridging social, political, and intellectual differences, and his interests above and beyond science ensured diversity among those who met at his hotel room or table. Besides technical people, lawyers, writers, gardeners, doctors, painters, priests, military officers, governmental officials, and political leaders were to be counted among his friends. By all accounts his human qualities--his tact, friendliness, tolerance, and interest in each person as an individual--were equally profound. Simon Ramo summed it up aptly: "Von Kármán, despite who he is, talks with any of us . . . as though that man were von Kármán and he, von Kármán, the one learning from the master."(14)

Of all his achievements, von Kármán's contribution as a teacher of engineering may well be judged the most significant.(15) To emphasize his pedagogical approach, he was fond of recalling the words of the famous British engineer, W. J. M. Rankine:

> When somebody told him that a practical engineer does
> not need much science, he said, 'Yes, what you call

a practical engineer is the man who perpetuates the errors and mistakes of his predecessors,' Although the definition is somewhat hard on many good practical engineers, it is correct in the sense that engiineering education should not only transmit experience from generation to generation, but should be based at all times on the old and new developments in fundamental sciences.(16)

He carried this message everywhere. Today his students can be found on five continents; together they make up what is affectionately known as "the von Kármán Circus." Attesting to the coherence of this group, William Pickering has observed that in his own travels around the world, word of GALCIT or Caltech was enough to cause von Kármán's students to materialize as if by magic to inquire after the Professor or to discuss old times together with new theory and developments.

At an International Astronautical Federation Congress shortly before his death, von Kármán's peers accorded him what was perhaps the greatest of personal tributes. Penned on a scrap of paper by Leonid Sedov, Chairman of the Department of Hydrodynamics at Moscow State University, and signed by twenty-two colleagues from nearly as many different countries, it read:(11) "We are truly happy to send greetings to Professor von Kármán. On any occasion when specialists in applied mechanics gather together, they warmly remember and send their greetings to Professor von Kármán, who is our father and classic of our time."(17)

Von Kármán affected profoundly the teaching of engineering, made lasting contributions to the basic scientific disciplines of aeronautics and astronautics, and organized effectively international scientific unions and projects. The von Kármán Circus, representing the fruitful marriage between science and engineering he helped create, still proceeds--a living testament to a most remarkable man.

REFERENCES

1. William H. Pickering with James H. Wilson, "Countdown To Space Exploration: A Memoir of the Jet Propulsion Laboratory, 1944-1958," in R. Cargill Hall, ed., Essays on the History of Rocketry and Astronautics: Proceedings of the Third through Sixth History Symposia of the International Academy of Astronautics, Washington, D.C., NASA Technical Memorandum to be published in 1977.

2. Von Kármán's collected works, published by Butterworths Scientific Publication (London) in 1952, contain 111 papers ranging over elasticity, structures, fluid mechanics, gas dynamics, aerodynamics, theory of specific heats, combustion theory, airship design, rocket propulsion, mathematical methods in engineering, etc. A Supplementary Volume of Von Kármán's Collected Works between 1952 and 1963 has been published under the auspices of AGARD through the Von Kármán Institute for Fluid Mechanics.

3. Frank J. Malina, "Kármán, Theodore von" in the fifteenth edition of the Encyclopedia Britannica, 1974, p. 405.

4. Theodore von Kármán with Lee Edson, The Wind and Beyond, (Boston: Little, Brown and Company, 1967), pp. 24,27.

5. Hugh L. Dryden, "Theodore von Kármán," in Biographical Memoirs of the National Academy of Sciences. (New York: Columbia University Press, 1967), Vol. 38, p. 350.

6. Von Kármán and Edson, The Wind and Beyond, pp. 140-145.

7. "Von Kármán, Theodore," in Current Biography 1955, pp. 622-264; also Frank L. Watterdorf and Frank J. Malina, "Theodore von Kármán, 1881-1963," Astronautica Acta, Vol. 10, 1964, p. 88.

8. Von Kármán and Edson, The Wind and Beyond, pp. 202-298.

9. Quoted in Will Jonathan, "The Wind's Deputy, Dr. Theodore von Kármán of Almost Everywhere," Saturday Review of Literature, February 2, 1963, Personality Portrait LXXXII.

10. See the Von Kármán Bibliography in Dryden, Biographical Memoirs, pp. 381-384.

11. See the honors listed in Ibid., pp. 366-368.

12. The von Kármán Gas Dynamics Facility was dedicated on October 30, 1959. Remarks of the speakers on this occasion can be found in the von Kármán folder 3-20, JPL History Archive.

13. Watterdorf and Malina, Astronautica Acta, p. 81.

14. Quoted in Dryden, Biographical Memoirs, p. 352; see also Dryden on this subject, quoted in Frank J. Malina, "Memorial, Theodore von Kármán," Technology and Culture. Vol. V, No. 2, Spring, 1964, p. 246.

15. Hugh L. Dryden, "Theodore von Kármán," Year Book of the American Philosophical Society, 1963, p. 166.

16. Theodore von Kármán, Aerodynamics: Selected Topics in the Light of Their Historical Development. (Ithaca, New York: Cornell University Press, 1957), p. 167; for a further elucidation, see Theodore von Kármán, "Engineering Education in Our Age," an Address to the IAS Northeastern Student Conference, Cornell University, April 4, 1959 (unpublished), in the von Kármán papers, Cornell University file, Caltech Archive.

17. Sedov note, n.d., in the von Kármán papers, Box 121, Caltech Archive.

HUGH LATIMER DRYDEN
(1898-1965)

Eugene M. Emme*

The legacy bequeathed by Hugh Latimer Dryden to the history of the conquest of space transcends the esteem accorded to this wise, humble, and dedicated scientist and leader by his contemporaries in many fields. This incomplete biography mainly gives recognition that a full-fledged treatment of this great American, whose life work is a distinguished part of the history of aeronautics and astronautics, remains to be written.

INTRODUCTION

Hugh Latimer Dryden was a Federal employee his entire career. Yet he became a scientist with an international reputation as well as an effective research coordinator and administrator with the U.S. National Bureau of Standards. After almost thirty years with the Bureau, he became the second Director of the National Advisory Committee for Aeronautics (NACA), 1947-1958, and the first Deputy Administrator of the National Aeronautics and Space Administration (NASA), 1958-1965. Honors received by Dryden are legion.

He early made basic contributions to transonic and boundary layer aerodynamics in the 1920's. Dryden became a member of the NACA Committee on Aerodynamics in 1931. During World War II, he managed the development of an operational guided missile, and was the Scientific Deputy of the first Scientific Advisory Group, under Theodore von Karman, of the U.S. Army Air Forces in 1945. Dryden played a central role in the genesis of the national Unitary Wind Tunnel Plan and of the joint research airplane programs. After the first sputniks, his was the seminal role in the national decision-making process leading to the creation of NASA, with the NACA as its nucleus. This record was prologue to further service.

Dryden served the first two Administrators of the non-military space agency of the United States. It was Deputy Administrator Dryden who insisted upon retaining the vital role of the research laboratories. He also helped to enlist the support of the military services and other governmental, academic and industrial activities as the new space agency carried through inherited space projects, continued the NACA aeronautical research while increasing attention to the needs of space technology, and formulated a long-range plan for the exploration of space. Dryden had already set in motion in NACA the R&D steps leading to the first manned satellite program later called "Mercury." Selection of the
----

*NASA Historian since November 1959. It should be pointed out that the author never worked with Hugh Dryden in mainline NASA actions. Dryden's foreword in Aeronautics and Astronautics, 1915-1960, published by NASA in 1961, did, in effect, constitute the genesis of the NASA Historical Program. He also endorsed participation in the founding of the History Committee of the International Academy of Astronautics.

manned lunar mission in early 1960 as the goal to follow Mercury, found Dryden's explanations persuasive in helping to gain White House and Congressional approval in 1961. Over the years, what Dryden told the Congress was intended to be accomplished was, in fact, achieved. Some considered Dryden conservative.

In 1962, President Kennedy designated Dryden to work out cooperative space projects with representatives of the U.S.S.R. Dryden's structure of principles for basic research and systems development served NASA exceedingly well as he contributed to helping make the Apollo goal a reality. He always openly shared his knowledge and his example inspired others.

Today, Hugh L. Dryden's bust stands outside of the National Academy of Sciences' auditorium constructed by monies given by his many friends. Many awards, scholarships and fellowships are named after Dryden. In March 1976, NASA dedicated the Hugh L. Dryden Flight Research Center at Edwards, California, as a living memorial to his life work. And, at the library of his John Hopkins University in Baltimore, Maryland, is also a tangible monument formed by the numerous papers of Hugh L. Dryden. They were assembled by others but they are a rich treasure of a disciplined mind and a sensitive humanist dedicated to the betterment and fulfillment of the destiny of mankind.

EARLY LIFE

Hugh Latimer Dryden was born on July 2, 1898, in Pocomoke City on Maryland's eastern shore.(1) In 1907, his family moved to Baltimore. Dryden saw his first airplane when he was twelve years old. It was an Antoinette monoplane flown by Hubert Latham, who won a $5,000 prize for making the first flight over Baltimore. Dryden often remarked meaningfully years later, "the airplane and I grew up together."

Honor Man At City College

Hugh Latimer Dryden
in 1913

(From Dryden Papers, Johns Hopkins Univ. Library)

If Hugh Dryden did not come from a scholarly family, he was certainly gifted with rare native intelligence. It was to be always utilized, and never flaunted. He completed his bachelor's degree with honors at John Hopkins University in 1916. His major professor was Joseph Ames, chairman of the department of physics and a member of the National Advisory Committee for Aeronautics (NACA). Ames recommended him for summer employment at the National Bureau of Standards as "the brightest young man I have ever had without exception."(3)

RESEARCH SCIENTIST

Dryden began his distinguished career as an inspector of gauges in the National Bureau of Standards in June 1918. He had planned to return to graduate school in the fall, but was to be transferred to the wind tunnel section. Professor Ames gave advanced courses for his Hopkins' graduate students at the Bureau so that Dryden was able to use its pioneering U.S. wind tunnel after working hours. Thus he could continue his studies while being employed full time. In the spring of 1919, Dryden received his Ph.D. in physics and mathematics just before his twenty-first birthday--the youngest person ever to receive a Ph.D. from Johns Hopkins University. His dissertation was entitled, "Air Forces on Circular Cylinders." It was based on measurements of drag and distribution of air flowing around cylinders. It marked the beginning of a brilliant career in a frontier of aerodynamical science, and which brought him international recognition. In 1920, Dryden became head of the newly-formed Aerodynamics Section of the Division of Mechanics and Sound. That same year he also married Mary Libbie Travers, who was to be his lifelong helpmate and partner in raising a family of three children.

With Lyman J. Briggs, who later became Director of the Bureau of Standards, Dryden became engaged in propeller-tip aerodynamic studies in 1924. They discovered that at 450 mph the flow of air separated from the air foil when compressibility was reached. Briggs and Dryden attained considerable fame for their NACA reports published in 1926, making known the results of research on a three-stage, turbine-driven compressor at General Electric in Lynn, Mass.(4) Dryden performed all data reduction and computation. Later they ran tests at Mach .5 to Mach 1.08, and observed a shock wave one-half inch in front of the air foil. They had used a compressor plant at the Army's Edgewood Arsenal, and had fitted airfoils with pressure orifices to record pressure distribution. They observed the now-familiar separation of flow, drop in lift, and increase in drag which occurred at transonic speeds. They next began a comparative study of 24 airfoil sections, discovering that thin airfoils--those having a low thickness-chord ratio--retained their effectiveness at higher speeds.

    Dryden and other Bureau colleagues, notably A. M. Kuethe, worked out methods of measuring turbulence in wind tunnels, and its effects on flow measurements. Using an electrically-compensated hot-wire anemometer, Dryden's work led to creation of low turbulence wind tunnels of critical importance as transonic speeds were attained--transition from laminar to turbulent flow in the boundary layer. In short, Prandtl's early theories were verified. Dryden became a member of the premier technical committee of the NACA, the Committee on Aerodynamics, in 1931. (5)

    In 1934, Dryden became the chief of the Bureau of Standards Division of Mechanics and Sound. In 1936, he delivered a noted paper on laminar flow on flat plates to the 4th International Congress of Applied Mechanics. And in 1938, he was the first American to deliver the Wilbur Wright Lecture. It was a classic survey of all that was known on "Turbulence and Boundary Layers." Between 1924 and 1947, Dryden produced ten major technical papers for the Bureau of Standards, a dozen NACA technical reports, and numerous articles in professional journals. He edited the Journal of the Institute of the Aeronautical Sciences for a decade. As the ever-improved airplane came to exert increasing influence in commerce and in international affairs, Dryden's duties increasingly included administration and coordination. With his ever-inquisitive mind and broadened horizons for his voracious reading habits, few appreciated the acceleration provided by his photographic eye and instant recall known by his colleagues. Many recall, even in NASA later, that Dryden used the unabridged dictionary regularly, ever seeking precision in the meaning to language.

RESEARCH AND DEVELOPMENT

    Dryden was elected to the National Academy of Sciences in 1944, and he was to become Home Secretary in 1955. He also became a Life Trustee of the National Geographic Society.

    In late 1944, General Henry H. Arnold, commanding the Army Air Forces, persuaded Theodore von Karman of the California Institute of Technology to form a Scientific Advisory Group to assess the impact of wartime technical progress on the conclusion of the war as well as the postwar peace. The appearance of the vastly superior German jet and rocket aircraft in operations and the revolutionary V-2 ballistic rocket were of great concern, even if the Germans had not developed an atomic bomb. Arnold knew that "today's weapons are tomorrow's museum pieces." Von Karman selected Hugh Dryden as his "Deputy Director, Scientific." In April 1945, von Karman, Dryden, Frank Wattendorf, and H. Tsien proceeded to Paris and thence to occupied Germany to examine the supersonic wind tunnel discovered at Volkenrode and other impressive R&D facilities, as well as interview German scientists and engineers.

In the fall of 1945, the classic and classified so-called "Von Karman Reports" were drafted and edited. The preliminary Where We Stand, and the overall Towards New Horizons, supplemented by twenty technical reports authored by Dryden, Tsien, and others. Dryden was the principal editor of the entire series, for which he received the Medal of Freedom. The reports forecast long-range ballistic missiles and earth satellites whenever they would serve useful purpose and were feasible. Most importantly, they laid out fundamental principles for long-term utilization of science in the R&D spectrum from basic research to operational systems. (7)

Dryden, Ludwig Prandtl, Theodore von Karman, and H. S. Tsien on the steps of the Kaiser Wilhelm Institute in Goettingen, Occupied Germany, May 14, 1945. Dryden is wearing without insignia an Army officer's uniform.
(From Dryden Papers, Johns Hopking Univ.)

RESEARCH POLICY AND ADMINISTRATION

Hugh Dryden's nearly three decades with the Bureau of Standards in war and peace, with all of his other service, well schooled him for his decade as the second Director of the National Advisory Committee for Aeronautics. Before he was elected, Dryden assessed the spectrum of research and development in aeronautics and its related disciplines in the generation of operational flight systems. He pointed out that once the atomic bomb was feasible with knowledge of implosion, development and production was possible. He related his principles to the large-scale inter-disciplinary group which developed the V-2 at Peenemuende. While Dryden's unpublished lecture was drawn from Towards New Horizons, his elaboration of four of the principles upon which he based his thoughts and actions in the NACA, and later NASA, seem worthy of broad examination here:(8)

> "1. An adequate national program for extending the frontier of knowledge in various fields of science is a necessary adjunct to the maintenance of the security of the nation.... Whereas applied research and development can be expedited by the application of more money and manpower, fundamental research contains many unpredictable elements and uncontrollable factors...."

"2. Fundamental scientific research is dependent on an
   atmosphere of freedom from immediate specific goals
   and timetables, freedom to discuss and exchange ideas,
   and freedom from controls and restriction. The discovery of new knowledge is an intellectual activity
   of the highest type and flourishes only in the atmosphere of greatest freedom...."

"3. The results of fundamental research furnish the broad
   base from which all applied research and industrial
   developments proceed. This principle has many implications...(and) the authorities should foster fundamental
   research...(and) they should in no way dictate or control
   the use of the results."

"4. As scientific research and development proceed from
   the state of pure research and to development, free
   enterprise and initiative should be maintained and
   centralized and dictatorial control of projects, funds,
   and facilities, should be avoided. Ideas and skills
   are widely dispersed. Even though the goals of applied
   research and development are concrete and specific, the
   attack on these goals should be diversified..."

It is not difficult to note how close in thought and spirit Federal civil-servant Dryden was to Professor von Karman, and how each, in their distinctive spheres and with wholly different lifestyles, were of stellar service to their countrymen in the shrinking aerospace world. (9) Neither rarely neglected to state their views nor said what their audience wanted to hear, which, in Dryden's case, included Congressional committees upon several occasions.

When Dryden joined the NACA in September 1947, its simple but basic legislative charter was already a part of his lexicon--"to supervise and direct the scientific study of the problems of flight, with a view to their practical solution." While the Director of NACA responded to the guidance of the main Committee, it was Dryden's task to lead the work at the laboratories. His first major task was to prepare a truly national Unitary Wind Tunnel Plan. It was a major contribution insuring that Federal monies for aerodynamic research faciltites in NACA and elsewhere in the government built with tax dollars were not duplicative.(10)

In 1949, Dryden gave a Wilbur Wright Lecture, and it was en-titled, "The Aerodynamical Research Scene--Goals, Methods, and Accomplishment." A masterful survey, it revealed more of his basic principles. It was one of his basic beliefs, when he stated: "The most important tool in aeronautical research, even more that the large wind tunnel, is the human mind." With regard to common goals enhancing basic research and practical development, Dryden said: "It (research) should not of its entirety be limited to exploratory research or to coordinated theoretical and experimental work, or experimental situations where complete understanding of basic phenomena is the principal goal. The need of designers of systematic surveys of basic areas and for research in support of development must be recognized by frequent and close contact between designers and research workers. The selection of some common advanced technical development as the goal of both groups has proved to be an excellent means of promoting cooperation and of channeling research into directions permitting early application, without sacrificing the values inherent in the personal enthusiasm of individuals, and freedom of the research worker." This concept of common goals but dissimilar contributions of scientists and engineers was to be instituted in NASA a decade later when it was created out of a reconstituted NACA. It was also a source of the "hands on" philosophy of NASA program management.

But it was well understood by NACA engineers who, later in NASA, were to be thrust into the management of space programs. (11)

Dryden in April 1949, enroute to deliver the 37th Wilbur Wright Lecture before the Royal Aeronautics Society.

As Home Secretary of the National Academy of Sciences in 1955, Dryden became involved in reviewing the satellite program proposed for the International Geophysical Year. It appeared very costly to the Director of NACA, the annual appropriation for which was less than $100 million. Once a structure of scientific experiments was worked out and the President had approved the program, Dryden endorsed the creation of the Satellite Vehicle Group under William J. O'Sullivan at the Langley Laboratory. This initiated fundamental research on spacecraft structures, trajectories, guidance, and other aspects of space technology additional to the rocket research work of the Pilotless Aircraft Research Division. To the NACA, and particularly the manned vehicle research to follow the X-15 rocket research program just getting underway, "Round Three" involved many elementary aspects of technology for which basic research had been by-passed by the priority programs, based on existing materials, guidance, life support, control systems, re-entry heating, and other problem areas. It was Dryden who best knew the strengths and the gaps in the competence of each of the NACA laboratories. James H. Doolittle, the Chairman of the NACA, was also Chairman of the Scientific Advisory Board of the Air Force. He had become the Chairman of the NACA in 1955, as the pace of R&D had enforced a need to expand its program. When the U.S.S.R. launched Sputnik I on the fire-tail of an ICBM rocket on October 4, 1957, this was no surprise to Dryden or the research engineers in the NACA.

SPACE POLICY: ACHIEVEMENT AND INTERNATIONAL COOPERATION

Hugh Dryden's finest hour was his "strategy" evolved in the post-Sputnik environment to insure that national needs for fundamental research were to be served by the NACA. The outcome was that in the process of nation-wide dialogue and decision-making by the White House and the Congress, the NACA was to be selected as the nucleus of a non-military space agency dedicated to contributing to "the peaceful purposes" in outer space. A complicated story, the role of Dryden appears a highly effective one when all aspects are considered.(12)

Dryden served with NACA Chairman Doolittle on the President's Science Advisory Committee (PSAC) under Jamed R. Killian. Initially, the NACA position voiced by Dryden was one of a minimal role to do for space technology what it had done for aeronautics for over four decades--that is, working with all governmental agencies with industry and with academe. NACA had a demonstrated record, one increasingly committed to advanced flight technology. In January 1958, under Dryden's leadership, the NACA issued carefully drawn position papers, and announced that it had created a Special Committee on Space Technology under H. Guyford Stever of M.I.T. Selected members included experts of national reputation such as James A. Van Allen, Wernher von Braun, Randolph Lovelace, William Pickering and others. (13) The Atomic Energy Commission also considered it could exercise civil control of the U.S. space venture. The U.S.S.R. had demonstrated rocketry with the sputniks for an intercontinental ballistic missile armed with thermonuclear firepower. The R&D problem was of highest urgency to all concerned with American priorities.

After the first U.S. earth satellite was at last flown, Explorer I on January 31, 1958, the question was whether to place the space program under the Advanced Research Project Agency (ARPA). Both the Congress and the White House had reservations about this. It became the recommendation of PSAC to President Eisenhower that a "reconstituted

Dryden with Chairman John W. McCormack and members of the House Select Committee on Astronautics and Space Exploration in April 1958. NACA-developed inflatible satellite prototype later used for Echo provides background. (Dryden papers, Johns Hopkins)

NACA" offered an existing nucleus for getting a space program underway with minimal disruption, leaving responsibility for national security in the Department of Defense. Eisenhower approved this recommendation on March 5, 1958. It had been drafted by Killian, Edward Purcell of Harvard, and Edwin Land to state the basic purposes of a national space program. It was well received nationwide. Dryden also made direct inputs in the drafting of the Administration bill for a "National Aeronautics and Space Agency" by the Bureau of the Budget. After Eisenhower sent it to the Congress early in April, it was Dryden and Doolittle who had to defend it before the special committees of both houses of Congress.

The National Aeronautics and Space Act of 1958, signed into law by President Eisenhower on July 29, 1958, was a remarkable national decision. It declared: "It is the policy of the United States that activities in space should be devoted to peaceful purposes for the benefit of all mankind." Powers and responsibility of the new civilian space agency had been beefed up by the Senate committee under Lyndon B. Johnson and the House committee, including provision for international cooperation. In the meantime, Hugh Dryden had initiated steps to get NACA moving into NASA. Robert Gilruth and colleagues at Langley began development of an initial ballistic manned satellite program using current knowledge. Abe Silverstein was brought in from Lewis to structure the NASA program, and Ira Abbott chaired a committee on an organization for NASA. White House intent to nominate Dryden for first Administrator of NASA was informally rejected by the House committee. Dryden testified candidly that a flood of money and the best of intentions could not buy the basic research that should have been accomplished yesterday. Dryden was not opposed to manned space flight, but he came to be regarded as too conservative to direct a vigorous space program by some. When T. Keith Glennan was offered the post of NASA Administrator, he accepted only if Dryden would serve as his Deputy.

Dryden receiving his commission as Deputy NASA Administrator from President Eisenhower, with T. Keith Glennan, August 19, 1958 (NASA)

Hugh Dryden's contribution as Deputy Administrator on NASA until December 1965, difficult to summarize briefly here, spans a sequence of interrelated and complicated activities symbolized grossly by the expansion of the 8,000-person NACA to a nation-wide assembly of talent including contractors of nearly 400,000. One Harvard Nobelist, not an aerodynamicist, commented that "Hugh Dryden was a great, a very great man to have lived so close to the NACA and be able to help make it go into the new world (of space)."(14) James R. Killian observed that NASA came to represent the greatest collection of engineering talent ever. Many do comment that the high cost of NASA was probably excessive but it was the one national achievement, after the Apollo lunar missions, that every American was proud to contemplate. Future historians will have their say. But one of the important things about NASA from its take-off was that it stated its business in the open for all to hear and see, in turn enforced a quality of performance with novel and complex systems. The list of public speeches and publications of the meaning of the space program by Dryden, to all types of audiences, is probably not exceeded in their impact by anyone.(15)

278

The beginning of the take-off of NASA in September 1958 was an event-jammed process for the NASA leadership. Working relationships with the White House and the Pentagon on the transfer of projects and money to NASA was coupled with hiring of only the best of the scientific and managerial talent out of the thousands of applicants. Initially, Glennan sought to infuse into headquarters key staff persons to leaven the dominance of those from the NACA. He initially asked Dryden to oversee those areas of greatest pertinence--the generation of the NASA space science program, the conduct of Project Mercury, and an upgraded NASA University Research Program.(16)

While Glennan actively cultivated outside consultants and advisory groups to gain policy perspectives not found in NASA, Dryden gave special attention to maintaining the viable NASA Research Centers, the former NACA laboratories. The Space Task Group at Langley running Mercury, the Jet Propulsion Laboratory in Pasadena transferred from the Army, to be given responsibility for the lunar and interplanetary mission, and the inherited Vanguard team from the Naval Research Laboratory for the continuation of the IGY momentum in space science--these were conceived to be three operational program areas best assigned to the new Space Flight Center at Greenbelt, Maryland. It was a major concern that the Small Task Group could digest all of the support from the Langley Research Center. By the time the Goddard Center got underway, the Space Task Group and the Langley Research Center had been divorced, and the Jet Propulsion Laboratory likewise was reporting to headquarters and not the Director of Goddard Space Flight Center.

The postponement for over a year of the transfer of the Development Operations Division of the Army Ballistic Missile Agency, under Wernher von Braun, to NASA, proved a major organizational problem involving the White House, the Pentagon and the Congress so that NASA would have a rocket systems development competence. Glennan found Dryden's counsel and support generous and helpful. Dryden had long-standing cordial working relations with the military services, the R&D establishment, universities, the aerospace industries, and professional societies. In hindsight, Glennan'a AEC experience matched well Dryden's NACA background since all of NASA's programs involved coalescence of resourses from many sources. Glennan's directness and Dryden's logic meshed well.

Dryden's reputation also provided NASA with scientific integrity. In 1959, he was appointed to assist Ambassador Henry Cabot Lodge at the first meeting of the United Nation' Conference on the Peaceful Uses of Outer Space. By December, NASA's International Program as specified in the Space Act was beginning to function and a proposal for the conduct of joint space research with other nations in a mutually-beneficial manner was announced. NASA's own program was moving.

On January 7, 1960, the manned lunar mission was approved by T. Keith Glennan as the planning goal for manned space flight after completion of the initial orbital demonstration by Project Mercury. It was a historic day. This far-out goal was one requiring White House approval to fund as an ongoing program. But the planning "goal" was needed for the "research goals" in order that basic research could provide the knowledge to provide the guidance for subsequent systems development. This had been recommended by the "Research Goals" Committee under Harry Goett. This step for manned space flight was similar in process to other portions of the NASA program projected for the next ten years in NASA's Long Range Plan. This plan had taken all of 1959 to prepare, and it was briefed to the Congress in January 1960. It was also necessary in order to generate sound initial programs in all classes of missions. But it was symbolic, that it was Hugh Dryden who announced at the NASA Industry Plans Conference on July 28, 1960, before 1300 people assembled, that

Project Apollo was the name of the manned space flight program to follow Project Mercury. The building-block modules for a coherent evolution of manned space flight from orbital to circum-lunar to lunar-landing missions were existent in concept. NASA inherited no plans. It planned from scratch.

The coming of Robert C. Seamans, Jr., as the Associate Administrator to serve as "general manager" of NASA reflected maturation in the organization and operation of a viable space agency. NASA's take-off had progressed into a trajectory with direction and purpose. Initial successes in 1960 of a number of Explorer scientific satellites, the Pioneer V interplanetary probe, the Tiros I weather satellite, and the Echo passive communications satellite visible to the naked eye around the world gave some credence to this. In its first two years, NASA had already turned out to be more expensive than expected by the White House, more institutionally independent than expected by the scientific community and the Pentagon, and more strongly supported by its Congressional Committees than some of its creators had intended. Its future achievements were to result from programs it had initiated after its creation. Scientists complained they were not running NASA's science program and Mercury should not be a priority NASA project. Some military enthusiasts felt that NASA's large launch vehicle development and manned space flight best be restored to Pentagon management. Elements of both of these views were presented to President-Elect John F. Kennedy by an ad hoc space panel, chaired by Jerome Wiesner. The report was submitted on January 10, 1961, reprinted in the New York Times along with announcement that Dr. Wiesner would be Kennedy's Science Adviser. It fell upon Hugh Dryden, whose resignation was neither accepted nor rejected by the Kennedy staff, to preside over NASA until a new NASA Administrator was selected.

It was NASA's good fortune that James E. Webb was selected as NASA's second Administrator. He was sworn in on Valentine's Day. He had not sought the post but was persuaded by the President to accept it, which he did if Hugh Dryden would continue to serve as Deputy Administrator.

Within seventy-five days after Webb became Administrator, the national decision had essentially been made to accelerate the NASA program to include a landing by Americans on the surface of the moon before the end of the 1960's. The catalyst had been the first flight by man in space, the world-circling flight of an unknown Major of the Soviet Air Force, Yuri Gagarin, on April 12, 1961. Three weeks before, Dryden had stated emphatically in a budget meeting in the White House, that the new President would be forced by the march of history to accelerate the NASA program.(17)

The dynamic "troika" in NASA's top management, as it was to be called, was the Webb-Dryden-Seamans team. James Webb explained it as follows: "The three of us decided together that the basis of relationships should be an understanding that we would hammer out the hard decisions together and that each would undertake those segments of responsibility for which he was best qualified. In effect, we formed an informal partnership within which all major policies and programs became our joint responsibility, but with the execution of each policy and program undertaken by just one of us...." This remarkable arrangement, although Webb as Administrator had full responsibility, was to work well during the incredible and complex months during the build-up phase of the Apollo program. Dryden is not known to have at any time reservations about this, and he continued to carry his share of Congressional and other duties.(18)

NASA's top management team: Dryden, James E. Webb and Robert C. Seamans, Jr., February 14, 1961 - December 2, 1965. (Dryden Papers, Johns Hopkins)

In October 1961, after surgery, Hugh Dryden learned that he was afflicted with an incurable malignancy. He merely increased his schedules in NASA and was unflagging in his speaking engagements supporting mounting the Apollo effort. In 1962, President Kennedy assigned him the task to conduct negotiations with A. A. Blagonravov of the U.S.S.R. for instituting initial cooperative projects for space experiments in meteorology and communications. This was a deliberate process, one that could have frustrated someone other than Dryden perhaps, but it was a beginning.

A great dialogue about the validity of Apollo erupted, one hinged upon whether there was truly "a race to the moon." This happened in mid-1963. President Kennedy had suggested in an address to the United Nations that perhaps the United States and the U.S.S.R. should go to the moon together--a "joint expedition to the moon." It resulted in a NASA budget cut for FY 1964, which was 11 percent below NASA's request, and 5 percent below NASA's authorization. Dryden and Webb were both outspoken in their battle against this. It certainly demonstrated the difficulty of sustaining a high priority R&D effort in a democracy during peacetime.(19)

IN RETROSPECT

Despite increasingly severe treatments requiring hospitalization, Dryden had so effectively maintained his public performance that his death on December 2, 1965, was somewhat of a surprise. His absence left a gap in the NASA leadership which was never to be filled.

President Lyndon B. Johnson said of Dryden's passing that it was "a deep personal loss and a reason for national sorrow. No soldier ever performed his duty with more bravery and no statesman charted new courses with more dedication than Hugh Dryden. Whenever the first American sets foot on the moon or finds a new trail to a new star, he will know that Hugh Dryden was one of those who gave him knowledge and illumination." (20)

If Hugh Latimer Dryden left a gap in NASA leadership, his structure of principles brought into the NACA and NASA contributions that he made served his successors who were conscious of them well. Deputy Administrator George M. Low expressed appreciation for Dryden's principles. His recent successor, Alan M. Lovelace, did not hesitate to put it, as follows: "In Hugh Dryden, the first Deputy Administrator, NASA was favored with the brilliant blending of spirit and pragmatism that motivates the individual to fully use his talents and material resources. Cheerfully, with tact and consideration, Dr. Dryden imbued this agency with a true sense of reason for being. His philosophy was built on a structure of principles that I believe in. I have discovered that I have long suffered an acute case of 'Drydenism'..."(21)

Hugh Dryden was greatly honored with degrees and awards during his life. On February 10, 1966, President Johnson presented to Mrs. Dryden the National Medal of Science, posthumously for Hugh Dryden. The citation said: "For contributions as an engineer, administrator, and civil servant for one-half a century, to aeronautics and astronautics which have immeasurably supported the Nation's preeminence in space."(22)

Will humble leadership by logic and performance, known by those who worked with Hugh Dryden, become a model highly regarded by future generations? His scientific curiosity, logical method, broad horizons, and initiatives for practical achievements were not divorced from his humanist regard for the welfare of individuals or society as a whole. Airplanes and spacecraft were amoral, just machines made by man useful for practical purpose including expanding knowledge. A perfect machine could be abused by imperfect men. As one biographer put it, Dryden "dedicated his intellect to solving turbulence in the science of flight as in the affairs of men."(23)

Hugh Latimer Dryden once said: "The opportunities of this age are their own inspiration. They provide us with the potential of travelling into space, of investigation of the moon and Mars, then more. But these individual projects are only segments of the real mission--to contribute to the welfare of mankind."(24) Hugh Dryden accomplished his mission.

REFERENCE NOTES

1. In preparation of this biography, the following works were most helpful: Richard K. Smith (ed.), The Hugh L. Dryden Papers, 1898-1965, A Preliminary Catalogue, Baltimore: Milton S. Eisenhower Library, Johns Hopkins University, 1974, 164 pp., which contains a biography, pp. 19-33; Shirley Thomas, "Hugh L. Dryden," Men of Space, vol. II, Philadelphia: Chilton Pub.,1961, pp. 64-88, which was edited by Dryden; Jerome C. Hunsaker and Robert C. Seamans, Jr., "Hugh Latimer Dryden," Biographical Memoirs (NAS), vol. XL, 1969, pp. 35-63, which contains a list of honors and publications; Raymond L. Bisplinghof, "Hugh Latimer Dryden," Applied Mechanics Review, vol. 19, January 1966, pp. 1-5, which focused on his scientific and engineering contributions. Hereafter citation to these are abbreviated.

2. Smith, pp. 19-20. Dryden kept all of his personal papers from his earliest years, which were added to the Johns Hopkins' collection by Mrs. Hugh L. Dryden.

3. Smith, p. 20. Joseph Ames served as Chairman of the NACA from 1927 to 1930, and was also Provost of John Hopkins in 1926, and became its President, 1927-1939.

4. Hunsaker and Seamans, pp. 37-38; Richard Hallion, Supersonic Flight, NY: Macmillan, 1972, pp. 9-10, 24, 27; R. C. Cochrane, Measures for Progress, Washington: Dept. of Commerce, 1966, pp. 182,314,368-69.

5. Bisplinghof, P. 2. Role of NACA Technical Committee seems greatly unappreciated. They gave national focus to specific areas of flight technology which went beyond the resolutions submitted to the main NACA, whose resolutions then were guidance governing the work of the NACA laboratories. The NACA laboratories had their own interface nationally with governmental agencies and industry on specific work, which, in turn, could be brought to the attention of the main and technical NACA committees.

6. Thomas, pp. 60-61.

7. Towards New Horizons and its supplemental volumes are now unclassified. Useful work is Thomas Sturm, The USAF Scientific Advisory Board: Its First Twenty Years, 1944-1964, USAF Historical Division Liaison Office, 1967.

8. Hugh L. Dryden, "Science and the Air Forces," lecture to the Air Command and Staff School, Air University, Maxwell AFB, Alabama, January 10, 1947, 23 p. One might ask why Dryden prepared such a sophisticated lecture for an audience of Majors and Lt. Colonels. Actually, this audience was the equivalent of a graduate school level and Hugh Dryden always did his utmost to elevate knowledge.

9. See above paper in this symposium on Theodore von Karman by R. Cargill Hall. Von Karman also drew Dryden into two other organizations he founded: AGARD, and the International Academy of Astronautics.

10. Jerome C. Hunsaker, "Forty Years of Aeronautical Research, 1915-1955," Smithsonian Report for 1955, pp. 241-271.

11. A. M. Lovelace, "Message from Washington," NASA Activities, September 1976, p. 3.

12. Hugh L. Dryden, "NACA-NASA Transition," memorandum to Eugene Emme for the History Files, September 3, 1965. The creation of NASA after the sputniks is a fascinating history which the author is attempting to document fully, one that was intended to be a chapter but could well be a book. Much hindsight legend has been generated about the impact of the sputniks and the political science version of its consequences. Killian was the sparkplug for the White House action, supported by most of the members of PSAC. But the President had strong ideas of his own on the "technological elite" and "open skies," which was reflected in his decision on the IGY satellite program in keeping space peaceful.

13. Doolittle, Dryden and Stever very carefully selected each member of the Special Committee on Space Technology. They represented leading specialists in various aspects of space technology who would be, if they were not already, foremost in future developments. Additionally, von Braun represented the Army, and Colonel Norman Appold, USAF, represented General Bernard A. Schriever. Its formal reports were not submitted until October, but its meetings and interactions with each of the NACA laboratories got things in NACA headed into space.

14. From an exerpted interview.

15. See Smith's catalog of the Dryden papers, passim.

16. Robert Rosholt, An Administrative History of NASA, 1958-63, NASA SP-4101, 1966, p. 65.

17. R. Rosholt, pp. 186-88; E. M. Emme, "Perspectives on Apollo," *Journal of Spacecraft and Rockets* (AIAA), vol. 5 (April 1968), pp. 369-82; and, John Logsdon, *The Decision to Go to the Moon*, Cambridge, M.I.T. Press, 1970.

18. James E. Webb, "Foreword," in Rosholt, p. iv. This essay foreword was Mr. Webb's first explanation in hard type concerning his philophy of NASA management, which was required reading in NASA for many years. Also see Webb's *Space Age Management: The Large-Scale Approach*, NY: Columbia Univ. Press, 1969.

19. See Rosholt, pp. 286-88.

20. *Astronautics and Aeronautics, 1965*, NASA SP-4006, 1966, pp. 534-35.

21. Alan M. Lovelace, "Message from Washington," *NASA Activities*, September 1976, p. 3.

22. *Astronautics and Aeronautics, 1966*, NASA SP-4007, 1967, p. 50.

23. Smith, p. 33.

24. Thomas, p. 83.

WILLIE LEY: PIONEER PUBLICIST FOR SPACE EXPLORATION
(1906-1969)

Mitchell Sharpe*

Willy Ley was born in Berlin on October 2, 1906. His father Julius was a wine merchant of Koenigsberg, in East Prussia; and his mother Frieda, the daughter of a Lutheran sexton, was a native Berliner. Young Ley attended the Fichte-Realschule Nr. 5, in Berlin, graduating with a good foundation in Latin, French, and English. In 1924, he entered the university at Berlin and studied physics, astronomy, mathematics, zoology, comparative anatomy, and paleontology. To help finance his way, he worked as a bank clerk for two and a half years during the day while attending classes at night. Ley's first ambition was to become a paleontologist. In 1927, the family simply could not make ends meet financially, and Ley transferred to the university in Koenigsberg for a few semesters. Again hard times dogged him, and after three years of university, he was forced to abandon any thoughts of completing his studies and taking a degree.

It was during his university days, in 1926, that he first became aware of the possibilities of space travel. Earlier in the realschule, he had read such diversionary works of science fiction as the books and stories of Jules Verne, H. G. Wells, Edgar Allan Poe, Kurd Lasswitz, and Maurice Renard. But in 1926, he read the book Die Rakete zu den Planetenraeumen (By Rocket Into Planetary Space) by Hermann Oberth. Despite the formidable mathematics and heavy-handed prose, Ley was fascinated with it. The book marked a turning point in his career. While he would never lose interest in paleontology, he dedicated himself to the future through rocketry rather than the past through the study of fossils.(1,2)

The effect of Oberth's work upon Ley has been described by Beryl Williams and Samuel Epstein as:

> "But it had never occurred to Ley that space travel could be anything but pure fantasy. And although he delighted in fantasy, he carefully distinguished, in his own mind, from fact. To him a fact was, for example, the footprint of some prehistoric animal found embedded in hardened clay. The footprint was mysterious, of course, just as the face of the moon was mysterious, because it hinted at the unknown. But the footprint also represented literal truth, and the mystery it suggested could be probed by careful study . . . . But in Oberth's book he found a potential means for probing that other greater mystery, the mystery of the moon and the whole realm of outer space. Oberth had translated into scientific reality the very dreamworld of which Lasswitz had written. He had suggested that a trip to the moon need not be imaginary--that it could actually be taken by matter-of-fact human beings, and that it could be taken fairly soon."(3)

At about the same time, Ley also read Der Vorstoss in den Weltenraum (Dash Into Space) by the very popular science writer Max Valier. It was based, with Oberth's consent, on his book mentioned above.
----

*Historian, Alabama Space and Rocket Center, Huntsville, Alabama.

However, Ley--despite the book's clarity and obvious appeal to the man in the street--found it lacking. It had too much mathematics and too many attempts at levity that seemed out of place. With all the brashness of youth, the 20-year old Ley felt he could do a better job; and he did!

In 1926, Hachmeister & Thal, in Leipzig, published his book Die Fahrt ins Weltall (The Voyage Into Space). The publisher gave him an advance of what amounted to $110 on a book that eventually sold 9000 copies in a time when such books were not selling well at all. Even more importantly, the book established Ley as a science writer of international reputation.(4)

As a result of the book, Ley met Valier, who sought him out while on a lecture tour in Berlin. Valier told him about plans for organizing a rocket society in Breslau, and Ley at once became enthusiastic. The Verein fuer Raumschiffahrt (Society for Space Travel) was formed by Valier and Johannes Winkler on June 5, 1926. While Ley was not a founding member, he was one of its very early recruits. He later became vice president of the VfR and, for a short while, editor of its publication Die Rakete (The Rocket).

"In the beginning it was simply a society of interested scientists and laymen who discussed rocket problems in meetings, papers and by correspondence. Experimental work was not begun earlier than 1929 . . . . In short, there was much activity but no rocket had yet been shot. It was only theoretical and a bit of preliminary experimental work," he later recalled.(5)

Within a year of its founding, the new organization had grown to a membership of more than 500. Included in its ranks were men of international stature in the fledgling science of astronautics and technology of rocketry. They included Robert Esnault-Pelterie, France; Prof. Nikolai Rynin, USSR; Guido von Pirquet, Austria; Hermann Oberth, Romania; and Walter Hohmann, Germany. It also attracted several enthusiasts in the USA: Hugo Gernsbeck, the publisher of science fiction magazines, and his editor David Lasser, who would later found the American Interplanetary Society (now the American Institute of Aeronautics and Astronautics) and G. Edward Pendray, science editor of "Literary Digest" and a leading member of the American Interplanetary Society.

Ley early convinced the members of the VfR that one way to attract the attention of the public, and hopefully money from it, would be to publish a book on spaceflight written in terms the public could understand. The result was Die Moeglichkeit der Weltraumfahrt (The Possibility of Space Travel) which appeared on 1928. It contained chapters on various topics by the leading members of the VfR; and it was edited by Ley, who also contributed a chapter.

Indirectly, the book helped turn the VfR from what was fundamentally a discussion group ihto a research and development activity.

It caught the eye of movie director Fritz Lang. Lang's wife, the actress Thea von Harbou, wrote a script based upon it and called it "Frau im Mond" ("Woman in the Moon"), which her husband produced and directed. In doing so, he called upon the VfR for technical assistance. Oberth was hired as a consultant, and Ley was taken on as a publicist for the film. He was contracted for a series of 12 articles that were released to the press as the film progressed.

Lang later recalled meeting Oberth and Ley for the first time. Ley had brought Oberth to the director's office to introduce him. Having performed that amenity, Ley turned around, walked to a door and opened it, carefully closing it behind him. Lang looked on with surprise when Ley did not immediately reappear because he had walked into a closet. After a minute or two, Ley did come out as though nothing had happened and left the office in a more conventional manner. He later said that while he was in the closet he had begun thinking about a problem and simply stayed until he had resolved it. Later, Lang inspected his movie set and found Oberth supine upon the floor with stage hands stacking lead weights upon his chest. The technical consultant told the director that he wanted to experience the effects of rocket acceleration at lift-off. Lang later recalled, "At that time I began to have serious doubts about the sanity of the rocket technicians I had hired."(6)

One member of the cast had no misgivings about Ley and Oberth. Gerda Maurus, the leading lady, became so fascinated with the theme of the film that she joined the VfR!

The film also contributed a technique that later became a standard practice in rocket launchings. Lang in order to build suspense for the launching scene had a "countdown" flashed on the screen: "10 seconds to go . . . 6, 5, 4, 3, 2, 1 --FIRE!"(7)

As a part of the agreement between Lang and the VfR, the film company making "Frau im Mond" was to finance a liquid-propellant rocket that would be launched on "opening night" of the movie. The rocket was designed by Oberth, who later summed up the results: "The affair, by the way, was nevertheless disgraceful. First, I was not a trained mechanic; and Henry Ford was right when he said that one should not invent an engine if one could not assemble it with one's own hands. Let me tell you, that man was right."(8)

After the film was released and the hurried attempt to design, build, and launch a liquid-propellant rocket within four months, the VfR undertook a serious and well-planned research program. It ceased publication of <u>Die Rakete</u> and used the money previously spent for it in designing and building rockets. The rockets were designed by Oberth and later Rudolph Nebel. One such motor developed and tested was named the "Aepyornis Egg" by Ley because of its resemblance in shape to the egg of an extinct ostrich-like bird.

While Ley was becoming more involved with publicizing rocketry and the VfR than in assisting in the more technical aspects of developing rockets, he did provide some answers for engineering problems. Toward the end of 1931, Klaus Riedel began considering the use of alcohol as a fuel for liquid-propellant rockets because it could be mixed with water to control the temperature within the combustion chamber of a motor. While Riedel was pondering the problem of what the best proportion of alcohol to water would be, Ley consulted with a man who was an expert in the field of alcohol if not in rocket motor design. He asked his father, a distiller of spirits and merchant of them in East Prussia. The reply was that 40 percent alcohol was the minimum mixture that would ignite.(9)

During the early 1930's, Ley began publishing articles abroad. Typical of the magazines for which he wrote were "The Bulletin of the American Interplanetary Society," "Cosmology," and the "Journal of the British Interplanetary Society." Soon, however, the economic depression in Germany made itself felt upon him as a free-lance journalist. Lectures and commissions for articles and books came less and less. He had to take a job as an office manager and accountant.

With arrival of the Nazis to power in 1933, the days of the VfR were numbered. (In June, 1933, the VfR was legally bankrupt and abolished. However, some of its members organized the E. V. Fortschrittliche **Verkerstechnik** with Major Hanns-Wolf von Dickhuth-Harrach as president and Ley as secretary.(12) So far as Ley was concerned, it was a short-lived office.) Nebel had earlier attempted to interest the government in rockets as possible weapons, with no success. However, the Nazis saw the utility of such weapons, especially since they did not violate the letter of the Treaty of Versailles, which was still technically in effect. One day in 1933, von Dickhuth-Harrach, last president of the VfR, called a meeting of the officers of the organization in his home. He very plainly spelled out the fact that the group had absolutely no future under the new political party and, indeed, several of its members probably would be safer if they left the country altogether because of political beliefs or ethnic origins.

Ley understood perfectly as did fellow member Herbert Schaefer. In the fall of 1934 they began making plans to leave Germany, and Ley got in touch with friends in The Netherlands and England requesting their help.(10) On January 7, 1935, he boarded a train carrying a visitor's visa to the USA and a letter written on his company's letterhead stationery stating that he was going to England for a holiday and would return in 10 days. (He had written the letter himself.) Ley stayed in England for two weeks as the guest of P. E. Cleator, president of the British Interplanetary Society, and then left for America by ship. He arrived on February 15, 1935.(11)

If things were economically bad in Germany, they were not much better in the United States.

Members of the American Interplanetary Society provided a bond for Ley to enter the country, and he spent his first month or so as a guest of G. Edward Pendray at his home in Crestwood, New York. He then moved into a room rented for him by Peter van Dresser, a member of the society, in Greenwich Village in New York City.

Since he was on a visitor's visa, Ley could not undertake a profession. However, Pendray arranged for him to evaluate a book manuscript in German for a publisher, who paid Ley $25. Realizing that he would have to earn his living by writing, Ley began studying colloquial English at the expense of everything else, including experimentation in rocketry. Slowly, he began to move ahead. His first sale was a translation of a French article for "Natural History," and articles such as "Some Practical Aspects of Rocketeering" were sold to Aviation in 1936.

In this article he made a prediction that seemed to cast doubt on his ability as a prognosticator in astronautics: "The most appropriate application for rocket motors are for altitude rockets to carry instruments into the layers of the atmosphere that cannot be reached by other means. Although these so-called 'meteorological rockets' may lead to mail-carrying rockets, it is doubtful whether rocket motors will ever be able to replace the internal combustion engine for ordinary aircraft."

Other articles appeared in "Sportsman Pilot," and he wrote for French and Swiss aviation periodicals as well. During his first year in the USA, he managed a precarious living, subsisting on an average of $12 a week.

Despite the necessity for concentrating his energy on writing for a living, he did not altogether foresake his interest in rocketry. With Herbert Schaefer, his fellow refugee from Nazi Germany, Ley co-authored an article on the use of rockets for studying the weather.(13) Unknown to them, it upset plans of Frank J. Malina and Apollo O. M. Smith, of the Guggenheim Aeronautical Laboratory of the California Institute of Technology, as Malina explained in a letter he wrote home: "Smith and I were much disappointed last week when we found a French paper with a study similar to ours. Have decided not to send ours to France (for the Robert Esnault-Pelterie-Hirsch prize competition in astronautics)."(14)

Ley also found time, briefly, to recapture the thrill of participating in the launching of rockets that he had known several years earlier at the VfR's Raketenflugplatz, Berlin. In 1936, he was the operations director for a series of experiments, financed by F. W. Lessler, that took place on a frozen lake at Greenwood Lake, New York. Two aluminum-bodied, rocket-propelled airplanes with 15-foot wingspans were launched under his direction. Propelled by liquid oxygen and ethyl alcohol, they were an early attempt to develop a means of delivering the mail. Neither rocket performed in a manner that would persuade postal authorities to abandon the airplane, train, or motor vehicle. Indeed, the rockets proved less reliable that the dependable horse in the days of the "pony express."(15,16)

In an attempt to gain a living, Ley even tried his hand at science-fiction, publishing "At the Perihelion" in the February, 1937, issue of Astounding Stories, under the pen-name of Robert Willey. F. Orlin Tremaine, editor of the magazine, suggested that Ley might better consider nonfiction. Ley obliged with an elementary description of the fundamentals of rocketry entitled "The Dawn of the Conquest of Space" for the March, 1937, issue. Since articles on rocketry and space travel reached only a rather parochial readership, Ley turned to articles on paleontology, a science in which he was equally at home. He sold articles on the topic to magazines such as "Astounding Science Fiction."

Such a precarious living could not go on indefinitely. In February, 1937, Ley officially set up a residence in Havana and then legally emigrated to Florida in the following month. Then followed a meager, hand-to-mouth existence for three years. In 1940, Ralph Ingersoll started the newspaper PM, an evening tabloid that was doomed from inception. Ingersoll believed that the public would pay 10 cents for a newspaper that had no advertising while its competitors, with advertising, sold for 3 cents. Ley became science editor for the noble experiment, a position he held for a year.

While PM proved to be a failure, it did prove to be a boon to Ley. While working on it, he met Olga Feldman, a svelt, young dancer, who was the newspaper's physical culture columnist. She, too, was a refugee--but from Russia, not Germany.

"I want you for Christmas," Willy said to Olga in 1941. They were married on December 24. Later, they became the parents of two daughters, Sandra and Xenzia.(17)

Daughter Xenzia, several years later demonstrated that she had inherited her father's devotion to scholarship. While leafing through a copy of his book "The Conquest of Space," with the young son of a neighbor, she was shocked to see the lad point to a picture of Mars and call it Jupiter. In indignation, she said to her mother, "That silly child didn't know Mars from Jupiter!"(18)

In March, 1944, Ley gained his U.S. citizenship and began a five-year position as an instructor at the Washington Institute of Technology, at College Park, Maryland, while continuing his writing. His book Rockets, the Future of Space Travel Beyond the Stratosphere also appeared. Little did he realize at the time that over 18 revisions later the book would be titled Rockets, Missiles, and Men in Space and become the standard history of astronautics and rocketry, inspiring a new generation of rocket historians to continue where he left off.

Ley as a prophet of events in space exploration was curiously uneven.

In the fall of 1944, his old friend in the British Interplanetary Society, A. V. Cleaver, visited him in New York. Early in September, German V-2 rockets had begun descending upon London. For a few weeks, their arrival was not noted in the press, but Cleaver had not only been aware of their arrival but had also been interviewed by British Intelligence officials concerning them.

Later remembering the meeting, Cleaver wrote: "When I met Willy in New York, I was astounded to find that, for some reason, he had decided that the rumors were a lot of nonsense. He spent much time and effort assuring me that his ex-countrymen were most unlikely to have developed such a weapon, which would be inaccurate and uneconomical, and probably impossible to achieve at that date, in any case. I argued weakly against these conclusions, but being very conscious of war-time security and of my own youth and junior position, I forebore to tell him that I had personally heard the 'rumors' arriving, with their characteristic double bangs, or that I could describe the rocket to him if he would only listen."(19)

From 1944 until 1947, Ley was engaged in a project that once again returned him to the developmental aspects of rocketry.

In October of that year, Ley moved from New York to Atlanta to begin work on a rocket that could reach a maximum altitude of 100,000 feet and radio back data on meteorological conditions as it ascended. After eight months, a turn-over in the ownership of the company required him to move to Washington, where he lived until 1947. The project was terminated because of the rapid proliferation of commercial air travel which made the random and frequent firing of such rockets impossible.(20)

Following World War II, Ley's rank as a science writer rose almost annually. In 1951, his book The Conquest of Space, illustrated by Chesley Bonestell, sold some 84,000 copies. He became, at the same time, along with his former colleague in the VfR, Wernher von Braun, a technical consultant to the Walt Disney Studios. Also, he performed similar services for the television show "Tom Corbett, Space Cadet" and a newspaper comic strip by the same title.

Ley's career during the 1950's is perhaps best summed up by Samuel Moskowitz, who wrote in 1966, "Everywhere one turned, Willy Ley's name was on a book, in a magazine, in the newspapers or on a catalog endorsing a rocket toy. His face peered out from the television screen; his voice, instantly identifiable, seemed always on the radio; and posters announced his lectures at major cities across the country."(21) One of his most important efforts in publicizing space during this period was a series of articles that appeared in the March 22, 1952, issue of Collier's magazine, entitled "Man Will Conquer Space Soon." Ley contributed an article on the manned space stations of the future. The collective authors of the series, including Wernher von Braun, Joseph Kaplan, Heinz Haber, and Oscar Schacter, placed a modest price tag of only $4,000,000,000 on the conquest of space.

In 1953, Ley was named guest of honor at the 11th World Science Fiction Convention, in Philadelphia, the first non-fiction writer ever to be so singled out.

From 1957 until 1958, Ley held a teaching position in the science department of Farleigh-Dickinson College (Madison Branch). It was during this time that the USSR launched the first artificial satellite of the earth. With reference to that event, Ley wrote in an introduction to Albert Parry's book Russia's Rockets and Missiles, in 1960: "It might be useful to explain here that I spent about half the time that has gone by since the date of Sputnik 1 in lecturing all over the United States.

Doing this, I probably met more 'public' than anybody who is tied down to a routine job, even if that routine job should be that of a newspaper reporter. The question asked, in innumerable versions and variations, always was: 'How could the Russians . . .?" Ley with his tremendous knowledge of the history of rocketry proceeded to tell them how the Russians could and did become the first country in the world to send an artificial satellite about the earth. In doing so, he predicted that the United States would send a man into space within six years (only four were actually required). A later prediction, in 1961, placed the first landing of an American on the Moon in 1968, a date Ley missed by only one year.

In addition to his writings in both space science and astronomy, Ley found time to teach, in the mid-1960's, at the New School (Mahattan) and Long Island University (Brooklyn). The subjects ranged from popular courses in astronomy to graduate courses in the history of astronomy and space sciences.

On June 25, 1969, less than one month before the first men set foot on the Moon, Willy Ley died. As a consultant to the National Aeronautics and Space Administration, he had planned to be at the Manned Spacecraft Center (now the Johnson Space Center) for the Apollo 11 mission. Following his death, Dr. Robert S. Richardson, a renowned astronomer, suggested that one of the larger of the many craters on the far side of the Moon be named for him. Apparently he was unaware of a similar honor that had been accorded Ley in Robert Heinlein's novel **Beyond the Horizon**, published in 1942:

> "Leyburg, Luna. Diana's Playground, long touted by its promoters as the greatest amusement enterprise ever undertaken off earth or on, was invaded by the first shipload of tourists at exactly twelve thirty-two, Earth Prime. These old eyes have seen many a pleasure city, but I was surprised! Biographers relate that Ley himself was fond of the gay spots--I'm going to keep one eye on his tomb while I'm here; he might show up--"(22)

A bazarre but wholly appropriate funeral service for Ley was proposed by Robert P. Crowley, editor of **Popular Mechanics**: "It's not too late, however, for America to honor this kindly, self-effacing scientist who convinced us the moon was within our reach. **Popular Mechanics** thinks it would be fitting and proper to honor this great man by scattering his ashes on the moon, and we urge NASA to take them there on Apollo 12."(23) NASA, with its characteristic lack of imagination in public relations, declined.

In summary, perhaps the best tribute to Willy Ley is that of his old friend P. E. Cleator, of the British Interplanetary Society: "From 1926 onwards, Willy Ley, in one way or another, was so much a part of the astronautical scene that it is difficult to visualize it without him. But in his writings, and in the writings of others, his name will live on as one of the pioneers of that once supposedly fantastic enterprise, the making of a journey to another planet."(24)

REFERENCES

1. "Willy Ley," **Die Rakete**, Nr. 8,15 August 1928, S. 126.

2. S. Moskowitz, "The Willy Ley Story," **Worlds of Tomorrow**, Vol, 3, No. 7, May 1966, p. 32.

3. B. Williams and S. Epstein, The Rocket Pioneers on the Road to Space, New York, Messner, 1958, pp. 116-117.

4. Moskowitz, op. cit., p. 33.

5. W. Ley, "Rocketeering or the Hunting of a Canard," Flight, Vol. 31, No. 1483, May 27, 1937, p. 534.

6. Anecdote related to author by Lang during a visit to the Marshall Space Flight Center, Huntsville, Alabama, c. 1965.

7. W. Ley, Rockets, Missiles and Men in Space, New York, Viking, 1968, p. 259n.

8. H. Oberth, "My Contributions to Astronautics" in F. C. Durant III and G. S. James, eds., First Steps Toward Space, Smithsonian Annals of Flight No. 10. Washington, Smithsonian Institute Press, 1974, p. 139.

9. H. Gartmann, The Men Behind the Space Rockets, London, Weidenfeld and Nicolson, 1955, p. 89.

10. "Willy Ley, 62, Prolific Writer on Scientific Subjects, is Dead," New York Times, June 25, 1969, p. 24.

11. Moskowitz, op. cit., p. 33.

12. "News of German Experimenters," Astronautics, 1934, n.p.

13. "Les Fusees volantes meteorologiques," L'Aerophile, Vol. 44, 1938, pp. 228-232.

14. F. J. Malina, "On the GALCIT Rocket Research Project, 1936-38," in F. C. Durant III and G. S. James, op. cit., p. 121.

15. W. von Braun and F. I. Ordway III, History of Rocketry and Space Travel. New York, Crowell, 1975, p. 82.

16. C. W. McNash, "Proposed Altitude Rocket Hops," Popular Aviation, Vol. 19, October, 1936, p. 24.

17. Moskowitz, op. cit., p. 40.

18. S. Bland, "Sky Guy, Willy Ley at Home Among Space Ships," The Philadelphia Inquirer, September 9, 1951, p. 10.

19. A. V. Cleaver, "Tribute to Willy Ley," Spaceflight, Vol 11, No. 11, November 1969, p. 408.

20. Moskowitz, op. cit., p. 41.

21. Ibid., p. 42.

22. p. 20.

23. R. P. Crowley, "Send Willy Ley's Ashes to the Moon!", Popular Mechanics, September 1969, p. 10.

24. P. E. Cleator, "Tribute to Willy Ley," Spaceflight, Vol. 11, No. 11, November 1969, p. 408.

WILLY LEY (1906-1969)

ANDREW GALLAGHER HALEY; A BIOGRAPHICAL SKETCH
(1904-1966)

Stephen E. Doyle*

This biographical sketch contains a summary of the principal roles played by A. G. Haley in the early development of national and international non-governmental aerospace organizations, and the promulgation of early concepts of space law.

INTRODUCTION

Andrew Gallagher Haley was an exceptionally talented man in many disciplines. His place on the list of famous men in the history of modern astronautics is justified by any of several contributions he made to that history. He was a lawyer, who helped shape and interpret emerging aspects of space law. He was an industrialist, who was instrumental in the founding and early management of a major rocket manufacturing company. He was an organizational leader of national and international aerospace and space-related organizations. He was a journalist, who reported and interpreted current events in astronautics, and he was a tireless researcher and author who produced hundreds of articles and papers describing, recommending and anticipating developments in the law, technology, and sociology of space. He wrote an historical book of renown on rocketry and space exploration(1) and, later, an award-winning book on space law and government(2). He was an educator who taught or lectured at more than thirty colleges in the United States and abroad(3). He was a privately modest philanthropist and publicly a benefactor of astronautical foundations and organizations.

Andy Haley had an unalterable dedication to the spirit and concept of international cooperation, and he devoted his time, energy and resources to stimulating and encouraging improved international understanding. The influence of Haley's life has been felt well beyond the boundaries of the United States, because while he was a great American, he was as much a citizen of the world.

THE MAN

The personal traits, the character, the intellect and personality of Haley were as complex and multifaceted as his professional capabilities. He had relentless drive and boundless energy. He pushed himself to his physical and intellectual limits, which were well above average, and he expected those around him to follow. He admired intellectual ability and despised laziness. He was an emotional man, capable of both temper and tenderness. He often suppressed his inherent compassion, which he may have considered a weakness. He was acutely sensitive, with
----

*Deputy Assistant Administrator for International Affairs, NASA: former law clerk (1962 and 1964) and associate member (1965) of the law firm Haley, Bader and Potts, Washington, D.C. Paper presented by Mrs. Eilene Galloway, Special Consultant to the Senate Committee on Aeronautical and Sciences; Member of the Board of Directors, International Institute of Space Law; President, Theodore von Karman Memorial Foundation.

love for his friends and family that he sometimes found difficult to convey.

Haley was politically astute and active, and socially aware, although publicly asserting his independence and individuality. His public behavior was sometimes enigmatic to observers, but it would be a mistake to assume that it was ever without purpose.

THE YOUNG LAWYER

Born in Tacoma, Washington, in 1904. Andrew G. Haley went to Washington, D.C., in 1924, where he enrolled at Georgetown University, earning his law degree there in 1928 (4). While a student, he was employed by Senator C. C. Dill, supporting research for and drafting of the Federal Radio Act of 1928. From 1929 to 1932 he practiced law in Tacoma, Washington, returning to the nation's capital in 1933. In 1934 he received his Bachelor of Arts degree at George Washington University in Washington, D. C.

In 1933 Haley became an attorney for the Federal Radio Commission and in that capacity contributed to the formulation of the Communications Act of 1934. He remained in government service as an attorney with the Federal Communications Commission until 1939, when he entered private law practice in Washington, D.C. In these active and constructive early days he built an enviable record of achievement in communications law which became the solid foundation of his later contributions to space communications law.

During the 1920's Haley did substantial research and writing on questions of immigration law, publishing several articles on this subject in Catholic magazines (5). His experience and knowledge in this relatively rare legal specialty first brought Andrew Haley into contact with Theodore von Karman in 1941, when Haley was instrumental in obtaining a permanent visa for von Karman's sister to remain in the United States (6). That contact opened a new chapter in Haley's life, taking him out of Washington, D.C., and, for the first time, out of the active practice of law.

THE PIONEERING INDUSTRIALIST

In January 1942, at von Karman's request, Haley went to California to advise the leaders of the Guggenheim Aeronautical Laboratory of the California Institute of Technology (GALCIT) concerning the legal intricacies of establishing a new corporation to conduct rocket manufacturing. Frank J. Malina, Martin Summerfield, Edward S. Forman and John W. Parsons met with Haley and von Karman in an organizational meeting at von Karman's home (7). The result was the incorporation, in March 1942, of the Aerojet Engineering Corporation. Theodore von Karman was President, Malina was Treasurer, and Haley was Secretary; Summerfield, Forman and Parsons were Vice Presidents. The initial capitalization was established by these founders, but the largest share was Haley's. Although he considered Haley the "owner" in a financial sense, von Karman had full management control through a voting trust of the stock, which gave von Karman the voting power of all the original investors (8).

Having completed the essential legalities of Aerojet's incorporation in early 1942, Haley returned to Washington, D.C., where he was called to active military duty in the mobilization that followed the December 1941 attack on Pearl Harbor (9). But von Karman, according to his own account, was himself not an administrator, and his corporate executives were too busy in the laboratories and at test sites to run the corporation. Consequently, within six months, the Air Force, which had given the novice corporation one of its first contracts, decided to terminate its relationship with Aerojet. Upon inquiry, von Karman learned that the principal reason was the Air Force's lack of confidence in Aerojet management. By direct appeal to General "Hap" Arnold, von Karman secured Haley's release from active military duty, to become President and General Manager of Aerojet. Meanwhile, von Karman assumed the position of Chairman of the Board. The result, in von Karman's words, was that "Haley became Aerojet's second President on August 26, 1942. He proved to be an incredible administrator. He held the company together by guts and audacity "(10).

Haley remained president of Aerojet throughout the war. In 1944, he began negotiations with the General Tire and Rubber Company in order to expand the capital base of Aerojet. After lengthy negotiations, an arrangement was reached for large block stock purchases (11). Control of Aerojet, known today as Aerojet-General, passed in 1945 to General Tire. Haley continued as president throughout the transition period, returning to Washington, D.C. and private law practice in 1946. During this period (1945-46) Haley also served as consultant to the U.S. Senate Special Committee established to assess the peacetime role of the defense industries (12). The completion of his diligent advisory work for the Senate, 1946, concluded the essentially industrial stage of Haley's professional career.

THE ORGANIZATIONAL LEADER

Haley's early organizational efforts in the American Rocket Society (ARS) anticipated his many later contributions to the American Institute of Aeronautics and Astronautics (AIAA). During the 1940's, Haley had vigorously argued the potential of rocketry and astronautics, when public and governmental opinion had little sympathy with such views. He was the first chairman of the ARS Spaceflight Committee, created in 1952, and drafted a special report from that committee to President Eisenhower encouraging full governmental support of astronautical research well before the IGY. He served the American Rocket Society continually during the 1950's with increasing responsibility, progressing from Director to Vice President, to President, and then Chairman of the Board and Fellow. He also served as legal counsel to the Society and contributed substantially to the complex and demanding task of merging the American Rocket Society and the Institute of Aeronautical Sciences in 1963 to form the American Institute of Aeronautics and Astronautics. Haley became legal counsel to the AIAA and was soon elected a Fellow. He organized and served as chairman of the AIAA Technical Committee on Law and Sociology (13).

At the Second International Astronautical Congress, in London, during September 1951, the IAF was formally established. At that time Haley was elected one of two Vice Presidents of the IAF and served in that role for three years. Haley accepted administrative and committee assignments whenever asked. In 1956, the IAF met in Rome, where Haley presented to the Congress a program for IAF relationship with UNESCO. The program was adopted and Haley was elected official IAF representative to UNESCO.

The Eighth International Astronautical Congress was held in Barcelona, Spain, in 1957, in the immediate wake of the USSR's launching of Sputnik 1. Haley was elected President of the IAF that year. Also, in Barcelona, at the suggestion of the United Kingdom delegation, a special committee was created to review and revise the constitution of the IAF. Haley was named Chairman of the review committee. In 1958, at the Ninth Congress, in Amsterdam, Haley was reelected IAF President. At that

Congress he was also instrumental in obtaining adoption of a resolution creating an IAF Permanent Legal Committee, the precursor organization to the International Institute of Space Law (IISL).

In a pioneering effort to stimulate international collaboration in the development of space law, Haley organized and was chairman of the First Colloquium on the Law of Outer Space, held at The Hague in 1958. He was convenor of the Second Colloquium at Lincoln's Inn, London, in 1959. He served as editor or co-editor of the proceedings of the colloquia each year from 1959 until his death in 1966.(14) While serving as General Counsel to the IAF, in 1959, Haley accepted the role of Executive Secretary of an ad hoc Organizing Committee for creation of the IISL. Thus, he was instrumental in drafting and presenting the Constitution of the IISL (as a sub-organ of the IAF) which was formally adopted at the Eleventh IAF Congress in Stockholm in 1960.

During the 1950's Haley worked continuously to obtain official status for the IAF in liaison with UNESCO, the International Telecommunication Union (ITU), and the International Council of Scientific Unions (ICSU). (15) Within the United States, he was instrumental in the establishment of the American Bar Association's Committee on the Law of Outer Space in 1957, serving first as its Vice-Chairman and in later years as its Chairman.

A sampling of leadership roles filled by Haley, just prior to his death in 1966, shows clear evidence of his ubiquitous involvement in national and international organizational activities in support of astronautics. In 1966 Haley was a Trustee and Executive Vice President of the Theodore von Karman Memorial Foundation, Chairman of the ABA Committee on the Law of Outer Space, Counsel and Academician of the International Academy of Astronautics, Counsel of the International Astronautical Federation, Legal Counsel of the American Institute of Asronautics and Astronautics, Counsel and Director of the International Institute of Space Law, and Vice President and Director of the Axe Science Fund. During his latter years Haley maintained active membership status in more than fifty professional and fraternal organizations of astronautical, legal, industrial and religious concern. But, while conducting his law practice in Washington, D.C., and serving in diverse leadership roles nationally and internationally, Haley relentlessly continued to write and lecture.

THE JOURNALIST AND LECTURER

Haley's first professional writing job was as a reporter on the Tacoma News Tribune, immediately after his graduation from high school in the early 1920's (16). He was interested throughout his life in writing articles, comments, monographs and books. In the 1920's and 30's he contributed often to the Commonweal and other Catholic magazines and wrote selected law review articles on immigration and radio regulation (17). He wrote relatively little during the 1940's, but began again in the 1950's writing on communications. From 1950 forward he also wrote substantially about aspects of astronautics and space law. One of his first space law papers, "Jurisdiction Beyond the Earth", was presented as an address to the Rotary Club of Charlotte, North Carolina, in June 1955. Throughout the 1950's he contributed to the ARS magazine Jet Propulsion. In the 1960's he continued journalistic writing for the AIAA magazine Astronautics and Aeronautics (18).

A total count of Haley's written speeches and articles exceeds two hundred; all were characteristically authoritative and informative. (19) Although drafts of some of the more scholarly legal articles were prepared or contributed to by law clerks or other lawyers in his firm, they all bear evidence of Haley's imagination and perspicuity. His writing was always intended to inform, but in his astronautical works Haley also wrote to stimulate interest and inspire support. He was like a prospector of minds, constantly searching for new ideas. After his Aerojet years, he was always the apostle and apologist of astronautics.

From the mid-1950's Haley lectured often and extensively on astronautics and space law. He spoke at major universities on five continents and in dozens of cities in the United States. His articles and lectures have been published at various times in more than half-a-dozen languages.

THE RESEARCHER AND AUTHOR

Evaluating Haley as a researcher and author requires knowledge of more than his latter years of work as a lawyer, organizational leader, lecturer and speaker. The foundations of Haley's scholarly interests were laid at Georgetown University in the late 1920's where, as a student, he did research and studies under James Brown Scott, one of America's greatest legal scholars.

During this period Haley was impressed profoundly by the early Spanish philosophers Vitoria and Suarez. He was a staunch supporter of the natural law school and decried the more pragmatic empirical sociological approach to international law (20).

After he left Georgetown University Haley continued to study and do research in support of his constant writing. When his personal library was catalogued, in 1966, it contained classic histories, biographies of statesmen and scientists, great works of literature, and a wealth of scientific, legal and engineering volumes. That library reflected the balance and comprehensiveness of Haley's approach to writing. He related Hammurabi's Code to the philosophy of anthropocentric jurisprudence to explain his concepts of Metalaw, (21) as comfortably as he discussed the need for IAF sub-organs to organize and pursue studies of implications of modern aerospace technology. He summarized and evaluated the results of an IAF Congress as accurately and succinctly as he introduced the von Karman Primary Jurisdictional Line, a proposed upper limit on national sovereignty at about fifty-three miles (22).

In the earliest of his major papers on basic concepts, Haley laid heavy stress on traditional international law terms such as "consent" and "sovereignty" (23). By 1955 he had written a formal legal rationale for the view that outer space is free for use by all nations for peaceful and scientific purposes (24). He declared in 1955 the need for controls on satellite radio communications because of the critical role radio must play in all spaceflight activities (25). He foresaw the need for an international treaty or convention on liability arising out of space activities long before the subject was taken up by the United Nations. (26).

Haley's most profound contribution to space law is one of his least known and certainly least understood concepts - Metalaw (27). Haley believed that there is a great mathematical probability of the existence of other sentient life forms in the universe (28). He evaluated traditional concepts of anthropocentric law and found them generally limited, because they were based essentially on the notion that one should do unto others as he would have others do unto him. But in different worlds, different life forms might not be able to survive being treated as we would want to be treated. To some forms of life, an oxygen-rich atmosphere could be deadly. Therefore, Haley reasoned, in matters involving inter-species contacts with other sentient life forms, man should be governed by the rule that we must do unto others as they would have us do unto them. From this precept he deduced a series of principles to guide man in exploring the universe. This set of principles, he named "Metalaw".

PHILANTHROPIST AND BENEFACTOR

During the later years of his law career in Washington and particularly in connection with his astronautical activities, Haley was frequently contributing his time and money to various organizations and causes. He selectively helped students and deserving young people interested in astronautics or space law to travel and attend conferences. He assisted funding administrative costs of the International Academy of Astronautics, the International Institute of Space Law (IISL) and the Theodore von Karman Memorial Foundation (29). In 1954 Haley established and funded the ARS (later AIAA) Astronautics Award (30). From 1960 to 1965 he contributed funding to the IISL to support the Andrew G. Haley Gold Medal Award, which was conferred upon outstanding contributors to the development of Space Law (31). These activities were not intended to obtain personal gain or reputation for himself; they were intended to be stimuli to dedication and excellence in the astronautical and space law fields. In large measure, Haley's philanthropy supplemented his unselfish dedication of personal time and energy to support international cooperation and to improve international understanding.

INTERNATIONALIST AND DIPLOMAT

Haley attended and participated in IAF Congresses regularly from 1951. In addition to the organizational leadership roles that he filled, Haley often undertook representational roles as well. In 1956 he represented the IAF at the VIIIth Plenary Assembly of the ITU International Consultative Committee on Radio (CCIR) in Warsaw, where he introduced the topic of spectrum allocation requirements for astronautical radio service. He attended CCIR Study Group meetings in Moscow and Geneva in 1958, and served as a U.S. Delegate to the ITU World Administrative Radio Conference in Geneva in 1959, where the first space radio frequency allocations were made. His legal work with the ITU bagan when he served as legal adviser to the 78-nation ITU Plenipotentiary Conference at Atlantic City in 1947. He also served as adviser at the Fourth Inter-American Radio Conference in Washington in 1949, and was legal adviser to the ITU Secretary General at the Plenipotentiary Conference of the ITU held in Montreux in 1965. He served continually on U.S. delegations to CCIR working groups, remaining active in this field until his death in 1966.

In recognition of his many efforts in the international arena Haley received numerous special awards and medals. Among early awards, in 1958 he was given the Grotius Medal of the International Grotius Foundation for the Propagation of the Law of Nations, in recognition of merit in the field of international law, for lecturing at more than a score of U.S. European universities on space law and international cooperation. In 1952 the Republic of San Marino conferred upon him the Cross of Commander of St. Agata, for his role in advancing the cause of peaceful space exploration. Also in 1962 he received the Medal of the British Interplanetary Society for outstanding contributions to astronautics. But one of his earliest awards possibly best characterizes his life--in 1954 he was given an American Rocket Society Special Award for "distinguished service and untiring efforts".

NOTES

1. Haley, A.G., <u>Rocketry and Space Exploration</u>, 334 pp., Illus., D. van Nostrand Co., Princeton, 1958.

2. Haley, A.G., Space Law and Government, 584 pp., Appleton-Century-Crofts, New York, 1963. In 1964 Haley was given the AIAA G. Edward Pendray Award, based on this book, for "a pioneering contribution to the analysis of governmental and international legal questions arising from the rapid development of space travel and space exploration".

3. In one tour, in the Fall of 1957, Haley traveled with Welf Heinrich Prince of Hanover across the United States (lecturing at thirty educational and scientific institutions in seventeen states) and then to Western and Eastern Europe and the USSR, on what Haley called an educational effort undertaken on behalf of the IAF. See <u>Space Law and Government</u>, <u>op. cit</u>., note 2, at 355. Welf Heinrich recently memorialized that tour in "Eine Reise in Sachen 'Weltraumrecht', Eindrücke und Erlebnisse einer Vortragsreise durch die Vereinigten Staaten von Amerika im Jahre 1957", in <u>Beiträge zum Luft - und Weltraumrecht, Festschrift zu Ehren von Alex Meyer</u> 385, Karl Heymanns Verlag KG, Köln, 1957.

4. At that time it was academically possible to enter law studies and take a bachelor's degree after four years. A general biographical review of Haley's early life is found in Vol. 16, No. 9, <u>Current Biography</u> 24, October 1955. A more comprehensive and interpretive biographical survey is contained in Thomas, S., Vol. 7, <u>Men of Space</u> 136, Chilton Books, Philadelphia, 1965.

5. See, e.g., Haley, A.G., "Debating the Immigration Question". IX:11 <u>Nat'l Catholic Welfare Conference Bulletin</u> 8, April 1928; or "America Welcomes the Immigrant", XLI:7 <u>America</u> 155, May 25, 1929.

6. Credit is given to Haley for obtaining Pipö's permanent visa in <u>The Wind and Beyond: Theodore von Karman</u> 256, (an autobiography written with the help of, and completed from rough notes after von Karman's death by Lee Edson.) Little, Brown and Co., Boston 1967. Martin Summerfield also recounts this story in "Haley and von Karman - A Beginning in Astronautics", <u>Astronautics and Aeronautics</u>, November 1966, p. 61.

7. <u>The Wind and Beyond</u>, <u>op. cit</u>., note 6, at 257. See also Haley's accounts in <u>Rocketry and Space Exploration</u>, <u>op. cit</u>., note 1, at 157 and <u>Space Law and Government</u>, <u>op. cit</u>., note 2, at xi.

8. <u>The Wind and Beyond</u>, <u>op. cit</u>., note 6, at 317.

9. <u>Men of Space</u>, <u>op. cit</u>., note 4, at 140. Haley was assigned the post of Chief of the Military Affairs Division, Office of the Air Judge Advocate.

10. <u>The Wind and Beyond</u>, <u>op. cit</u>., note 6, at 259.

11. <u>Ibid</u>., at 317.

12. Haley made substantial contributions to a report entitled "Investigation of the National Defense Program", Additional Report of the Senate Special Committee Investigating the National Defense Program (Aircraft), Rept. No. 10, Pt. 6, 79th Cong. 2d Sess., June 7, 1946, GPO Wash., D.C. This effort is summarized by Shirley Thomas in <u>Men of Space</u>, <u>op, cit</u>., note 4, at 142.

13. See historical origins of the Committee discussed by Haley in <u>Space Law and Government</u>, <u>op. cit</u>., note 2, at 389-90.

14. Dates and places with citations of proceedings for the first ten IISL Colloquia can be found in Lay, S.H. and H.J. Taubenfeld, The Law Relating to Activities of Man in Space 305, University of Chicago Press for the American Bar Association, Chicago, 1970.

15. Haley devoted considerable effort to recording his efforts on behalf of IAF, see e.g., chapter 11 of Space Law and Government, op. cit., note 2; and chapter 10 of Rocketry and Space Exploration, op. cit., note 1.

16. See Current Biography, op. cit., note 4, and an article by M. Shannon, entitled "Hovercraft Launch to the Homecoming," a social feature article about Haley, in the Tacoma News Tribune, July 23, 1971.

17. See, e.g., Haley, A.G., "The Broadcasting and Postal Lottery Statutes," 4:4 Geo. Washington Law Rev. 475 (May 1936); "Note-Radio Broadcasting and the Public Interest Involved", 22:4 Georgetown Law J. 844 (May 1934). See also articles cited in note 5 above.

18. Haley submitted news articles, current events reports and evaluations, and for several years did an annual state-of-the-art report on space law for Astronautics and Aeronautics. See cumulative index for each year, published with final number for each year in the 1960's.

19. A nearly complete bibliography of his works is contained in Space Law and Government, op. cit., note 2, at 529-539. An eight volume, bound compilation of Haley manuscripts and reprints is maintained at the Washington law offices of Haley, Bader and Potts, 1730 M St., N.W., Wash., D.C. 20036.

20. The best and latest written summary of Haley's jurisprudential philosophy is contained in chapter 2 of Space Law and Government, op. cit., note 2. Haley extended the traditional concepts of international law with his concepts of Metalaw, discussed below.

21. The best, most-comprehensive, and latest statement of Haley's proposed regime of Metalaw is contained in chapter 12 of Space Law and Government, op. cit., note 2.

22. The von Karman Primary Jurisdictional Line, as Haley called it, has been widely commented upon. It is presented with substantial commentary in chapter 4 of Space Law and Government, op. cit., note 2.

23. These terms later became the core of chapter headings for his book on space law and government.

24. Haley, A.G., "Basic Concepts of Space Law," a paper presented at the 25th Anniversary Annual Meeting of the American Rocket Society, Chicago, Nov. 14-18, 1955. This paper was published later in Jet Propulsion, November 1956, p. 951.

25. Ibid. See also, Haley, A.G., "International Cooperation in Rocketry and Astronautics," Jet Propulsion, November 1955, p. 627.

26. Haley, A.G., "Space Law and Metalaw - A Synoptic View," Harvard Law Record, November 8, 1956. This same paper was presented to and published in the proceedings of the VII International Astornautical Congress, Rome, September 17-22, 1956.

27. Haley's concepts of Metalaw evolved and expanded. His earliest mention of it is in the 1955-56 articles cited in notes 24, 25, and 26, above. See also note 21.

28. In one early article Haley wrote that lawyers may obtain evidence from many sources and witnesses, including mathematicians and scientists. He based his argument for the existence of other sentient life forms on works such as Shapley, H., *Of Stars and Men*, Beacon Press, Boston, 1958 and Carl Sagan, "Direct Contact Among Galactic Civilizations by Relativistic Interstellar Spaceflight" (undated).

29. Most of his contributions to these organizations were unpublicized but essential to the continued functioning of the organizations.

30. See *Rocketry and Space Exploration* 228-29, *op. cit.*, note 1.

31. The recipients of the IISL Gold Medal are listed in the proceedings of the XIII Coloquium on the Law of Outer Space, held October 4-10, 1970, in Constance, Germany, p. 377, Fred B. Rothman and Co., South Hachensack, N.J. (1971).

FRITZ ZWICKY
(1898-1974)

R. Cargill Hall*

Brilliant, assertive, indefatigable, and fiercely independent, Fritz Zwicky made important contributions to physics, astronomy, rocket engineering, and international scientific cooperation. He probed the crystal structure of solids, discovered supernovae, conceived of neutron stars, catalogued compact galaxies, patented rocket engines and exotic propellant combinations, invented coruscatives, and launched the first man-made object to Earth-escape velocity.

An implacable scientist first and last, he deplored unequivocally the jumbug artists, the social or religious dogmatist, the "fat cats" and the fatuous who hindered man's progress and curtailed freedom. If he was sometimes outspoken to the point of abrasion, Zwicky's Weltanschauung was always rooted firmly in the traditions of rationalism, humanism and internationalism. Scientists, he believed, ought to use the results of their research for the enrichment of the world community. The political, economic, and social leaders of that community ought to create the conditions in which everyone could discover and develop the genius that is his.(1)

THE ZWICKY BEGINNINGS

Fritz Zwicky was born in Varna, Bulgaria, to Swiss parents, Fridolin and Franziska Zwicky, on February 14, 1898. His father managed the Swiss Credit Bank in Varna and served variously as the Consul General of Norway, Russia, Germany, France, and England, depending on Bulgarian allegiances of the period. Though counting only 35,000 inhabitants, Varna was a departmental capital and important Bulgarian seaport and commercial center on the Black Sea, enriched by Jewish, Greek, and gypsy communities. As a youth Zwicky learned the value of cultural diversity, knew his Swiss heritage, and mastered Slavic, French, and German languages.

In 1912 Zwicky was sent to school in his native Switzerland to prepare for a career in international banking and commerce. But he soon demonstrated great promise in mathematics and physics, and in World War I, Zwicky went on to the Swiss Federal Institute of Technology in Zurich. He received his B.S. degree in 1920, and, under Professors Debye and Scherrer, the PhD in the physics of solids in 1922.(2) For the next few years Zwicky served as a Research Assistant in physics and the Institute while he explored the secondary structure in crystals and pursued a pastime he had come to love, mountaineering. The heights of the Eiger and other Swiss peaks helped provide the distance and perspective for contemplating the marvels of nature and the follies of man. The cataclysmic way and the manner in which French, English, and Belgian scientists sought to restructure international societies and humiliate scientists of the former Central Powers in the postwar years caused the young man to eschew mindless nationalism. In its place, he fashioned an independent international outlook that he would never forsake.

----

*Jet Propulsion Laboratory, California Institute of Technology. Presented at the International Space Hall of Fame-International Academy of Astronautics Symposium SELECT BIOGRAPHIES OF OUTSTANDING SPACE PIONEERS, Alamogordo, New Mexico, October 9. 1976.

In August 1925 Wicliffe Rose and Professor Trowbridge of the Rockefeller Foundation visited the Institute. Zwicky guided them, and at the end of the tour they inquired whether he might be interested in going to America on a Rockefeller International Education Board Fellowship. Indeed he would. Where would he like to go to, his guests asked? "Where there are mountains," Zwicky replied. In short order the three men agreed on the California Institute of Technology in Pasadena. With the approval of Robert A. Millikan, Chairman of the Executive Committee at Caltech, Zwicky was on his way to the United States and a long and productive scientific career.(3)

FROM PHYSICS TO ASTRONOMY

Named an Assistant Professor of Theoretical Physics at Caltech upon completing his fellowship in the spring of 1927, Zwicky's interests expanded from the physical properties of liquids and solids to embrace the high energy physics of cosmic rays. Fascinated by the physical processes at work in distant galaxies, he was before long "dabbling in astronomy." (4)

At this time Zwicky evinced two characteristics that would become his hallmarks. First, no matter accepted scientific truth, he would analyze a subject from every angle and approach--"Morphological research" as he later termed it--"if a possibility existed, nature would have filled it and scientists should discover it."(5) For example, he and Walter Baade in the early 1930's realized that there was a class of exploding stars far more luminous that ordinary novae, which had been seen for many years. In doing so, they succeeded in making the important distinction between novae and supernovae--whose enormous luminosities and great brilliance would make them detectable at extreme distances. Second, Zwicky would undertake systematic and very large programs of observation of long duration. Unable as a physicist to qualify for the time he thought necessary on the telescopes at Mount Wilson, he spent his evenings between 1934 and 1936 on the roof of Robinson Hall at Caltech with a Wollensack camera strapped to a 12-inch reflector in a futile search for extra-galactic supernovae. Did they really exist beyond the Milky Way? Logic said they did; what was needed were better observational techniques and instruments.

In 1935 Zwicky learned of just such an instrument: the Schmidt wide-field reflector-refractor telescope. Though he actively promoted the new invention, few astronomers of the day expressed interest. As a last resort, Zwicky presented his case to George Ellery Hale, the doyen of American astronomers. Hale agreed that supernovae probably did exist, "but that nevertheless the search for supernovae was like searching for gold, and people searching for gold were not really very respectable in scientific circles," Zwicky later recalled. "So I almost gave up and I said: 'Dr. Hale, it all depends on who searched.' Dr. Hale laughed and replied, 'Okay, $25,000,' and we built the 18-inch Schmidt in one year"(6) With the Schmidt telescope in operation at Palomar Mountain on the night of September 5, 1936, Zwicky began "beating the 'tar' out of the sky." He discovered his first supernovae in March 1937, and, on August 26, 1937, the brightest one of this century.(7)

Continuing the supernovae patrol throughout his life, Zwicky would personally discover and catalogue more supernovae than any other single individual. He proposed classification of them that still dominates the subject, and made estimates of their frequency (1942) which have only been improved by more recent finding. Zwicky devised a theory to explain how supernovae produced cosmic rays (1934), and how stars might implode and collapse to the neutron-star state (1934)--just two years after the discovery of the neutron. In 1937 J. Robert Oppenheimer and George M. Volkoff would develop the mathematical theory of neutron stars, and in the 1960's the discovery of pulsars would provide strong evidence of their existence. Zwicky pursued the idea of stellar collapse still further, contemplating controversial "pygmy stars" and "object Hades" (the latter now called "black holes").(8)

Always skeptical of the large recession velocities claimed for distant galaxies, Zwicky took care to refer to "symbolic" values of redshift and "indicative" redshifts. Driven by this skepticism, a fascination with extreme types of cosmic objects and the underlying high energy physical process at work in the universe, and a remarkable determination, he set out to catalogue conspicuous clusters of galaxies and individual galaxies down to the apparent magnitude 15.7 over most of the sky visible from the northern hemisphere. In the late 1930's he individually discovered many clusters of galaxies and even suggested that all galaxies exist in clusters. After his appointment as Professor of Astrophysics in 1941, collaborating with other astronomers, he discovered and catalogued tens of thousands of galaxies and clusters of galaxies eventually leading to the six-volume Caltech Catalogue of Galaxies and Clusters of Galaxies (1968). With Milton Humason, Zwicky also studied and catalogued "faint blue stars" far from the galactic plane (1947), and later, with others, the relatively dense, high-surface brightness "compact galaxies." (Figure 1) He perceived the relation of compact galaxies to the Seyfert and N-type systems, published a large Catalogue of Compact Galaxies (1971), and had another in preparation.(9)

Fig. 1 Fritz Zwicky, 1949

FROM ASTRONOMY TO ASTRONAUTICS

During these astronomical endeavors, Zwicky also took time out to consider rocket engineering and astronautics, and the implications these subjects held for "experimental astronomy." In the early 1930's Theodore von Karman, the famous aerodynamicist who directed the Guggenheim Aeronautics Laboratory at the California Institute of Technology (GALCIT) supported Zwicky in his efforts to introduce more physics and engineering into astronomy. Zwicky had already begun lofting cameras high into the stratosphere for astronomical observation free of atmospheric disturbances; now, with some colleagues at the Mount Wilson Observatory, he "started to hallucinate about observatories on the Moon."(10) The von Karman connection soon provided him the opportunity to explore directly the ways and means to realize that goal.

In 1936 the GALCIT Director encouraged some of his students in an investigation of solid and liquid-propellant jet propulsion engines. To produce these engines for the armed services when war broke out in Europe a few years later, von Karman, his attorney Andrew Haley, and some of his former students founded the Aerojet Engineering Corporation (now the Aerojet-General Corporation). When the company began to grow, they hired astrophysicist Zwicky as Director of Research.(11)

At Aerojet (1943-1949) in nearby Azusa, California, Zwicky attacked the problems of jet propulsion research in his customary synoptic fashion. He created a "morphological box" to survey all possible jet propulsion propellants and power plants, including hydroturbojets, hydrofuels, and terrajet engines. The latter engine was designed to move through solid earth, a function he determined to be vital not only for terrestrial uses, but indispensable for operating underground on the Moon and other bodies in the Solar System. For future lunar colonies engaged in research in astronomy, physics, chemistry, and engineering, Zwicky designed solar furnaces and generators that could be used to produce water, oxygen, nitrogen, foodstuffs, propellants, and electrical power. Meanwhile, he acquired numerous patents for exotic propellants--including free radical and metachemical propellants--heat detonators, coruscatives, and novel launchers for ultrafast particles.

Zwicky had long wanted to practice experimental astronomy, that is, investigate cosmic bodies directly to test hypotheses. The advances in rocket technology achieved during World War II made that dream possible. Beginning in 1946 he formulated an ambitious five-phase plan for "man's march into space." First, launch ultrafast particles (artificial meteors) at all heights in the Earth's atmosphere to explore its physicochemical composition; second, launch artificial meteors from great heights with velocities greater than 11.2 km/sec into interplanetary space; third, bombard the Moon on the dark side of the terminator with ultrafast particles containing reducing metals such as Al, to determine whether water of crystallization was present in the Moon's surface material; fourth, launch man into space; and fifth, restructure the Solar System so as to make the planets and their satellites habitable.(12)

In March 1946 Zwicky proposed to Army Ordnance that one of the V-2 rockets captured in Germany be used to launch artificial meteors in the Earth's upper atmosphere. Army Ordnance approved. Zwicky began preparing shaped charges of cast and putty explosives that would produce ultrafast particle jets made luminous even at extreme altitudes by air friction. Tested on the ground at night, the shaped charges produced a luminous trail of particles with velocities of nearly 10 km/sec as measured by cameras equipped with rotating hexagonal mirrors.

The first night launching of a V-2 in the United States occurred without a hitch on the evening of December 17, 1946, at the White Sands

Proving Ground in New Mexico. Thirty cameras and telescopes tracked the missile exhaust and glowing graphite vanes to a height of 190 km. But the cone shaped charges, to fired in groups of three at 36, 48, and 60 km altitude, failed to detonate. The failure of this experiment--later traced to the timing mechanism--severely disappointed Zwicky, but its repercussions for future research proved even more detrimental. Other American scientists, eager to place experiments aboard the V-2 and other available sounding rockets derided artificial meteors as impractical and without scientific merit Zwicky could not secure governmental approval for any further test shots.(13)

Though prevented from experimenting in space directly, Zwicky continued experiments at Aerojet with artificial meteors. He recognized that the problem in phase two of his space exploration plan involved producing self-luminous particles that could be detected even when ejected into a vacuum. He solved the problem by inventing coruscatives, or heat explosives. Coruscatives are solids that detonate with the application of sufficiently strong shearing stresses for instance, but in contrast to common explosives, they generate virtually no gas, and the reaction products are solids. After testing a number of highly compressed fine powders of powerful reagents, Zwicky settled on powders of titanium and carbon which, as a compressed conical insert in a shaped charge, could be collapsed fast enough to react explosively, resulting in the ejection of TiC droplets at almost 6000°C with speeds up to 15 km/sec.(14)

More time passed; then, in August 1957, Knox Kilsaps, Chief Scientist at Holloman Air Force Base in Alamogordo, New Mexico, notified Zwicky that one cubic foot of space would be available on an Arobee sounding rocket scheduled for launch in October. Zwicky and industrial chemist Joseph Cuneo quickly prepared the shaped charges and arranged for the photographic coverage. On the evening of October 16, 1957, the Aerobee sounding rocket with Zwicky's artificial meteors rose from its launch tower at Holloman Air Force Base. This time the shaped charges detonated properly 91 seconds after liftoff, at a height of 85 km. One very luminous pellet consisting mostly of TiC and some Al_2O_3 ejected with a velocity of 15 km/sec, considerably in excess of the escape velocity of the Earth, and became the first man-made object to be shot into interplanetary space--a tiny artificial satellite of the Sun. Its luminous escape trajectory was photographed by a number of instruments, including the 48-inch telescope at Palomar Mountain, 1000 kilometers to the West.(15) Zwicky had accomplished the first two phases of his plan for man's march into space.

But by now, large, government-supported teams of engineers and scientists pursued the exploration and utilization of outer space. Though he would never be harnessed to such large-scale efforts, Zwicky's enthusiasm for space exploration nonetheless would remain as avid as ever. The technology required to initiate manned lunar operations, Zwicky observed one month before President Kennedy proposed Project Apollo, was already at hand; "imagination and an indomitable spirit are all that are needed for success."(16) In the 1960's and 1970's he would participate as an officer and trustee in the operation of the International Academy of Astronautics, contribute to the deliberations of the International Astronomical Union and International Institute of Space Law, and propose ways for realizing his proposed reconstruction of the solar system. (Figure 2)

THE ZWICKY HERITAGE

On arriving in Pasadena in 1927, it is said, one of Zwicky's first questions to Robert Millikan concerned the availability of mountains to climb. When Millikan pointed toward 5,700-foot Mount Wilson nearby,

Fig. 2 Fritz Zwicky, 1968

Zwicky replied: "Ja, I see the foothills." Recounting this anecdote in 1950, People Today commented, "That's his attitude toward the impossible in science."(17)

Zwicky attempted, often with great success, to scale the biggest mountains wherever he found them. Well educated in the classics as well as in the physical sciences, he considered no subject foreign to his interests. Over 76 years he wrote 10 books and over 300 papers and articles ranging from "Zue Theorie der heteropolaren kristalle," to "Reynold's Number for Extragalactic Nebulae." He spoke fluent German, French, Italian, English, and Russian, and many of these works were first published in Europe. Not only did he "dabble" successfully in physics, astronomy, rocket engineering, and astronautics, he wrote about ways to solve such problems as smog in Los Angeles and the potential conflict of national activities in space. He developed a forceful epistemological case for problem-solving in general (co-founding the Society for Morphological Research). He exhibited a Renaissance flavor in the 20th century.

Zwicky deplored the "self-centered" men of research unable or unwilling to modify scientific paradigms in the light of new data. He judged government "experts" to be mostly "'yes men' and promoters . . . chosen because of their pliability, rather than because of their independence of thought and . . . indomitable will to make the truth prevail without compromise."(18) Having seen his own theorizing about supernovae, neutron stars, artificial meteors, and compact galaxies at first rejected by established scientific journals, he concluded that many scientists and laymen in our time had fallen prey "to the overbearing and presumptuous belief that scientific truth is absolute and science omniscient. This fallacious notion [has] caused, or is associated with, the emergence of groups and organizations of men of research . . . who successfully work on problems of limited scope, but who unduly promote each other to the exclusion of the free and lone inventigator."(19) The scientific process itself, he implied, could be dulled until it failed to plow new ground. It had to be constantly checked, honed, and resharpened.

Seemingly in the eye of one controversy or another throughout his life, the intercession of influential friends was often necessary if Zwicky's talents and ideas were to be utilized effectively. Zwicky recognized this and the reasons for it:(20)

> While [von Karman] was Director of the Scientific Board of the Air Force (on my standing with my colleagues I would have never been on that), he insisted that I too should be on it. So, he pushed that through, and I am indebted to him for that, and also later on for having pushed me into the International Academy of Astronautics, and so on. And it would have been quite impossible if all the hierarchy in power would have had their say, because they can not really admit a non-conformist like myself. On the other hand, he had his little jokes with me. He thought I was treating people too abruptly, too roughly, and it would be better not to be that rough; but to commemorate this abrasiveness he said, 'Now we have an occasion to get you into history, and we must devise a unit for the roughness of airplane wings, the surfaces of missiles, and so on. The proper thing will be to name this unit a Zwicky.' But then on second thought, he added, 'There is no such thing as a whole Zwicky except you--that's far too excessive--so the practical unit will be a micro-Zwicky!'

Zwicky displayed a fine sense of humor, and, despite his mordant barbs, remained a humane and kindly man. At the end of World War II he organized a lengthy project to restock war-ravaged libraries, and, with colleagues at Caltech, spent his weekends for a number of years collecting, crating, and dispatching books and periodicals to Europe and the Orient. In the 1950's he served as chairman of the board of trustees of the Pestalozzi Foundation of America devoted to the support of orphanages. He assumed good works to be a scientist's obligation. Scientists, he believed furthermore, were obligated to forestall the use of science for anti-social ends by all governments, however constituted. Men of research were "accorded special privileges by their fellow men and in return they have a duty to pursue the social implications of their work." (21) (Figure 3)

Busy with preparations for a meeting of the Trustees of the International Academy of Astronautics and at work on two books, Fritz Zwicky died in Pasadena on February 8, 1974. To the end he remained one of the "Fifth Swiss," as he liked to say, those from Switzerland who worked and resided abroad. True to that heritage, he had declined to become a naturalized citizen of the United States, and thus forfeited a security clearance, the directorship of research at Aerojet, and a chance for honors by American scientific societies. Nonetheless, for his technical achievements Zwicky had been honored in 1949 with the Medal of Freedom from President Truman. Near the end of his long career, in recognition of his singular contributions to astronomy, in 1973 the Royal Astronomical Society had awarded him its gold medal.

Every era has need of the lone wolf, independent, free to inquire, discover, and chasten to bodies scientific and political. Our era was fortunate to have shared Fritz Zwicky.

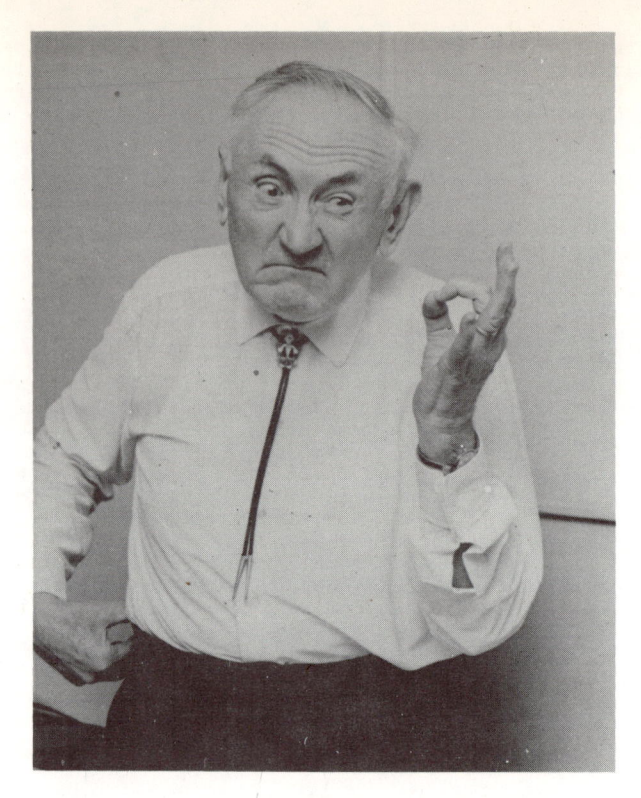

Fig. 3   Fritz Zwicky Lecturing, 1971

REFERENCES

1. See Fritz Zwicky, Discovery, Invention, Research, (New York: The MacMillan Company, 1969), also Zwicky, Jeder Ein Genie, (Bern: Herbert Lang Verlag, 1971).

2. Current Biography 1953, p. 677.

3. Fritz Zwicky, interview with the author, May 17, 1971, JPL History Archive, folder 3-561, pp. 1-2.

4. Ibid.

5. Jesse L. Greenstein, "Fritz Zwicky--Scientific Eagle," Engineering and Science, March-April 1974, p. 16. This procedure involves not only the "totality of all the possible aspects and solutions of any given problem," Zwicky explained, but "it insists on exploring all the possibilities for practical application of the results achieved to insure the continued stabilization of human society and the enrichment and optimization of its activities."

6. Zwicky interview, p. 3.

7. Fritz Zwicky, "100 Supernovae--The Reward of a 40-Year Search," Engineering and Science, May 1973, p. 21.

8. Halton Arp, "Fritz Zwicky," Physics Today, June 1974, pp. 70-71.

9. Cecilia Payne-Gaposchkin, "A Special Kind of Astronomer," Sky and Telescope, May 1974, pp. 213-313.

10. Zwicky interview, p. 3; supporting Zwicky's recollection, see also the contemporary article by R. S. Richardson, a colleague at Mount Wilson, "Luna Observatory No. 1," in Astounding Science Fiction, February 1940, pp. 113-123.

11. See Theodore von Karman with Lee Edson, The Wind and Beyond (Boston: Little, Brown and Company, 1967), pp. 261-263.

12. Fritz Zwicky, "Not Every Stone That is Thrown Has to Fall," Neue Zuercher Zeitung, August 17, 1946, p. 6; see also Zwicky in Observatory, Vol. 68, 1948, pp. 121-143; and Zwicky, Morphology of Propulsive Power, (Pasadena, California: Society for Morphological Research Monograph No. 1, 1962), pp. 145-149.

13. Zwicky, "A Stone's Throw Into the Universe: A Memoir," in R. Cargill Hall, ed., Essays on the History of Rocketry and Astronautics: Proceedings of the Third through Sixth History Symposia of the International Academy of Astronautics, Washington, D.C.: NASA Technical Memorandum to be published in 1977.

14. Zwicky, Morphology of Propulsive Power, pp. 179-181.

15. Zwicky, "A Stone's Throw Into the Universe."

16. Fritz Zwicky, "Possible Operations on the Moon," Aerojet Pamphlet, April 1961, JPL History Archive, folder 5-1143.

17. People Today 1:16, June 20, 1950.

18. Zwicky, Morphology of Propulsive Power, p. 3.

19. Zwicky, Discovery, Invention, Research, pp. 15-16.

20. Zwicky Interview, p. 15.

21. Zwicky, Discovery, Invention, Research, p. 24.

DR. EUGEN SAENGER BIOBRAPHY
(22 September 1905 to 10 February 1964) Austria-Germany

by Mrs. Irene Saenger-Bredt

Condensed and partly translated by Miss Tanestin Boubel, San Antonio, Texas, and translation completed by E. A. Steinhoff*, Alamogordo, New Mexico.

    The realization of any technological concept appears to be possible only after some developmental stages are satisfied during which the "originators" of the idea are called "dreamers", "inventors", "builders" and "doers". The originators within two interim stages may be considered "pioneers" of the particular new technology.

    Much before mankind becomes aware of, or even begins to realize, the evolution of a new technological ideal and concept, the potential idea already begins in a visionary way, to occupy the minds of poets and writers as an elusive, etheric mirage, for which the term "science fiction" has been coined. At the threshold of all pioneering technological advances unavoidably stand wishes and vague dreams.

    Stimulated by the creative thoughts of such poetry and science fiction, gifted individuals, "so-called inventors", begin to believe in the realisability of this or that science fiction project by evolving and matching requirements of technical performance of their "project" with the existing body of basic or fundamental physics, to mold and dress these new ideas in words or sketches, and to begin to vigorously pursue these without initially involving themselves in, or justifying, excessively quantitative dimensions, interrelations of details and practical realizability. These, by their contemporaries often under-estimated and seldom believed, genial individuals who can be termed as "first generation pioneers".

    They are followed, in a world which oppresses new and daring projects, generally by pioneers of the "second generation". Engineers and natural scientists, who on the basis of thorough studies and reflections, evolve all technological details of the plans of their concept, its development potential and of its rational use of the realizable product. In this effort they consider factors of environment of use with this project, which for the first time will yield applicable designs and approaches to their achievement. In this way, as researchers or "engineers", they "blaze" the trail towards the realization of their initially almost elusive goals for their "realization", who under consideration of in-dustrial potential and capacity of existing political realities as well as available or predicatable technological possibilities and needs of the society, conduct all the studies and planning necessary to drive forward all the phases of such a new project from specifications throught design, experiments, testing all the way to eventual production, until his achievement becomes obvious to the world in general and becomes recognized as such.

    Within the rocket and space technology, the most classical example of the career of a "realizor", who began as a pioneer of the second generation, is the life of the German physicist and rocket designer, Wernher von Braun, whose work was crowned by the achievement of much more than a thousand-year-old dream of humanity, the landing of a manned spacecraft on the Moon. To the pioneers of the "second generation" in the realm of spaceflight and rocket technology, who forwarded the

----

*President of the New Mexico Research Institute; Director of the International Space Hall of Fame Dedication Conference

particular building blocks towards the success of this technology, one has to count research engineers of many nations of our Earth, as e.g. the Russian Sergej Pavlovich Korolev, the American Frank Malina, the Austrian Eugen Saenger, the Chinese Hsue-Shen Tsien and others. They follow the pioneers of the first hour, men like Konstantin E. Tsiolkovski (1903), Robert H. Goddard (1914), Hermann Oberth (1913), Max Valier (1924), and Robert Esnault-Pelterie (1927).

Among the pioneers of the second generation space technology, Eugen Saenger occupies a particular position, insofar as he himself did not deal with the technology of ballistic rockets, which at that time, thanks to the vision of a Jules Verne, and the preparations of Tsiolkovski, Oberth and other pioneers of the first generation, became popular, and which did lead, in 1969, to the first landing of an American space capsule on the Moon. His work concentrated on the investigation of problems which correspond approximately to today's post-Apollo Program and possibly beyond these.

From the data of Eugen Saenger's life and action, it can be seen that his merits as trail blazer of spaceflight lie in four fully independent planes. They concern:
1. The development of space propulsion units
2. The conceptual design of spacecraft
3. The evolution of systematic theory of spaceflight,
4. The promotion of the spaceflight idea as a supra-national objective of humanity.

As the main propulsion unit for ballistic launch-vechicles and space transports, Saenger designed and tested <u>on teststands regeneratively cooled liquid rocket engines</u>. These were designed for pressure chambers of up to 1500 psi, and thrust of up to 100 tons, and according to patents of Saenger, already equipped with bell-shaped short nozzles, so that these, in principle, are comparable or a sample of the J2 power plants later developed by Rocketdyne, and then installed into the upper stages of the Saturn vehicles.

The reactants used by Saenger were liquid oxygen as oxidizer and hydrocarbons or diesel fuel as metal dispersions. Already in 1934 Saenger reached with more modern combustion chambers of this design, at two kilopounds of thrust, an effective exhaust velocity of about 3000 m/sec. (Remarks by Dr. Steinhoff - I can only tell you that at Peenemunde, in 1939, we reached somewhere around 2000 to 2300 m/sec exhaust velocity, and you see that Saenger's project data are substantially above what we did at that time, however, with much larger rockets, and thrust of 55,000 lbs in old units.)

Besides rocket power plants, Saenger, since 1941, designed novel ramjet power plants and tested these in towed flights. These power plants were initially designed for use as main propulsion units for rapidly climbing subsonic interceptors. Some years later, a ramjet power plant was used on Saenger's advice to envelope a central turbo-jet power plant, ATAR 101, and installed in the French supersonic "Griffon" fighter. Beyond this, Saenger, however, has developed plans and submitted patents in the last decade of his life, which visualized ramjet power plants in enveloping turbojet power plants by ramjet power plants, and at first stages of ballistic launch vehicles of space transports. That latter use is not a ramjet power plant, but a <u>combined ramjet-rocket</u> power plant of first stage ballistic launch vehicles, and of space transport, useful and operating in the transition between the aeronautical and spaceflight zone.

As a hypothesis, he eventually developed the concept of photon and jet power plants which were expected to fly in near light-velocity to collect their fuel in the form of interstellar matter within their neighborhood.

Since 1955, static test stands and mobile vehicles were used to develop and test, eventually, hot water or steam power plants which could serve as economic jet assists for space transports and unmanned flying units particularly designed as power plants for launch sleds for the takeoff of horizontally launch transports.

Saenger's most interesting contribution to spaceflight was without doubt the concept of a design of a single stage manned re-usable liquid fuel rocket-propelled space transport to be used for roundtrips and payload transport between Earth surface and Earth-orbit space stations, respectively for short duration missions of near satellite type character, also useful for objectives as these were simulated within the application range, and corresponded at that time to the early development phase of the Space Shuttle. The design, alreay published in 1944, visualized a space frame of 100 tons weight, of which 90 tons of fuel and payload was achieved. After a horizontal takeoff with the aid of a 500-meter per second launch sled (corresponding roughly to Mach 1.5), this space transport, the fuselage and wing sections of which were selected for hypersonic flight above Mach 5, and which could accelerate to speeds of in the order of Mach 10 within a propulsion period of a few minutes. It then could continue in a semi-ballistic flight or skip trajectory, (in German Rikochetier Flug), and possibly cover large distances and land in a glide. Within a skip phase of its flight, flight altitudes of more than 300 kilometers would be reached, (based on computations which Saenger personally conducted), flying like a flat rock thrown across a water surface and ricocheting repeatedly from the surface of a denser medium (water or low atmosphere), and after this undulating flight, finally landing after a straight, purely aerodynamic flight. With a specific impulse of 400 seconds at fuel cutoff, such a space transport would be expected to reach orbital speed with one ton payload, or to carry four tons of payload once around the Earth, or transport eight tons of payload halfway around the Earth. (Dr. Steinhoff: I use here Saenger's numbers which he used, I would say, 40 years ago, or more than 40 years ago).

From 1963 to 1964, Saenger projected for EUROSPACE, 307 pages of project study for a stepwise realizable development plan of a space transport of a launch weight of 180 tons and 2.5 tons pure payload for a flight into a 300 kilometer orbit, launched with the assist of a steam catapult into a horizontal takeoff with 20 seconds launch time and 300 m/sec release velocity. (Dr. Steinhoff: I have now abbreviated Mrs. Saenger-Bredt's text for time reasons, with the complete text in the final proceeding.

Saenger's contribution to specialized theory concerned scientific treatises in individual areas of specialized theory for the introduction, organization and promotion of specialized theory as a recognized scientific discipline to be taught at technical universities and comparable national and international schools of higher learning. The early intoduction of terms like _effective thrust_, _free thrust_, _nozzle exit velocity_, _theoretical exit velocity_, _maximum exit velocity_, are terms frequently found in publications of Saenger in the 1930's. In 1933, Saenger, as a first, published a scientific textbook on "Space Technology" in which all contemporary knowledge on thrust forces, aerodynamic forces and projectories of liquid rockets were presented in a systematic way, which was later translated into numerous foreign languages particularly Russian and Japanese.

In 1954, Saenger was asked to plan development, and make development plans for the leadership of an "Institute for the Physics of Jet Power Plants" at Stuttgart, Germany. This institute, represented under this colorless and vague name, became the first scientific institute for research in space power plants. Research themes were subdivided into disciplines, as e.g. "thermo-dynamics and flow", "steam and hot water rockets and jets", "molecular jets", "plasma jets", and "fuel chemistry". During the semester of 1954/1955, Saenger received a call to the Stuttgart Institute of Technology in Germany to teach "Fundamentals of Jet Propulsion Systems". In the semester of 1962/1963, he became full professor with a new chair, "Elements of Space Flight", which, at least in Europe, became the first research institute which was exclusively devoted to space flight.

Saenger's contributions to promote the space flight idea as a supranational task of humanity did not restrict themselves into the publication of treatises concerning the motives and consequences of spaceflight, or organization and development cost of spaceflight. As a co-author and member of advisory committees, he influenced the creation of recognized spaceflight journals and book series, as e.g. "Astronautica Acta", (1954); "Missiles and Rockets", (1956); "Flugwelt", (1958); "Advances in Space Science and Technology", (1958); and "Atompraxis", (1960), by his own contributions.

Beyond this, he was participating and active in the founding of numerous spaceflight organizations, and supported these by his advice, or by becoming an officer of them. In this way, he became Chairman of the International Bureau for the preparation of the founding of the "International Astronautical Federation" (IAF), and participated in the founding of the "International Academy of Astronautics", and became a member of it, and from 1955 to his death in 1964, was President of the 'Deutsche Gesellschaft Fuer Weltraumforschung" (GFW), and of the successor organization of the GFW, and served as a member until his death.

His active participation in organizing the budding spaceflight field includes:

1. Membership in the Council of German Federal Transportation Secretary (1956)

2. Director of Section XI of the National Committee on the IGY (1958)

3. Advisor on the "Select Committee on Astronautics and Space Exploration on the U. S. House of Representatives" (1959)

4. Member of the Committee for "Spaceflight Objectives" in the Air-council of the German Federal Transportation Department" (since 1960)

5. Member of the Expert Council for the guidance for the establishment of the "European Satellite Launch Program", and the German delegate for the founding of ELDO (1961)

6. Chairman of the German team of the subgroup of the scientific-technical working group of COPERS, the precursor of ESRO (1961), of the founding of the ESRO, or European Space Research Organization.

7. Member of the Joint Commission for Spaceflight Technology of the federally supported "Deutsche Gesellschaft Fuer Flugwissenschaften (DGF)", and of the "Eighth Federation of the German Society of Aeronautical Engineers (BDLI)".

Eugen Saenger's work was not limited to Germany, but also influenced U. S. Space Developments in their early phases, as e.g. the ROBO project of Bell Aerospace Corporation, and some early contributions to the Space Shuttle idea.

EUGEN SAENGER (1905-1964)

WILHELM THEODOR UNGE
A BIOGRAPHICAL SKETCH
(1845 - 1915)

Wilhelm Theodor Unge was born in Stockholm, Sweden, in 1845. He graduated from the College of Technology, and he started his military career in 1866. As a very promising young officer he was appointed to the Military College. Afterwards, he was attached to the General Staff. Soon his technical education became predominant, and he started a career as an inventor in the field of military technology. His first patented invention was a telemeter, in 1887, and in a short time he patented several improvements for an automatic rifle.

In the late 1880s Unge became interested in artillery. He regarded rocketry as a possible way to improve artillery, and he concentrated on improvements of rockets at a time when military rocket use and technology advances were at an all time low.

From 1891-1896, Unge was sponsored by Alfred Nobel, the millionaire inventor of dynamite, and for the following five years by the Nobel Estate. In 1892, Unge formed the Mars Company to improve the performance and accuracy of rockets, and to develop, manufacture and sell Unge's inventions. He first tried to increase range and accuracy by firing a rocket from a cannon for first-stage velocity, and then igniting the propellant charge for second-stage velocity. Though the technique was sound, and would be proven so in practice in the 1960s, when Unge made his tests in the early 1890s he met with failure. The first rocket, tested in 1892, was made of brass, with a diameter of 20 mm (0.8 in.), 1 mm wall thickness (0.4 in.), and a length of 150 mm (6 in.). The conical burning area was placed with the base at the top of the rocket, which meant that the gas had to turn 180° in order to accelerate the rocket forward. The turning of the gas was achieved by a cupola at the top of the rocket. The great disadvantage of this rocket was the heating of the cupola, and even the body, when the gas stream was forced to turn.

The next two types of rockets were very like those of William Hale, but they had only two instead of three exhaust pipes at the rear end. In one, the exhaust pipes were cut along the center axis opposite each other, and bent at the ends to form a "spoon" which would cause rotation of the rocket when the gas streams passed through. The rotation was not great enough to stabilize the rocket, and making two combustion chambers inside the rocket didn't help either. To improve the rotation in the initial part of the trajectory, the launch tube was replaced by a rotation gun. This and other methods were not sufficiently effective.

New designs were tried, and one of the more significant details was oblique exhaust orifices. These were at first uniformly thick, but soon Unge tested rockets with conical orifices. The conical orifice is the first step to the final solution to the problem of stabilization by rotating the rocket.

A few years earlier, in 1888, Gustaf de Laval had made the first sketches of the later well-known Laval nozzle, and in 1892 the approved patent was published in a paper. This new idea, which showed how to get maximum force out of a high-pressure gas stream, was soon adopted by Unge. He improved the lateral stability of rockets by designing an exhaust gasturbine, which first used de Laval nozzles to accelerate combustion gases shockfree through the nozzle exit, reaching higher exit velocities than ever before. Turbine reaction torques imparted spin to the projectile, later augmented by spinning the launcher itself, avoiding directional instability at the launcher exit. By 1905, launcher-rotated rockets were reaching five-mile ranges with accuracies that competed with the rifled artillery of the day.

With the encouragement of Nobel, Unge also worked toward improving propellant powder. No work had been accomplished during the 19th century toward developing more powerful powders, which had remained essentially as they had been in the 17th and 18th centuries. Unge took maximum advantage of Nobel's work in double-base ballistite smokeless powders, which consisted of nitrocellulose and nitroglycerin, and soon came up with a controlled burning mixture that produced a higher exhaust velocity than traditional powders. To Nobel's double-base powder, Unge added a stabilizer (which served to retard chemical decompostion during storage), and a plasticizer (which increased the propellant's plasticity or workability). Later, in co-operation with the Skanske Bomullskrufabriks company, he developed a binder that gave greater mechanical strength to the propellant grain. Unge's first ballistite-powered rocket flight took place on September 12, 1896.

In 1907, Unge began experimenting with lifesaving rockets. The work was based upon two patents, one for a new ignition system, and the other for "improvements in or relating to the means for connecting lines, cables or the like to rotatory projectiles for conveying them through the air". Test launchings were made not only in Sweden, but also in England. Unge managed to sell some of these life-saving rockets to England, India, Australia and Greece.

Unge spent a lot of effort on improving manufacturing methods. A way to make a more inexpensive rocket body was introduced in 1912. Another idea tried with great success, was the manufacturing of a very inexpensive turbine out of clay.

New ideas for the use of the rockets were developed by Unge when he started to calculate how the heavy guns on armored vessels could be replaced with the batteries of his aerial torpedoes. He also suggested surface-to-air versions to be used to knock down enemy balloons. Ship-to-ship types would prove invaluable during naval engagement, he said, while ship-to-shore missiles would assist conventional cannons in softening up fixed positions during a general naval bombardment.

One of Unge's later ideas was a system to propel and guide rockets, aeroplanes, and airships by using the reaction force of a jet of gas. Unfortunately, this idea will be secret forever, because it is only to be found in a 1909 patent application, which Unge did not carry through; the application is therefore marked secret, according to the patent law of that time.

Wilhelm Theodor Unge, who retired as a lieutenant colonel from the Army, died in 1915.

WILHELM THEODOR UNGE (1845-1915)

CLINTON P. ANDERSON (1895-1975)

CLINTON P. ANDERSON: Life as Statesman, Pioneer of U.S. Space Objectives, Goals and Legislation, as well as Executive Manager

Abstract on Anderson pylon in Space Hall display

CLINTON P. ANDERSON (1895 - 1975), remarkable Statesman, U.S. Legislator and ardent believer in U.S. future in space undertakings. Educated at the Dakota Wellesleyan University and the U. of Michigan, moved to New Mexico in search of cure from severe ailments which treatened his life, in 1917, gradually rebuilding health, but never to become fully cured. Being Journalist, Managing Editor of the Albuquerque Journal, Banker and Insurance Executive, served as Rotary International President 1932/33, which brought his active interest in international affairs and led to the beginning of his unique 40 year public service career in 1933, becoming State Treasurer, New Mexico Relief Administrator 1935 and U.S. Secretary of Agriculture 1945-48, U.S. Senator from 1948 - 1973, when he retired, a very sick man, whose sickness was never premitted to interfere with his far-ranging goals and objectives. Before becoming Secretary of Agriculture, he served in the U.S. House of Representatives from 1941 - 1945. He served a cumulative 27 years in the following three Senate and Joint Senate House Committees: a) on the Joint Committee on Atomic Energy, b) Interior and Insular Affairs, and c) the Senate Committee on Aeronautical and Space Sciences from 1958 thru 1973, most of this time being Chairman of one or more Committees at a time. His sponsorship of far-reaching, beneficial and successful legislation is legend, and his power to convince and to compromise, and his ability to advise our national leadership has become legendary, and his achievements for the Nations longterm space objectives, and policies are:

o He believed in Development of the Resources of Space as one of the great movements in Human Destiny, in which the U.S. should push ahead

o He successfully resisted efforts in Congress to redirect National efforts away from the Development of new Space Resources

o He viewed the success of the Apollo Program a triumph in Management, Technology and outstanding Engineering, which united Government, Industry and Universities in one common undertaking

o He saw in Man's landing on the Moon a beginning and not an ending of Human efforts to increasingly use Space Resources to secure the Nations future

o He pushed for a well balanced Space Program with adequate resources to support current and future goals of National Significance

His dedicated efforts to achieve these unique legislative records and statesmanship, his effective Senate leadership, his belief in International Co-operation, and his great vision of Man's future in Space made him one of the most outstanding National Leaders, this Nation has benefitted from in times of severe tribulations and challenge to legislative Statesmanship. He uniquely responded to national needs and always was there when the right man was sought to undertake an unusual legislative mission.

by Ernst A. Steinhoff *

*Dr. Steinhoff made the abstract from material by Dr. Frank C. Diluzio.

WERNHER VON BRAUN
IN MEMORIAM

F. C. Durant III

"Come to America. We are going to the Moon!" Wernher von Braun cabled that short but stirring message in the late 1950s to a young German scientist torn between accepting an attractive industry job at home or going to the U.S. to join the von Braun team.

And to the Moon we did go. More than any one other person, Wernher von Braun was responsible for the NASA Saturn series of rockets that propelled 17 astronauts into Earth orbit and 24 to lunar orbit, making possible the exploration of the Moon's surface by 12 Americans.

Most people dream of what they would like to see happen in the future. Few make dreams come true. Wernher von Braun did. He dreamed of man leaving the Earth and exploring the Moon and planets. As a boy he grasped the significance of rocket power to space flight, and he dedicated his life to making space flight a reality. He prepared himself academically, earning a doctorate in physics, and early established his brilliance as an engineer. In 1932, as a schoolboy of 16, he wrote, "As soon as the art of orbital flight is developed, mankind will quickly proceed to utilize this technical ability for practical applications." In his 20's he became a leader in the development of the rocket technology required to achieve satellite velocity.

During the past twenty-five years, von Braun probably moved more young--and young-minded--people than anyone else to recognize that the ancient dream of space flight was possible in their lifetime. He wrote and spoke with enthusiasm and grace, and had the rare skill of explaining complex subjects in simple understandable terms. He was a popular witness before Congressional Committees and greatly sought after as a public speaker.

In the early 1950s, von Braun led a group of experts who wrote a series of articles for Collier's magazine on the possibilities of space stations and manned flights to the Moon and Mars. Magnificently illustrated by Chesley Bonestell, these articles subsequently took book form.

As a manager, von Braun had an uncanny ability to inspire others to give their best, to excel themselves and to be a team member. In 1956 (Missiles and Rockets, Vol. 1, No. 1), he stated the following factors to be indispensable to a successful scientific and technical team: "Maximum delegation of authority; efficient and continuous system of communications, from top to bottom and bottom to top; loyalty; honesty; justice."

He was patriotic. Von Braun justified his involvement in the development of the German V-2 rocket with a nationalistic credo: "In times of war a man has to stand up for his country, whether as a combat soldier or as a scientist or engineer, regardless of whether or not he agrees with the policy the government is pursing." Despite the fact that during World War II he fully dedicated himself to the development of the long-range rocket as a weapon, he was arrested and imprisoned by the Gestapo, accused of really having space flight in mind.

He was a philosopher. "Science, by itself, has no moral dimension. The drug which cures when taken in small doses may kill when taken in excess. The knife in the hands of a skillful surgeon may save a life, but it will kill when thrust just a few inches deeper. The nuclear energies that produce cheap electrical power when harnessed in a reactor may kill when abruptly released in a bomb. Thus it does not make sense to ask a scientist whether his drug...or knife or his nuclear energy is 'good' or 'bad' for mankind."

He was a humanitarian. "We are now coming into an era of space research that one might call the humanitarian era in which man will use the tools and capabilities of space. It will be an era when we set out to solve many of the problems that we haven't been able to solve any other way. If a free society cannot help the many who are poor, it cannot save the few who are rich."

He was religious. "The universe as revealed through scientific inquiry is the living witness that God has indeed been at work. Understanding the nature of the creation provides a substantive basis for the faith by which we attempt to know the nature of the Creator."

An Honorary Fellow of AIAA, von Braun received the Institute's Goddard, Louis W. Hill Space Transportation and Haley Astronautics Awards. He served on the Board of the American Rocket Society in the Fifties, and chaired the ARS Space Flight Report to the Nation in 1961.

Wernher von Braun inspired millions to look to space as an expanding frontier of terrestrial enterprise. Future historians may well note this century (or millenium) as significant because mankind then took its first tentative steps into space. In these steps Wernher von Braun was a preeminent leader. He not only had a dream, but he made his dream come true for all of us.

Reprinted with approval of the AIAA stating: From Aeronautics-Astronautics, Volume 15, No. 7, July/August 1977.

Dr. von Braun explains the Applications Technology Satellite 6 (ATS-6), which is currently positioned such that it provides India with educational TV services covering 4500 villages with ground receiving facilities. Dec 1975

WERHNER VON BRAUN (1912-1977)

CHAPTER VIII

EVOLUTION OF FLIGHT CONTROL SYMPOSIUM

"We desire to open the planetary
world to mankind."
   Dr. Wernher von Braun, 1950
   (Germany-U.S.A.)

THE EVOLUTION OF FLIGHT CONTROL OF THE APOLLO MISSION

Dr. Maxime A. Faget*

The purpose of this paper is to recount how the various Apollo flight control techniques and systems were first conceived and how they evolved. However, as I started tracing early events and attempted to recall to memory the motivation for the things we did and didn't do, it became increasingly clear that the question at the time was not how man may fly to the Moon, but could it be done with adequate safety. And one of the dominant considerations concerned the feasibility of navigation and flight control. Thus, the evolution of flight control of the Apollo mission is best seen as part of the history of American manned space flight.

Shortly after NASA started the Mercury Program to make experimental orbital flights of a man-carrying spacecraft it also started considering more ambitious manned space missions. Although the Mercury team, called the Space Task Group, was the focal point for most of this planning, virtually every NASA installation was involved in some sort of advanced mission study involving manned space flight. Furthermore, the USAF, the Army's ABMA and a number of industrial teams were also promoting or participating in this sort of activity. In the Spring of 1959 NASA formed a Research Steering Committee on Manned Space Flight. This committee became NASA's first forum for coordinating the various planning studies then underway. The most obvious and appealing prospect was for some type of lunar mission. Three types of manned missions were considered. They were in order of increasing difficulty, circumlunar flight, lunar orbit, and lunar landing. Our initial considerations for these missions in retrospect may appear quite naive. However, they were based on a very narrow experience base and a conservative assessment of technology. Although the Mercury mission of manned orbital flight had not yet been achieved, the basic approach had been firmly set, and hardware development was sufficiently mature to demonstrate that we were on the path to success. The Mercury design and flight operation philosophy had a dominant influence on lunar planning.

Compared to flying to the Moon and back, the Mercury Program requirement of orbital flight could be met by relatively crude flight control hardware. Basically the Mercury spacecraft was to be inserted into a low earth orbit using the launch vehicle guidance system. Once in orbit, no further velocity change maneuvers were required until it was time for descent. Return to earth was accomplished by firing a cluster of three solid rockets. This deflected the flight path to one which entered the earth's atmosphere. Since the spacecraft was designed to produce no lift, it followed a highly predictable ballistic entry trajectory. Consequently, the time at which the retro-rockets were fired primarily determined the location of splash down in the ocean. It was recognized that there would be a fairly large dispersion about the planned landing location. However, consideration for emergency descent or aborts during launch which could result in a landing anywhere along the flight track made survival for a period of time on the water after landing and extensive use of location aids a basic design requirement anyway.

---
*Dr. Faget is Director of Engineering and Development, NASA-Johnson Space Center, Houston, Texas

Flight control equipment onboard the spacecraft was needed only for attitude control and firing the retro-rockets. These functions could be done both manually and automatically. When in automatic flight an autopilot and a horizon scanner were set up to maintain the vehicle in a fixed attitude with respect to local vertical, there was also a timer onboard which initiated the retro firing sequence after commanding the attitude to the optimum one for the maneuver. This timer which was started at lift-off could be corrected during the flight by command signals from the ground. The astronaut could also control attitude with a hand controller using either an attitude indicator on the instrument panel or by looking through the window. He could also override the automatic initiation of the retro-rocket sequence. The astronaut could crudely determine his position in orbit by comparing his view of the earth with a clock driven replica of the earth's globe.

The control of the mission was carried out on the ground. Communications with the spacecraft and tracking data were obtained from a network of stations along the path of the first three orbits. There were 16 different stations so located to allow a maximum of 10 minutes without communication contact. The data from these stations was sent to the Mercury Control Center at the launch site in Florida. It was on the basis of this processed tracking data that orbital ephemeris was determined. The location of intended splashdown was predetermined to accommodate post retro-firing tracking, thereby enhancing final location for recovery.

Since it required the least amount of propulsion, the least sophistication in navigation and guidance equipment onboard the spacecraft, and it clearly seemed the least hazardous, the first mission seriously considered for manned lunar flight was simple circumlunar navigation and return. This seemed to be modest extension of orbital flight - in fact, circumlunar flight is achievable by a highly eccentric earth orbit of the proper parameters. However, the gravity field of the Moon creates a major influence on such orbits. Consequently, even the smallest error in state vector at the time of translunar injection could not go uncorrected if a safe entry into the earth's atmosphere was to be made at the end of the mission. It was clear that flight path corrections would have to be made. The question was how to determine the error and how accurately could the corrections be made.

There was real concern by a number of NASA leaders, particularly Dr. Harry Goett, Chairman of the NASA Research Steering Committee on Manned Space Flight, that a safe entry into the earth's atmosphere at lunar return velocity might be beyond the guidance and navigation "state-of-the-art" technology. At these velocities, during the initial stages of entry, the spacecraft must pull negative lift to skim along a very narrow corridor of the upper layer of the earth's atmosphere. If the upper boundary of this corridor were exceeded, the spacecraft would skip out of the atmosphere back into a highly eccentric orbit and perhaps expend its supplies before re-entering the atmosphere a second time. On the other hand, if the corridor boundaries were missed on the lower side, the spacecraft would exceed the heating or load limitations of its structure. An associated concern was entry and landing point location. Since the mission was not very well understood, it was conjectured that the time of return might vary greatly from the planned time, and because of the earth's rotation the geographical position of the entry may have a large dispersion. For these reasons configurations with a fairly high lift-to-drag ratio appeared desirable. In summary, the thinking in 1959 was that from a flight control standpoint the circumlunar mission would be flown using ground-based navigation obtained from tracking data. Guidance instructions would be transmitted to the crew for the necessary mid-course corrections. A budget for velocity changes as high as 500 ft/sec., was initially estimated for this purpose. Tracking and communications would be maintained to the entry interface. The entry guidance would be done with an inertial measuring system and an onboard computer that would receive a navigation update prior to entry. The aerodynamic configuration of the entry vehicle was estimated by some to need a lift-to-drag ratio of better than one to provide a large maneuvering footprint while safely staying within the entry corridors.

Concurrent to this activity the Deep Space Information Facility (DSIF) was being defined by JPL. This would consist of three communication and tracking stations located at approximately 120 degree intervals around the earth, and thus could continuously track and communicate with any spacecraft on a lunar or interplanetary mission. This development subsequently led to the Manned Space Flight Network (MSFN) which was used to support all the Apollo missions. This consisted of three 85' diameter antennas dedicated to the Apollo program but located at the same sites as the DSIF. I will come back to the use of MSFN later.

During 1960 enough studies had been carried out by NASA and industry to achieve a fairly good understanding of the implications of various manned lunar missions. Lunar orbit missions with a total flight duration of 2 weeks received a great deal of interest. Such a mission did not appear to be a great deal more difficult than circumlunar flight, but would provide much more scientific data. Furthermore, it would provide the means of gaining significant flight operational experience and reconnaissance data that would support a future lunar landing. The biggest hindrance to enthusiastic support for lunar landing was the enormous size of launch vehicle that would be required. Another consideration was that features of the lunar surface were known to a resolution of no better than one kilometer. Thus, the roughness and soil properties of the surface upon which a landing would have to be made were woefully unpredictable.

Meanwhile NASA had moved out with an unmanned space flight program to explore the planets of the Moon. The Ranger spacecraft was a probe that transmitted a few closeup images of the lunar surface just before colliding with the Moon at high velocity. The Surveyor was a soft-lander that made five successful landings on the lunar surface. Subsequent to landing, the Surveyor transmitted pictures of the surrounding features of the moonscape providing extremely useful information on surface roughness as well as the quantity and size of rocks. Just as important, engineering data obtained from the Surveyor landings were extremely valuable in verifying the firmness of the lunar surface for the landing of the Lunar Module. The unmanned Lunar Orbiter flights, however, were every bit as valuable to the Apollo Program. They provided high resolution photographs of the lunar surface that were extremely useful in selection of landing sites. The cartographic quality of the photographs was more than sufficient to make accurate maps of the lunar surface that could be used for orbital navigation and for visual recognition by the astronauts in the terminal phase of their descent. Just as important, analysis of orbital tracking data greatly improved the accuracy of the lunar gravitational constant and provided valuable data on lunar gravitational anomalies - all of which facilitated translunar and lunar orbit navigation on the first missions.

Before any of these unmanned missions had been successfully accomplished, our concern with the suitability of the lunar soil to support the Lunar Module led us into a study of the use of penetrometers for this purpose. The concept was that a number of penetrometers would be carried aboard the Apollo. The Apollo would be put in an orbit that would pass over the chosen landing site a number of times. One or more penetrometers would be released, whereupon most of their forward velocity would be checked by a retro-rocket. Upon hitting the surface, the penetrometers would telemeter a deceleration signature that could discriminate the suitability of the soil to support the Lunar Module upon landing. Happily, information obtained by the successful landing of the Surveyors eliminated the need for these penetrometers.

There was also great concern with the accuracy of navigation in lunar orbit and that the landing might be made in an unexpected and unsuitable location, for instance, on a rugged mountain slope rather than a smooth plain or valley. Since the radius of the Moon is comparatively small and because the surface features are quite rugged, the astronauts would be completely committed to the general area of the landing before it came within view. For this reason it seemed highly desirable to have a radar beacon at a known location relative to the landing site as a terminal navigation aid. Like the bell on the cat, this was considered a wonderful thing without a practical means of accomplishment. Design studies were made of beacon-carrying hard-landers that could be deployed from the Apollo spacecraft near the desired landing location. Fortunately, the cartographic maps and knowledge of the gravitational figure of the Moon resulting from the Orbiter missions coupled with confidence in improved navigation techniques terminated this work before too much effort was wasted. In fact, the accuracy of navigation techniques that were ultimately developed made it possible for Apollo 12, the second landing mission, to come to rest within short walking distance of Surveyor III that had landed on the Moon two and one-half years previously. The Surveyor's location had been identified on a Lunar Orbiter photograph by patient and meticulous study.

I would like to go back again to 1960 when the concept of the mission itself was still being established. We at the Space Task Group visualized that, regardless of the ultimate goal, the first flight would be circumlunar with the spacecraft passing within several hundred kilometers of the lunar backside. After this, orbital flights would be made using the same outbound and homeward navigation techniques proven in the circumlunar missions. Finally, a lunar landing would be made by descending from lunar orbit. By this scheme of things each mission would be an extension of the previous one, thus, the overall difficulty of achieving the final goal would be divided into a number of incremental steps, each with a greatly reduced exposure into the unknown. Nevertheless, the first time a plan to make a manned landing by descending from lunar orbit was outlined to NASA management, several in the audience severely questioned the wisdom of not taking advantage of the experience that would be obtained from Surveyor, which was designed to go directly from the earth straight down to the lunar surface. Clearly, they had not considered how thrilled the crew would be during a landing that started at hyperbolic speed in a near vertical direction and would be fully committed before they knew if the landing propulsion would fire up. I only mentioned this incident to illustrate that at the same time that a manned lunar landing was seriously debated, the basic understanding of the venture was still quite primitive.

As conceptual mission plans firmed up it became clear that both lunar orbit and lunar landing would require sophisticated onboard navigation and guidance capabilities. It was not considered feasibile to provide this function from the ground. The Instrumentation Laboratory at MIT had been studying both lunar and Mars missions for some time and had established themselves as the leaders in deep space navigation. A contract was therefore negotiated with MIT making the instrumentation Laboratory a partner to the Space Task Group in studying manned lunar missions. Subsequently, when the Apollo program was implemented, the

Instrumentation Laboratory became the program's first contractor when they were given the responsibility for the onboard navigation and guidance hardware which subsequently included the digital autopilot. Dr. Hoag's paper presents a historical account of these and related systems.

The Apollo Program got its official blessing and start when on May 25, 1961, President Kennedy said "....I believe that this Nation should commit itself to achieving the goal, before this decade is out, of landing a man on the Moon and returning him safely to the earth." This precipitated a great number of decisions. The guidance and control precision for atmospheric entry at lunar return velocity was sufficiently well established to commit to an entry configuration that would have an L/D of 0.5 instead of a value of one previously mentioned. The selected 0.5 L/D value was compatible with the use of a semiballistic entry configuration design. Such configurations could achieve the relatively low entry heating of high drag ballistic designs with a modest amount of lift. Furthermore, these features could be embodied in an axisymetrical shape which simplified a number of design, manufacturing and test considerations. The design chosen for Apollo was a derivative of the Mercury shape. By offsetting the center of mass a distance of 7½ inches from the center line, Apollo would trim at about 33 degree angle of attack which was sufficient to produce the desired L/D of 0.5. With this much lift Apollo could be confidently guided to land 5000 miles downrange of the entry interface.(1) On the other hand, the landing point could be limited to only 800 miles downrange without exceeding 4 g's deceleration. However, an interesting thing happened between preliminary design and final assembly. As various equipment was stuffed in the entry capsule, the center of mass inexorably moved toward what one of our engineers called the "idiot point." That is the center of volume. Consequently, the center of mass ended up displaced only a little over 5 inches from the center line and the resulting L/D was 0.35. However, this was more than sufficient. Planned splashdown for all missions actually flown was set for 1400 miles downrange with the never used capability to either decrease it to 800 miles or increase to 2200. I should mention here that only once in all the returns from the Moon was it felt desirable to move the pre-planned landing point. It was moved 500 miles further downrange to avoid the possibility of predicted bad weather at the initially intended landing point. However, the decision to relocate was made early, and the change was accomplished by a propulsive maneuver during transearth coast a day prior to entry. Thus, the actual flight was a standard one with the nominal downrange distance.

Returning once again to 1961, the decision by President Kennedy to make a lunar landing the principal space effort of the decade precipitated the famous debate on the mission scheme to be employed. NASA, industry and the USAF at that time were studying a great variety of launch vehicles. The most mature of these studies was by the Marshall Space Flight Center dating back to the days when they were still part of the Army. Marshall's principal effort was devoted to the Saturn series of launch vehicles. Dr. Haeussermann is presenting a paper in this session which will recount the history of the development of the Saturn's guidance and navigation system. The NASA also was studying a larger launch vehicle, called the Nova, but this was planned to come after the Saturn.

The lunar landing missions studied by the Space Task Group required a launch vehicle more powerful than the largest Saturn under consideration. This was so since it was at that time envisioned that the entire spacecraft would land on the Moon. Since there was insufficient confidence that a Nova class launch vehicle could be built, attention was turned to employment of rendezvous in lunar orbit or earth orbit as alternatives that would fit the mission within the capability of the largest feasible Saturn class launchers. Without getting into the multitude of considerations that finally settled this sticky situation, lunar orbit rendezvous was chosen. Thus, Apollo became two spacecrafts; the Command and Service Module manufactured by North American, and the Lunar Module manufactured by Grumman.

From a mission planning and guidance and navigation standpoint, lunar orbit rendezvous was completely compatible with all the work that had taken place up to the time of that decision. The requirement of rendezvous in lunar orbit during the mission, of course, had a major impact on onboard equipment and operational techniques associated with rendezvous navigation. More than anyone else, Dr. Gilruth was greatly concerned with the high increase in difficulty that the Apollo missions represented when compared with Mercury. He, therefore, convinced NASA that the Gemini Program was a necessary interim step that would, among other things, provide a means for gaining experience and building up the organization needed for Apollo.

Gemini was extremely valuable as a tool for developing flight control techniques and procedures for orbital rendezvous. Furthermore, the general philosophy of the interplay between the Mission Control Center in Houston and the astronauts in the spacecraft was developed during the Gemini Program. This is particularly true in dealing with the critical problem of mission navigation. Gemini also had its center of mass displaced from the centerline to produce a weak but sufficient amount of lift to control its flight path during entry. The basic scheme used in Apollo of rolling the lift vector about the stability axis to steer the flight during entry was proven in the Gemini Program.

Both the MSFN and hardware onboard Apollo were able to produce highly accurate navigation data. Data from both sources were checked against one another prior to making any velocity change maneuver. Also, immediately after the maneuver was made, data was again cross-checked. Navigation done on the ground had the benefit of a large complex of powerful computers. Furthermore, at least two S-band trackers were always available as data sources. On the other hand, onboard navigation was a necessity in the event communication equipment failed. Ground processed navigation data was transmitted directly to the spacecraft computer. However, as a crew option, it could be held out in a separate register for display prior to insertion into the memory.

The data from the S-band tracker was extremely accurate. In addition to providing a doppler count for velocity, the carrier signal was also phase-modulated with a pseudo-random noise (PRN) code for range measurement. This digital signal which was non-repetitive for $5\frac{1}{2}$ seconds, was turned around and re-transmitted on another carrier by a transponder on the spacecraft. Distance measurements with an accuracy of about 10 meters could be obtained. Velocity measurements were much more useful. High powered data processing techniques could produce an accuracy better than a millimeter per second by smoothing doppler data over a period of one minute. With such data, extremely accurate state vectors could be obtained not only on translunar and transearth flight, but also while Apollo was in lunar orbit. This was extremely important since lunar gravity anomalies and venting from the spacecraft continually perturbed the orbit. Computational techniques were developed to the point where tracking data obtained from the Lunar Module during its landing descent burn could be processed to serve as a sufficiently accurate "tie-breaker" in the event that onboard primary and back-up computations produced inexplicable differences. This highly sophisticated computation technique was developed by Bill Lear of TRW.

The general approach to mission planning was to break the mission down into a number of discrete events and periods. Each of these was analyzed in great detail and a complete model of the mission to great precision was constructed before flight. The missions when flown would usually duplicate the plan to exact detail. A feature of the planning was the inclusion of time allowances for unexpected events so that the preplanned schedule could be maintained. The advantage was that almost every event or phase of the basic mission was extremely well understood and exercised. In addition to the basic mission plan there were a great number of contingency plans that would cover every rational problem.

All missions were planned to accommodate mid-course corrections both outbound and on return. There were specific times set aside for these maneuvers. Outbound there were four times set for mid-course correction events, whereas on the transearth leg there were three. However, if the error to be corrected was sufficiently small, the maneuver would not be made and, as a matter of fact, many missions used only one corrective maneuver each way. When we first started considering translunar flight in 1959, we budgeted 500 ft/sec for mid-course corrections. The estimated need was down to 300 ft/sec when we actually put the program into gear several years later. When we actually started flying the 3σ estimate was 78 ft/sec. As it turned out most flights required less than 20 ft/sec. For example, on the last flight, Apollo 17 executed only one correction maneuver each way: translunar it was 10.6 ft/sec, and for transearth only 2.1 ft/sec was needed.

Optimization of trajectories, improved precision and other sophistications in the guidance, navigation and control systems can greatly reduce the quantity of propellant needed for any space mission. This was particularly true for the Apollo missions where a large number of large velocity change maneuvers were required. A basic planning problem was the quantity of reserve propellant to carry for worst case flight control performance. A particularly bothersome consideration was that secondary or back-up systems usually did not have the precision of the primary system. Consequently, in planning missions expendables, all return maneuvers were usually based on the worst case performance of the poorest system in the redundancy chain. Nevertheless, as successive missions were flown, it was usually found possible to increase the load carry capacity in terms of instruments or returned lunar material without compromising safety. However, the impressive fact is that nine missions were flown to the Moon with almost flawless performance of the guidance, navigation and control systems.

Reference

(1) An arbitrarily chosen altitude (400,000) where atmospheric encounter was presumed to first begin.

CHAPTER IX

SPACE STATIONS SYMPOSIUM

ISHF

"How beautiful is our Earth!"
 Yuri A. Gararin, 1961 (U.S.S.R.)

SPACE STATIONS-SYMBOLS AND TOOLS
OF NEW GROWTH IN AN OPEN WORLD

Krafft A. Ehricke*

INTRODUCTORY REMARKS

The line of approach in this address is based on a set of propositions which provides the frame of reference for an overview assessment of the space station as a symbol and toll of New Growth in an Open World. The term New Growth is meant to characterize the advance of mankind into a state in which science and technology are indispensible determinants, an advance whose potential for human development is virtually limitless, whose potential risks are severe, but nevertheless a historically unavoidable process. Open World identifies the physical frame of reference in which this potential for New Growth can--indeed, must--play itself out.

In this frame of reference, terrestrial and extraterrestrial environments become one integrated whole. Its capacity for growth is determined by human scientific insight and technological powers, giving humanity a material foundation for continued growth in it. The Open World outgrows the closed cycle--and, therefore, closed--biospheric world of the terrestrial environment; and it overcomes the dichotomy between Earth and space implied in the recently so fashionable notion of Earth as an "isolated island in hostile space."

The first step in building the Open World is the productive and cost-effective use of environments beyond Earth--in short, space industrialization or exoindustrialization. In this context, a space station has basically a residential, a productive, and a supportive function. However, in a broader sense, a large, complex automated productive facility, requiring human attendance--a light station, a communications station or a power station--is also a space station. Celestial mechanically, they are satellites of Earth or any other body. Functionally, they are way stations of human activity in space. Therefore, both types of stations are considered in the following observations. But, for reasons of clarity, the latter type, without residential provisions, will be referred to as a space facility.

Space stations, like all technical means are tools. They receive meaning through the goals and objectives to which they are put. These, in turn, determine their functions.

Thus, my main point of departure--as often before--is the <u>social function</u> of the space station. In this context, space stations must first of all be related to societal needs--that is, to the needs of national societies and, because of the global scope of space stations, to the needs of the society of nations.

With recognition of its national societal relevance, the laboratory function of the space station is readily broadened to include industrial production and support in the construction of space facilities and other space stations. Once the relevance to the society of nations is recognized, the cooperative aspects (between station and countries) and the
----

*Keynote address to Session I: International Space Stations, of the International Space Hall of Fame Dedication Conference, Alamogordo, N.M., October 5-9, 1976.

participatory aspects (multi-national crews) are strengthened. This, in turn, will essentially serve to reduce the practical importance of differences between "pure" and "applied" research, since a wide range of research--on station as well as Earth-related--will be covered, if the needs of different industrialized and of developing nations are considered. Thus, on both counts, the useful functions of the space station can be broadened and its social impact enhanced.

IMPACT OF THE SPACE STATION

Next to space transportation, the space station is the most important feature in all space operations involving human participation and/or maintenance. Within the scope of this review, it suffices to sketch in barest outline some of the most important assets that make the space station indispensable. They provide the background and rationale for its functions in regard to space industrialization, research, exploration and other missions involving extended stay times in space or on other surfaces (Figure 1)

Practicality

The space station is a precondition for most industrial ventures designed to realize the productive potential of space, commensurate with the capability of the space transportation systems available. The space station is needed to conduct tests of extended duration and a level of complexity requiring in-situ human presence. The space station, in suitable orbit, also is a precondition for the operational practicality of constructing and maintaining numerous, large and complex space industrial facilities on with the functioning of an increasing segment of important terrestrial activities and supplies will come to depend.

Economy

If sized appropriately (i.e., if the personnel to be accommodated is based on practical and economic considerations rather than oversized), the space station reduces the cost of most space industrial facilities both in terms of construction cost and operational costs. Even with extensive automation, in situ personnel are needed for several purposes. Being the first fully assembled and operational <u>unit</u> on the construction site, Man must be employed to assemble and test the automated units,

• PRECONDITION FOR MANY SPACE INDUSTRIAL VENTURES	• REQUIRED TEST FACILITY • OPERATIONAL PRACTICALITY FOR CONSTRUCTION & MAINTENANCE
• REDUCES THE COST OF MOST SPACE INDUSTRIAL FACILITIES	• CONSTRUCTION COST • OPERATING COST • NECESSARY OPERATIONAL BASE FOR ADVANCED, MORE COST-EFFECTIVE INTERORBITAL TRANSPORTS
• CAN PROVIDE THE STEPPING STONES NEEDED FOR PRIVATE VENTURE CAPITAL ENGAGEMENT	• PROVIDES TEST LABORATORIES • PROVIDES HOUSING FOR EMPLOYEES • PROVIDES CENTRAL UTILITIES • PROVIDES BASIS FOR GRADUAL EXPANSION IN RESPONSE TO MARKET DEMANDS FOR PRODUCT
• PREREQUISITE FOR IN-DEPTH RESEARCHING & PROBING IN NEW ENVIRONMENT FOR NEW SP. IND. OPPORTUNITIES FOR FULLEST EXPLOITATION OF SP. IND. POTENTIAL	• LONG-TERM RESEARCH AND DEVELOPMENT ACTIVITIES IN EXTRATERRESTRIAL ENVIRONMENTS
• PROMISES A NEW ROUND OF TECHNOLOGIES AND SPIN-OFFS BENEFITTING TERRESTRIAL PRODUCTIVITY	• PRIMARY INDUSTRIAL PRODUCTS • CLOSED CYCLE TECHNIQUES

Figure 1. Space Station

feed them with the materials or structural members needed, monitor their
operation, and maintain, repair, or exchange these units. In addition,
Man will undertake key tasks in handling their products - tasks which
are relatively too infrequent, too complex, and too critical to warrant
full automation (such as interconnect structures, quality control, electric or optical interconnections, etc.). Considering the importance
of transportation economy, the space station also contributes indirectly
to cost reduction by serving as operational base for advanced (electrostatic, electromagnetic, delicate solar-thermal powered and nuclear)
interorbital transports which are no longer structured for return to the
Earth's surface or are designed for intra-orbital taxiing (e.g., between
points in the geosynchronous orbit).

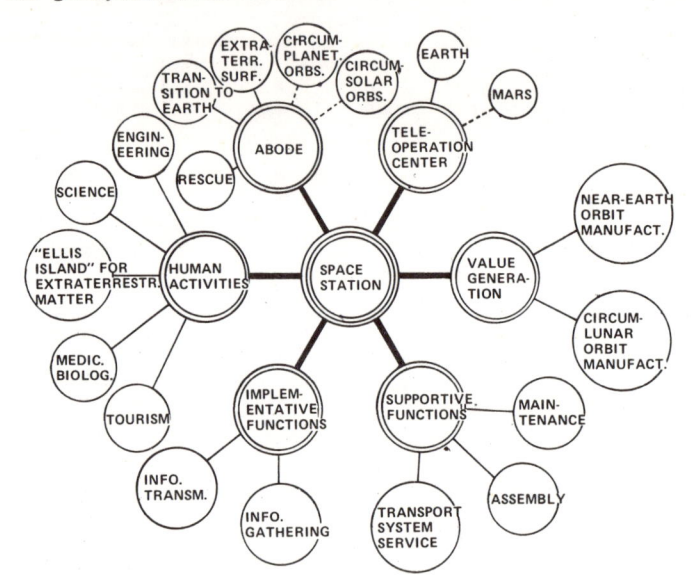

Figure 2. Space Station – Functional Survey

Commercial Significance

     An important economic consideration is the flexibility offered by
the space station. Through this flexibility, the space station is an
essential means, if not a prerequisite for providing the appropriately
sized investment stepping stones needed to attract private capital to
space industrial ventures, if other conditions (e.g. transportation cost)
are right. The space station provides capacity for a company's test laboratories, housing for its employees, and central utilities (power, life
support, basic equipment) for developing and marketing one first product.
Subsequently, it provides the basis for gradual expansion of a successful
product line, until private space industrial capital formation is strong
enough to procure individual modules and eventually entire facilities.
Even at that level, private enterprises will continue for some time to
depend on governmental space transportation services (for which, in turn,
a space station capability may be required as mentioned before); and they
may depend, as on Earth, on basic utilities (power, water, air, etc.)
which may or may not be government-provided. But for getting private
capital engaged beyond small automated satellites (as in Comsat or Intelsat), a space station capability is about as important as cost-effective
space transportation.

Innovation

The space environment as an industrial asset can hardly be expected to be fully understood at this time. In analogy to the discovery of the New World, we are still at the beginning of the 16th century when few if any of the eventually important or decisive socio-economic potentials and opportunities of the New World are grasped. The space station is a prerequisite for indepth researching and probing of the New Environment for new scientific, technological, and space industrial opportunities not yet, or not yet fully understood -- for example, fusion power plants, efficient pulsed energy transmission at shorter, more convenient (from a structural standpoint) wavelengths in space, the proven inevitability of accidental "discoveries," etc. They will have an innovative impact on space industrialization as well as on terrestrial industries, productivity, and services.

New Spin-Offs

Manufacturing in space stations will furnish new products for the terrestrial industry such as pharmaceuticals, near-perfect silicon crystals, high-strength permanent magnets, high-quality alloys, etc.

Another potentially significant terrestrial spin-off from space station development is likely to be the spreading of closed-cycle operations. Such operations must necessarily be advanced to a high state of perfection, if space stations and other industrial space facilities are to function economically. On Earth, closed-cycle modes of operation, where economically feasible, are environmentally benign and conserve resources.

Thus the space station is essential if space technology is to generate a commensurate impact on the furtherance (a) of development objectives on Earth, (b) of space industrialization, and (c) of space exploration. Along with energy conversion systems -- especially propulsion and power generation -- and the ability to process extraterrestrial materials anywhere, the space station as a habitat constitutes the third cornerstone on which the space sector of new growth and the open world will rest.

FUNCTIONAL VERSATILITY

Perhaps it is not an exaggeration to say that, while propulsion makes the Open World of space possible, the space station makes it real.

Most basically, the reason is that it integrates unique human abilities with unique capabilities of fully automated and cybernetic systems. It is creation by an intelligent life form through which life propagates itself into the post-planetary phase of its evolution. The already polyenvironmental versatility of the intelligent life form within the terrestrial environment -- on land, in the sea, and in the air -- is enhanced thousandfold, its senses and its computer-enhanced mind power are increased millionfold, and so is its ability to technologically metabolize energy and matter by gaining access to the inexhaustible wealth of primordial resources -- the energies and the stuff that made the solar system.

On a more pragmatic level, the space station promotes international space ventures because it offers high and useful participatory visibility to nations with limited space efforts. It combines practical benefits with technological and scientific advances; and it provides enhanced versatility of response in complex activities as well as versatility of operations.

In short, the possibility of integrating a wide plurality of function gives the space station almost unlimited functional versatility. It appears useful to divide the functions into six categories which are selected primarily from the standpoint of determining design charteristics and power requirements of the space station which either specializes in those functions or incorporates them as part of its overall activity pattern, as will likely be the case with larger stations (Fig. 2).

In the <u>human activities</u> category the space station accommodates basic and applied research as well as laboratory type engineering testing and developments. It can serve medical purposes -- eventually as orbiting hospital -- and space tourism as a potentially profitable facility. Since a space station can be perfectly isolated from Earth -- at least in principle -- genetic research, including gene transplants and modifications, leading to new life forms, could more safely be conducted there than on Earth. In fact, the space station may offer the environmental growth potential needed to advance life research beyond the limits acceptable on Earth, because this field of research is probably more pregnant with surprises than any other, except extraterrestrial life. Only after certain advances are fully understood (i.e., essentially risk-free) would they be fed back into the terrestrial biosphere. For the same reason, a space station in the longer term, is a natural "Elis Island" for examining extraterrestrial matter returned from other celestial bodies by automated retrievers, if the presence of organic matter is suspected.

In the field of electronic services a potentially important category of <u>implementative functions</u> exists which includes a variety of information gathering activities. In both cases, human judgement in conjunction with the likely superior data processing, data display, and even sensory equipment of the space station can complement and implement the operations of automated satellites which expectedly will continue to carry the main load of information service activities. The extent to which the space station can complement and implement their services depends on the space station's orbit. In near-Earth space its potential appears comparatively more restricted than in geosynchronous orbit.

With increasing sophistication and capability of unmanned Earth service satellites -- in the immensely varied area of communication, data transmission, and Earth observation -- more and more reliance may be placed on their operations, thus increasing the economic and social sensitivity to failures. At one point, the stationing of a maintenance crew in geosynchronous orbit will pay off, depending, as almost always, on transportation costs.

Maintenance of space facilities in general, is an important component of the space station's <u>supportive functions</u>. Other supportive functions include transportation system servicing and assembly of large structures. Transportation system service will include repair, maintenance, and propellant management -- a very interesting area which I cannot pursue further in the frame of this discussion. The space station is indispensible for the assembly of large structures -- ranging from advanced information processing and Earth observation platforms to light and microwave reflectors as well as generation and transmission facilities (Fig. 3). Also, once established, a space station can facilitate and economize the assembly of a new space station in its vicinity. With advancing space manufacturing, more parts will be produced in space, for large space facilities and space stations. Existing facilities can be updated. Eventually, entire facilities will be produced in space.

Space manufacturing is the essence of the <u>value generation</u> category. The science and technology of material processing in space will be devel-

Figure 3. Support Function of Space Station in Assembly of Large Structures

oped near Earth where it also will achieve its first industrial production application, based on terrestrial material supplies. Thereafter, it will not be too difficult to transfer such operations to circumlunar orbit where indigenous materials can be supplied from the lunar surface, at less energy consumption. For this to translate into superior economic competitiveness, other conditions related to lunar industrialization must be met. These developments will determine the time and place of the transfer.

In space manufacturing the outlines of a wide product spectrum become apparent. They meet growing needs, rather than just stimulating new but not vital demands. In the biomaterial sector they include new vaccines, livestock sperm control, perhaps genetic control of plant and animal characteristics, breakthroughs in cell and enzyme pathology and, in general, a deeper understanding of biochemical and life processes. In the material sector, the products point to improved and advanced data processing and cybernetic systems, higher efficiency in thermodynamic processes and long-distance power transmission (both of which conserve energy and reduce environmental burdens), improved electric equipment, greater efficiency in the exploitation of oil wells, important improvements in transportation (bearings) and to superior materials and structures for transportation systems and facilities in space or on the Moon.

Commercial manufacturing for the extraterrestrial market holds considerable promise (Fig. 4). In large space structures, the use of foam metal could result in reduced mass requirements without loss in structural rigidity. The cost of mass transports from Earth would be reduced. Also, for space use, the gas pressure in the bubbles of foam metals could be kept very low, thereby further enhancing the insulation characteristics. The presence of many low-pressure bubbles permits "internal" expansion and contraction of the material, thereby reducing or eliminating the coefficient of external thermal expansion -- a vitally important feature for large space structures that are exposed to extreme temperature variations. Solar-powered spacecraft with large structures will benefit, as will other space transports, through the reduction of their mass fraction.

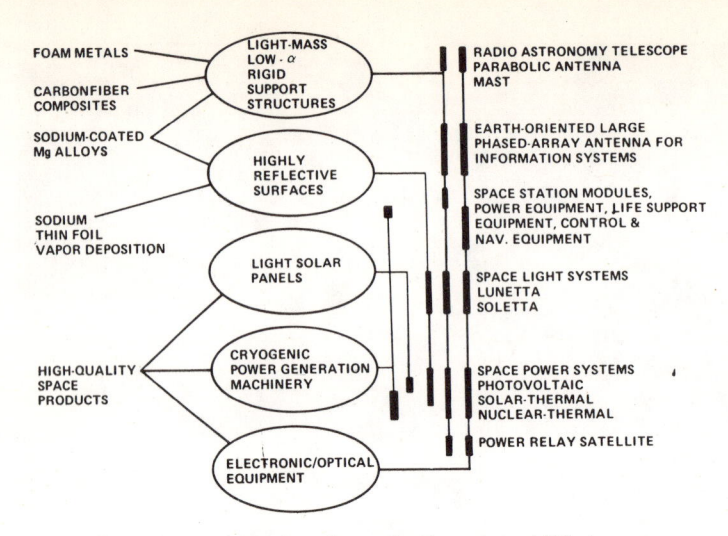

Figure 4. Typical Product Groups for Extraterrestrial Market

Another interesting prospect is the industrial use of sodium for the extraterrestrial market. Sodium is an excellent reflector material, more closely approaching the superior optical characteristics of silver than does aluminum, while being both significantly lighter and less expensive than either. On Earth, the manufacturing of sodium foils, or the spraying of sodium on a suitable substrate, and their transportation into space would be extremely difficult because sodium oxidizes readily. In space, these manufacturing problems are eliminated. The reflectance of a good evaporated coating is always higher than that of a polished or electrolytically produced surface of the same material. One of the main conditions for preparing a high-quality reflection coating is evaporation in high vacuum and a high rate of deposition or fast evaporation of the metal.

If experience is a guide, additional possibilites will be discovered and developed with growing familiarity and skills in the new processing environment of orbital manufacturing facilities. But already on the basis of recognizable benefits, the orbital manufacturing facility promises products and results that bear favorably on disease control, food production, on fuel and power conservation, and on environmental protection at continuing industrialization. Space manufacturing can thereby directly (i.e. apart from technology transfer benefits) strengthen the export capacity of the industrial world, open new markets and create new job opportunities on Earth. At a time when developing countries enter increasingly into the processing and manufacturing phases of industrial production, as they must in the course of needed industrial growth, it becomes of growing importance for the advanced industrial nations to move forward also. Advancing into the neo-environmental processing and manufacturing of space and the Moon is as much a part of industrial growth as are advances in gaining access to new resources beyond the Arctic Circle, under the sea, or at greater depths on Earth and in the sub-atomic world.

With recognition of these possibilities in orbital manufacturing must go a realistic approach to the expectations. Many basic tests and much capital are required to raise the possibilities to the level of feasibilities and even then, many problems associated with the establishment of competitive product lines must be overcome. This fact adds urgency to the availability of Shuttle and Spacelab by 1980, if the development of product lines is to commence around 1985, and if near-Earth space manufacturing is to generate an impact -- albeit a selective one -- in the terrestrial market by 1990. Only demand will trigger major private investments. Demand is stimulated by new, superior quality of space products within a commensurate price bracket. To assure those qualities, experiments must be carried out in space to understand the processes instructure, the cost of transportation, of investment in capital goods on the station, and the recurring operating costs must be assessed with reasonable confidence. This is a long and uncertain way for private capital to go it alone.

In West Germany, leading developer of the Spacelab, about 400 experiments have been submitted by industry and research institutes; but there is almost no willingness to spend private capital. In view of this reluctance, I understand, the German government has established the general rule that only those who pay at least 50 per cent of the cost of development are entitled to a profit from the given product line. I believe that to stimulate ventures, this threshold should be lower, even if it is only on a loan basis, retrievable from future profits if and when the development is successful. To advance into these new fields of industrial production is in the general interest of the industrialized nations whose economic strength rests largely on material processing and exports.

It appears that the use of the space station as <u>teleoperation center</u> holds promise, depending, of course, on many ancillary factors which cannot be discussed here. Remote control techniques can be applied from space to Earth. The control loop would have to be centered in a space station, preferably located in geosynchronous orbit. In stationary position, continuous contact and almost real-time control of ground facilities is possible (Fig. 5).

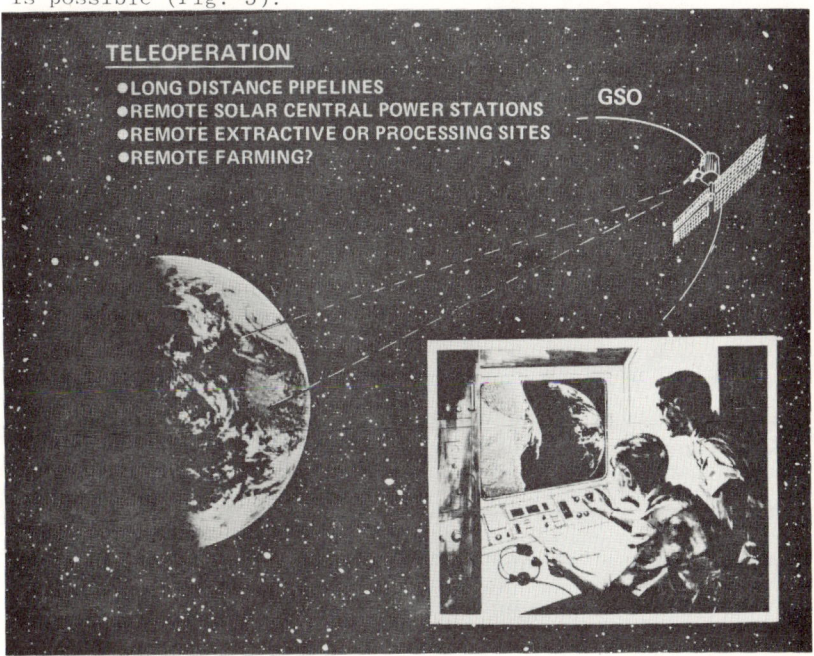

Figure 5. Teleoperation

For a command transmitted to a ground installation from a control center in equatorial GSO, the path delay is 0.13 second for a point 60 degrees from the subsatellite point. Remotely controlled facilities on Earth can be equipped with sensors and built-in controls to enable them to accommodate delays of a fraction of a second in automatic feedback operations. The two-way path delay is in the order of 0.26 second. Fixed delays within the remotely controlled system can probably be kept within 0.15 second, resulting in a total delay of 0.42 second or less. Depending on the type of ground operation, it appears that even higher values, caused by longer fixed delays, could be accommodated (Fig. 6).

Figure 6. Teleoperation

The objective of "reverse teleoperation" is to increase terrestrial productivity where feasible and economically advantageous. A preliminary assessment of possible applications suggests several categories, subject to further studies:

1. Very expansive facilities, whose operations and routine maintenance can be automated relatively readily, such as large solar power stations or long-distance pipelines. In those cases ground supervision and control is expensive. Local damage occurence or attempts at sabotage are difficult to monitor. A plant manager in geosynchronous orbit easily overlooks a 1000-square mile solar plant or a 2000-mile pipeline.

2. Operating remote and unfavorably located industrial facilities at minimum on-site manpower requirement. It reduces the cost of accommodations and services. Emergency and maintenance crews are brought in on a temporary basis. Examples are large nuclear parks and strip mining for coal or metal ore in remote, undesirable locations.

3. It appears possible to cultivate land areas for food production without extensive settlements, with sowing, irrigation, monitor-

ing, pesticide spraying, harvesting, and transportation to collection points all remotely operated from GSO.

The world population is to exceed 6 billion by the turn of the century. Increase in good production is the most urgent task confronting mankind. Should the global climate worsen during the next 50 years, present resources will be strained beyond capacity. Close crop supervision from air and space becomes necessary to maximize soil productivity and minimize losses -- down to such fine points as close control of soil moisture at various stages of plant growth and its relation to light intensity, length of day, air temperature and winds, as well as supplies of fertilizers and pesticides.

More land needs to be cultivated to increase food production and compensate for the occupation of land for urban/industrial uses. Most of the prime agricultural areas already are developed. Expanding food production must, in most cases, attempt to utilize land, not necessarily of low fertility, but located in more difficult climatic environments, such as deserts, steppes, and tundras. A complex of semi-automated circular fields near the Kufra Oasis (Lybian Dessert) was observed by Landsat (1000 km). Each field, measuring about 1.5 km in diameter, is irrigated by a center-pivot rotating irrigation system fed by a large underground water reservoir (Fig. 7).

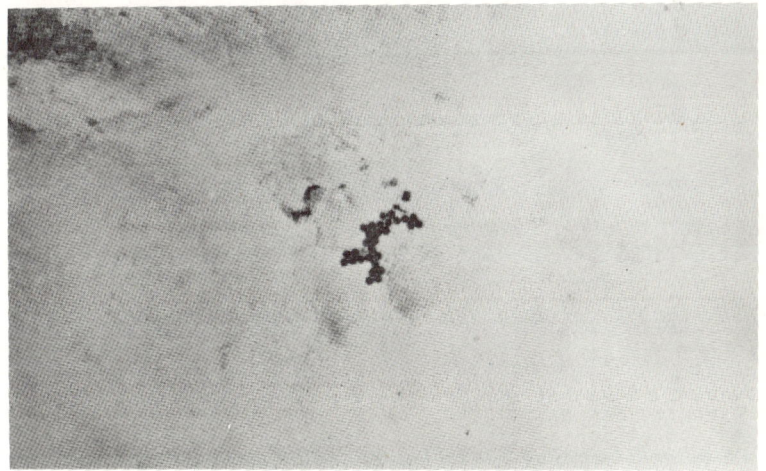

Figure 7. Circular Fields Near Kufra Oasis (Libyan Desert) as Observed by Landsat

Under such conditions, close control of the above-mentioned parameters is even more important. In teleoperated agricultural facilities of large, economic size, close monitoring of these parameters and corresponding close regulation of irrigation and pesticide dispensation become an integral part of the teleoperating system which must rely on data from ground sensors as well as on observation from space.

Since governments and industry increasingly roam the planet in search of scarce resources and for places where industrial processing does not interfere with already strained environmental conditions in densly populated areas, further studies may reveal additional applications of teleoperations to terrestrial activities.

Finally, and this is the sixth category, the space station has functions as the abode of personnel, not only near Earth and in geosynchronous orbit, as mentioned before. If capable of generating artificial gravity, the station can train and condition teams destined for another g-environment -- for example, the Moon. Conversely, it can serve as receiving and reconditioning station of personnel returning to Earth after long dwell time at different gravity levels; and, wherever it is, the space station becomes a natural center for rescue operations in its greater environment. Space stations provide habitats far from Earth in lunar space, in the orbits around other worlds, on their surface, and on the far-flung cruises between planets. That is to say, the orbit-locked space station of today, whose rudimentary forms have been realized already in Skylab and Salyut, will become the mobile habitat of tomorrow, carrying human crews through the solar system and providing mining camps anywhere from the lunar surface to heliocentric orbits in the asteroid belt.

Fig. 8 illustrates the use of space station type modules at the docking port of a lunar manufacturing base.

Figure 8. Docking Port of a Lunar Manufacturing Base

Fig. 9 depicts a manned Mars transport whose front section is designed to provide artificial gravity during the transfer and become part of a space station in circum-Martian orbit. The radial distance of the living and work modules from the main body is about 30 meters. Completing about three rotations per minute, the g-level in the modules ranges from 0.3 to 0.4g, conditioning the crew to the gravity level on the Martian surface. The radial tunnels are so constructed that they can be angled during powered maneuvers at Earth departure and Mars arrival. The thrust acceleration acts in the same direction as later the centrifugal effect. Using two vehicles of this type, the two front sections are mated. The tunnels and modules can be coupled to form a larger rotating subassembly of great diameter (lower space station configuration). More simply, the two payload sections are decoupled from their respective main propulsion modules and mated back to back to form a space station with two counter-rotating sets of arms (upper space station configuration).

SPACE STATION CONFIGURATIONS

The architecture of space can evolve into a great variety of forms and configurations. Launch vehicles, choice of electric power sources, artificial gravity, function, and size are among the strong configurational drivers.

Figure 9. A Space Station for Mars

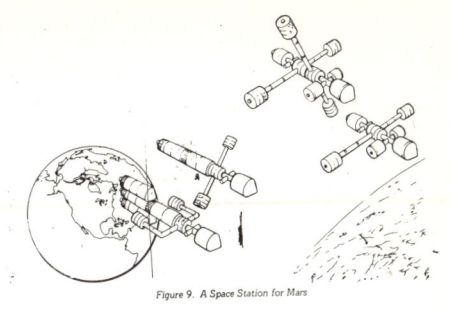

Illustration No: 9A : Planetary Surface Base of the Future!
This artists redition shows a potential Mars Base as conceived by U.S. Space Planners. In the foreground, we observe living and working quarters of the Base Staff. These could be industrial research laboratories, processing plants, schools, recreational facilities. In the background we observe Mars landers, which transport crews to the Androcell and to the departure facilities for interplanetary destinations, of space freight transports and orbital logistics systems. We see solar powerplants, energy distribution systems and surface vehicles. Surface facilities are covered with surface soil, heavy insulated to conserve energy and provide protection against radiation incidence.

In any case, however, modular construction offers many advantages. Fig. 10 illustrates a possible progression from the mini-station -- one step beyond the Spacelab -- to a very large modular configuration (Astropolis). The first two stations offer only zero-g environment. The third station has an axis of rotation. It is designed for energy-intensive operations. The spin axis offers null-g space. A nuclear Brayton cycle power station is used as counter-mass to the modules. The living and work modules are arranged on the opposite end, in the manner of rungs on a ladder, except that they are mounted on the outside of the two radial conduits. This facilitates the installation and removal of modules.

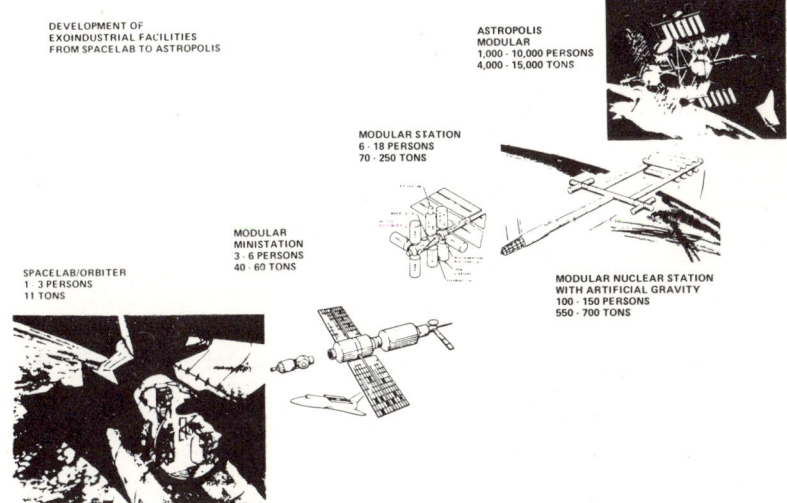

Figure 10. Development of Exoindustrial Facilities From Spacelab to Astropolis

343

The presence of two individual modules at the same distance from the hub doubles the volume at a given g-level but keeps the modules smaller. Thereby, the range of compatibility with the Shuttle or Shuttle-derived transports is extended to larger space stations. The arrangement also provides a dual egress for faster removal or installation of equipment and for emergencies. The nuclear power plant, which signifies the higher productive capacity of the facility, is located so that the modules and the entire width of the hub are within the reactor's radiation cone. The darker cross-bar sections at the end of the hub do not participate in the rotation (or counter-rotate), thereby providing additional null-g environment. Their size can be extended by adding null-g modules primarily in the directions parallel to the radial conduits, since this would safely keep them inside the shielding cone. The non-rotating end points of the hub provide docking ports for space transports. A crossed ladder configuration reduces the tendency to wobble and provides superior means of wobble control. The crossed-ladder configuration is the basis of the Astropolis concept.

SPACE STATION EVOLUTION

The European Spacelab is a significant and promising start. The Spacelab can grow into a free flying mini-space station. The mini-station may be envisioned as consisting of a core module, a shelter module, and a discretionary module for scientific or technological experimentation.

Out of these modular developments can emerge the large modular space station, multi-functional and capable of accommodating a larger number of value-generating modules. Again, it appears cost-effective (primarily for reasons of standardization, extended Shuttle compatibility, and optimum orbit for its functions) to build more than one medium size station and to proceed to a larger station size only when such station offers functional and economic advantages over a cluster of smaller stations.

CONCLUSION

Let me summarize the principal conclusions.

The space station is an exoindustrial facility in its own right, combining unique human abilities with the unique capabilities of cybernetic systems in performing Earth-oriented services, processing materials in new ways, and manufacturing new products for terrestrial markets.

The space station is indispensible for the construction of other exoindustrial facilities -- such as large information systems, space light and power facilities -- as well as for lunar industrialization.

But the space station is more than an exoindustrial facility and a construction/operations base. It is the means for providing the appropriate environment for Man in Space.

Thereby, the space station renders human permanency in space possible.

Beyond that, the space station enables Man to traverse the enormous distances between planets and to drop anchor on new worlds.

Finally, the space station in all its manifestations and growth versions makes possible human growth beyond Earth in numbers. Medical and biological sciences will receive powerful new impulses. In its larger versions the space station will bring behavioral and social sciences into space and open new dimensions for space law as the relations between individuals and groups, the rights of individuals and those of the community must be thought through and weighed anew. Thereby the space station leads up to the Astropolis and the Androcell about which I will talk this afternoon.

Indeed, space stations are tools as well as symbols of New Growth in an Open World.

SPACE STATIONS AND THEIR PROLOGUE

J. F. Madewell*

This paper addresses the potential evolution of space stations. Past studies of flight hardware and accomplishments are assessed and the objectives and constraints for future systems are discussed. Scenarios for future space station programs are discussed and the potential early application for a system derived from the present Space Shuttle and Spacelab systems is evaluated. Low earth orbit and geosynchronous orbit space station configurations and their operational characteristics are described.

INTRODUCTION. If man is truly to use the unique vantage point of space to enhance the quality of life on earth, he must be capable of three things. First, he must be able to go into space and return at will with full safety and adequate support equipment. Second, he must be able to stay in space in a routine way for a long period. Third, he must be able to perform complex tasks while in space. Space Shuttle puts us very near the attainment of the first capability plateau. A permanent manned space station will bring the next capability plateau within grasp, and it will provide for development of the third capability.

Previous space station program options and scenarios have emphasized laboratory-type experimentation. Newest program options have emphasized the commercial applications of space stations and the building of large, useful structures in earth orbit.

This paper surveys previous studies and flight programs, looks ahead at the objectives for future flights and the constraints and development requirements for new systems, and discusses the importance of maximizing the capabilities and achievements of the pre-Space Station era.

BACKGROUND AND PERSPECTIVE. Future space stations must be viewed from the aspects of where we have been and where we want to go and what we have as a starting base.

In December 1968, we were all treated to the spectacular total view of our tiny earth (Fig. 1) from the cabin of Apollo 8. The impact was immense. Throughout the world, statesmen and poets, philosophers and technicians all acknowledged the implication. We were all voyagers on a tiny space station. It is clearly our international space station. It glides effortlessly in orbit about the sun, collecting the waves of energy, and broadly spread solar collectors and dissipating the excess thermal energy collected with giant radiators to the black abyss of space. Except for that slender thread of energy, Spaceship Earth appears otherwise to be a closed ecologic system.

----

* Director, Advanced Systems at Rockwell's International Space Division.

Figure 1. The Earth as Seen From Apollo 8

    In July 1969, one of the truly remarkable achievements of mankind occurred (Fig. 2). The result was monumental. The capability implied by this technical accomplishment will undoubtedly be the standard by which **all of mankind's problems (from power to pollution) will be measured for decades to come.**

Astronaut Harrison Schmitt on Taurus Littrow excursion and reconnaissance.

Figure 3. Skylab in Flight

The launch of Skylab (Fig. 3) in May 1973 displayed the early gains that can be expected from a Space Station. Four key functions were performed: observation of earth, observation of space bodies (comets to corona), space processing of materials, and space repair and construction. In addition, we were allowed to observe and assess man's ability to perform in space for extended durations. As he has shown before, man's ability to adapt was remarkable.

The Apollo Soyuz Test Project (ASTP Fig 4) broke ground in resolving several international technical and operational incompatibilities. These include docking-unit compatibility, where we first developed androgynous elements; the means of providing spacecraft ranging and rendezvous; the spacecraft habitability module atmospheric system incompatibility; the ground flight control system; and the methodology compatibilities such as mission trajectory analysis.

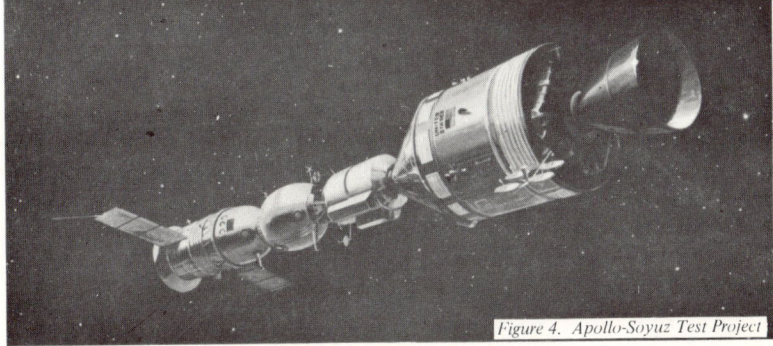

Figure 4. Apollo-Soyuz Test Project

Even as the ASTP flight was being designed and tested, and well before flight, a second space station was conceived. Throughout 1972, the NASA Marshall Space Flight Center (MSFC) and the European Space Research Organization (ESRO) conducted definition studies of a sortie can, which eventually became Spacelab.

347

The future for space flight activities clearly requires a sizeable, continuously manned space station in earth orbit. Probably several such stations will be appropriate to manage the significant tasks and functions to be performed. The scope of this work is depicted in Fig. 5.

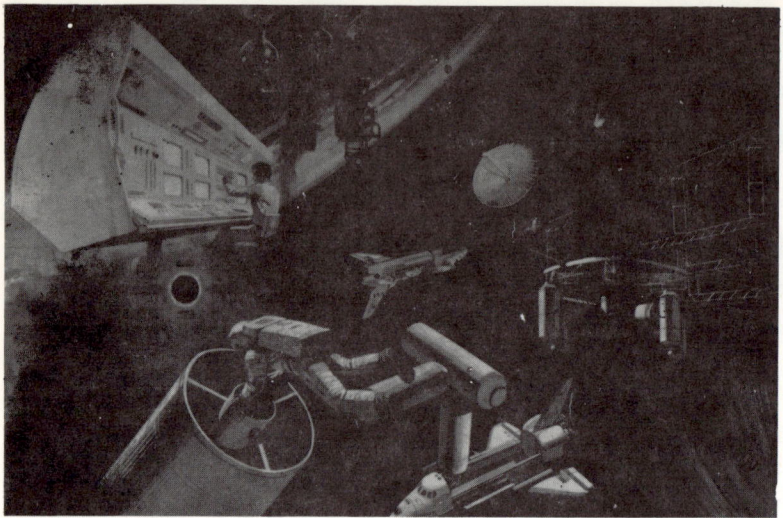

Figure 5. The Evolving Scope of International Space Stations

The next decade will find us conducting various activities in earth orbit which transcend national boundaries. These include:

- Monitoring of earth's dynamic processes: weather, moisture patterns, urban sprawl, river bed changes, disaster assessment, and crop status among many others.

- Repair of satellites on orbit and preparation and possibly fabrication of automated systems on orbit, thereby reducing even further the rigors of the ascent flight.

- Searching for additional locations of scarce natural resources.

- Conducting materials research to develop commercially competitive products for terrestrial consumption.

- Fabricating large structures for collection and transmission of energy from space to earth.

- Assembling large systems to assure continuous direct point-to-point or person-to-person communication around the globe.

In the short span of less than two decades, a strong base of space station background has been developed. Certainly, many studies have been conducted on the subject. Additional studies are underway to define the operational requirements and characteristics of a space station system capable of performing these functions.

The role of man has grown steadily in past flights, as noted in Fig. 6. In the early flights, man was a passenger and observer or, at first, even the "observee". Once it was clearly established that he could make the trip, he was put to work as an operator and an experimenter. To date, the apex of man's role in a space station was demonstrated in the Skylab program. In the role of repairman and innovator, he successfully rescued a crippled space station and went on to log more than 12,000 man-hours of the most useful and effective spaceflight time to date. Perhaps the most effective man-hours were spent by Pete Conrad in the repair of the Skylab solar panel.

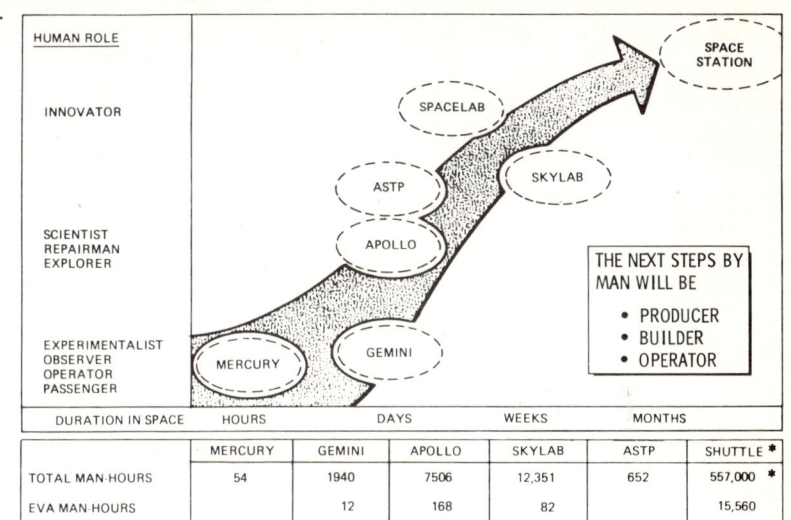

DURATION IN SPACE	MERCURY	GEMINI	APOLLO	SKYLAB	ASTP	SHUTTLE *
TOTAL MAN-HOURS	54	1940	7506	12,351	652	557,000 *
EVA MAN-HOURS		12	168	82		15,560

*PROJECTED FROM MISSION MODEL

Figure 6. Growth of Man's Role

In the early planning stages of Skylab, a fundamental objective was to demonstrate the utility of man in an expanded role. Although the timeline did not finally evolve as it was initially perceived, the actual results achieved greatly exceeded those anticipated. The question of man's role was put to rest. Further, the concept of manned and unmanned activities was finally set aside and the approach became "How can the job be done best?" Present plans project more than 500,000 man-hours on orbit in Space Shuttle and in Spacelab in the first decade of its operation. More than 15,000 man-hours of EVA time are presently envisioned. In the next decade, man's role will evolve to producer, builder, and operator in space.

FUTURE SPACE STATIONS

Program constraints. In developing the character of future manned space stations, it is appropriate to establish clear objectives, derive requirements, and evaluate options to accommodate these requirements. It is equally important to acknowledge the existence of key constraints to the system. In the case of all past stations, the transportation system (Fig. 7) was the most significant constraint. This situation can be expected to continue for future space stations.

The Mercury, Gemini, Apollo, Skylab, and ASTP space stations were driven by the limits and availability of the transportation systems; in this case, Atlas, Titan, and Saturn class launch vehicles. Such basic characteristics as vehicle diameter, shape, and weight were driven by the transportation limits rather than the requirements. Only on the Apollo Program, where total mission objectives were considered early, was

transportation limit eased. Even in this case, however, the shift to
the Lunar Orbit Rendezvous mission mode in 1962, and the decision for a
single-launch mission concept, put the transportation system back into
a constraining role.

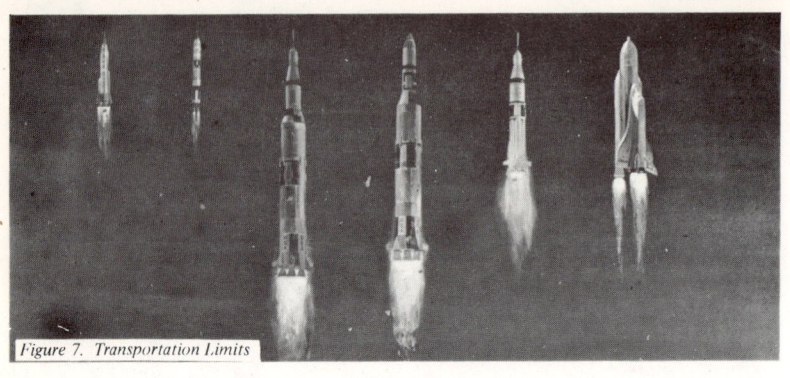

Figure 7. Transportation Limits

The Space Transportation System (STS), with Space Shuttle and Spacelab, will be fully operational in 1980. Space station designs for the
early subsequent years must accept the characteristics of these systems
as strong constraints. This suggests that modular configurations compatible with Shuttle cargo bay limits on size and weight will be foremost
in the space station designer's handbook.

Man's ability to perform on orbit is another key factor on the constraint side of the space station equation. Fig. 8 depicts some performance parameters projected from Skylab flight data. Several key factors
emerge. A nine-to-ten-hour work day is a good choice for long-duration
missions. If tasks are fairly repetitive in nature, man has a significant rate of performance improvement. This rate varies somewhat with the
nature of the task but averages along the 85-percent learning curve.

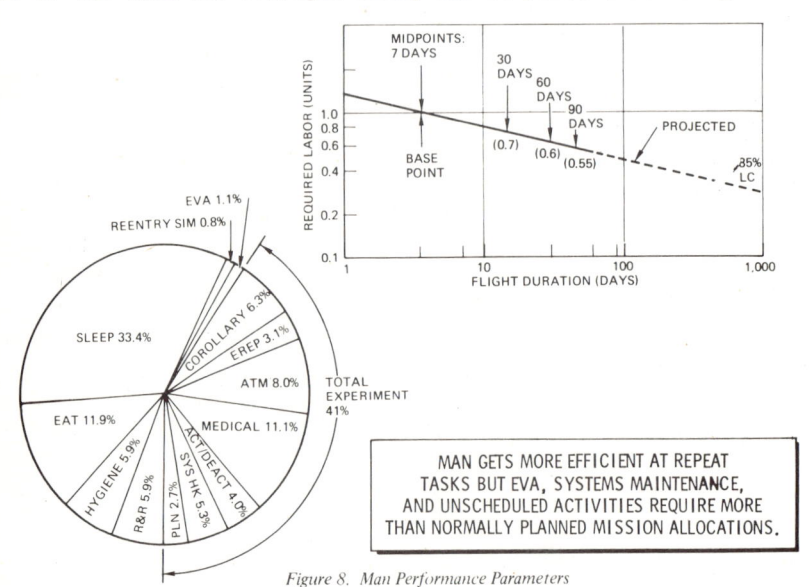

Figure 8. Man Performance Parameters

MAN GETS MORE EFFICIENT AT REPEAT TASKS BUT EVA, SYSTEMS MAINTENANCE, AND UNSCHEDULED ACTIVITIES REQUIRE MORE THAN NORMALLY PLANNED MISSION ALLOCATIONS.

Thus, a task such as assembling a large structure might take 12 hours on a first attempt, but after a 90-day period that task could be cut to six or seven hours. Skylab experience also warns that EVA activities and system maintenance will require more effort than an initial operational timeline allocates.

Program Objectives. Space experimentation has always been an integral part of space flight. A continuing series of Flight Opportunities Announcements yield proposals from universities, industry, and government agencies. These proposals cover the fields of medicine, astronomy, space physics, atmospheric sciences, earth sensing and monitoring, and space applications and technology. The orbital environment at altitudes of 100 to 300 nautical miles offers the opportunity of performing unique experiments over a wide spectrum of disciplines that are either difficult and costly or impossible to accomplish on earth. A space station facility can provide (1) ultra-high vacuum, (2) a high degree of sterile cleanliness, (3) low-vibration environment, and (4) near-zero-g loading. These factors enable the experimental study of physical processes normally obscured by terrestrial disturbances. Experimentation utilizing the greater weight, power, energy, and accommodation capability of a permanent space station are certain to reflect even longer experiment operational time. These experiments, however, are merely the prelude to the real payoff - space industrialization.

Space Industrialization. Space industrialization is the utilization of space to provide goods or services of commercial value on earth. It differs from space colonization in intent; to improve the quality of life for a vast number of people on earth, rather than the movement of a few people into a space colony with a goal of eventual self sufficiency. Fig. 9 describes key opportunities in the area of space industrialization.

INFORMATION SERVICES
- GLOBAL COMMUNICATIONS
- GLOBAL NAVIGATION
- INFORMATION ACQUISITION FOR:
  - AGRICULTURE
  - DISASTER DETECTION
  - RESCUE
  - BORDER SURVEILLANCE
  - WEATHER
  - OCEAN MONITORING

✓ INEXPENSIVE LONG-DISTANCE CALLS AND DATA TRANSFER
✓ TIME TO MILLISECONDS, POSITION TO < 10 METERS
✓ CROP PREDICTIONS, WATER MONITORING.

LIGHT & POWER
- SOLAR POWER FROM SPACE
- POWER RELAY SATELLITES
- ILLUMINATION FROM SPACE

✓ SPACE MAY PROVIDE THE TECHNOLOGY FOR ABUNDANT ELECTRICAL ENERGY THROUGHOUT THE 21ST CENTURY THAT IS CONTINUOUSLY RENEWED, ECONOMICAL, AND ENVIRONMENTALLY SAFE.

MATERIAL PRODUCTS
- ELECTRONICS & POWER DIST.
- VACCINES/DIAGONISTICS
- METALS
- GLASSES
- NEW PROD.

✓ SOME OF THE EXPERIMENTS ON SHUTTLE/SPACELAB WILL BE IMMINENTLY SUCCESSFUL. WHOLE INDUSTRIES CAN DEVELOP QUICKLY ON A NEW MATERIAL OR PHENOMENON. EFFICIENCY IMPROVEMENTS IN AGRICULTURE OR ENERGY COULD BE QUICKLY EXPLOITED.

PRODUCTS OR SERVICES OF COMMERCIAL VALUE ON EARTH

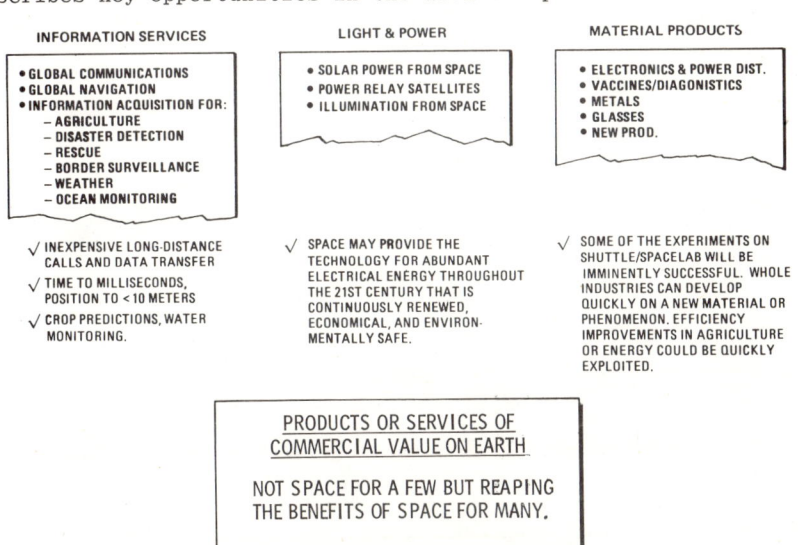

NOT SPACE FOR A FEW BUT REAPING THE BENEFITS OF SPACE FOR MANY.

Figure 9. Space Industrialization (The Payoff)

The vantage point of space is already being used commercially for information services. Communications via a geosynchronous orbit satellite is cheaper than corresponding ground systems and is fostering an exponential growth in relationships between geographically distant people and machines. Space can also provide the best position for many surveillance and information acquisition services. Small improvements in agriculture, security systems, and weather predictions can quickly result in major commercial and/or humanistic value. Navigation also can benefit from the vantage point of space; systems presently are being built that will give world-wide, three dimensional position information to within 10 m (32.8 ft).

Most of today's processed materials use the earth's gravitational force as an integral part of separation, distillation, and other processes in their manufacture. Extending the gravity spectrum available down to zero opens up materials processes unavailable except in space. A wealth of experiments is presently planned that will result in new pharmaceuticals, metals, glasses, and electrical/electronic processes. With the current expanding need for food and energy, any of these that result in efficiency or productivity improvements would be quickly incorporated on a wide scale.

To an average citizen, space industrialization can provide the payoff of space research and development directly beneficial to his daily life. An example of a direct daily impact is shown in Fig. 10, which describes the fundamental elements of a personal communication broadcast system. The parts of the system include a hand-held, or perhaps even wristwatch size, communicator which enables point-to-point communications with the fidelity of a citizen's band radio system. The key to this industrialization system is a large space structure. A multibeam antenna, about 70 to 100 m (228 to 328 ft.) in diameter at geosynchronous orbit, provides a 60-dB gain for the system.

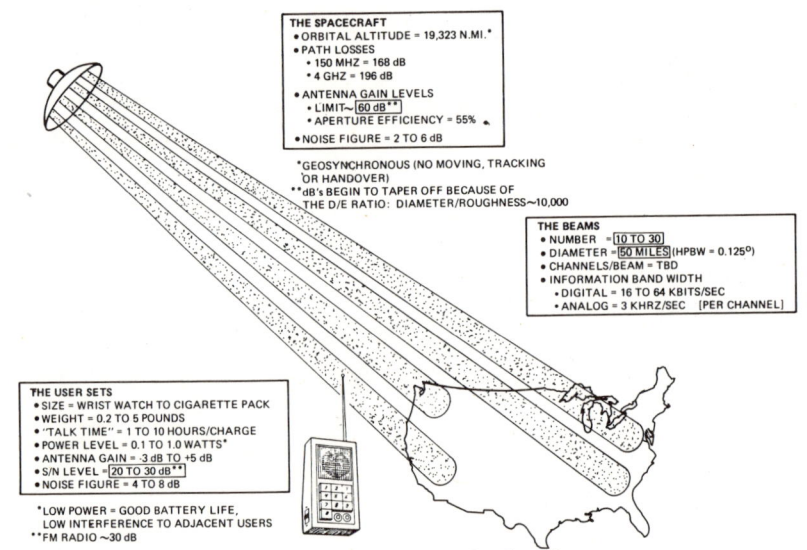

Figure 10. Personal Communication Broadcast System

Parametric link margin calculations (Fig. 11) were used in sizing the antennas and in determining the corresponding transmission frequencies. In this analysis, the maximum power level of the portable communication unit was set at 0.5 watt (ERP) and the signal-to-noise ratio of the system was taken to be 30 dB. The other critical assumptions are given in the figure. The results show that a transmission frequency of 900 MHz would require an antenna aperture of 225 feet with a spot diameter of 130 miles. At 4GHz, the antenna aperture would be 145 feet and the spot diameter would be 46 miles. If the signal-to-noise ratio could be reduced from 30 to 20 dB, the antenna aperture could be reduced to about one-third this size. Specifically, at 900 MHz a 75-foot antenna would suffice; at 4 GHz, the aperture would be 50 feet in diameter. Though promising, this 10-dB reduction in the S/N value could seriously impair the fade margin of the system under certain conditions.

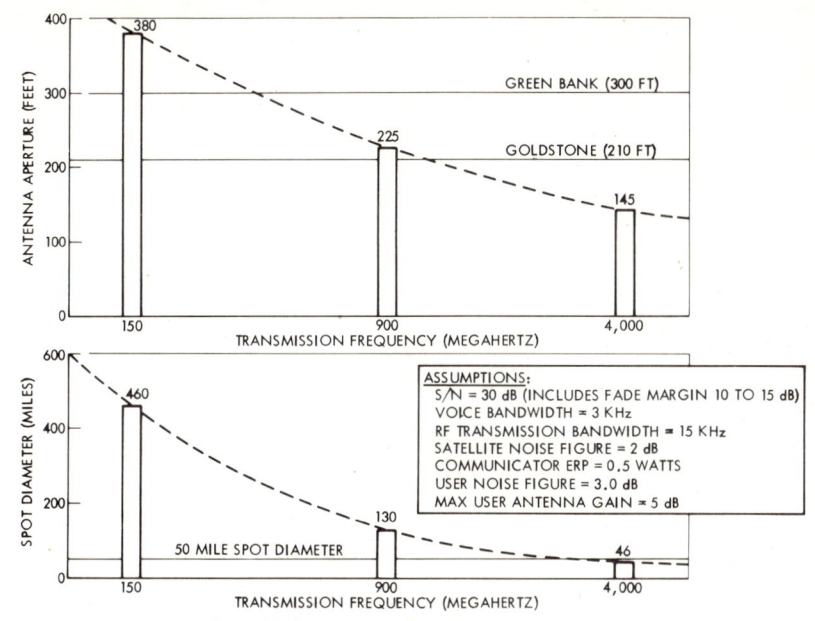

Figure 11. Link Margin Calculations (S/N = 30 dB)

NASA has had a series of studies underway to investigate the optimum construction technique for large structural elements. Fig. 12 shows a construction approach for a slightly larger structure, a microwave radiometry reflector, approximately 400 m in diameter. This approach was developed by Grumman. It shows the use of a support module (space station), three fabrication elements (antenna mast, antenna ribs, and the circumferentials), and a jig. The antenna reflection mesh is prefabricated, fastened to the hub, then unrolled and attached to the gores. Surface characteristics are maintained by tension ties.

Space Power. Probably the most exciting space possibility is the utilization of solar energy in space to generate power, both for space industrialization itself and to help satisfy the growing terrestrial demand for electricity. Satellite power systems avoid the thermal and atmospheric pollution of even advanced electrical power plants using hydrocarbon fuels. Solar energy, both space and terrestrial, can be complementary to nuclear power. In addition to electrical power, light for large areas could be directly reflected from space, eliminating the cost of lighting installations and the efficiency chain of fuel-through-electricity-to-light.

The technical feasibility and economic viability of collecting, converting, and transmitting solar energy for use on earth is currently under study by ERDA and NASA. If such a program comes to fruition, it will impose major requirements on space station facilities, first during its developmental phase and later in its operational phase. Early testing for many of the development objectives shown must be conducted by the Space Shuttle and Spacelab.

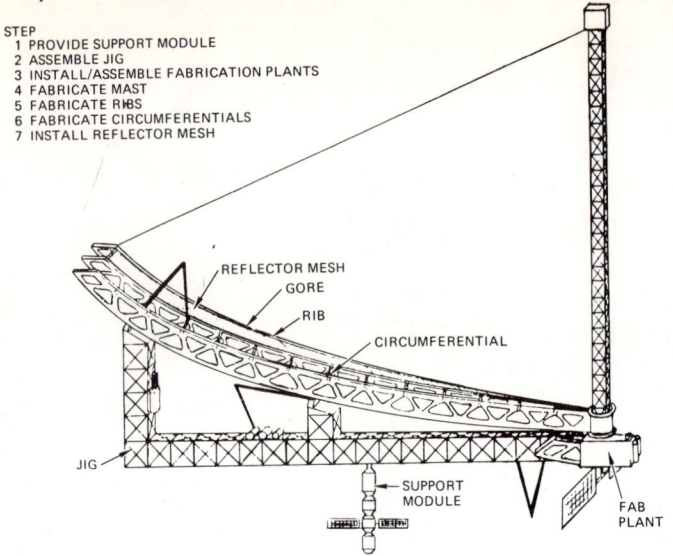

STEP
1 PROVIDE SUPPORT MODULE
2 ASSEMBLE JIG
3 INSTALL/ASSEMBLE FABRICATION PLANTS
4 FABRICATE MAST
5 FABRICATE RIBS
6 FABRICATE CIRCUMFERENTIALS
7 INSTALL REFLECTOR MESH

Figure 12. Microwave Radiometry Reflector Assembly

Eventually, however, the physical sizes, power demands, and crew complements and durations required will justify investment in a low earth orbit test facility. This facility will serve to demonstrate technology advancements, operational processes, and scaled prototype systems not only during the initial development phase of a program, but on a continuous basis to test improved systems and man-machine operations.

From early studies, it has been concluded that economic viability, in part, will require the deployment of large numbers of satellite power systems delivering hundreds of gigawatts of electrical power to earth. The low earth orbit test facility will be supplemented with a staging base to be used as a fuel depot for checkout of payloads launched from earth and as a crew and materials transfer station between earth and geosynchronous orbit (GSO). As implied in Fig. 13, large facilities will be required in GSO to house the crews needed for SPS assembly and checkout and for control and maintenance and repair of the operational units. For satellite power systems, many "space stations" will be required and each becomes a program-dedicated operational support facility.

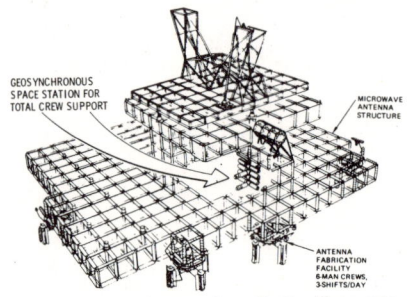

Figure 13. Geosynchronous Space Station (Supporting Construction of SPS Antenna)

In examining manned assembly operations in GEO, Rockwell International has included a space station located at the GEO construction site, as shown in Fig. 13. The station is composed of Shuttle cargo-bay-sized modules. The fabrication facilities are also identified in the figure. Each of the fabrication facilities requires a crew of 9 to 15 for three-shift operation. As the structure increases in size, the number of fabrication facilities working simultaneously is increased. Eight fabrication facilities can work on the section of structure shown. The space station houses all of these crew when not on shift, plus support personnel.

Space Station Configurations. Space station configurational concepts have been and probably will continue to be largely determined by the delivery capability of the available space transportation system. The concepts shown in Fig. 14 are all compatible with the deployment and support capabilities of the present STS. The variations in the concepts primarily reflect a distinction in the application and cost.

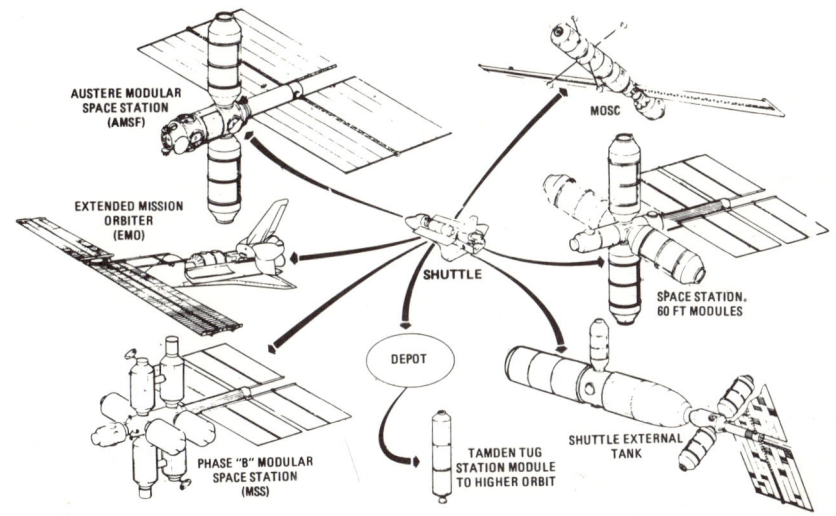

Figure 14. Configuration Evolution - Shuttle Launch Vehicle

The Rockwell International austere modular space facility is a recently defined concept for a minimum-cost, initial configuration for a crew of six with growth capability to 24. It is derived from the previously defined NASA JSC Modular Space Station (MSS) concept modified by simplification and incorporation of a significant amount of developed orbiter hardware. The concept could become an integral element of a program philosophy which emphasizes exploitation of Shuttle and Spacelab hardware. The NASA MSFC four-crew-member MOSC (manned orbital systems concept) also is a low-cost modular configuration. By using shorter modules, two modules can be placed in orbit with a single Shuttle launch.

The MSS is a full-capacity modular station with integrated mission element functions. The modules are approximately 12 m (40 ft) long, leaving cargo bay length for a tunnel docking adapter of an orbital maneuvering system (OMS) kit for delivery to higher energy orbits. Eighteen-meter (60-ft) length modules represent maximum volume deliverable in the Shuttle cargo bay.

Shuttle payload-bay-sized space stations require few, if any, technological advances to become operational. Shuttle Orbiter and Spacelab subsystems and technologies are largely applicable. Growth can be accommodated by more modules, and life support subsystem closure can be increased as required to reduce logistics costs. Beyond a certain point, however, this size module may be excessively restrictive, both structurally and operationally. The Shuttle external tank (ET) offers an initial large-module alternative, utilizing a premodified, spend ET and Shuttle payload-bay-sized modules.

Fig. 15 shows space station configurations alternatives with increased flexibility and capability launched by a heavy lift launch vehicle (HLLV). The concepts illustrated are representative of a series of Rockwell-developed 8.2-m (27-ft) diameter HLLV-placeable space facilities which provide for crew sizes ranging from 20 to 108. This class of facility has been named OASIS to signify its flexibility for observation, assembly, staging, and industrial support missions. Delivery of these modules to higher energy orbits would require development of a heavy payload orbital transfer vehicle (OTV), man-rated for transfer of large crew sizes.

It is generally conceded that long-range mission requirements can best be accomplished by permanent space stations, capable of continuous manned orbital operations with periodic resupply. Early availability, low-cost concepts, however, are always useful to support a transition to the next era.

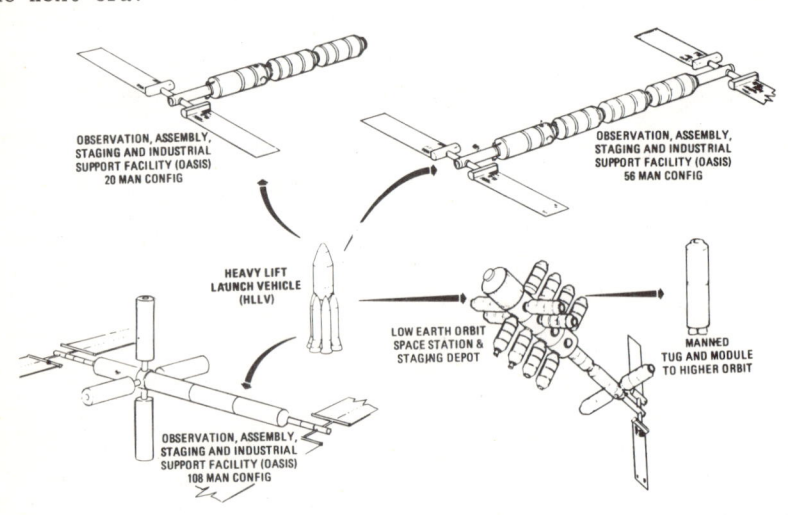

Figure 15. Configuration Evolution - Heavy Lift Launch Vehicle

<u>Full-Capability Space Facilities</u>. The new space initiatives of high current interest, such as the satellite power system (SPS), require the fabrication of structures in space which are so large that their dimensions are measured in kilometers. Space station requirements to support the structural fabrication are a major departure from requirements for missions of traditional interest. Millions of kilograms of materials must be space-assembled. Estimates for construction crew size for this range up to 500 with operations occurring at both low earth orbit (LEO) and geosynchronous equatorial orbit (GEO). Shuttle-compatible space stations for operation in LEO have been the subject of many detailed preliminary designs, as shown in Fig. 14. Fig. 16 is repre-

sentative of previous designs, all of which have been assemblies of varying numbers of Shuttle orbiter-deliverable modules. Prior to serious consideration of SPS, the requirement for a manned space station at GEO was not clear; therefore, no preliminary designs were specifically identified for GEO operation. Recently the LEO designs have been evaluated for GEO service and have been found to be well-suited for that application with modifications for the radiation environment. An OTV with adequate delivery (and crew/crew module return) performance is required.

Figure 16. Low Earth Orbit Space Station Concept

LEO Space Power Development Laboratory. The buildup of huge structures in space for the SPS will require development support of permanent LEO manned facilities. NASA JSC has suggested an LEO Space Power Development Laboratory (SPDL) where the technology of fabricating and operating these systems can be investigated and developed. One Rockwell concept for Shuttle-compatible LEO SPDL is shown in Fig. 17. It consists of a hub module and four-radial modules, and a fabrication platform attached to one end of the hub.

The four modules consist of one crew habitation, one crew support, and two fabrication/test support modules. The crew habitability and support modules are described later. The two fabrication/test support modules house the facilities necessary to support the antenna assembly and other test operations, and consist of repair shops, equipment stowage, test control stations, and stowage of test consumables. They also provide the docking location for the cargo modules during their unloading period.

The hub consists of a multiple crew member airlock at one end and a multiple docking adapter at the other. The platform is attached to the airlock end, which provides for egress of suited crew members to the platform work area and/or a docked manned maneuvering module (MMM).

The laboratory is used to develop assembly concepts for large structures. It will also be used to test the strength and dynamic characteristics of candidate structural designs in the space environment, and to develop and evaluate microwave power transmission concepts.

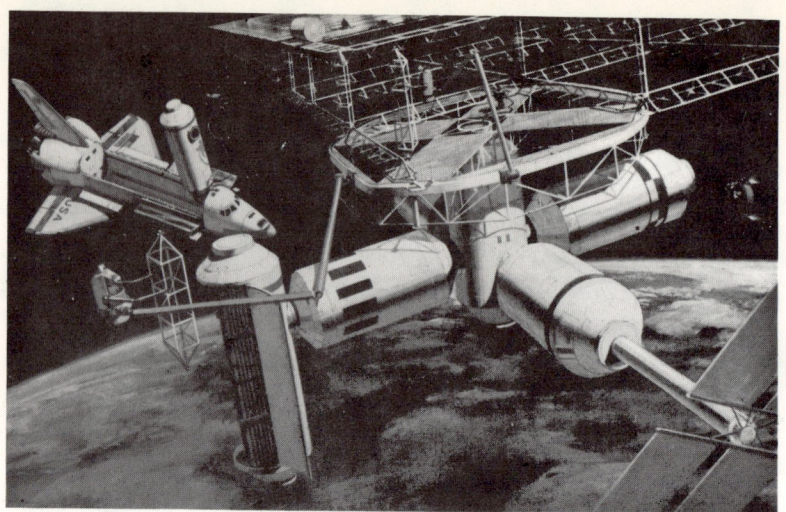

Figure 17. Low Earth Orbit Space Power Development Laboratory

Construction Crew Habitat. Living accommodations for SPS fabrication work crews must be appropriate for orbital staytimes of 90 to 180 days. These must be provisions for some degree of privacy, recreation, food, uncramped living spaces and medical care, in addition to provisions for support of the fundamental living and stationkeeping requirements. Figs. 18 and 19 are habitat modules which have been designed to operate together to provide extended mission duration crew support.

The crew habitability module (CHM), Fig. 18, provides facilities for sleeping, food preparation, eating, personal hygiene, and redundant subsystems for complete environmental control and life support. All facilities and subsystems are sized for a crew of six plus ECLS provisions for airlock and EVA operations and for temporary support of docked modules. It is self-contained and can maintain six crew members without benefit of the support module (SM).

The SM, Fig. 19 provides a large recreation space, a medical facility, an office, and the module control center with backup capability for total habitat control. It also provides a second hygiene station identical to that in the crew habitability module, a shower, a large freezer, emergency rations, and a small emergency food preparation station. A general-purpose workshop and photo-processing facility is located on the lower deck of the two-deck area. A redundant set of subsystems, identical to those in the crew habitability module, is located below decks. Emergency sleeping restraints are carried in the SM. The SM can provide for all the facility crew requirements on an emergency basis, with the CHM isolated or shut down for repairs. The SM is large enough to support a 12-man station, thus each combination of one SM and two CHM's constitute extended mission living facilities for a crew of 12.

The crew habitability and support modules are used in conjunction with a power module and with special-purpose modules that are required for specification applications.

EVA Considerations. The Apollo lunar and Skylab missions have demonstrated the need for good EVA support facility design and EVA operational techniques. Experienced EVA crews have identified the requirement for a dedicated EVA module which is an appendage (not a passageway) to

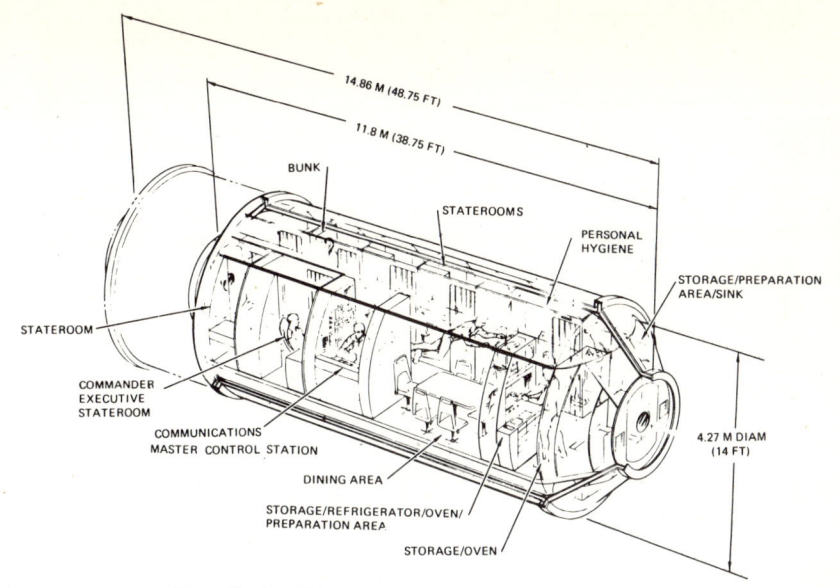

Figure 18. Crew Habitability Module for Construction Habitat

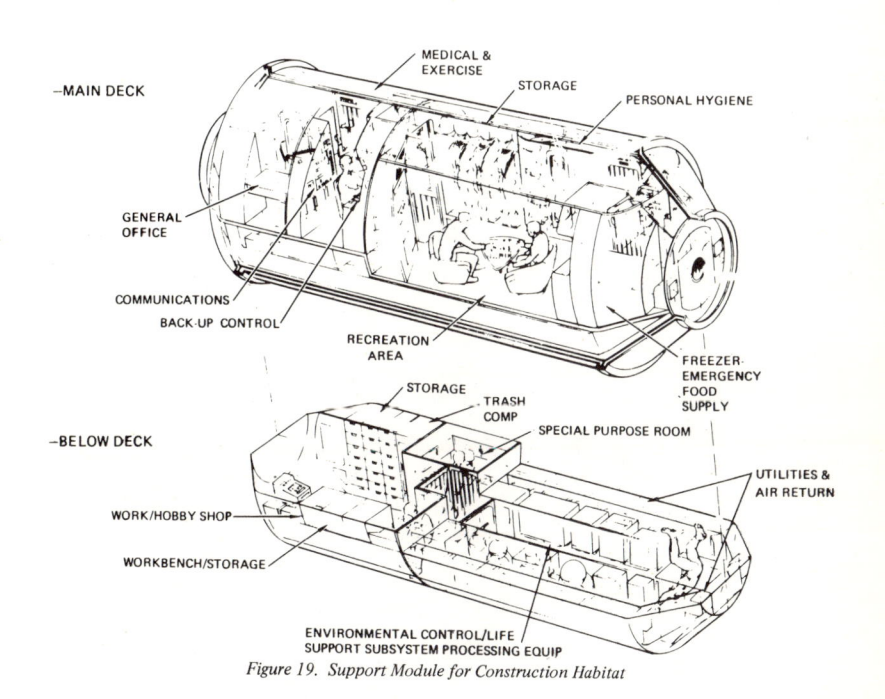

Figure 19. Support Module for Construction Habitat

the space station, in which there is ample space for permanent suit stowage, EVA planning and preparation, and checkout. The airlock design should preclude papers and other loose items from rushing to the outlet port during depressurization. Those features were deemed necessary even for missions involving a crew of three. Future missions involving fabrication of large structures in space are projected to require three-shift operation with each shift being in excess of 24-crew members.

Purging of nitrogen from the blood prior to each EVA imposes special design and operation requirements where large crews are involved. NASA data show that crew members living in an oxygen-nitrogen space station environment of 101.3×10^3 N/m^2 (14.7psia), with a partial pressure of oxygen (22.1×10^3 (N/m^2 (3.2 psia)), must breathe pure oxygen for three hours prior to entering the spacesuit environment of 27.6×10^3-N/m^2 (4-psia) pure oxygen total pressure. Increasing suit pressure to 55.1×10^3 N/m^2 (8 psia) permits EVA without prebreathing, and development of a suit that operates at that pressure is being investigated by NASA.. Another approach is to maintain a reduced-pressure EVA module where each shift spends several hours prior to EVA, possibly using it as sleeping quarters.

Fig. 20 depicts one Rockwell International concept for a dedicated EVA station/airlock to support shift-type EVA operations. The module provides crew sleeping compartments; stations for extravehicular mobility unit (EMU), manned mobility unit (MMU) and pressure suit stowage, servicing and repair; fit-up and checkout stations; and a hygiene facility. The EVA module atmosphere is 55.1×10^3 N/m^2 (8 psia), the remainder of the station being maintained at 101.3×10^3 N/m^2 (14.7 psia) with partial pressure of O$_2$ at 22.1×10^3 N/m^2 (3.2 psia). Entrance from the space station to the EVA module is via the rotating airlock on the left. Meals for the EVA crew are prepared in the space station proper and passed through the same airlock. The rotating airlock on the right takes the new shift (two at a time) outside and simultaneously brings the retiring shift into the EVA module. The atmosphere in the airlock is pumped from the egress sector to the ingress section. The sectors have been made small to provide for very rapid pump-down/pump-up, thus providing rapid crew transfer and minimum gas loss. The low differential pressure and pressurized rotating seal also helps minimize overboard loss of cabin gas. The design has been sized for eight, but can be extended to accommodate larger crew sizes.

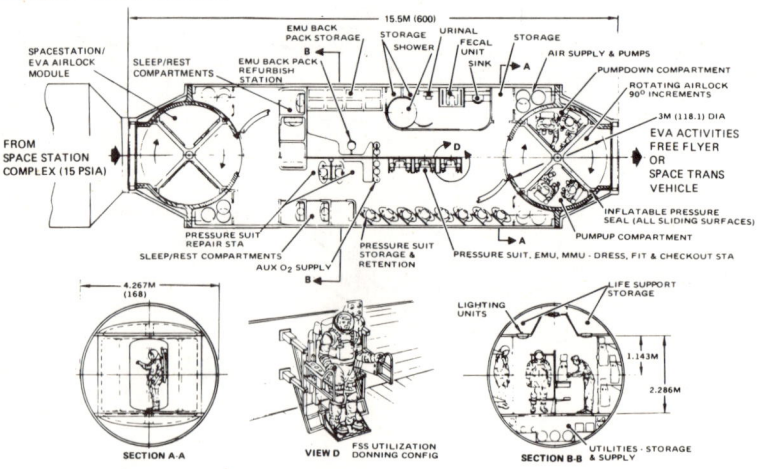

Figure 20. EVA Station/Airlock Concept

Subsystem Evolution. Subsystems for space stations are the subject of refined trades for changed applications and for integration of new technology on a continuing basis. Technology advancements are directed toward increased performance, lower weight, lower cost, and reduced logistics requirements. Basically, however, the technology exists now to design and build a full-capability space station.

Fig. 21 lists subsystems that have a major influence on a space station's physical characteristics and logistics requirements. These are the subsystems where technology advancements can have the most pronounced effect.

The present cryo tanks for the Shuttle have heat leak rates that limit their usefulness to about 50 days. For longer missions (extended-duration orbiter) high-pressure gas containers must be utilized. Improvements in cryo tankage are expected that will allow a 90-day resupply cycle.

	INITIAL OPERATIONAL CAPABILITY EXTENDED MISSION ORBITER (1981 – 1983)	LEO STATION (1985 – 1990)	GEO STATION (1990 – 1995)
LIFE SUPPORT			
ATMOSPHERE STORAGE	CRYO/HI-PRESS	IMPROVED CRYO	IMPROVED CRYO & WATER
WATER SUPPLY	STORED	RECLAIM	RECLAIM
CO_2 REMOVAL	REGENERATIVE (5 MM CO_2 PP)	REGENERATIVE (3 MM CO_2 PP)	REGENERATIVE (3 MM CO_2 PP)
CO_2 REDUCTION	NA	NA	SABATIER/ELECTROLYSIS
ELECTRICAL POWER GENERATION	SILICON SOLAR ARRAY *	IMPROVED SILICON SOLAR ARRAY	LIGHT WEIGHT GAILLUM ARSENIDE SOLAR ARRAY
STORAGE (ELECTRICAL)	N_1H_2 BATT	N_1H_2 BATT	N_1H_2 BATT
DATA PROCESSING	• ON-GROUND DATA REDUCTION • DATA BUS/CENTRAL COMPUTER	• ON-BOARD DATA REDUCTION • DECENTRALIZED SYSTEM	• ON-BOARD DATA REDUCTION • DECENTRALIZED SYSTEM

*FUEL CELLS USED FOR NON-ORBITAL FLIGHT PHASES

Figure 21. Subsystem Evolution

The degree of closure of the life support subsystems is a major factor in space station logistics requirements. Two areas that provide the most improvement potential are reclamation of water and oxygen (O_2) recovery from carbon dioxide (CO_2) (Sabatier/electrolysis). Water reclamation is the more dominant, comprising nearly 65 percent (versus 12 percent for oxygen) of the logistics requirement for a six-person space station with an open-loop design. It is expected that water reclamation will be cost-effective for all space stations and that the recovery of O_2 from CO_2 will have potential only for the space stations in high earth orbits, where the logistics costs are considerably higher than for space stations in orbits accessible with the Shuttle.

Regenerative CO_2 removal systems are required for cost-effective extended-duration operation, and there is a strong desire to maintain the CO_2 partial pressure below 506 N/m^2 (3.8 mmHG) in permanent space stations. The baseline Shuttle uses an lithium hydroxide (LiOH) CO_2 removal system which is capable of maintaining CO_2 partial pressure below 400 N/m^2 (3.0mmHg), but it is not regenerable and therefore imposes a significant weight penalty on long missions for the consumable LiOH. Skylab used a molecular sieve system which is regenerable by exposure to space vacuum, but it cannot maintain the cabin CO_2 partial pressure below 670 N/m^2 (5.0 mmHg). Solid-amine systems are being developed by NASA which are designed to maintain 506 N/m^2 (3.8 mmHg) CO_2 partial pressure. Both molecular sieves and solid amines are under consideration for the extended mission orbiter (EMO), although the lower partial pressure is not critical for the EMO. As solid amines and other concepts (i.e., electrochemical concentrators) are developed, trades will continue to be made to select the best system for the application.

Lighter weight and higher efficiencies are expected improvements in silicon solar arrays. Lightweight, radiation-resistant gallium arsenide cells with solar concentrators is a developing technology that is expected to replace silicon cell arrays at some future point, probably for the GEO station.

Nickel-cadmium (Ni-Cad) batteries have been the common storage battery for past space programs. Nickel-hydrogen is a relatively recent development in alkaline storage batteries and has some distinct advantages over Ni-Cad. Prime advantages are improved energy density, potential for longer life, and safe operation at low depth of discharge. Prime disadvantage is volumetric energy density.

The continued advancement in sophistication and miniaturization of electronic equipment makes feasible the concept of placing system complexity in space so that the user equipment on earth can be simpler and less expensive. Space stations are expected to use this approach and provide for greater on-board data reduction, and decentralize more autonomous subsystems.

This describes some key future elements for Space Station systems. An important part of this future is to maximize the productivity of the pre-Space Station era, the era of Space Shuttle and Spacelab.

SPACE SHUTTLE AND SPACELAB

<u>Baseline Shuttle</u>. Fig. 22 depicts our next space station on orbit. This figure shows the Shuttle payload bay loaded with the airlock, tunnel, Spacelab, and a pallet loaded with an experiment payload. This station has a 4- to 10-man crew complement and a 7- to 30-day mission duration. The payload is launched to orbit aboard the orbiter and will be returned to earth intact once the mission is complete. After injection into the operational orbit, the orbiter payload bay doors are opened, exposing the payload sensor equipment and allowing the orbiter thermal rejection system to operate. Experimenters enter the Spacelab from the orbiter cabin via an airlock, then into a transfer tunnel, and finally into the Spacelab itself. The flight commander remains on station in the orbiter while experimentation is underway in the Spacelab.

Initial launch of the Space Shuttle system begins in 1979 from the Kennedy Space Center. The program is proceeding well, with orbiter roll-out having just occurred. A second launch site, Vandenberg Air Force Base, is planned for operation in 1982. At that time, launch into orbits with inclinations greater than 56 degrees will be available for payload systems.

Figs. 23, 24, and 25 describe key Space Shuttle payload interfacing data regarding payload weight-to-orbit capability, payload location limits, and power and energy accommodations.

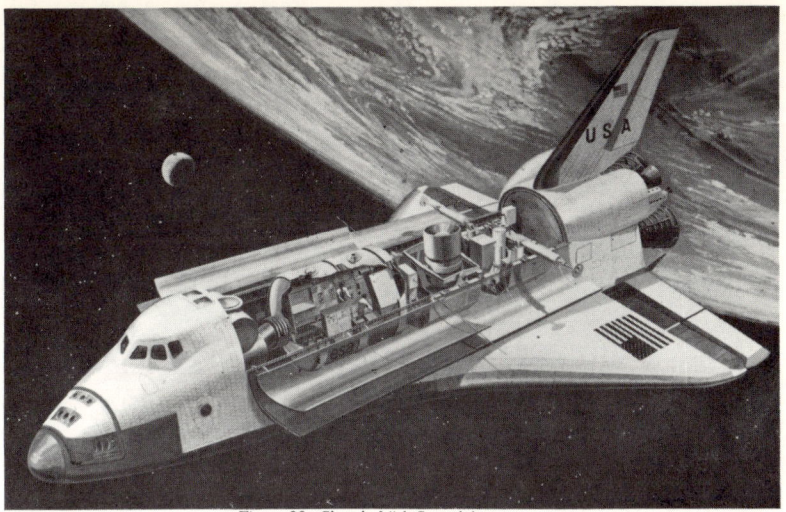

Figure 22. Shuttle With Spacelab

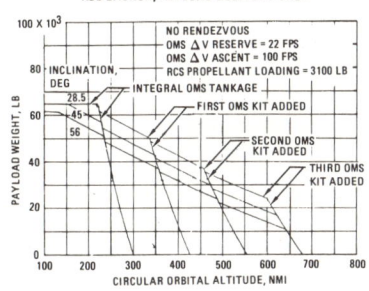

SPACE SHUTTLE PAYLOAD TO CIRCULAR ORBIT
KSC LAUNCH, PAYLOAD DELIVERY ONLY

SPACE SHUTTLE PAYLOAD TO CIRCULAR ORBIT
VAFB LAUNCH, PAYLOAD DELIVERY ONLY

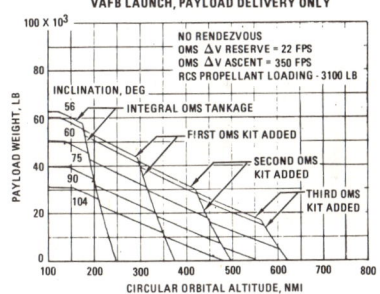

SPECIFIED CAPACITY COULD BE MODIFIED IN THE FUTURE

- FLIGHT EXPERIENCE
- OPERATIONAL MODIFICATIONS
  - INCREASE DURATION OF SSME HIGH THRUST
  - SELECTIVE INSTALLATION OF (HIGH ISP) SSMES
- SHUTTLE DESIGN MODIFICATIONS
  - WEIGHT REDUCTION
  - SSME PERFORMANCE IMPROVEMENTS
  - HIGHER PERFORMANCE BOOSTER

Figure 23. Shuttle Payload Delivery Performance

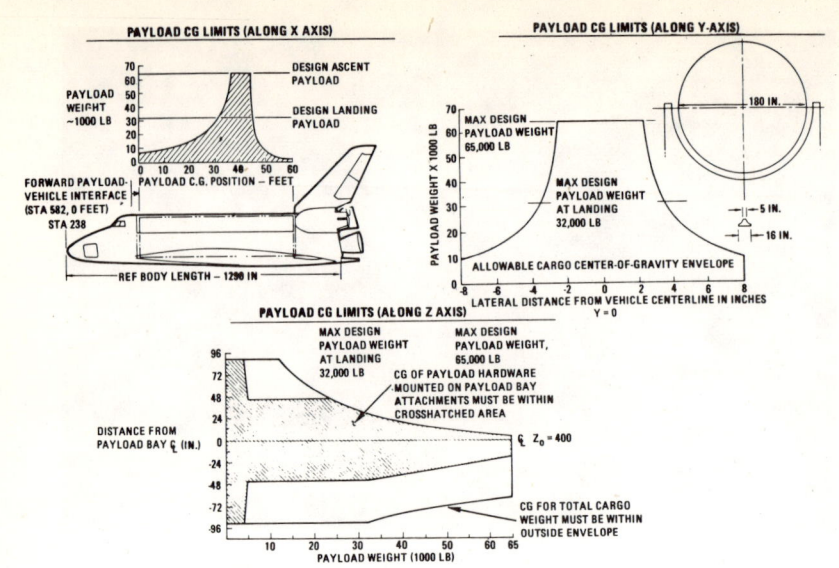

Figure 24. Shuttle Allowable Payload c.g. Location

	NOMINAL PAYLOAD MASS (KG)	PAYLOAD VOLUME	PAYLOAD POWER	PAYLOAD ENERGY (KWH)
	5500	MODULE 7.6 M^3 PALLET 102.2 M^3	3.8 KW	391
	5500	22.4 M^3	4 KW	422
	8000	182.8 M^3	5.2 KW	609
	9100	182.8 M^3	5.2 KW	609

NOTE: 1. HEAT REJECTION (ALL CONFIGURATIONS) 8.5 KW
2. COMMAND AND DATA MANAGEMENT IS MISSION DEPENDENT

Figure 25. Spacelab Payload Accommodation Summary

As these missions begin, it is critical to note the fundamental differences in this new era. Past manned space efforts have been discrete steps. Each system was marked by a limit in the number of flight attempts. The STS provides us with what is essentially a limitless number of flights. Where extreme design and operational guidelines were needed to assure success in a single flight, these can be relaxed in the continuing STS flight program. The Shuttle-Spacelab system provides a flexible, continuing experiment program. The next step beyond must be to provide continuous manned on-orbit capability. This is the prelude to space stations for the future.

Extended Mission Orbiter. Rockwell International has analyzed concepts for extending the on-orbit capability of the Shuttle-Spacelab through the use of add-on modification kits to provide a very low cost interim space station capability. The result of this activity has been the definition of the extended-mission orbiter described in Figs. 26, 27, and 28. The major departure from the baseline orbiter configuration is the addition of the solar array to provide electrical power for the orbiter, Spacelab, mission equipment, and crew during the orbital stay period. This releases payload weight and cargo space which would have been required for fuel cell reactants if the baseline Shuttle fuel cell were used during orbital operations. The array is deployed upon reaching orbit, and is retracted and stowed for post-orbital mission phases. The baseline orbiter fuel cells provide power for the non-orbital phases of the mission. Other modifications are indicated in Fig. 27 and include addition of nickel-hydrogen (Ni-H$_2$) batteries to provide on-orbit power during solar occultation, stored water, and a regenerative CO$_2$ removal system.

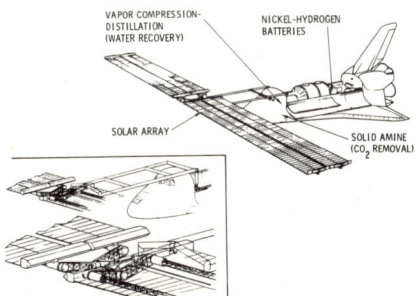

Figure 26. Extended Mission Orbiter

Fig. 27 also shows the increased mission capability provided by the EMO configuration when constrained by the 14,500-kg (32,000-lb) landing weight ground rule. It will support a crew of four with a 10,000-kg (22,000-lb) cargo, and provide 7-kw minimum continuous power to the cargo for an orbital staytime of 60 days. Longer missions are feasible (i.e., 90 days with a 9000-kg (20,000-lb) cargo) from a cargo mass standpoint. Spacelab mission payload equipment and materials accommodation provided by three alternative configurations of Spacelab when installed in the EMO cargo bay are defined in Fig. 28. The minimum mission equipment weight accommodation is 4100 kg (9,000-lb) and minimum power available to the mission equipment is 3.7 kw.

The EMO is highly attractive as an interim manned orbital facility. It maximizes the utilization of Shuttle and Spacelab systems with minimum hardware modification. It provides early long-duration manned mission capability. The potentially lower cost allows more funding for payload development to prove utility of manned space operations to develop requirements for a permanent manned operational facility.

BASED ON 14,500 KG (32,000 LBS) LANDING WEIGHT CONSTRAINT
4 CREW MEMBERS

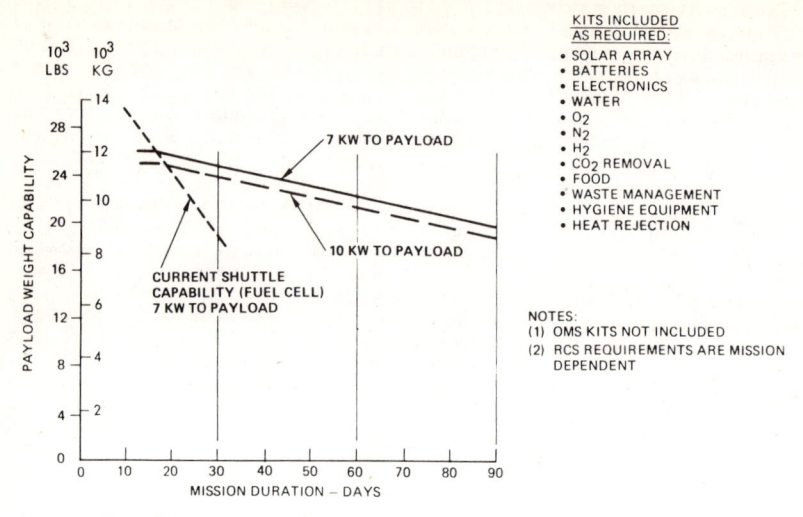

KITS INCLUDED AS REQUIRED:
- SOLAR ARRAY
- BATTERIES
- ELECTRONICS
- WATER
- O_2
- N_2
- H_2
- CO_2 REMOVAL
- FOOD
- WASTE MANAGEMENT
- HYGIENE EQUIPMENT
- HEAT REJECTION

NOTES:
(1) OMS KITS NOT INCLUDED
(2) RCS REQUIREMENTS ARE MISSION DEPENDENT

Figure 27. Extended Mission Orbiter Parametric Payload Performance

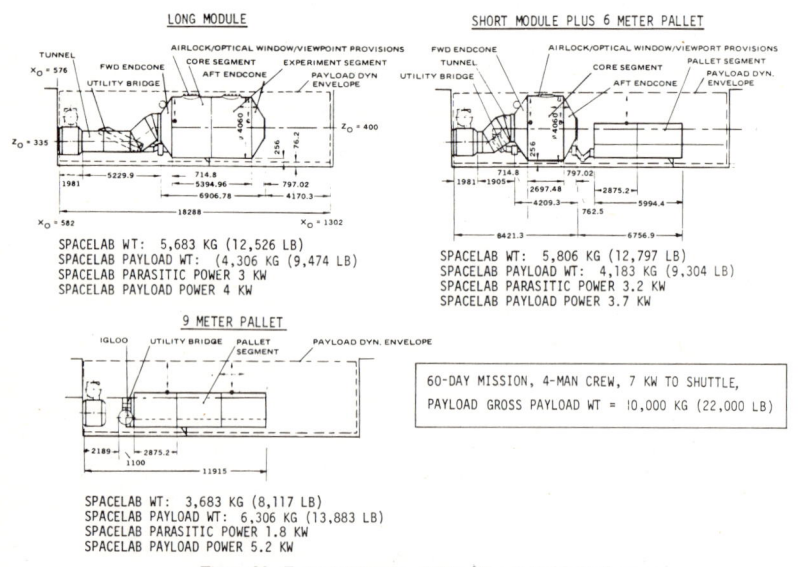

LONG MODULE

SPACELAB WT: 5,683 KG (12,526 LB)
SPACELAB PAYLOAD WT: (4,306 KG (9,474 LB)
SPACELAB PARASITIC POWER 3 KW
SPACELAB PAYLOAD POWER 4 KW

SHORT MODULE PLUS 6 METER PALLET

SPACELAB WT: 5,806 KG (12,797 LB)
SPACELAB PAYLOAD WT: 4,183 KG (9,304 LB)
SPACELAB PARASITIC POWER 3.2 KW
SPACELAB PAYLOAD POWER 3.7 KW

9 METER PALLET

SPACELAB WT: 3,683 KG (8,117 LB)
SPACELAB PAYLOAD WT: 6,306 KG (13,883 LB)
SPACELAB PARASITIC POWER 1.8 KW
SPACELAB PAYLOAD POWER 5.2 KW

60-DAY MISSION, 4-MAN CREW, 7 KW TO SHUTTLE,
PAYLOAD GROSS PAYLOAD WT = 10,000 KG (22,000 LB)

Figure 28. Extended Mission Orbiter Plus Spacelab Performance

Minimum-Cost, Permanent, Man-Tended Concept. A low-cost concept for a permanent orbital facility to support man-tended payloads which require very long mission durations or minimum disturbance during operation is the free-flying power/support system (FPSS) concept of Fig. 29. The FPSS consists of a solar array power module and a subsystem support module with docking provisions for the supported payloads, the Shuttle orbiter, and other specialized modules that may be required. The docking section provides for pressurized crew transfer between docked elements. The entire support module is pressurizable for shirtsleeve maintenance. The structure of the subsystems support module is derived from Spacelab structure. The module provides power conditioning and distribution, heat transport and heat rejection, data management and communications, stationkeeping, and consumables distribution for attached payloads. Spacelab payloads designed for operation in the Shuttle orbiter could operate indefinitely from the FPSS by the addition of a docking ring and minor modifications.

The FPSS is deployed and its operations initiated by the Shuttle orbiter. Subsequent operations are semi-autonomous with provisions for ground updates and data dumps, and for orbiter control when attached. Spacelab/payload configurations, previously designed for orbiter-supported operation, are modified to provide a docking interface. These are delivered to the FPSS by the orbiter where they remain for the duration of their mission. The Spacelab/payload hardware elements are returned to earth by the Shuttle orbiter upon mission completion.

A very important additional aspect of this concept is that the FPSS can provide power to the Shuttle orbiter while it is attached. Since the orbiter fuel cell consumables are the major constraint to extending Shuttle orbital staytime, the FPSS establishes the potential for extending the length of time the orbiter can maintain man's presence on orbit. The advantages of use of developed hardware, discussed with respect to the EMO, also prevail with the FPSS concept.

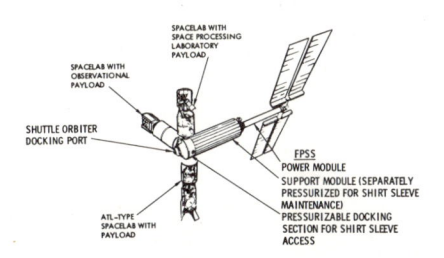

Figure 29. Free-Flying Power/Support System

<u>Available Test Systems</u>. The technology development program for the required subsystems and large structures should include early flight development testing. The Advanced Technology Laboratory system presently being defined by NASA Langley Research Center provides a vehicle for such testing. This system is described in Fig. 30. An example of specific development/experimentation aboard the ATL program is shown in Fig. 31; this summarizes the generic characteristics of large space structures composed of both erected space girders and expandable-type structure. Both types of structures are compressed for transportation in the Orbiter payload bay; in orbit they are then erected or expanded to their full shape, the folding or hinged joints are rigidized, and in the case of the erected structure the assembly into a larger space structure is begun. All of these steps require astronaut participation and, perhaps, mechanical aids such as the orbiter's manipulator.

A key early test for these Shuttle systems will be the assembly/test of large structure elements. A concept for assembling long truss structures composed of individual members is shown in Fig. 32. This entails the use of the RMS to retrieve stick elements from the cargo bay and systematically form pentahedrons on a holding fixture positioned to the side of the cargo bay doors. After a pentahedron cell has been completed, the beam can be moved one cell length along the fixture and the process repeated. This process can be pre-programmed to install the 9 sticks in the same position on the holding fixture. A cell span/length will be constrained by the reach limitations of the RMS.

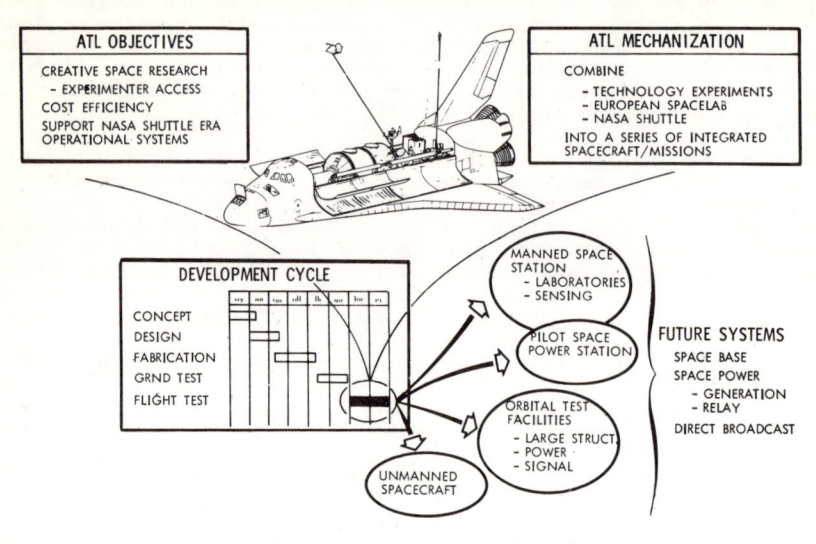

Figure 30. Advanced Technology Laboratory Experiment System Concept

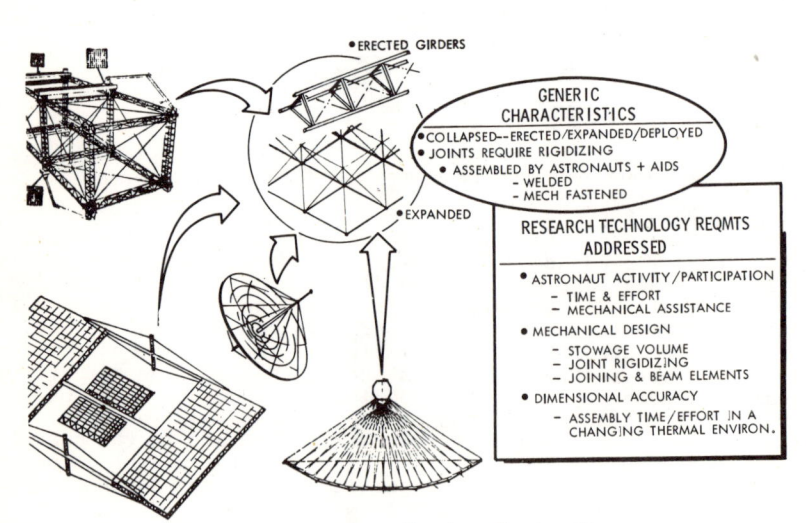

Figure 31. Structure Element Experiment Requirements

The fixture as conceived in Fig. 32 is ideally suited to fabricating long truss subassemblies. An additional re-indexing of the total truss beam one cell forward across the fixture width would allow the building of the second row of cells parallel to the first. Systematically moving the trusses forward one cell width at a time would produce larger area subassemblies. Another holding fixture concept would allow the turning of corners with the cell indexing to assemble hexagonal planforms composed of numerous pentahedron cells.

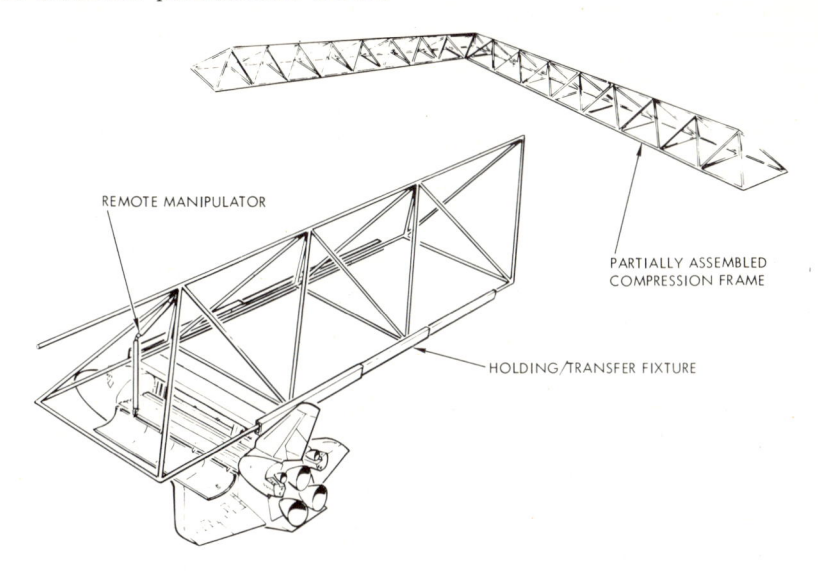

Figure 32. Assembly Concept for Compression Frame - Tension Web Configuration

The holding fixture would be transported to orbit in the Shuttle bay and deployed and set up in orbit. If the basic elements do not exceed the weight limit of the orbiter, the holding fixture can be transported each flight and returned with the Shuttle. It would be installed at the side of the bay in a similar fashion to the manipulator but on the opposite side of the bay. A second approach would be to leave the fixture in space attached to an unfinished subassembly and dock the Shuttle to it when the next cargo of structural elements is delivered.

An alternative approach for onboard Shuttle assembly is shown in Fig. 33, where the assembly is constructed over the front of the orbiter. Assembly operations can include the attached teleoperator, a free-flyer teleoperator, and EVA.

The off-Shuttle assembly would involve the mating of partial subassemblies, away from the confines of the cargo bay. At the elemental level, this could be considered as using the manipulator or construction boom to position and connect a single structural element (e.g., girder) to a partially completed subassembly.

This concept might require some form of stablilizing device. A preliminary estimate for such a stabilization module could weigh less than 10 percent of the weight of tne subassembly structure required to be stabilized. Such a device would have to be low-cost and reusable.

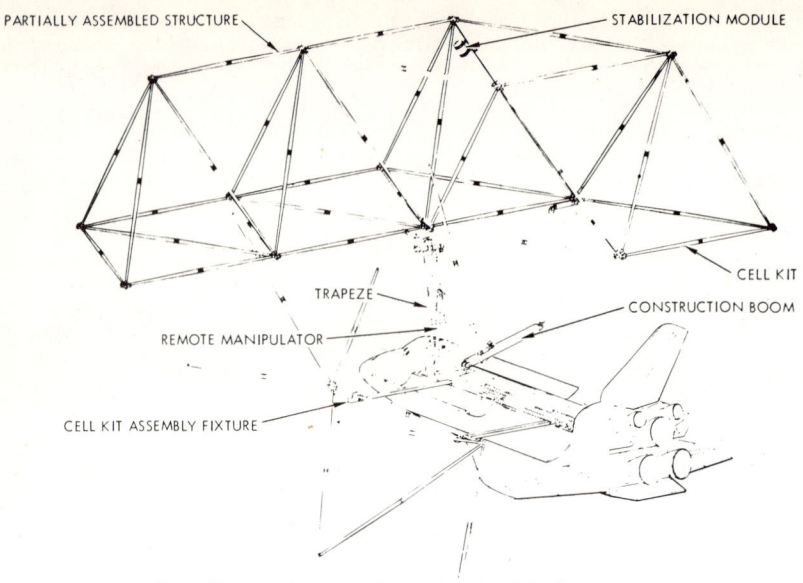

Figure 33. Assembly Concept for Tetrahedral Grib Configuration

Fig. 34 presents a time-phased development program emphasizing the year-by-year installation of the major elements in the power-oriented scenario for the space industrialization program. The installation of a pilot SPS will be supported by the early activities involving the Space Shuttle, the Spacelab, and a low-altitude space station. Test programs involving large structural elements will provide particularly important background data on the assembly procedures, attachment methods, alignment fixtures, natural vibration modes, and control systems development, and the detrimental effects of thermal gradients. These early tests will provide an orderly development program leading to the efficient installation of the pilot SPS. The early assembly and test operations will involve single Shuttle launches: later materials from multiple launches will be incorporated into the program.

Figure 34. Space Station Program Evolution

370

CONCLUSIONS

The definition of the next generation of space stations will continue for some time. Even now, however, we can conclude the following:

1. We have developed a broad depth of experience in hardware definition, design, test, and mission operation for space stations. Man's unique capability has been clearly established. Since the requirements and mission experience exist, it is reasonable to conclude that there will be future space stations.

2. The constraints imposed by transportation system characteristics appear to be stronger configuration drivers than those derived from the program objectives and requirements. All past station designs were strongly driven by the payload weight, volume, location, and density constraints of the Atlas, Titan, and Saturn launch vehicles. Space Shuttle must be assumed as the basic launch vehicle for the next generation of space stations. As such, a station made up of modules less than 4.6 m (15 ft) in diameter (18.3 m (60 ft) long and weighing less than 29.484 kg (65,000-lb), seems the most probable design.

3. Space Shuttle and Spacelab elements, systems, and subsystems can provide a significant technical capability. An orbiter, plus Spacelab, can be modified with solar panels, extension kits, and minor changes to allow for missions of 60 to 90 days. A modular space station with permanent on-orbit capability for 6- to 25-man crews can evolve from the systems and technology of Shuttle and Spacelab.

4. Very large space stations may be required to construct very large structures on orbit. The Construction approach is not fully defined. It may be an erection process such as large building construction. It may be more of an in-place fabrication process similar to street paving. The selection of the process will require the assessment of many factors including man's ability to do the job, surface flatness characteristics, thermal-stress, deflection, weight, attitude control, and dynamic frequency. An assessment of the assembly process on a proposed microwave power transmission system structure suggests the job can be done. Better solutions will be found but clearly, a viable solution exists.

5. The Space Shuttle and Spacelab, along with payload systems such as the Advanced Technology Laboratory, provide an excellent base to develop the required subsystems and support elements for future stations. Large structure experiments, proposed for flight on Shuttle/Spacelab/ATL missions, will provide early comparative data on structures that are erected, expanded, or fabricated on orbit. Proper development of these early missions is critical.

BIBLIOGRAPHY

1. Space Station Systems Analysis, Space Division, Rockwell International Corporation, SD 75-SA-0301 (6 Feb. 1976).

2. Modular Space Station Phase B extension, Space Division, Rockwell International Corporation, Contract NAS9-9953, NASA JSC.

3. Shuttle Orbiter and System Integration Program. Space Division, Rockwell International Corporation, Contract NAS9-14000, NASA JSC.

4. Microwave Power Transmission System, Raytheon Corporation, NAS3-17835, NASA Lewis Research Center, Report NASA CR-134866, December 1975.

5. Extended Mission Shuttle Feasibility Study. Space Division, Rockwell International Corporation, SD 75-SA-0071 (Sept. 1975).

6. Austere Modular Space Facility, Space Divison, Rockwell International Corporation, SD 75-SA-0105 (Sept. 1975).

7. Observation, Assembly, Staging and Industrial Support (OASIS) Facility, Space Division, Rockwell International Corporation, SD 75-SA-0106 (Sept. 1975).

8. Manned Orbital Systems Concepts Study, McDonnell-Douglas Astronautics Company - West, NAS8-3104, NASA MSFC.

9. Manned Assembly of SPS Structure, Space Division, Rockwell International Corporation, SD 76-SA-0071.

10. Space Station Concepts, Space Division, Rockwell International Corporation, SD 76-SA-0072.

11. The Skylab Results. A compendium of papers presented at the 20th American Astronautical Society Annual Meeting, August 20-22, 1974.

12. Study of the Commonality of Space Vehicle Applications to Future National Needs. (Unclassified Portion), The Aerospace Corporation, ATR-75(7365)-2, 24 March 1975.

13. K.A. Ehricke, Exoindustrial Productivity, The Extraterrestrial Imperative of Our Time. Rockwell International, North American Space Operations, May 1975.

14. NAVSTAR GPS Space Vehicle System Segment, Air Force Contract, AF 4701-74-C-0527, Space Division, Rockwell International Corp.

15. Geosynchronous Space Station Study, MSFC-PA-7-75, July, 1976, NASA MSFC.

16. Outlook for Space, NASA SP-387, January, 1976, Scientific and Technical Information Office, National Aeronautics and Space Administration.

17. J.F. Madewell and R.E. Sexton, "Space Stations for the International Future," Technical Paper presented to TGRB, Bonn, Germany, June, 1976.

18. Orbital Construction Demonstration Study, Grumman Aerospace, Johnson Space Center Contract NAS9-14916, first quality report (June, 1976).

ASTROPOLIS AND ANDROCELL -- THE PSYCHOLOGY AND TECHNOLOGY
OF SPACE UTILIZATION AND EXTRATERRESTRIALIZATION

Krafft A. Ehricke*

Today the social and environmental frame of reference of nations and mankind is still determined exclusively by the terrestrial scene. The advent of the space station and the growing impact of space industrialization will mark the beginning of a profound change. Activities in never before utilized environments will bring many improvements to the social and geo-environmental scenario. They will affect industrial developments, food production, the economy of nations.

Their social impact will include vastly expanded communication with its special provisions of better education and better health service for billions of people, as well as new jobs. Work habits will change as more people will communicate rather than travel or drive to work over more or less great distances. The savings in gasoline, the reduction in rush hour traffic, pollution and accidents could be considerable already in the 1990's in highly industrialized countries. The gradual spreading of an industrial lifestyle will generate a social environment conducive to smaller families. Thereby, the acceptance of contraceptives and family planning around the globe will be facilitated and the prospects for slowing down population growth will be enhanced greatly.

The environmental impacts will include better pollution monitoring, reduced pollution, long-range (up to 30 days) weather forecasting and the beginnings of climate forecasting.

The common feature of the above effects is that they represent improvements and problem solutions. These, of course, are critically important and highly desirable. Nevertheless, they do not include the changes brought about by the emergence of human life into space. The peoples of Earth will continue to make history. But new branches of history will begin their time line or lines beyond Earth. There is likely to be feedback, yet this will no longer be the whole story.

Space industrialization still has the primary purpose of contributing to the before-mentioned improvements on Earth. The emphasis is on maximum productivity and cost-effectiveness, which implies small, specialized teams of professionals and technicians. But the existence of a space industrial capability unavoidably produces a new consequence world, that is, a changed and broadened option bank: options to apply space industries to make the existence of larger communities possible beyond Earth. But new motivations are necessary to exercise these options. To these and to their technological implementations my observations are directed.

The Socio-Psychological Scenario. To develop a better grasp of future needs, motivations and opportunities, it is useful to develop a likely scenario. Its main features are representative of the trends and determinants that can logically be expected to assert themselves in the time period under consideration -- for present purposes, within the next 50 to 100 years.
----
*Director, Advanced Planning, Space Division, Rockwell International Inglewood, California.

In this context, permit me to fall back onto a scenario a colleague and I developed in connection with our study entitled "The Extraterrestrial Imperative" (1) (2) (3). Back in 1970 our work had reached the point where we felt it could be helpful in counteracting the then rising emotional anti-technology and anti-space moods. No book publisher would then touch a manuscript in this field that did not deal with gloom, doom, ecocatastrophies, limits to growth etc. Official encouragement was unavailable. So we offered to a film studio an outline for a TV series we thought could be both exciting and informative.

The series, which we gave the working title "The Olympians", was novel for those days in that it did not deal with an expedition to Mars or into the galaxy. It focused on the involvement of medium-sized and large space stations, in different orbits, in the affairs of Earth, and those of a lunar industrial colony. By international consent, the Olympians were given five principal mandates: (1) to assist and provide services to all nations in the fields of Earth observation, communication, data storage and others; (2) to be concerned with spaceship Earth, that is, its international domains, climate and ecosystem; (3) to present and symbolize a new unifying human factor; (4) to support orbital industrial activities, and the lunar industrial colony; and (5) to further and manage the continued development of space communities.

While many episodes dealt with external factors and adventure, a good number of them reflected the socio-psychological aspects involved in this scenario. These aspects have four main components.

One component is the social impact of the international space station as a viable and beneficial institution on the national interfaces and on the awareness of human commonality as a species or life form.

The second component is the reverse of the first -- the impact of terrestrial affairs on the orbiting communities. Terrestrial needs, problems, conflicts and confrontations are reflected within the orbital communities whose members are drawn from virtually all nations, even though on a selective and compatibility oriented basis. Their reflection causes internal stresses and problems. Their solution models offer new approaches to terrestrial behavioral problem solving -- a kind of socio-psychological spin-off from space.

The third component involves the intra-community relationships and the emergence of new cultural traits under the impact of prolonged orbital existence.

The fourth component of this scenario, finally, comprises the sociopsychological relationships and differences between the orbiting communities and the comparatively less confined lunar community.

The lunar activities were to include inter alia, the construction of a large space community, numbering in the thousands to tens of thousands, capable of navigating into heliocentric space as a planetella -- to Mars, into the asteroid belt and into the Jovian satellite system. Its socio-psychological aspects were to reflect the integration of orbiting and extraterrestrial surface existence, and the ultimate socio-psychological independence from human society on Earth. We called this community a human cell or Androcell released from the mother planet into the limitless ocean of space. The largest of the Earth orbiting stations we called Astropolis.

Needless to say, the series was not accepted in this form. But it did lead to a pilot film organized around an Astropolis. The pilot, entitled "Earth II" ran on ABC-- I believe in 1972. Our background studies continued, particularly as regards the identification of major evolutionary phases -- space industrialization and beyond -- and of the priorities and objectives characterizing each phase.

The reason why I mention this event of 6 years ago is because such a scenario is still valid; and to point out that there is a very important qualitative difference between the industrialization of extraterrestrial environments and the issues involved in populating them. The problems and the objectives are different. The placement, for its own sake, of larger populations beyond Earth rather than the smaller and highly specialized teams needed for optimum cost-effective space industrial productivity is feasible and realistic only with the experience of space industrialization behind us.

<u>Ability and Motive</u>. Fundamentally, two detriments are involved in progression: ability or capability and motive arising from need, preference or desire.

To begin with a very long look, I believe it is useful to recognize some relevant aspects of our position in the evolutionary chain of life as it advanced on this planet. In "The Extraterrestrial Imperative" we found it useful to percieve the particular human form of intelligence -- in distinction from the intelligence and learning capability of dolphins, for example -- as information metabolism. A central feature of human intelligence is the ability to metabolize random information into its basic or generic context and to polymerize the information building blocks into coherent systems of knowledge and know-how or skills enabling us to interact more efficiently, hence on a growing scale, with the environment and also to gain access to environments that would otherwise remain closed to us for reasons of physical and/or sensory limitations.

As far as interaction with the environment is concerned, perhaps the most outstanding fact is that information metabolism is the first metabolic technique since photosynthesis enabling a life form to interact directly with primordial energy and material resources. On this planet it shares this distinction only with certain forms of chemosynthesis (life's original metabolism and retained to this day in some forms, such as the

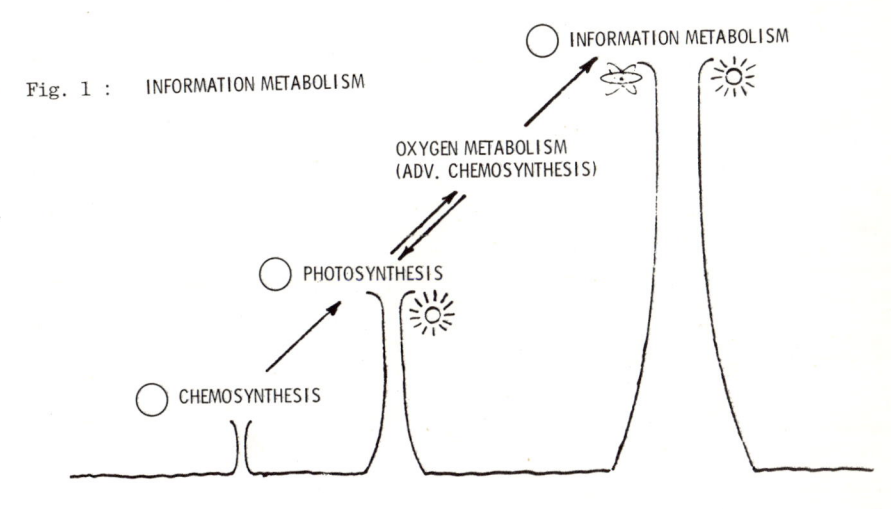

Fig. 1 : INFORMATION METABOLISM

INFORMATION METABOLISM

OXYGEN METABOLISM
(ADV. CHEMOSYNTHESIS)

PHOTOSYNTHESIS

CHEMOSYNTHESIS

PRIMORDIAL MATERIAL/ENERGY RESOURCES

important nitrogen fixation, but also gangrene) and photosynthesis (Fig. 1). Oxygen metabolism, a greatly improved form of chemosynthesis is nevertheless a photo-synthesis-dependent technique, not an "umbilical" metabolism.

As umbilical metabolisms we designate those which interconnect with the primordial (inanimate) resource world whose energy and material wealth is truly infinite.

In the fact that information metabolism is an umbilical metabolism lie the roots of our present environmental and resource confrontations as well as the abilities to overcome the problems and crises arising from those confrontations. Advances in umbilical metabolic technologies always share the common characteristic of first confronting their "native" environment, then superseding it. In the case of photosynthesis the process of superseding -- i.e. the formation of the biosphere -- was accompanied by the destruction of the native environment -- the reducing character of the terrestrial environment was destroyed by the accumulation of large amounts of free oxygen. The human life form is not desirous to destroy its native environment -- the one created by photosynthesis Because information metabolism renders extraterrestrial resources accessible, destruction fortunately is not necessary, although modifications occur continuously, at least since the agricultural revolution some 6000 years ago.

All umbilical metabolic processes must create their own Open Worlds in equilibrium, since they are working systems. Energy input must equal energy output, material input must equal material output either in the form of recycling or by expanding the physical environment. Bio-industry being confined to the terrestrial environment had to rely entirely on recycling in creating its work-performing equilibrium, the biosphere -- a specific planetary environmental niche interfacing with an extraterrestrial energy source (Sun) and an extraterrestrial energy sink (space). The chlorophyll molecule acts as the principal material flow "gate" determining the size and limits of the biosphere.

The equilibrium system of information metabolism we named __Androsphere__. Its energy and material processing gate is the human brain. Because of broader access to primordial resources, including extraterrestrial ones, the Androsphere can supersede the biosphere without destroying it. In turn, the Androsphere must be transplanetary so that accommodation of its terrestrial sector with the biosphere becomes possible, while meeting the particular needs of the information metabolic life form continuing to reside on Earth.

Therewith a fundamental ability and a basic need are established.

On a less fundamental level, of course, more pragmatic motives assert themselves in our time. They are divided into terrestrial preferences and space or integral space/Earth preferences (e.g. communication, Earth observation and material processing and manufacturing in space for terrestrial uses). Some of the most important preference criteria either way employed by industrialized and developing nations are listed in Fig. 2.

They are related, in effect, to space industrial requirements. Space industrialization has very clear cut generic, and more specific, goals and objectives which are the reason why it is desirable to proceed with its stepwise realization.

CONSIDERATIONS IN EVALUATING COMPARATIVE MERITS OF TERRESTRIAL AND SPACE OPTIONS

	WHAT MAKE TERRESTRIAL OPTIONS DESIRABLE	WHAT MAKES SPACE OR EARTH/SPACE OPTIONS DESIRABLE
INDUSTRIAL NATIONS	• EXISTING INVESTMENT PATTERNS • LABOR INTERESTS • VENTURE RISK CONSIDERATIONS • GROWTH POTENTIAL OF EMERGING TERRESTRIAL TECHNOLOGIES (ADV. POWER TECHNOL., SEA BED MINING ETC.) OPENING NEW RESOURCES ON EARTH • NEED FOR FUEL • COST OF SPACE TRANSPORTATION AND SPACE JOBS (ECONOMIC COMPETITIVENESS PROBLEMS)	• IMPROVED SERVICES • IMPROVED INVENTORY RESOURCE MANAGEMENT • ACCESS TO NEW RESOURCES • DEV. OF NEW INDUSTRIES (IMPROVED ECONOMIC STRENGTH & BALANCE OF PAYM'TS) • ENVIRONMENTAL CONSIDERATIONS • ASSISTENCE TO DEV. NATIONS
DEV. NATIONS	• GREATER LABOR INTENSITY[1] • INDUSTRIAL DIVERSIFICATION • BALANCE OF PAYMENT CONSIDERATIONS • ENHANCED SELF-SUFFICIENCY • NEED FOR SHORT-/MEDIUM-TERM RETURNS AND THEIR RAPID RE-INVESTMENT TO CREATE EMPLOYMENT [1] TOWARD THE END OF CENTURY, LABOR FORCE IN DEV. COUNTRIES IS INCREASED BY ~ 400M, THAT OF INDUSTR. COUNTRIES BY ~ 35M	• IMPROVED SERVICES (SHORTCUTS) • IMPROVED INVENTORY & RESOURCE MANAGEMENT • NATIONAL EDUCATION & HEALTH SERVICES • DEVELOPMENT OF NATIONAL RESOURCES • MAY SUPPORT TRANSFER OF LABOR FROM TRADITIONAL TO MODERN SECTOR OF ECONOMY (HIGHER PRODUCTIVITY, CAPITAL FORMATION) • TRANSITION TO INDUSTRIALIZATED LIFE STYLE

Fig. 2   Rockwell International Space Division   96PD123093

SPACE INDUSTRIALIZATION IN PERSPECTIVE

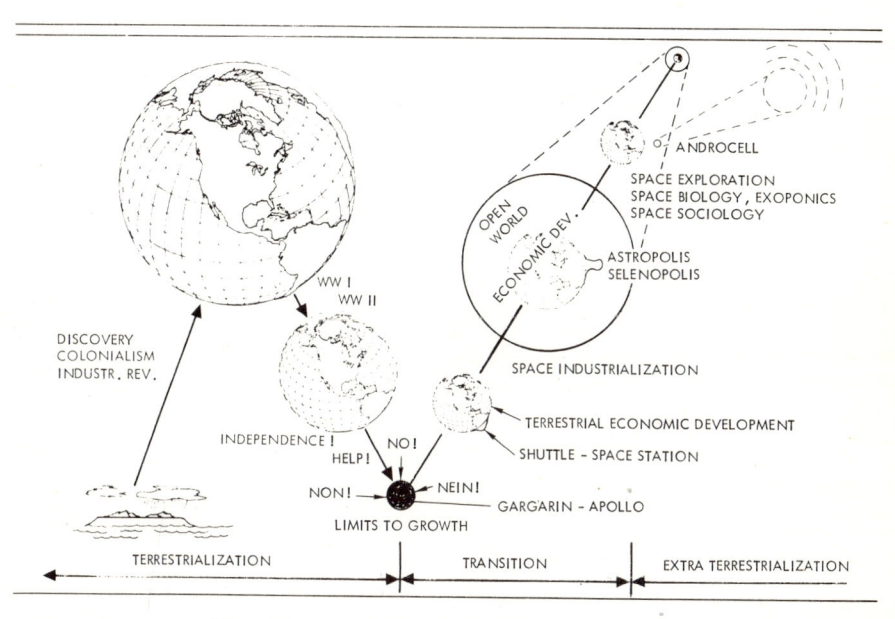

Fig. 3 Space Industrialization in Perspective

Space Industrialization in Perspective. Realization of the Renaissance and its consequence world (the Age of Discovery and the scientific revolution) as a brand new beginning permits one to place space industrialization in perspective (Fig. 3.).

Through the Age of Discovery the sundry parts of humanity were eventually knitted together. Man became aware of his planetary existence. This terrestrialization opened a seemingly inexhaustible world to a human activity. The greatest migrations in all history and an unprecedented quickening of scientific and technological progress took place. The great wars in the first half of this century, however, ushered in a further acceleration in resource demands. The world began to shrink. Many newly formed nations after World War II soon realized the need for economic assistance from the wealthy industrialized nations.

In the early 1970's, long-term prospects for this aid seemed to become doubtful by anti-technological movements and no-growth-laced environmentalism (UN Conference on the Environment, Stockholm 1974). The 1970 depression and political factors raised additional early warning signs. The upshot of these developments was the determination of many developing countries to establish their own growth plans based on higher prices for their indigenous raw materials. The oil producing and exporting countries (OPEC) spearheaded this New Economic Order. The sharp increase in oil prices led to inflation, recession and sharply reminded the wealthy nations of the vital need for continued technological and economic growth -- on new frontiers as well as along existing channels. Now the Earth begins to "grow" again, as new energy resources and new material resources (e.g. on the sea bottom) come into economic view. An additional resource growth potential is offered by the geolunar environments through space industrialization.

In this perspective, space industrialization represents an advance of terrestrial industry for terrestrial purposes. Shuttle, space stations, even lunar industrial facilities are Earth related. The men and women who make up the industrial teams operate in space for more or less limited periods. They are and remain terrestrials.

The same is true for tourists and for many who may avail themselves of the by then better defined therapeutic benefits of space hospitals and reconvalescent facilities in larger satellites of the Astropolis type.

Throughout this process, the space industrial capability level and the skills of productive space utilization will have risen. The number of humans living in space for greatly extended periods (major fraction of their life span) will increase. These people will no doubt develop new preferences as to g-levels and life styles no longer necessarily compatible with terrestrial conditions. Space stations and lunar abodes will have become their home -- Earth a place to visit or, better even, to be experienced holographically in the comfort of their gravity environment. Their space-born offspring may migrate back or they may stay and, in that case, will recede socio-psychologically still further away from the ways of terrestrial mankind.

Their willingness to achieve organizational and resource independence will grow as both motivation and capability to create new worlds in their totality. The tool will be Androcell. It will be a new beginning in the Open World, as back on Earth conditions move toward a demographic and industrial equilibrium.

Open World Evolution. In the perspective of the evolutionary thrust of the Open World of Earth and space, therefore, exoindustrialization may be regarded as the first phase of extraterrestrialization.

Extraterrestrialization proper -- the Androcell phase -- is likely to occur following an intermediate phase (exo-urbanization). Over-simplifying a bit, it may be said that exoindustrialization puts the machines into space; exo-urbanization introduces the human and biological elements; and extraterrestrialization integrates the two components into whole new worlds.

Thus, each of the three evolutionary phases is justified by clearly identifiable primary purposes as well as by their impact in changing the consequence world (Fig. 4).

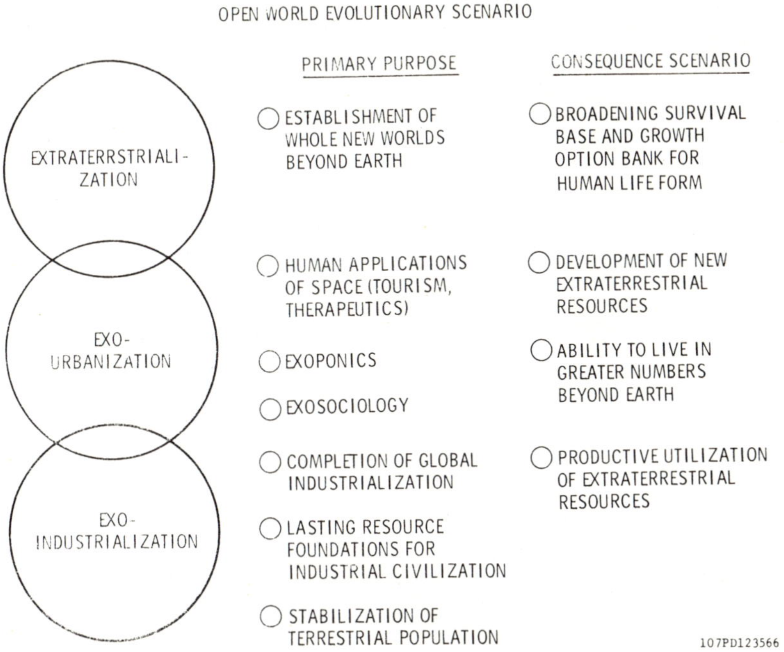

OPEN WORLD EVOLUTIONARY SCENARIO

	PRIMARY PURPOSE	CONSEQUENCE SCENARIO
EXTRATERRSTRIALI-ZATION	○ ESTABLISHMENT OF WHOLE NEW WORLDS BEYOND EARTH	○ BROADENING SURVIVAL BASE AND GROWTH OPTION BANK FOR HUMAN LIFE FORM
EXO-URBANIZATION	○ HUMAN APPLICATIONS OF SPACE (TOURISM, THERAPEUTICS) ○ EXOPONICS ○ EXOSOCIOLOGY	○ DEVELOPMENT OF NEW EXTRATERRESTRIAL RESOURCES ○ ABILITY TO LIVE IN GREATER NUMBERS BEYOND EARTH
EXO-INDUSTRIALIZATION	○ COMPLETION OF GLOBAL INDUSTRIALIZATION ○ LASTING RESOURCE FOUNDATIONS FOR INDUSTRIAL CIVILIZATION ○ STABILIZATION OF TERRESTRIAL POPULATION	○ PRODUCTIVE UTILIZATION OF EXTRATERRESTRIAL RESOURCES

Fig. 4

It is through this positive feedback that each phase contributes to the capability and motivation in the consequence world to progress to the subsequent phase. In the third phase we leave the harbor and emerge into the open sea of space. In the trackless infinity of space and time, human history will henceforth pulse through many arteries which lose themselves beyond the horizon of our perception.

The civilization of the Androcell is truly three-dimensional and becomes four-dimensional as it spreads through interstellar space, whether in the relativistic or Newtonian flight regime. Through exoindustrialization (Space Station), exo-urbanization (Astropolis) and extraterrestrialization (Androcell), the human life form returns to the three-dimensional origin of all terrestrial life. The two-dimensional existence on Earth's land surface ultimately becomes only a brief (by evolutionary standards) interim phase. This phase is wedged between the finite oceanic womb from which life rose to the brightness of consciousness and the infinite cosmic womb in which it will, or at least can, rise to a level beyond our understanding. The growth potential of all life forms (not only human) capable of emerging into this infinity exceeds all comprehension. It relegates all its previous history to the proverbial first step in a journey that lasts a thousand miles.

In returning to our time, we note that at no point in the evolution from exoindustrialization to extraterrestrialization, do the primary purposes include relief of terrestrial population pressure through mass emigration into space. The impracticability of this approach is apparent from Fig. 5. The top line shows a representative growth curve of world population between 1970 and 2100. This growth may be overrun by as much as five to seven billion by 2100. But it will hardly be underrun by as much as three billions. The growth curve, therefore, is conservative for the present reasoning, since a stronger growth would only intensify the point. The point is that the practicality of population pressure relief when it counts -- i.e. in the next 50 years -- depends on an impractically high growth rate of emigration into space. The dashed curve shows the annual increase in world population for the given growth curve. The increase shows a rise from an average of 96 million/year in the 1970 to 1975 period to 103 $\bar{M}$/yr in 1980-85, followed by a gradual decline to 44 $\bar{M}$/yr in the 2095-2100 period. For the next 50 years the growth rate does not drop below 80 $\bar{M}$/yr. This is the period when relief would be most urgent, because of the as yet unfavorable relation between population growth and economic as well as resource (esp. energy) development. The latter conditions will be far better a century from now, barring a major catastrophe. For this reason and because of the concomitant reduction in population growth rate (the top curve corresponds to a population doubling time of 27 years in 1970 and 199 years in 2100), demographic conditions will be under control beyond 50 years from now.

Relief 100 years from now through emigration would require the capability of removing 10 to 30 million people annually between 2080 and 2090 (curve B), causing the annual increase in global population to decline along curve B[1]. Such capability may be possible by then, but demographic relief will no longer be the issue. To be relevant to this issue, the emigration rate would have to follow curve A, at least. This buildup would require a concentration of resources that is quite out of proportion to the priorities urgently needed by the some 6 billion people then living on Earth without any chance of leaving the planet, not to mention the socio-psychological unknowns involved for whose exploration no time would be available.

A growth rate of emigration which causes the extraterrestrial population to double every 10 years (a 7% growth rate, not counting extraterrestrial births), from a starting base of 1000 peoples in 2000, is shown in curve C. This growth rate, though vigorous, would not have a significant influence on the annual world population increase in the 21st century.

Demographic pressure can be alleviated by family planning and by a superios industrial infrastructure provided by space industrialization. Indeed, contribution to the stabilization of the terrestrial population through completion of the global industrialization and by providing lasting resource foundations for a (global) industrial civilization, is the key purpose of exoindustrialization, as indicated in Fig. 4.

Fig. 4 expresses the proposition that the advent of extraterrestrial communities will primarily be the result of the new exoindustrial capabilities, coupled with the intrinsic human desire to develop new social

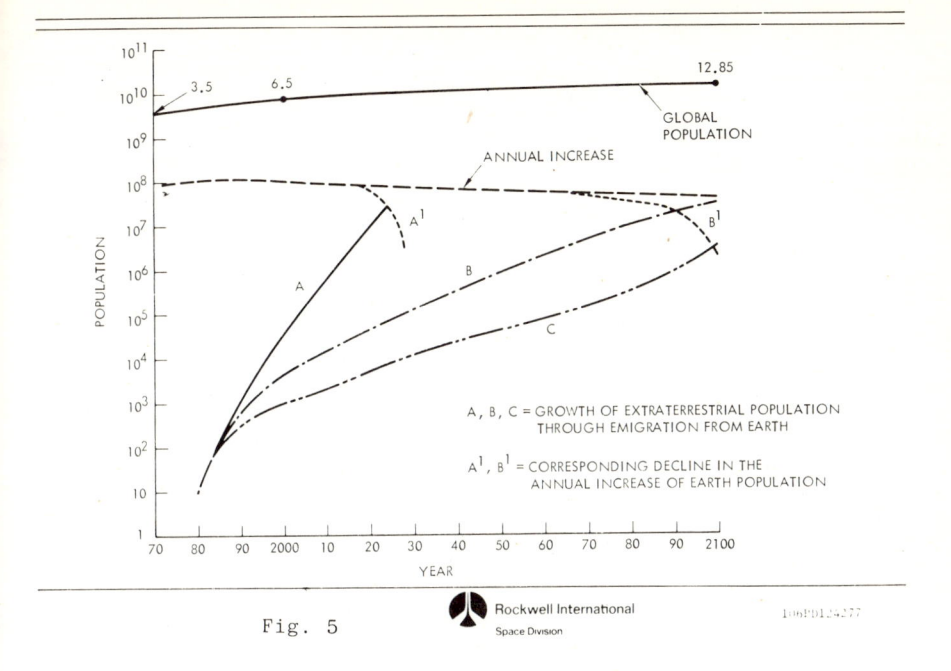

EFFECT OF DIFFERENT RATES OF POPULATION
REMOVAL FROM EARTH ON THE ANNUAL INCREASE OF GLOBAL POPULATION

Fig. 5

and cultural choices in the wake of these new capabilites (4). It appears logically profoundly inconsistent and therefore most unlikely that any civilization could possibly advance scientifically and technologically to the point where it could build whole new worlds on other surfaces and in orbits without having achieved civilized forms of population stabilization. It is not impossible, of course, but it seems to be about as probable as to assume that a civilization has developed TV but never heard of electric light.

In accordance with the primary purposes and the impact on the consequence scenario of human affairs, as proposed in Fig. 4, the expansion process and major achievement plateaus associated with each of the three phases are presented in Fig. 6.

Exoindustry will be born in geolunar space. All industrial operations beyond the Earth-Moon system will largely be applications and extensions of the foundations acquired in geolunar space. Through exoindustrialization a number of basic new capabilities will be developed and become an integral part of the consequence scenario. These include low-cost space transportation, permanent human habitats beyond Earth (Space Stations) and the ability to construct large facilities in orbits and on the Moon for the purpose of processing energy and materials as well as building new habitats. The new material resource base acquired in this phase will be the Moon.

Exoindustrialization will impart new and powerful stimuli as well as practical possibilities for space research and exploration which will benefit exo-urbanization and which will be indispensible for extraterrestrialization.

The foundations of exo-urbanization will also be laid in geolunar space. Exo-urbanization comprises the human advance in the wake of the industrial advance into space. Its principal achievements, therefore, are human-oriented and include space tourism, the utilization of space for therapeutic purposes, the construction of large habitats with urban character and comfort (Astropolis and the "Moon city" Selenopolis) and the development of relatively large-scale food production through "space plantations". The transport of a broader spectrum of people, especially of the sick in need of space therapies, and of permanently disabled and handicapped persons who would find a lower g-environment more desirable, requires special transportation systems. These systems are designed to lower the g-load during ascent below that of the Space Shuttle. For this purpose, either airbreathers or gas core reactor drives must be provided.

Exoindustrialization and exo-urbanization raise the level of achievements to where the advance to extraterrestrialization becomes feasible. The theatre of action is broadened now beyond lunar space, although lunar space will be its "womb". This phase will see surface colonies and orbiting colonies (Androcells). Mobile Androcells built with lunar material, equipped with terrestrial life and provided with propulsion will navigate into the solar system. Androcells are characterized by complete local resource autarky. For this they require fusion as energy base and, with it, the ability to carry artificial suns (Helioids) along into the outer solar system for the purpose of powering illumination, laser tools, material processing, propulsion and even for illuminating and temperature changes on other worlds in the asteroid belt and beyond. To the concept of reconstructing the solar system by transposing more planetary matter into the Sun's ecosphere (especially the Earth's orbit) as envisioned by Tsiolkovski, there is the alternative -- which could not be envisioned by Tsiolkovski -- of carrying the nuclear fire into the outer solar system along with Androcells. Redistributing energy is far more cost-effective, in the sense of achieving more for a given energy expenditure, than is the redistribution of matter. Therefore, the former is at least likely to precede the latter on a large scale.

EFFECT OF DIFFERENT RATES OF POPULATION
REMOVAL FROM EARTH ON THE ANNUAL INCREASE OF GLOBAL POPULATION

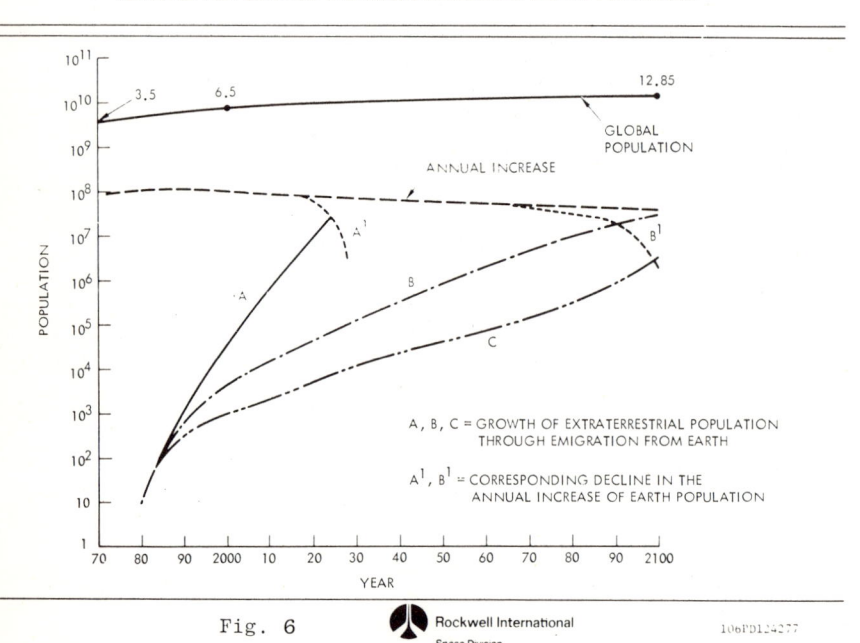

A, B, C = GROWTH OF EXTRATERRESTRIAL POPULATION THROUGH EMIGRATION FROM EARTH

A^1, B^1 = CORRESPONDING DECLINE IN THE ANNUAL INCREASE OF EARTH POPULATION

Fig. 6   Rockwell International
Space Division

The New Growth. Fig. 7 illustrates the New Growth process into the Open World. This buildup of orbiting industries from Earth (right) will provide the first benefits. To begin with, these benefits will include greatly expanded information services; material processing/manufacturing, producing important vaccines and such products as large silicon crystals and reflectors for the night illumination of remote industrial activities (e.g. above the arctic circle at winter), or metropolitan areas in industrialized as well as developed countries, and other purposes.

THE NEW GROWTH

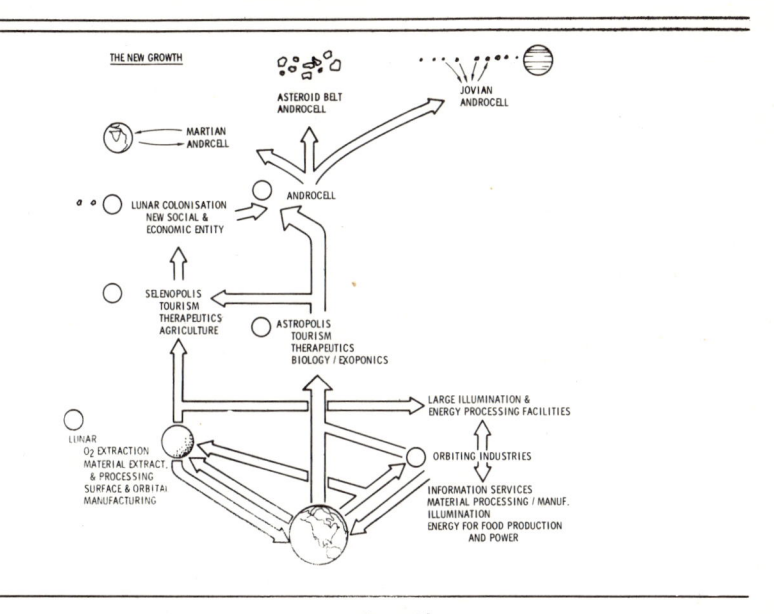

Fig. 7 The New Growth

Orbiting industries will support the buildup of an initial lunar capability which then becomes self-growing. Oxygen is an attractive initial product because it greatly reduces the transportation requirements from Earth to orbit to meet the oxygen demand for interorbital transportation between near-Earth orbit, geosynchronous orbit and circumlunar orbit.

The reduction in transportation requirement has economic advantages since some 20 tons can be lifted from the lunar surface for the same energy expenditure as that required to lift one ton from Earth to low orbit. There are also environmental benefits. For example, by delivering 2000 tons of lunar oxygen to a depot in geosynchronous orbit, 5000 tons of facilities can be delivered from near-Earth orbit to geosynchronous orbit without that the delivery vehicles need to carry along the oxygen for their return flight to near-Earth orbit. They just carry the payload and the hydrogen for the return flight. The associated oxygen, which amounts to about 5.5 times the hydrogen mass, is tanked in geosynchronous orbit. Thereby alone, the dumping of about 90,000 tons of O_2/H_2 combustion gases into the atmosphere is avoided.

Methods for cost-effective and large-scale extraction of lunar oxygen are described elsewhere (5,6). The resulting utilization of lunar oxygen facilitates the construction of very large facilities for illumination and energy processing in geosychronous orbit. Subsequent extraction of other materials contributes to the supply of structural material from Moon rather than Earth and to the delivery of products to Earth.

Orbital and lunar industrial capabilities form the basis for exo-urbanization -- a space city in near-Earth orbit (Astropolis) and a Moon city (Selenopolis). These exo-urban facilities add to industrial capabilities the human aspects for touristic, therapeutic and agricultural (exoponic) uses of space. While Astropolis would be constructed largely from Earth, Selenopolis would be built with lunar materials and supplied from Earth only with tools, equipment and biological materials not yet producible on the Moon.

Selenopolis represents the transition from the early industrial settlements to a more comprehensive colonization of the Moon. Lunar colonization offers an ideal testing and proving ground for rationally and effectively exploring and developing all aspects involving the building of whole new worlds. These aspects range from complex questions of establishing and balancing new biological ecosystems (not just recyclable life support systems), and the intricacies of human socio-psychological group behavior under extraterrestrial conditions to the vast technological and managerial infrastructure required -- material management, sanitation, fire protection, closed environmental purity control, microclimate control and emergency provisions associated with failures anywhere in the large array of integrated and interdependent processes. It is unrealistic in the face of past practical experience to believe such worlds can be built quickly and operated with "off the shelf" equipment. Out there Man has to go it alone without the forgiving biospherical environment. Creation is a "new trade" and, as all previous trades, requires an intensive learning process. Failures will occur. The Moon is more suitable for experimentation and comparatively more "forgiving" than orbital space.

Lunar colonization necessarily will be accompanied by a significant expansion of lunar industrial capability. From this level and with terrestrial support using Astropolis as its outer base, the construction of the first free flying world, Androcell, can be undertaken.

A detailed discussion of the placement of this Androcell exceeds the frame of this presentation. Three major options are available: near-Moon orbit, near-stable libration points L_4 or L_5 in the Earth-Moon system, and stable libration points L_4 or L_5 in the Earth-Moon system, and stable libration points L_4 or L_5 in the Sun-Earth system. Near-Moon orbit (well within the stable orbit limits) is most convenient from the standpoint of construction with lunar materials and the utilization of existing lunar surface and orbiting factories. Terrestrial and solar perturbances must be controlled out to some extent. Periodic passage through the lunar shadow does not represent as much, if any, of the problem as caused by the associated temperature gradient in the structure of a large unmanned facility. Practically all parts of the inhabited Androcell are heated internally. Experience gained with inhabitable lunar surface installations and with the Earth-circling Astropolis will be a guide in designing the Androcell.

Location at geolunar L_4 or L_5 offers easier (i.e. lower-energy) access from Earth without significantly increasing the access energy from Moon. Because of solar perturbation, the theoretical 3-body (Earth Moon, Androcell) stability of the Androcell is not met completely. The solar gravity influence must be navigated out at a velocity expenditure of some 1300 m/s per year. At continuous propulsion yielding a velocity increment of 3.6 m/s per day, a thrust acceleration of $4.25 \cdot 10^{-6}$ g is required, consuming 22.33 tons of propellant per year per 1000 tons of Androcell mass if the specific impulse is 6000 sec.

Stability at the geosolar equilateral libration points is much higher, because of the weaker perturbative influence of the outer planets and the great distance of the nearest stars. However, access from Earth and Moon requires more energy, depending on transfer time which, in any case, is much longer than within the geolunar system, although not prohibitively so (1 to 2 months).

In both cases of libration point placement, the Androcell is exposed to permanent sunshine. But individual parts nevertheless pass through shadow periods, because of rotation. External temperature gradients are unavoidable in all three cases, but their effect is not particularly significant for the reasons given above.

With the advent of the mobile Androcell, other parts of the solar system are opened to larger population groups. More about that later in this presentation.

<u>Social and Psychological Aspects.</u> With the transition from exoindustrialization to exo-urbanization, the extraterrestrial population will begin to grow at an accelerated rate. In the new environments beyond, there will take place not only physiological adjustments of the body chemistry to different gravity levels but also new psychological adjustments to the new conditions to which group behavior will have to conform. A new branch of psychology, exopsychology, and of sociology, exosociology, will evolve.

Essentially unimpeded freedom in a vast environment (in relation to the individual) since the dawn of existence has bred certain basic tendencies, instincts, mental desires, ways of life and comfort criteria into the human species. The human can be domesticated to constrained conditions, given time. But the transition to life beyond Earth is very profound. Nobody can predict at this time what the ultimate outcome of the adjustments will be in what may be the greatest socio-psychological challenge faced in human history so far. Glib optimism appears as unjustified as dire pessimism which virtually always turns out to be wrong. Therefore a stepwise progression from exo-urbanization to lunar colonization and to free flying Androcells appears the logical as well as practical course to take. Nevertheless, in projecting human individual and group behavior into these conditions we are not entirely without clues. In the following I will address myself to some of those which to me appear to be of particular significance to arrive at an understanding of how to best integrate human nature and the design as well as the mode of operation of Androcell.

The industrial civilization on Earth has led to a degree of urbanization without historical parallel. Mankind is still in the process of learning to shape cities and develop an acceptable, lasting urban lifestyle. People who for the first time move from country to city tend to suffer from a "life-in-a-box" syndrome, from confusion and from physical as well as cultural disorientation. New norms, habits, customs and traditions must be acquired. The environment cannot be mastered psychologically if it is too large. Neighborhoods develop -- modules of the city, cells in the body of the city at large -- or suburbs. Suburbs in particular are the combined result of modularization and of centrifugal forces which, in the terrestrial environment, can play themselves out.

In an orbiting community, the degree of containment is far stronger. It is virtually absolute if one considers that moving from one enclosure into another newly built enclosure does not alter the basic situation at all. It offers no relief for centrifugal pressures.

In 5 million years of evolution the human being has not advanced much beyond the point of seeking, and finding, emotional safety and security in groups generally not exceeding 10 persons. The structure of this group does not, as a rule, change much over the years. In such groups earlier Man shared caves or other abodes, went hunting and went to war. Living in close contact with numbers exceeding 10 at most, results in a keen sense of loss of privacy.

The human, like every higher animal, is extremely sensitive to a violation of its body zones. We surround ourselves with an intimate protective shield at about 20 cm distance, roughly twice the length of our middle finger. Short of brief periods in a crowded subway, this invisible shield can be penetrated only by violence or close personal relationships. A person also needs protection in the form of spatial privacy. Hospitals, prisons, garrisons, training camps and related open groups present notorious problems rooted in human nature because such installations deny people the opportunity to be alone. Humans are not ants or bees.

This suggests that an open, undifferentiated commune in an orbiting enclosure, even if large, is not a viable social structure. An amorphous population would break down into smaller groups. Territorial drives would assert themselves in the absence of a system design philosophy that provides enclosures and features that are agreeably behavior-channeling. Even if conditions in the enclosure are quite even -- i.e. if there are no particularly desirable and particularly undesirable regions -- the fact that minor differences cannot be avoided are prone to lead to contests or confrontations in an open environment. Airlines know this and provide seat selection in a controlled manner, although the difference between "best" and "worst" seats is very small and exposure time to conditions is but brief. To appreciate seat selection one has only to be exposed to the irrational pushing and shoving of crowds at the admission gate to an airplane at European airports where seat assignment is waived, <u>because of the brevity of most flights</u> at the short distances between terminals.

Large groups can be stabilized only by smaller groups. It appears logical to expect that an Androcell (and, before it, an Astropolis) design will have to take this fact into account rather than to assume that Man in space will be miraculously changed into some behaviorist's ideal. This means that socio-psychological design criteria must be taken into consideration along with physics and technology. This may not result in minimum mass or other technological optima; but these have little meaning if they are attained at the price of violating basic needs of human nature. The design should strive to minimize inducements of contests or confrontations or their alternative in the routine of everyday life. It should provide "neighborhoods" as well as "common ground" and privacy as well as opportunities for exposure.

The alternative to contests and confrontation is inactivity or some kind of aimless, bleached and atrophied communal lifestyle. If supplies and amenities are assured -- and they must be assured in a space enclosure, because of the much higher survival sensitivity of the community than on Earth -- other incentives and stimulations become all-important. This also follows from the fact that such a community is far more sensitive than a comparable terrestrial community asocial behavior, outright crime or even unemployment. In absence of incentives and stimulations either these asocial trends will assert themselves or desertification of the mind, dullness will result. In either case loss of individuality and eventually loss of socio-political freedom within the community will result. Attempts or limitations toward social and societal experimentation and evolution may then be abandoned.

From these considerations certain design criteria can be derived for the Androcell. It is further possible to state the following conclusions:
1. Strong purposes and objectives must be developed for Astropolis and, even more, for Androcell to develop a stable social structure.
2. Not only for material reasons but for socio-psychological reasons, Androcell needs a planetary or a large Moon or asteroid base. This base provides a sense of freedom and privacy (in addition to resources) for its human population that no artificial enclosure of foreseeable size can offer.

<u>Astropolis</u>. Astropolis symbolizes the exo-urban state of orbital architecture. It no longer is a space station or a construction base, but it is not yet an Androcell. However, its design features anticipate the likely design characteristics of Androcell -- neighborhoods/suburbs and common space; privacy and communal exposure. It is no accident that the large modern resort hotel and resort ships served as basic models for Astropolis which is neither an aircraft carrier, an amusement park, nor a factory, although it contains manufacturing facilities. Resort ships and hotels have developed to a high level the art of offering a community both privacy and pleasant public surroundings within limited space. Astropolis goes beyond the tourist aspects by incorporating a broader spectrum of human activities into this style -- hospital and research facilities as well as manufacturing and other space service capabilities.

Figures 8 and 9 show Astropolis designed according to modular principles (7). The particular configuration shown here has four Wings, attached in cruciform arrangement to a central axis of rotation. Each wing consists of three sections: a Residential Seciton (6 cylindrical Domiciles arranged radially), a Research Section (an OWE) and a Production Section, including the exoponic installations, that is, the non-radially arranged bio-ecological cylinders indicated on the radial tubes of two of the four wings.

Each Residential Section consists of radially oriented multi-story cylindrical modules attached to a large cylinder running parallel to the axis of rotation. The long cylinder provides a public "open space" with Coriolis force-free gravity for the residents. The Residence Modules offer apartments and privacy in the manner of high-rise apartment buildings on Earth. Each Residential Section is provided with a sausage-shaped multi-storage business and supply center. Each Residential Module contains annular floors, divided into apartments and tied to a central duct serving as ingress and egress path and an emergency shelter in case of major damage to one or several apartments. Each apartment has at least two access doors to the central duct for safety reasons. The central duct contains all supply lines (water, air, food, power, communication, emergency oxygen) and the waste lines leading out of the apartments. Being fully pressurized at the same pressure level as the apartments, the central duct constitutes a much larger air volume than does each apart-

L   = Residential Section
D   = Dynarium
OWE   = Other World Enclosures
PS-EI = Production Section -
        Exoponic Installations
S   = Sewage Treatment
H   = Hydroponic Farms
A   = Animal Farms
PS-II = Production Section -
        Industrial Installations
C   = Spin Control Masses
V   = Vernier Spin/Module
        Adjustment Masses
NRDB = Nonrotating Docking and
        Bearthing Facility
NDRB/PLC = NRDB for People
           and light Cargo
NDRB/IEC + NDRB for Industrial
           and Exoponic Cargo

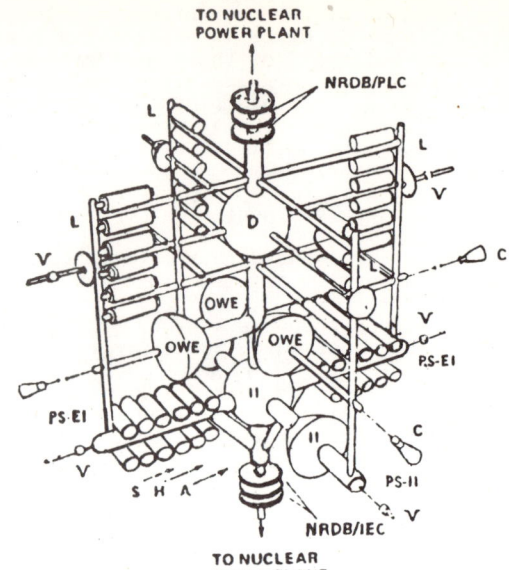

ANDROCELL (Fig. 8)

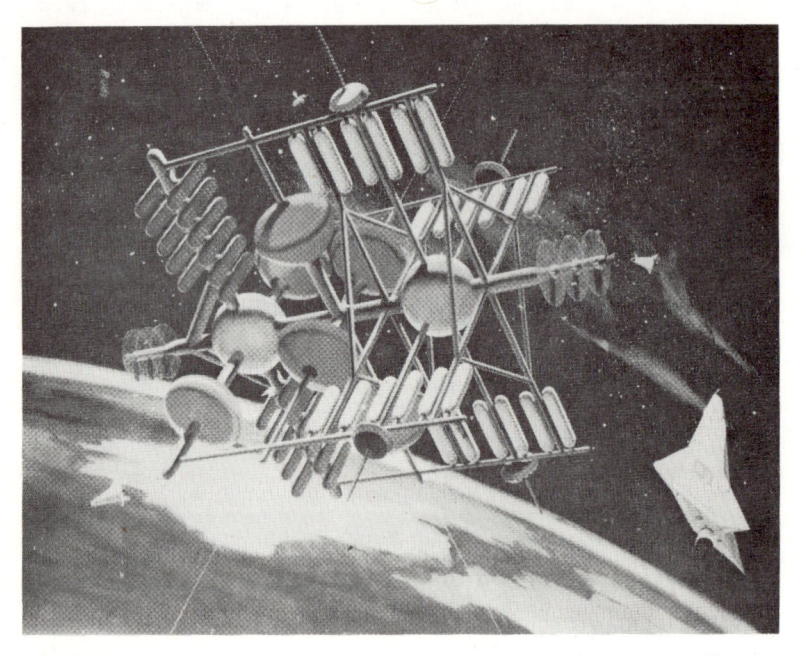

Rockwell International
Space Division

ASTROPOLIS (Fig. 9)

Rockwell International
Space Division

388

ment individually. In case of a major leak in an apartment, sudden decompression is averted, or at least greatly ameliorated by the automatic opening of emergency air vents connecting the stricken apartment to the air supply in the central duct. All central ducts are connected to the extensive system of pressurized radial and axiparallel ducts that constitute the structural framework of Astropolis. Thus, each central duct, in turn, is backed up by the vast air volume of the space city's network of tubular connectors. At the intersection of the radial access tubes to the four Residential Sections with the hub is a large spherical enclosure. Inside this outer sphere is a slightly smaller inner sphere which does not necessarily participate in the rotation of Astropolis. The sphere forms a large zero-g volume designated Dynarium. The residents of Astropolis use the Dynarium by tumbling and floating in its artificial air currents (Fig. 10).

Fig. 9. Astropolis - Conceptual Arrangement

Fig. 10. The Dynarium of Astropolis

The Research Section is organized for extensive and long-term utilization of the orbital environment for basic and applied research. On Earth, laboratories may simulate many environments, but they cannot simulate the correct combination of gravity, vacuum, temperature and radiation environment on the surface of bodies like our Moon, Mars, Mercury, or the surface of the Martian moons or the asteroids. This becomes possible on Astropolis.

Astropolis rotates very slowly. Because of the low angular velocity, the Coriolis force causes little disturbance, even at lower gravity levels closer to the hub. Therefore, the comfort range (at least in terms of Coriolis disturbances) on Astropolis is considerably wider than the comfort range on smaller, more rapidly spinning space stations. The

Residential Section most likely will be located in the Venusian to Martian gravity Range (Fig. 11).

Fig. 11 shows the variation of the g-level in a uniformly rotating system. For a radius at whose outer end the gravity level is one-g, the g-level corresponds to the fraction of the distance from the center line of the hub to the one-g level. For a spin rate of about 925 revolutions per 24-hour day, the radial distance to the one-g level is about 1200 ft. (366 m). The distance of a given g-level equals the product of g-level times radius (e.g. the radial distance to 0.4 g is 480 ft. (146.4m). Even the g-level of Callisto is still 100 ft. (30 m) from the axis of revolution, or about 15 m above the outer surface of the hub. It is, therefore, possible to provide relatively sizeable enclosures at various distances from the hub, in which the gravity environment of all significant sub-terrestrial worlds in this solar system can be simulated. Hospitals and research laboratories can be placed at desired g-levels, including null-g in the hub, most of which does not participate in the rotation of the Wings.

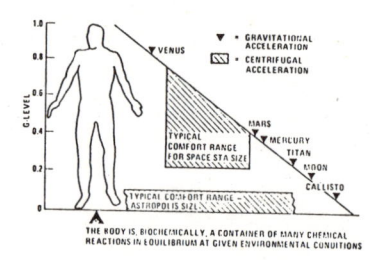

Fig. 11 G-Environments on Astropolis

The industrial installations of the Production Sections contain enclosures which house manufacturing facilities in a low-g and zero-g environment. The exoponic installations are the heart of the closed bio-ecological system of Astropolis. Gases, liquids and solid organic materials are fully recycled. This means that the organic waste from the residential zones must be fed into sewage tanks where bacterial breakdown reconditions the raw waste so that it become suitable as fertilizer for hydroponic farms -- plant farms that use nutrients resolved in water rather than in soil. Some of the plants raised on that basis are fed to animals, others will wind up on the human dinner table. The animal farms are located at the lowest gravity level to utilize the experimentally already established fact that in a low-g environment bone growth is minimized in favor of meat production. The waste of living and processed animals is also recycled.

An important advantage of modular design is the feasibility of replacing individual modules -- produced on Astropolis or elsewhere -- without discontinuing the rotation of the space city. Fig. 12 illustrates the method -- which is also assumed to be applied in the buildup phase (8). The hub is divided into several sections. The non-shaded sections, to which the radial tubes are attached, are part of the rotating system. In between them and at the far ends of the hub (crosshatched sections) lie sections which do not rotate (zero-g sections). Between the two outermost zero-g sections and the non-shaded rotating sections lie two (single-hatched) sections which can be spun-up to rotational speed and coupled to the rotating sections, or which can be decoupled and kept, non-rotating, in zero-g state. We will refer to these as variable-spin sections. A new module is placed first on the outermost zero-g section, then transferred to the variable-spin section in a zero-g condition. Once the module is properly fastened, the variable-spin section is slowly spun up until it is synchronized with and coupled to the rotating section. Thereafter the module is moved outward along the radial tube to its position at any radial distance from the hub cylinder.

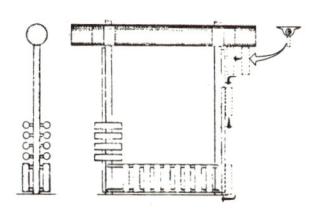

Fig. 12 Astropolis -- In-Rotation Module Replacement

Androcell. Man-made planetellas using centrifugal inertia in lieu of gravity are vastly more mass-effective than those formed of original stellar material, where only the surface layer is useful while an unproportionately large amount of mass contained in the interior is needed to provide gravity. In return for this, the entire surface of a sphere can be used.

When rotation replaces gravity, the "useful" mass of a planetella can be reduced to far less than one percent of the mass needed to generate the same weight gravitationally. But the usable surface is reduced also, namely to a cylindrical region surrounding the axis of rotation. In return for this limitation, however, the no longer mass-filled interior is accessible. This, of course, means that in the Man-made planetella all "gravity" levels below the maximum outer level are available, all the way to zero in the hub. In this modular design, installations can readily be placed at different weight levels.

Thus, the Androcell configuration may be envisioned as an extension of the modular Astropolis design. Instead of two rectangular units in cruciform arrangement as shown previously, many more rectangular units can be mounted onto a large axis, making the facility capable of housing a population of tens of thousands in a completely closed, self-sufficient eco-system. One such wing is shown in Fig. 13.

As with Astropolis, the design is chosen to maximize the productive utilization of different gravity levels by placing installations at various distances from the hub, as best suited, from the outer residential sections down to null-gravity within the hub. This goes for the residential, the industrial and the exoponic installations.

ANDROCELL WING

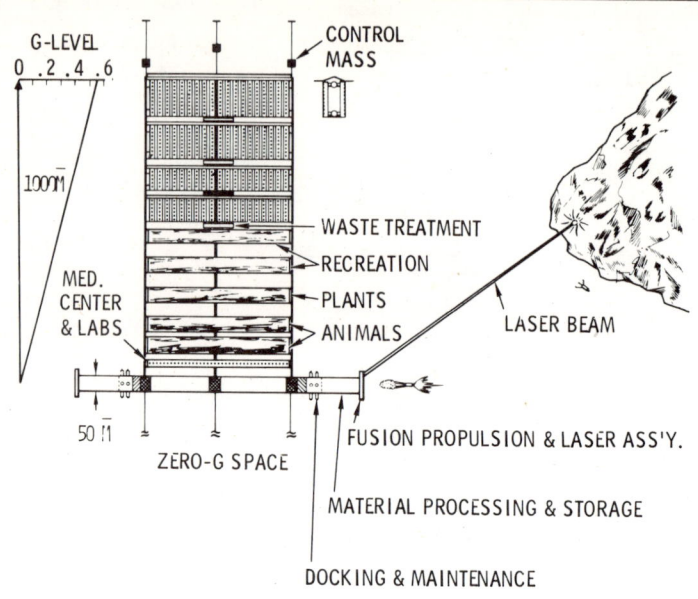

Fig. 13 Androcell Wing

The modules in the residential section which extends from 0.5 to 0.3 g are exchangeable by the same technique as illustrated in Fig. 12. The other enclosures are all cylindrical, in order to achieve greater standardization and maximum utilization of the low-g regions in each wing. The main recreational spaces are at the 0.29 and 0.23 g-level, but there are also zero-g Dynarium sections in the non-rotating part of the hub. These are available to all wings. The recreational cylinders also contain therapeutical low-g installations. The main medical center and certain research labs for each wing are located in a one-story, but sub-divided, axiparallel cylinder at the 0.05-g level. As in Astropolis, waste treatment facilities are at higher g-levels, with plantations and animal housing at progressively lower gravity levels (0.2 g and 0.13/0.1 g-range, respectively).

Near the entrance on both sides of the large hub cylinder are the main material pre-processing, processing and storage facilities (there are also smaller manufacturing facilities at other g-levels). The main facilities act as the "mouth" and "stomach" of the Androcell. Raw materials from the surface of celestial bodies can be delivered "as is", then crushed and sorted. Specific minerals and elements are extracted and stored separately for later manufacturing purposes. All parts and components of Androcell can be manufactured -- i.e. Androcell has complete regenerative capability.

The power source is fusion, providing heat, electric power, laser power and propulsive power. Laser beams are used to analyze surface materials from great distance and for loosening materials at the surface, where necessary, prior to the arrival of landing parties with cargo shuttles. If Androcell is in fixed orbit, unprocessed raw material with its indigenous element distribution may not be acceptable, unless all elements can be used, because waste cannot be released. In this case preprocessing will take place on the surface with laser beams from Androcell transmitting the necessary power. Fusion power generation, laser assemblies and fusion propulsion are located at both ends of the hub cylinder.

Androcell is designed not only as a modular, multi-g world structure but also for acceleration to more or less high speeds. The large hub cylinder provides a suitable thrust structure for propulsion systems mounted on both ends. Since the thrust acceleration is very low, the centrifugal acceleration in the rotating wings will be tilted only slightly from its radial direction. Even at a relatively high acceleration of 10^{-4} g, the deviation from the radially oriented "vertical" is only 0.06 degrees at the 0.1 g level and correspondingly less at the higher g-levels. Obviously, null-g conditions are not available while Androcell is under propulsion. However, acceleration levels of $5 \cdot 10^{-5}$ to 10^{-4} g are sufficiently close to null-g for most purposes.

A mobile Androcell will seek other resource bases beyond Earth-Moon system. It may move from one world to another, especially in the asteroid belt. Its people will have the surface of another celestial body to explore, on which to experiment and to enjoy greater freedom from confinement. Returning from an extended stay on Mars or on a Jupiter moon enhances the predeliction for perceiving Androcell as "home" world.

It is possible to preserve individuality and individual freedom and dignity of the people in Androcell, and yet to understand Androcell as a benign macro-life form interacting with the primordial cosmic environment through which it moves by means of two metabolic processes -- information metabolism whose nucleus is the human brain, and industrial metabolism powered by primordial energy that is controlled and processed by information metabolism. This is illustrated symbolically in Fig. 14. The experiences, the tasks and the achievement for this macro-life form and extraterrestrial Man in it are virtually limitless in the polarity of endless horizons on other celestial bodies and the security and comfort of Androcell and the mind power enhancing qualities of its computer and sensor systems.

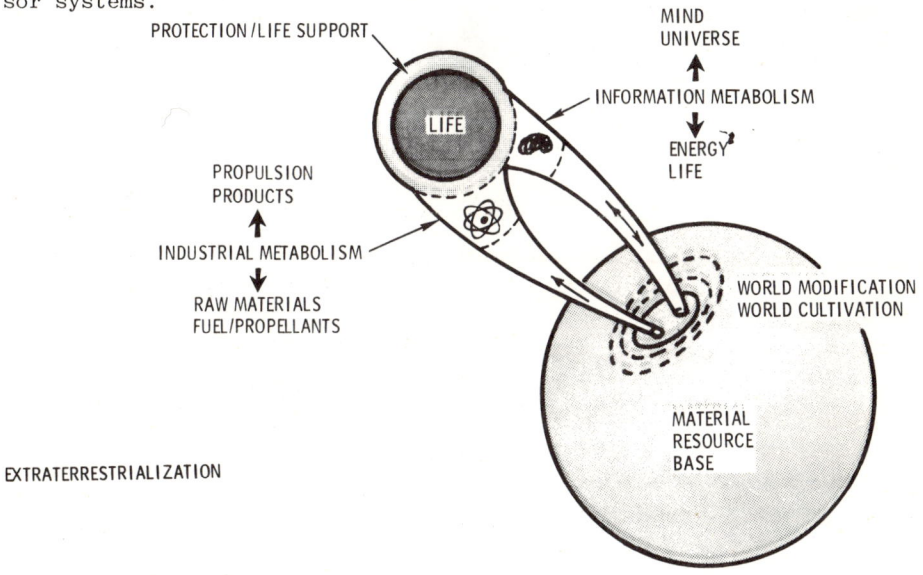

EXTRATERRESTRIALIZATION

Fig. 15 depicts an Androcell that has penetrated into the Jovian satellite system. An Androcell-attached Helioid system provides the natural sunlight furnished in the inner solar system. Fusion-powered sunlight-spectrum beams are emitted from a source near both ends of the hub cylinder to a suitably convex reflector for diffuse reflection. The Helioid system rotates with respect to the rotating wings at slightly different angular velocity. The difference is determined by the desired circadian rythm. For a centrifugal acceleration of 0.5 g (4.87 m/s^2) at the outer axiparallel cylinder of the residential section and a wing radius of 1000 m, Androcell completes 960 revolutions per day (40 rev/hr or 1.5 min/rev) at an angular velocity of 4 deg/sec. If the desired circadian rythm is 24 hours, the Helioids complete 959 revolutions per day.

ANDROCELL

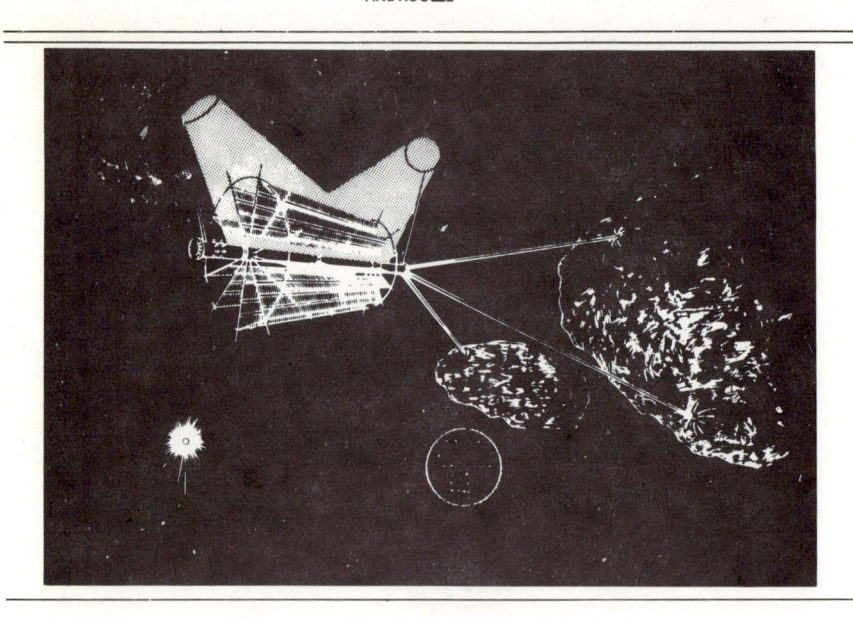

Epilogue. Thus, as before when an Open World was created on this planet by the biosphere, the genesis of a new Open World proceeds on the basis of information, controls, materials and energy, only this time around a higher level of intensity and cosmic mobility (Fig. 16).

OPEN WORLD - RESOURCE BACE

PHASE AND MAIN SYSTEM	MATERIAL RESOURCE BASE	ENERGY BASE
A SPACE INDUSTRIALIZATION		
A-1 EARTH ORBITING INDUSTRIES	EARTH/(MOON)	SOLAR
A-2 LUNAR INDUSTRIES	MOON/EARTH	NUCLEAR/SOLAR
B SPACE URBANIZATION		
B-1 ORBITING TOURIST FACILITY	EARTH	SOLAR
B-2 SPACE HOSPITAL	EARTH	SOLAR
B-3 ASTROPOLIS	EARTH/MOON	NUCLEAR
B-4 SELENOPOLIS	MOON/EARTH	NUCLEAR/SOLAR
C EXTRATERRESTRIALIZATION		
C-1 LUNAR COLONIZATION	MOON/EARTH	NUCLEAR/SOLAR
C-2 GEOLUNAR ANDROCELL	MOON/EARTH	NUCLEAR/(FUSION)
C-3 MARTIAN ANDROCELL	MARS/PHOBOS/DEIMOS	FUSION
C-4 ASTEROID BELT ANDROCELL	ASTEROIDS	FUSION
C-5 JOVIAN ANDROCELL	JUPITER MOONS	FUSION

ip. 16

107PD123563

<u>Conclusions</u>. In conclusion I summarize the main points of my presentation.

Evolutionary logic suggests three phases in Open World development. These phases overlap and coexist but they are nevertheless clearly identifiable as exoindustrialization, exo-urbanization and extraterrestrialization. They are identifiable by their purposes and by the problems that must be solved to render their realization possible. These purposes also justify the evolutionary logic.

We live in one of the rare times when fundamental evolutionary processes become more transparent than usual, because we become consciously aware of our own role in the evolution of life as an active agent, a cutting edge in the continuation of evolution -- from inorganic to organic to bio-metabolic and now to the information metabolic level. Evolution is growth through response to change -- not any response but consistently one that leads to progressively more powerful metabolic processes in terms of complexity, data processing, energy concentrations, material handling and mobility -- in other words, in the orchestration of resources and in cognitive assets.

Open World evolution turns on the human, i.e. information metabolic life form. Manned space flight will move ahead beyond the Shuttle level.

Space industrialization allows Man to progress in steps -- near Earth, then farther out; shorter periods, then longer ones.

Prolonged exposure to weightlessness leads to biofeedback processes by which the organism adapts itself to the new conditions.

Prolonged exposure to large-group existence beyond Earth will stimulate socio-feedback resulting in the emergence of social changes and new cultural traits, which is a fascinating process in its own right. At this point we do not know whether major social and cultural changes in any Androcell can occur in one generation or whether several generations will be required. But we shall see .....

References

1. K.A. Ehricke, The Extraterrestrial Imperative, Bulletin of the American Scientists, Nov. 1971

2. K.A. Ehricke, Estraterrestrial Industry, A Challenge to Growth Limitations, the Essential Resources Conference, The Conference Board, Washington, D.C., April 1973.

3. K.A. Ehricke and E.A. Miller, The Extraterrestrial Imperative, to be published by Mechanical Engineering, Publ., U.S.S.R.

4. K.A. Ehricke, A Long-Range Perspective and Some Fundamental Aspects of Interstellar Evolution, J. Brit. Interpl. Soc., vol 28, pp 713-734, 1975.

5. K.A. Ehricke, Lunar Industries and Their Value For The Human Environment on Earth, 23rd International Astronautical Congress, Vienna, Austria, Oct. 1972; Acta Astronautica, vol 1., pp. 585-682, Pergamon Press, 1974.

6. K.A. Ehricke, Industrial Productivity As a New Overarching Goal of Space Development, in publ., Brit. Interpl. Society.

7. K.A. Ehricke, Space Tourism, 13th Annual Meeting of the Am. Astronautical Soc., Dallas, Texas, May, 1967; Commercial Utilization of Space, vol. 23, Advances in Astronautical Sciences, 1968.

CONTRIBUTIONS OF SPACEFLIGHT AND SPACE STATIONS TO THE
SOLUTION OF SOCIAL AND ENVIRONMENTAL PROBLEMS

Dr. Leonard Jaffe*

I was asked to sort of key-note the question of social impact of Space Stations and of the new era of space transportation. I'd to do this a little differently than I would have normally done it. Normally, I show a number of slides and describe the various applications that are possible, that we visualize that they are possible, but I'm sure that you're going to hear a great deal of this during the course of this afternoon and you heard some of it this morning. I'm never really quite sure where to start with this question of how to bound the discussion of the contributions of space flight to our social and environmental needs. In the very larger sense, it's probably very important for man, human beings, to know that there are new frontiers, that he's not limited to the environment of Earth. There are many among us who, while assuming that we are constrained to the environment to Earth, have become purveyors of gloom and doom regarding our ability to further continue to avail ourselves of the resources that have made our lives what they are today, and to continue to improve our living standards here on Earth. But, fortunately, this near frontier of space has arrived and has given us a little bit of living room with which to expand our horizons. We're no longer constrained to think in terms of, "Boy, the last horizon is gone. We've now got to constrain our way of living." So, in the larger sense, I think that space and this new capability to go into space easily on a routine basis for you and I and perhaps the many of the students that are in the balcony right now, to go into space perhaps as easily as one gets on a 747 and goes around the world. This kind of recognition that we are no longer bound to mother Earth and that we have an unlimited horizon, a new frontier, I consider to be probably the most, the largest contribution of spaceflight to our social well-being here on Earth. But I'm sure you want to hear about something a bit more mundane than that, so I would like to talk about spaceflight and how it has contributed, and will continue to contribute to some of the down-to-Earth needs; needs that we have here on Earth.

It's in this area of space applications, as we call it, and the use of Space Stations to somehow provide for an Earthly use, or take care of an Earthly need, which I feel really more comfortable about talking. I needn't remind this group, that we've already realized over a decade of benefits from satellite systems that have kept track of our daily cloud cover and have assisted the world's weather forecasters, particularly, in the area of large storms such as cyclones, hurricanes. In the last half-dozen years, almost a decade, not a single hurricane or cyclone has gone undetected and not a single large scale storm has failed to be forecast to the human beings that they would later affect. This is an important contribution of space. I needn't remind you also that a new international industry has been established using satellites to tie the continents of the world together with highly reliable telephone and television communications. It should be noted that these services became operational in much less than a decade, both of them really about 5 or 6 years after the introduction of the first satellite, man-made satellite into orbit. This rapid introduction of new technology to operational service is really phenomenal in the history of mankind. For those of you who haven't followed how long it takes to introduce a new piece of technology into the market and into our homes, a minimum is ten

---

*Associate Administrator for Applications, National Aeronautics and Space Administration, Washington, D.C.

years and usually on the order of 15 to 20 years, before something that
is developed in the laboratory is realized in our daily lives. Here, we
have for the first time, I believe, introduced technology to the point
where human beings are making use of it in less than half a decade after
the first possibility of this introduction. This rapid introduction you
can take into account, or account for it, by two factors. One is that
the need existed. There was really no way to observe the global cloud
cover, and everyone recognized that meteorology is a global phenomena.
The weather in the United States is going to affect the weather in Europe
a few days from now. The weather in the Pacific is going to affect the
weather in the United States a few days from now. This was understood
so that the use of satellites was fairly obvious even to the doubters in
the early days. The use of communications to span the continents was
obvious. The need was there. We were limited in 1958 to a few telephone
channels worth of capability between the United States and Europe carried by a very marginal capacity undersea cable. The rest of the communications had to be carried out by radio waves which were highly unreliable.
So, the reason for the rapid growth and the rapid introduction, was the
fact that the need existed and we were expanding a service with which
the world was familiar. We were expanding a meteorological forecasting
service which existed. We were expanding a telephone service which existed but today we're talking about new applications of space, and new
services resulting from new data and new capabilities with which the
potential users are not familiar. I might digress for a second to tell
you a little story which indicates how sophisticated we really are in
talking about space but how naive perhaps the public is and the potential users of space even today. In 1960's when we went to Europe to talk
about communication satellites and to ask the Europe community to join
with us in some very early experiments to use satellites to transmit
telephone and television programs between the United States and Europe,
we had a very serious and very deliberate discussion over the subject,
a very lengthly discussion. People were very concerned about the investment that they had to make but they recognized the potential in some
intuitive way. We ended up signing an agreement and after we had signed
the agreement, in one of the countries, the minister of telecommunications had a little celebration that we had after the signing of the
agreement, gave me a little poke in the ribs and said, "All right, now
that we've signed this agreement, tell me very frankly, won't these
things fall out of the sky?" Now that kind of skepticism, that kind
of unfamiliarity with high technology, still exists today. We have
it to overcome with the new applications that are around the corner.

We will in the next few years develop the capability to measure in
quantitative terms the condition of the atmosphere anywhere and all over
the Earth. We will be able to measure the temperature, the wind velocity, the water content and other chemical constituents which make up or
pollute our atmosphere. What do we do with this new capability? A new
level of meteorological understandings must be created in order to take
advantage of these measurements. A better understanding of the physical
and chemical processes taking place in our atmosphere will eventually
lead to a better and longer term weather forecast and the ability to
protect us from the short lived storms such as tornadoes. But the science to achieve this is not here at the moment. Ten years ago we thought,
gee, all we need are the measurements and computers. Now we have the
measurements virtually around the corner, computers virtually around the
corner, however, the science is not here yet. The measurement capability
from space and the science of interpretation of meteorology must grow
hand-in-hand. They must pull each other up by the boot straps, so to
speak. It will take time. Those among us who are insistent upon in-

stant benefits today from each step in this difficult development will be disappointed. Is there any one among us who doubts the need or desireability of pursuing this capability? The visionaries of the world have joined together in a great undertaking called the "Global atmospheric Research Program" to cooperate in this effort. Space will play a key role. In the 1979 and 80' time period, almost all of the nations of the world have joined together in a grand and glorious experiment to make as many measurements as possible of our atmosphere for a period of a year and to these measurements that will be used to upgrade our theoretical understanding of the atmosphere and our ability to make mathematical models of the atmosphere. Only then will we be able to truly use the capability that we will have created in space to make better weather forecasts.

Satellites will be the main observational tool, as I have said, for this Global Atmospheric Research Program. In addition to just making measurements of the space and the environment that we are in, the Space Station that you've just seen a film on will provide us with a tremendous capability to do some experiments in space that will tell us more about how the atmosphere and the chemistry, the chemical constituents of the atmosphere affect weather, how they affect our environment. One of the things we don't really understand is how clouds are formed. The energy involved in clouds accounts for a major portion of our weather. In the physical laboratory, if you will, that you just saw a picture of - Spacelab, we will be able to carry out experiments in which single droplets can be suspended in the zero-gravity condition of space, and we can watch the formation of raindrops and determine what the energetic processes are.

Observations from space of the surface of the Earth on a routine and periodic basis are today beginning to be used for a lot of things - for updating maps, for monitoring our snow and ice cover, for keeping track of the extent and quality of our water and vegetation resources and to better define the geologic nature and structure of our planet. These applications are still in their infancy. Again, the data are of a new type. Very few people have had the opportunity to stand away from the Earth and observe the Earth rather than the trees so to speak. Its application and the institutions required to apply this data to produce information of real value to decision makers have yet to be developed. Until recently the concerns over maps, water, ice, snow and vegetation, etc. were of a very local interest. With the increased demand for resources caused by the increases in population and industrialization of the world, man must have an understanding of what is happening on a world-wide basis to fully understand his own environment and his own situation.

If our prospects for water or food indicate a short supply in one hemisphere, let's say the Northern hemisphere, or I'm experiencing a drought this year, isn't it important that I know what's going on in the Southern hemisphere where perhaps they can make up for any deficiencies that may be occurring in the North? The same can be reversed. Isn't it important that we know if there is a drought or a severe constraint in food production in the Southern hemisphere so that we can take remedial action? We believe that this can be done via satellites. The need for global observations and understanding of environmental trends using satellites will become more important with each passing year. Satellites observing our Earth and its environment can monitor our status and permit us to better understand the current situation as well as assist us in developing the capability to forecast our future.

Satellites exploring other planets in our solar system can help us relate those different environments - those which are different from ours. They, these other planets, may be experiencing an environment today which is related to our past, or perhaps to our future. In understanding how these planets have gotten to be what they are, as they are, we will learn to better understand what the Earth is and where it's going. Of course, the sun, which is the major source of energy for all of the Earth, must be monitored on a continued basis for any changes, even though minor, which can affect our Earthly climate. Is this all so far off? No. We can with our satellites today, such as our meteorological satellites and the LANDSAT satellites, observe all parts of the Earth. We've taken the first steps to monitoring its changing conditions. These satellites have been used by many nations to focus on their own resources' problems. In many of the developing areas of the world, they are producing the first reasonably accurate up-to-date maps. Canada, Brazil, Iran and India have already made committments to install ground receivers to read out the satellite data of their own region on their own soil, and we can expect many more areas of the world to avail themselves of this opportunity in the near future. The utility of these satellites in helping to assess the problems of flood, drought, of vegatation and stress continues to develop. Hopefully, soon we will be able to assess agricultural production, and hopefully optimize it so that the needs of the world population can be accommodated. We can expect this capability to grow with space laboratories where more sophisticated sensors will be developed to better and more efficiently provide the needed data and to permit observing the Earth through the now troublesome clouds which almost always cover at least 50% of our Earth. The latter, looking through the clouds, can be achieved using microwaves in both active radar techniques, or radar modes, and in passive instrumentation such as radiometers. When we have this capability, then we can truly observe the Earth on a continuous basis.

Soon also we will be able to measure the motions of our Earth's land masses, its continents and the plates which make up the continents. We will be able to measure the motions of these continents with centimeter precision. These measurements will assist in the understanding of our plastic Earth and the science of plate tektonics to the extent that earthquakes may one day be forecast with a degree of precision, which certainly today isn't available, and with a degree of precision that may, in fact, be found useful to our civilizations. The measurements will contribute to measurements of our ocean surface, its currents, its wave heights, so important to efficient shipping and so important to the optimization of off-shore structures which must be built in the hostile oceans if we are to continue to use our off-shore resources.

Let me return for a moment to an old application - communications. Have we done all that needs doing in this area? Absolutely not. To date communications satellites have been used largely to increase our capacity to handle familiar services, that is telephones and television. Communications satellites have mainly been used to provide inter-continental links, high capacity, of these conventional services. This is somewhat ironic because the real attributes of satellites are not in their ability to communicate between two places on the Earth's surface, but to communicate with the entire surface of the Earth, with entire regions, with many stations and with many small and cheap stations on the surface of the Earth.

New services, contributing to education, delivery of health care, teacher training, instant access to data bases such as those that are available in hospitals and libraries throughout the world, access to the best universities and the most current research procedures, will be available to anyone, anywhere on Earth. Provision of these services will require large high-power satellites in orbit which can focus their energy on local areas so that we minimize the cost of the many ground receivers that will be required. Space Stations will permit the development and construction of these very large satellites in orbit. I'm personally convinced that the greatest social contribution will result from such communication satellites of the future.

The just completed satellite educational experiment conducted in India during this last year using the Applications Technology Satellite (ATS) built by the United States is a forerunner in this area. I might just say I was interested in noting that in one of the newspapers here they referred to the Applications Technology Satellite as an atmospheric test satellite. It's nothing of the sort. It's a communications satellite - a very powerful one. We used this satellite in the United States for almost a year to provide communications with our sparsely populated regions, and we have them. We have sparsely populated regions of the country in Alaska. We have sparsely populated regions of our country in the Rocky Mountain regions and in the Appalachian regions. We have areas where the educational facilities are less than desirable. For a year we provided services on a test basis to the Alaskans and to educational facilities in portions of the Rocky Mountains and the Appalachian. We provided health care services to the Alaskans with a great deal of success. For the first time, they had a reliable means for reaching a doctor. Their only means for communicating with doctors in many portions of Alaska was high frequency radio which most of the time is unreliable. After a year's worth of experimentation, we shifted the satellite over to the area of the Indian continent and they used this satellite to test the utility of instructional television to deliver education to primary schools for teacher instruction and for information to over 5,000 villages on health care, hygiene, agriculture, and family planning. This was not merely a technical experiment nor an effort to see if India could mount an effort to build and maintain 5,000 television sets in a primitive environment, or to produce some 900 hours of programming. It was more than this. It was a program to determine how effective and how rapidly the much needed services could be supplied by satellites and how effective the social impact of these services in such an area is.

Satellites are unique in their ability to cover such large territories at once. One satellite can cover the entire sub-continent of India. It would take a decade, if not more, to cover the Indian continent with conventional ground communications. The world eagerly awaits the results of the great social experiment, and I might say that there will be a tremendous amount of data, thanks to the foresight of the people running this experiment. A great many sociologists have been collecting data in the villages, both prior to the onset of the experiment, during the experiment, and will continue to collect data for a period of 4 or 5 months after the conclusion of the experiment, which did conclude on August 1, 1976, to insure that they really have a valid set of data with which to weigh the effects. I look forward to the results of this experiment.

So far, we've considered only limited extensions of our current capability. If one considers that some day virtually unlimited communications can be coupled with an ability to observe the environment of the Earth with an ability to predict the weather, the winds, the sea state, and other conditions of our environment, when one couples all of this with the reliable future computational equipment - the computers - one can imagine whole new realms of space application. When I say this, I'm not so sure it's understood but let me give you an example which may bring it home. I'm not advocating what I'm going to use to demonstrate this point but let me throw it out anyway. We've all talked about expanded communications for telephone services but imagine if one had the ability to apply this plus weather forecasting, plus our ability to determine winds, sea state, etc., the conditions of our local environment, and we could communicate with unlimited capacity anywhere on Earth. What could I do about things if I looked differently? What could I do about transportation across the oceans? Well, it's rather interesting. We've been using sophisticated ships taking a lot of energy to transport lots of oil across the oceans. We've got an unlimited energy resource on the ocean called "wind." We've forgotten about sailing vessels. Sailing vessels have been put in the past because they're not fast because they have been overtaken by the energy-consuming ships that are modern, but who cares how long it takes to get a cargo of oil across the Atlantic? There's a continuous chain of ships going across the oceans at this particular time. All we care about, is that a ship lands at the desired port on a daily basis - not how long it took to get across. Suppose we could predict the winds with absolute accuracy over the entire oceans. Suppose we could provide automatic computational equipment which could control the ship, the sailing vessel, very accurately. Suppose we could observe the environment of that ship and its condition on a continuous basis. We could chart the course of that ship across the oceans automatically. We could predict where it should go. We could take advantage of the winds. We could know where the ship was instantaneously, governing its course from land in a very cost effective way provide for a transportation system unmanned, which does not now exist and which is not now possible because we don,t have this information. The point that I'm trying to make, and I'm not advocating automatized sailing vessels, but that we can begin to think about putting together all this tremendous capability for knowing precisely what's going on anywhere on Earth with a communications capability to transmit that information wherever we want to transmit it. With our ability to control automatically devices which are remote from our particular homes into a situation which will allow us to think much more different about how we use our environment and how we use our resources. Well, let me bring this discussion to a conclusion by discussing one final element of space which I think is going to be terribly important and terribly exciting.

We've been examining the utility of the so-called gravitational characteristics of space for the last couple of years. We've done a few experiments on the Spacelab, on the Apollo-Soyuz missions, on Skylab missions, and we have determined that, in fact, there is a value to manufacturing materials or causing materials to change form in the space environment, in the environment of zero-g. If we can remove gravity from the process of solidifying crystals or metals, we will be able to manufacture new and more uniform materials. When we can melt and solidify these materials while free floating in space unencumbered by a containment vessel, we will be able to manufacture purer materials. Almost anything you put a material in, a molten material in, will contribute some contaminate to the material itself. In zero-g, we can suspend the material in space without a container, melt it, allow it to solidify, untouched by anything. We've had some early successes in space already. Crystals have been grown of some materials which are larger than we've experienced here on Earth and alloys which were melted and solidified in space, showed a marked increase in uniformity of the material. This was due to the removal of conduction currents that are caused by the gravitational field of the Earth. We have separated some biological materials into various protein classes through the use of electro-freezes. This is using electric currents to move cells at different velocities depending on their charge. In a recent experiment, human kidney cells were electroferetically separated in space in an attempt to isolate a strain of the cells which are producers of an anticoagulant called urokinase. Urokinase is produced only by a small number of kidney cells - not by all of them - and it's a very difficult thing to produce. It was theorized that if we could separate out these cells that do produce urokinase, that we could bring them back on Earth and culture them and create more and more of this valuable anticoagulant as a result of this separation. The tests, which were conducted on the Apollo-Soyuz test project, did produce a cell separation which yielded several times the urokinase producing population than the average population of cells would produce. With this encouragement, we can look forward to a very great need for a laboratory in space in which solid state physicists, material scientists and biologists can work much as they do in laboratories on the ground. Their efforts may result in a requirement for manufacturing and manufacturing plants in space or it may result in increased understanding in the science of materials which will permit the production of new materials here on Earth. The prospects are many and they are real. The realization of these capabilities and the reduction to operations will take time. I'm afraid that if we have been spoiled by the rapid introduction to operations of communications satellites and meteorological satellites in the past, we might be a little disappointed in the time it takes to introduce this new capability into daily lives in the future. As I pointed out, daily experience in the past is that it has taken a decade or more. In many cases, the things that I'm talking about require new institutions, new social institutions to accommodate the full potential of these new tools. There is no one really responsible for global agriculture at this particular moment. There is no one responsible for the production of biologicals at this particular moment in space. I hope that the industrial group will rise to the occasion and manufacture their own institutions. My own guess is that the social problems have adjustment, and the development of new institutions will take longer to develop than the science and the technology itself.

Now, in closing I'd like to just tell you another story which relates to the first one and that is that we haven't gone as far as I would have imagined, as probably this group has imagined, in educating the public. I recently gave a talk to some people in a very esteemed institution and one of the benevolent grant institutions in the country at the Vice-Presidential level. They were very interested in the potential applications in space and what they might do to aid particularly developing countries in the realization of these benefits. After I got finished giving him a very lengthy two-hour demonstration of the potential benefits, again I was called aside by a young lady who was one of the Vice-Presidents of the institutions, and she asked me very privately, "Won't these things fall out of the sky?" So we have a long way to go. Thank you.

Left to right: Prof. Perek, Leonard Jaffe, Portuguese representative, Prof. Cid, Dr. Diederiks-Verschoor, Dr. Draper (Courtesy C. Machado, Lisbon)

"ABOUT THE NEED FOR SOCIAL INSTITUTIONS TO CHANGE TO ACCOMMODATE TECHNOLOGICAL ADVANCES AND CHANGE". COMMENTS ON DR. JAFFE'S AND DR. EHRICKE'S KEYNOTES

Dr. Warren Armstrong*

As I listened to Dr. Jaffe and Dr. Ehricke speak from their expertise, it became clearer to me that, as the afternoon wore on, there would be little in the way of expert knowledge that I could contribute. I was interested in a theme that I thought was consistent in the comments of both gentlemen, however. It was a theme I guess I could best describe as one of change and a need for social institutions to change to accommodate technological advances and changes. As I pondered that concept, I thought of the life of my grandfather Armstrong who died in 1968 at the age of 84. He had lived to see the locomotive convert from wood to coal, and be replaced by a diesel locomotive and the electric train. He had lived to see the perfection of telephone and other electronics means of distant communication. He had lived to see the development of the internal combustion engine, its adaptation to the automobile from the Model-T through modern automobiles with air-conditioning and automatic transmissions, power system steering and braking, etc. He lived to see the advent of television bringing into the home live scenes from distant continents by means of communication satellites. He had lived to see spaceflight from the very beginning in Kitty Hawk through the spaceflight that's commonplace today. Had he lived another 18 months, he would have lived to see a namesake, Neil Armstrong, be the first man to step from a space vehicle to the surface of the Moon. And as I think about his life, I cannot imagine that anyone's life could accommodate that much in the way of technological change and advance. It's simply mind-boggling that so much could happen and change the way of life in a nation in a span of one man's life. And yet reason tells me that the pace of change is accelerating, not decreasing, and that although so much of what I've just referred to is now past that the changes in our style of life in the future will be so great that if we were given the privilege to look back a century from now, we could hardly have imagined what will transpire in those next 100 years.

I'm reminded also of some comments that I heard by a bio-physicist who is now deceased but who taught for years at Michigan State University in East Lansing, Michigan. Leroy Augestein was his name, and perhaps some of you have heard of him or read some of his works. He was speaking as a biophysicist about the changes that technology had brought about in his area of special expertise, and he spoke specifically of genetic engineering as a future horizon that the American people needed to try to the very best of their ability to anticipate. In other words, the social institutions need to advance, to meet the changes that technology is going to bring to us, the capacities for determining our future that they are going to provide for us. I was fascinated by Dr. Ehricke's comments about the growth of the population on the face of this Earth, and the potential for accommodating that increasing population through the development of a frontier which we have just begun to touch - the frontier of space. Some of you, I am sure, have had studies in history in a formal way and you may remember that a frontier hypothesis was advanced to explain American history, and that the last frontier was supposed to have closed when the Northern Great Plains were settled in the late decades of the 19th century. The frontiers that are open to men and women of science are not limited by this terrestrial ball. I think all of us know that now. We've begun to develop knowledge of the seas and what lies beneath the waters in the seas of this globe, but we have a greater frontier that we've just begun to tap that's far beyond anyone's imagination at this time. While I have very little specific scientific knowledge of it, I do appreciate the opportunity of being included in the program this afternoon, and perhaps these comments from a layman may stimulate a question or two to the members of the panel who can answer questions from an expert's point of view.

----
*Dr. Warren Armstrong, President, University of Eastern New Mexico, Portales, N.M.

THE EUROPEAN SPACE AGENCY AND THE SPACELAB PROGRAMME

W. J. Mellors, ESA*

Ladies and Gentlemen, 1975 was a most important year for European Space activities in general. In May of that year the European Space Agency (ESA) was created out of the two former space organizations - ESRO and ELDO.

By the end of 1975 eleven European countries had signed the Convention, namely the ten original Member States of ESRO plus Ireland; Austria, Canada and Norway had been granted observer status with the Agency, and both Austria and Norway had joined specific programmes. It is perhaps worth noting that the new Convention endows the Agency with significant, new functions, which the previous organizations did not have. These are primarily designed to permit the development of a coordinated European space programme, which makes the maximum use of the resources of the Agency itself and of its Member States. This may not appear very exciting at first sight but - to take a historical perspective - the United States has been a single nation with a stable political system for two hundred years. Two of our member states, Germany and Italy, have been single nations for less than 150 years, and there has not been anything approaching a United States of Europe since the time of Charlemagne. Consequently, we believe the healthy existence of a European Space Agency is extremely encouraging. We also believe its existence is important - as an essential working prototype of the type of wider International Organization of which Dr. Chatel spoke in his paper.

The Agency was also given a new internal structure, of which one of the most important characteristics was the introduction of three Executive Directorates, which manage our different kinds of development programmes.

One of these is the Spacelab Programme Directorate which is responsible for the implementation of the development and production phase of the Spacelab Programme and for the coordination and integration of European Spacelab experiments. The present Director of the Spacelab Programme is M. Michael Bignier, who until recently was the Director of the French National Space Agency, CNES.

May I now turn to the Spacelab Programme itself. We, and we believe, NASA also, regard the Spacelab as an integral part of the Space Transportation System. It is being developed by ESA and is being jointly funded by nine of our member states together with Austria who has voluntarily associated herself with a contribution of nearly 1%. The major contributor is Germany followed by Italy, France and the UK. The total projected cost of the programme, at mid 1975 prices, is 396 millions of Accounting Units. Due to the comparative weakness of the dollar, this 396 MAU amounts to $515 millions in 1975 prices. Due to the rise in the value of the dollar over the past year, this figure is now somewhat less.

---

*European Space Administration, Washington D.C. Office, L'Enfant Plaza.

What is Spacelab? Basically it consists of a pressurized module in which scientists and technicians will be able to carry out experiments in a shirt sleeve environment, and a series of pallets on which will be mounted experimental equipment, telescopes, etc., which need to be fully exposed to space. Access to the module will be by means of a tunnel being developed by NASA. On any flight it will be possible to select, - dependent on the type of experiment to be flown -, a particular configuration from a number of combinations of module and pallet.

I should now like to mention the major events and decisions achieved so far during the life of the programme.

Between 1972 and 1974, following competitive Phase A and Phase B studies, the prime contract for the development design and manufacture was awarded to the German-Dutch firm of VFW-Fokker who worked with a large number of co and subcontractors throughout our member states.

Amoung the main milestones achieved in 1975 was the signing of the main development contract which had been awarded the previous year.

VFW-Fokker have established a competent industrial consortium to handle the task. As I said, the Agency has given the main contract to ERNO, and under ERNO, there are the various co-contractors and sub-contractors charged with various parts of the Spacelab.

The negotiation of the individual contracts between the prime contractor and his co-contractor is now complete. This, together with the completion of the Sub-system Requirements Review conducted in 1975, means that we have now completed the definition and documentation phase of the development and have entered the detail design and hardware development phase.

We have also made major progress in the management of the all-important interface with the Shuttle Orbiter. This particular interface was of great concern to us as a source of uncertainty in our control of the Spacelab design. Early this year NASA, ESA and their respective main contractors on both sides signed an Interface Control Document which places this interface under joint control.

Agreements between NASA and ESA were also reached this year in the definition of ground support equipment and the command and data management system.

By the end of 1975 the ESA contractor had produced most of the documentation necessary for manufacture of the hardware. The full scale soft Spacelab mock-up was completed on schedule and was on show this summer at the Bicentennial Exhibition at KSC; while the Spacelab integration building at Bremen was also completed.

Ensuing the effective operation of such a complicated management system involving companies in so many European countries, was an extremely difficult task. To begin with, as one might expect, there were weaknesses and difficulties largely in the areas of management coordination but, once they had been recognized, they were dealt with and, as the Director General was able to inform Dr. Fletcher, the NASA Administrator last month, we have recently had very considerable success in giving the whole complex industrial team including the co and subcontractors - the motivation necessary to make Spacelab a success. The industrial companies concerned are now all dedicated to the programme to an extent far beyond the mere fulfillment of contractual obligations. The manpower build-up in industry is proceeding very close to the original estimates and has practially reached its peak at between 1500 and 2000.

In the past 18 months several design-changes have been approved, all of which either improved the service to users or reduced costs. By way of example, a French computer with a higher performance than the baseline proposal has been selected. The concept of the remote control of sub-systems was introduced so as to make better use of valuable crew time. Further, it was arranged that the Orbiter will also provide the oxygen needed for the Spacelab environmental control. These and other agreed changes are all in line with the basic philosophy of Spacelab, namely to achieve a low cost (from the points of view of both development and operational) space research facility available to as wide a range of experimenters as possible.

Since then we have had some slippage, due mainly to the large number of design changes and the management problems to which I have already referred. However, the necessary corrective action has been taken, and the Preliminary Design Review will be completed this year - admittedly some months late - and we are confident that we will be able to deliver to NASA a Spacelab at the time they require it to be flown on the first Spacelab flight, which has been scheduled for the third quarter of 1980.

Now a few words about Spacelab operations and utilization. The programme cost ceiling agreed by the participating countries does not include any funds for the development or integration of European Spacelab experiments. These will be procured and funded separately from interested governments. The discussions with our Member States on this score have gone well, and the replies to our Announcement of Opportunity for the first experimental Spacelab payload, (which will be a joint NASA/ESA flight), have been most encouraging. The experimental objectives for the flight, and the constraints imposed by the verification testing have been formulated and agreed with NASA. In the past month we have agreed that the configuration for the first flight will be the long module plus one pallet.

In Europe the development and integration of the European experiments come under the direction of the Director of Spacelab, and a special ESA management group for this purpose has been set up. This group is located at the German Space Research Organization (DFVLR) near Bonn.

On the basis of the currently available statements of interest by European users, we have established a Spacelab Mission model for the years 1980-85. This includes European participation in several flights through 1983, and full European payloads for five flights during the years 1983-85. The major European interests for Spacelab utilization presently lie in the disciplines of space processing, earth observation and astronomy.

One further remark regarding the utilization of Spacelab. ESA's role will be as the mission planner for European payloads -- whether the experiments originate from universities, institutes, industry, or national agencies. In this way ESA will become the focus for harmonizing European ideas and proposals, whilst at the same time drawing on the reservoir of resources available throughout Europe for its functional support. In all these plans, ESA fully recognizes the role of the experimenter himself as being vital, and considerable responsibilities will be assigned to him for all phases of the mission. From experiment conception through instrument development, and payload integration, even to the equipment operation, and data handling during the Spacelab flight, and to post-flight operations.

How will Spacelab be used? The short film which will be shown immediately after this talk, will be better than any description of mine.

Along with the planning for the operational phase, planning for follow-on production, and procurement of additional Spacelab units are proceeding. To begin with ESA is acting only as the procurement agent for NASA vis-a-vis European industry, but we hope that as time passes many other customers will come forward.

Finally, I would like to emphasize that in Europe Spacelab is not regarded as an end in itself, but as a beginning. Any of you present who are familiar with the intergovernmental negotiations which lead to Europe's undertaking the Spacelab programme, will recall that the representatives of the European nations stressed the importance of considering this joint venture as a prelude to further joint space programmes.

We believe that in the next several years the Spacelab concept will evolve and grow. ESA intends to initiate appropriate planning to ascertain what would be necessary to increase mission duration, and provide autonomy from the Shuttle and various improved services to the user.

This we hope will be followed by our participation in programmes such as the Space Stations which are now being studied. Thank you.

16 years ago of the six space pioneers facing us, three were, and still are, well-known in their field, space biology - Al Mayo, Harold von Beckh and Hubertus Strughold, with space systems and technology planners Ernst Stuhlinger, Hermann Oberth and Wernher von Braun, assessing manned space exploration. Five of the above were either present at the Dedicatation Conference or were inducted into it. (Courtsey Ge-Be Foto)

CHAPTER X

LUNCHEON AND DINNER SPEAKERS AND BANQUET PROGRAM

LUNCHEON ADDRESS, 6 OCTOBER 1976

THE SPACE SHUTTLE ORBITER

Mr. George Merrick*

I have really enjoyed the time we have spent out here the last few days. I think I was particularly impressed yesterday when the honorees -- considering the stature of the honorees, and the depth of experience and history that they represent -- were announced. In that context, one feels kind of like a new kid on the block. I would like to say a few things that will kind of maintain the vein of history in considering what I would call the legacies of the Apollo and Skylab programs and how they relate to the Shuttle development.

But first, even though that news release had a certain amount of detail about what the Shuttle program is, I'd like to briefly describe it. I am sure most of you are familiar with it but I will be very brief. Probably most importantly the system will provide a significant increase in the capability to deliver large payloads including man, to near-Earth orbit and with additional stages synchronous and beyond and at significantly reduced cost. The secret of the cost reduction really lies in the ability to reuse the hardware and to make near-Earth space operations a routine matter. As an example, in a lunar program for every hundred pounds of structure we launched at Kennedy, we brought back three and those three pounds or that three percent were represented in the Command Module and today most of those command modules are in museums. In the Shuttle Program, for every 100 pounds of structure that we launch, we will bring back close to 90 and they will be reused routinely.

Let me describe the elements. The most known element of the Shuttle is the Orbiter which the Space Division at Rockwell is responsible for developing. The Orbiter is kind of like an airplane - is a spacecraft. It has got the length of a DC-9 and the girth of a DC-10. It carries a crew of 2 to 7 and it carries most of the system's avionics and subsystems. It has a payload bay of 60 feet long and 15 feet in diameter and will carry 65,000 pounds, so therein lies the issue of being able to carry a sizable increase in our capabilities today to near-Earth orbit. It has three main engines that are developed by the Rocketdyne Division of Rockwell, liquid/hydrogen/oxygen engines. They developed a thrust of almost a half a million pounds. The Orbiter is launched vertically on an external tank. The external tank is the one element that is really discarded each mission. It carries about a million and a half pounds of propellant. On the tank are strapped two solid-rockets. They weigh about a million and a half pounds apiece and they generate about $2\frac{1}{2}$ million pounds of thrust each. The whole stack when it is lined up is about 180 feet - that is about half of what the lunar stack height was. There is another element of the flight hardware that is kind of unique. NASA now is the proud owner of a 747. It is in the process of being modified and that configuration with the Orbiter mounted on top of a 747 much like it is mounted on the external tank, provides the very rationale for ferrying the aircraft or the spacecraft, the Orbiter, from one facility to another, be it final assembly to launch or recovery to launch. But the 747 also gives us the capability to accomplish a horizontal flight test program. And I will talk a little bit about that in a few minutes. The mission is classic in its start.

----

*President, Rockwell International Space Division, Downey, California.

The whole stack launches vertically with all engines firing. The solid rockets are staged at about 150,000 feet. The external tank is dropped off just prior to orbit and that is the last time the main engines on the Orbiter burn. There are small 6,000 pound thrusters that are on the Orbiter that gives it its final thrust to get it into orbit. It will stay there from 7 days to 30 days depending on how one configures the vehicle in orbit. It will deorbit and reenter the atmosphere at a high angle of attack to provide the drag to reduce the energy and then lands very much like an airplane. It lands at 180 knots, at a relatively steep descent angle - about 20 degrees. That is very much like the X15, an aspect ratio of 4½ - very comparable to that configuration in its flight path at landing. The X15 is certainly a small version of the Orbiter.

There are some kind of key technologies that I think are worth discussing that relate to the Orbiter or the Shuttle System. One is the engine itself. The engine represents a significant increase in efficiency of a liquid-hydrogen/oxygen engine. It has a chamber pressure of about 3,000 pounds. That is three times what the normal engines that we have burned or utilized previously. The Orbiter is an aluminum construction and the other technology that is kind of significant and certainly new to this system is the thermal protection system. The aluminum construction is good to 350 degrees. That is the rather conventional milskin, skin, and stringer - things that we know pretty much about how to do today. But to protect it from heat of reentry, it is covered with a solid silicone insulation. This is put on in six inch squares all over the vehicle. That amounts to about 25,000 tiles. They are bonded on the structure. The squares allow for deformation of the structure but it is good for about 2,500 degrees on the surface and for various thicknesses. For that kind of temperatures, the thicknesses are perhaps in the order of three inches and that insulation is efficient enough to keep the fuel lines at 350 degrees Fahrenheit. There is another area of technology that is kind of unique and that is the piggy-back configuration. I think that has certainly been the subject of a significant amount of analytical and test work. We have a reasonable high degree of confidence in that. We have a lot of testing left in front of us before we fly the first launch, but it does represent a relatively unique and different sort of technology from what we have been used to in launch vehicle configuration.

The last area that I think is worth noting in the technological sense is avionics. The hardware is baiscally off-the-shelf. The hardware that is on a subsystem basically is not unique in its technology. The task of putting it all together is formidable. We have a central processing capability with five computers, 64,000K of memory. They are pretty good size machines. They do all the computation in essence for the spacecraft. Those five machines represent a dedication to redundancy. The management and operation of that hardware in an integrated sense, including its software, is one of the bigger tasks that we are faced with today. I think we have a good handle on that and we have, as you can imagine, a tremendous amount of laboratory work in process working out the software. We are just coming up on the software we will be using this summer and are working diligently for the first manned orbiter flight software.

Now, I would like to spend a little time on what I consider are the legacies of the Apollo and Skylab programs. First is the general area of technology. I have commented to date that a lot of the things we are doing are things that we know pretty well how to do - materials, material properties, adhesives, for example fracture mechanics, flammability. All of those things were really formidable major development issues in the Apollo program. Although we have got our hands full in the quantity of that sort of activity, we have got an awfully good leg to start on in that area of technology on the Shuttle. Subsystems are the same way. The propulsion systems - we spent significant resources and time on the combustion stability, for instance, and hyperbolic engines that we use for the reaction control systems and the orbiter maneuvering engines. Those kinds of issues we haven't seen. They are kind of behind us. We know what that is all about and how those things have to be handled.

Power Generation. We are using the same kind of fuel cells. We are procuring them from the same source. Pretty much on top of what has to be done in terms of fuel cells and all based on the Apollo experience.

The environmental control system and heat rejection in orbit. The same sort of thing of most of the basic work behind us - implementation required on the Shuttle System.

Crew Provisions. A tremendous amount of effort was put into the crew provisions and the biomed area in general on the Apollo program and that is a good foundation and base to work on and has significantly reduced the kinds of issues in that area that we are faced with on the Shuttle program.

Avionics the same way. Most of the hardware is off-the-shelf as a subsystem as I said. Guidance and control is an example. The equipment itself is with us and we are not having to go out and work the problems on an individual basis, and as I said, the biggest issue is putting it all together.

Missions Operations. A formidable task in Apollo evolved from Mercury and Gemini and certainly the ultimate in the lunar missions. It is good to have that background and knowledge and understanding to know what sort of issues one would face in the Shuttle operations and they most likely will be far simpler than a task such as the lunar orbit, or the lunar mission.

Crew Training. Same sort of thing - operational and test facilities. They are all kind of honest so none of that planning is required - it is all there. A lot of our own test facilities, a lot of the NASA test facilities, facilities here in this area that we are using for the Shuttle, that were developed and in hand, and the processes and procedures for that are well-known.

The utilization of man in space. I think we learned a lot about what you could do with a man in the Apollo program, certainly far more in Skylab - what he can do for experiments. What we will certainly now be calling payloads. The capabilities of the man - there are far fewer questions now than existed ten years ago when we started out on the Apollo program. In the idea of what it takes to put payloads and experiments, and interface those equipments with the flight hardware, that was a very time consuming process in the Skylab and ASTP programs, and I think that background base will really see us through what will be a major part of the shuttle operational activity. Now in the past day we spent - there has been much discussion on the scientist. I noticed that last night - and the engineer and technology, and there is an aspect to programs of this size that I think got missed a little bit, and I would say it is probably one of the most significant Apollo legacies and that is the one of technical management. If you will consider Dr. Gilruth's remarks last night about President Kennedy's charge to put a man on the Moon and return him in that decade, that was precisely a defined goal but a fantastic, technological, engineering, and perhaps more significantly, management challenge.

A collection of government agencies, but headed by a relatively young NASA, would never have really tackled a task like that. A requirement for numerous prime industrial contracts for the various elements that went together to build that lunar stack, and almost uncountable subcontractors under each one of those prime subcontracts. A brand new task in the scope of putting all those things together - schemes had to be developed for planned technology that was not really in hand - the growth and development of that technology. And added on that, the rigors required for manned space flight that really were yet to be defined in the early days of manned programs.

The issue of how you interface hardware and control those interfaces, and be sure when you stack them all up at the Cape that they fit. That was a formidable task in the Apollo program days, and I think it is going to be formidable on the Shuttle, but I suspect that that Apollo experience will be and is very useful.

<u>The test planning</u>. The planning of minor component tests down to the black boxes, all the way up to the major test vehicles, and an understanding of what is required was all kind of evolved in the Apollo program, and that evolution and knowledge of what is really required is behind us on the Shuttle and we are using all that knowledge.

<u>Program schedule visibility and control</u>. Again, a massive task like as required to put that many together in an integrated effort - a very significant issue in the Apollo program. Again, much more in hand as far as the Shuttle program is concerned and a massive job of resource planning and control. Probably more emphasis on the Shuttle program as I recall, than on the Apollo. We have had some rather unique resource funding requirements. I think we have all been able to face up to that task, and again kind of based on the understanding that was generally developed in the Apollo days.

A couple of examples in the management of subsystems. I can see some change of command in the NASA, and the prime contractor and the subcontractor are the very same people that worked the same kinds of programs. The fuel cells for instance. The Program Manager at Houston, the Program Manager in our shop, and the Program Manager of Pratt & Whitney - the same people - the same interface. They kind of know what has to happen, and the real issue on the Shuttle program is to go make it happen, but they don't have to generate any techniques or at least management techniques.

The transition in the management program or the management issue is interesting to me too. The Apollo program - programs of that type phase out in an operational phase and the issues that come up while you are flying a spacecraft that goes behind the Moon periodically. Those issues demand immediate management reaction. I can remember Cris Kraft describing a problem to us and saying, "I want you guys back in 30 minutes with the answer." Well, training like that for a management team just cannot be beat. You just cannot sit on your backside and spend very much time thinking about it, because if you have not got it behind you, you do not have it. So I think the success of that program showed that the whole team had what they needed and were very responsive to all the problems, the little ones and the big ones, because you never knew when a little one was going to become a big one. That is a very good environment to launch off into a relatively new development program. You do not take the time to dawdle with the issues. You are really primed to go after them and go after them quickly.

I think the Apollo started with the "what" that did not really have a lot of the "how." The Shuttle, to my way of thinking, started with a good definition of what was to be done and had a good handle on how most of it was to happen. The NASA industry team, I think, when we began the Shuttle, knew what had to be done and did not really have to create a management system to start with.

Now let me talk just briefly about the status of the program. Last month we rolled out the 101 - the first Orbiter at Palmdale. That was rolled out on a schedule that was established over three years ago. That is in my mind kind of unique for aerospace programs. We actually also accomplished that for less money than I think we all expected it was going to cost us three years ago. That spacecraft has been through sybsystem testing. It has been through ground vibration tests. It will go up to Edwards Air Force Base in January along with a modified 747. That 747 is being modified at Boeing right now. The flight test program will start in December and it will be at Edwards in January. We will start captive flights in January. These will be inert without any power in the Orbiter or crew and that allows us to understand the relationships between the Orbiter and the 747 and also practice some procedural maneuvers for launching the Orbiter. In May, we will start manned captive flights and we plan a separation of the Orbiter from the 747 in July. That will allow us a few minutes of the last part of the mission phase. That separation will take place at about 20,000 feet.

We have a static test article that is in assembly now. That test article should be finished and in test next summer. We have a main propulsion test article which is really the back end of the Orbiter. That is also in final assembly. One of the major elements that adds fuselage about 15 feet in length in the back end and has all the propulsion feed systems. The propellent feed for the engines from the main tank to the main engines goes through the aft fuselage. We deliver about 60,000 gallons a minute. Those feedlines directly out of the tank are 17 inches in diameter. That ends up with a lot of plumbing in the back of that fuselage. That will go to Mississippi at the National Space Technology Laboratory next summer. It will be assembled with a full flight version of an external tank and three main engines and we will start firing that main engine system in the latter part of next year.

Orbiter 102 which is the second Orbiter that we are building on the DDT&E contract is in initial assembly. The crew module is an all-welded aluminum structure. We are finished with that structural assembly. There is a legacy. The welders by name and a lot of the welding equipment that we are using on the crew module are the same that were utilized on the second stage of the Saturn V. We will start final assembly on 102 in Palmdale next summer. We will put it into test at Palmdale in the latter part of the year and it will get delivered to KSC in 1978 in August and it will be the vehicle that accomplished the first manned orbital flight in the spring of 1979.

The main engines are at test in Mississippi. There have been three of them down there now. They have had like 600 seconds at 50% thrust and they are up to 95% thrust on a couple of the engines for shorter duration burns, but I think they are getting there. They have had their typical engine problems in development - nothing that is going to disrupt the schedule for that main propulsion test in Mississippi the end of next year.

<u>The external tank in the SRB</u>. The external tank is made by Martin and the SRB - the motors for that are made by Thiokol. Those are both in fab and the facilities across the country are really coming on line, and most of those facilities have been modifications of existing facilities from the Apollo program.

'77 is a big year. 101 horizontal flight testing, main propulsion testing in Mississippi, the Orbital static test, first firing of the SRB, and 102 assembly. There is probably more in front of us than is behind us. I am very gratified for the progress that has been made to date. As I say, we have got an awful lot of work in front of us to get to that first orbital flight. I think the program is really progressing very well and on a solid foundation of technology and management capabilities developed on previous programs.

I have enjoyed being here and I thank you very much.

LUNCHEON ADDRESS, 7 OCTOBER 1976

TRENDS IN SPACE COMMUNICATIONS

Mr. George Harter*

Dr. Draper, Dr. Steinhoff, other honored guests, space pioneers - and by that I think almost everyone here today qualifies - ladies and gentlemen. I am indeed honored to have the opportunity to be here at the Dedication conference for the International Space Hall of Fame and even doubly honored to have this opportunity to be invited to speak to you here at this luncheon.

I have chosen at the invitation of Dr. Steinhoff here to cover just a brief overview of the trends in space communications. I had two reasons for choosing this. One, I noted how many panels and experts they had here in guidance and control and I knew I better choose something which was as far away from that as possible. I also chose it more seriously, because of my company's emphasis in communications - communications satellites over the past number of years and also because of what I think is a critical importance of communications to any of the space missions that you are talking about here at this conference.

As a brief historical perspective relative to communications, I note that the modern space flight - not the ones back to 1940's - but modern space flight, at least in my era, was initiated just a few short years ago exactly on October 4, 1957 with the launch of Sputnik and followed very shortly by the Jupiter launch in January of '58. I'm sorry - which was termed Explorer and the Vanguard launch - I think it was in March of '58 and then followed by a series of the early Pioneers in October 1958 which my company built. In those first systems as far as communications were concerned, the command link and the telemetry link back were characterized by only processing a very few bits per second, and by bits for those of you who are not in communications, that is characters, and a few bits per second would be less communication transfer, I believe, than the manual Morse that is used by telegraphers and modern day Ham radio operators. Now less than 20 years later today, our systems can provide up to 300 million bits per second - 3300 megabits per second which is the capability that the NASA's Data Relay Satellite will be providing. This is something on the order of 100 million to one fold increase and the data transfer for one single link and when you come to think that the Data Relay Satellite will have several links of that type. We are talking about an extremely large increase in capability over those first very crude systems. Another way of looking at it, the early commercial communications satellites which were used for telephone links through a satellite, and the first one launched which was really operational which was Intelsat II in the mid-1960's - about 1965 - which is barely 10 years ago - provided 240 two-way voice channels. Since that time, the Intelsat group have launched Intelsat III and Intelsat IV. They are now working on Intelsat V which will provide 12 to 15,000 two-way links, so again we are talking in 10 years of about a 50 to 1 increase in capability, and even there not limited by the basic communications technology but limited more by the electrical power that can be provided on a satellite that size to the communication system.

The early satellites were designed with objectives of a few months of life. I can recall that even as late as the mid-60's, we were designing satellites where the objective was only six months. The hope

----

*General Manager, Electronics Systems Division, TRW, Inc.

416

was for more but the objective six months, and now it is routine to design these satellites for seven to ten years. There are satellites up there that were designed for six months, that I will relate later, which have been operating more than ten years, so it is now safe to expect that when we go into a new program.

As communications in the space arena have developed, there are kind of three basic types which have tended to develop and the trends have been to have them typed by a mission area. The first major type would be the communication satellites which are used for both the commercial comsets as you are aware of but also the military communication satellites. The second would be what I would call application satellites such as for Earth resource, weather applications, and the third of which much talk and discussion is taking place here, is for planetary spacecraft used for your landing or fly-by missions.

I would like to briefly dwell just on the characteristics of each of these that make them somewhat unique and make the design process a little different for each one. In the communication satellite area, they operate now at synchronous altitude so that they can provide full time day/night coverage for a single area. Now that may sound rather obvious that that is where they would operate, but I can recall back in the early '60's there was a great controversy. Many studies done, lots of heated discussion as to whether it should be a synchronous altitude, a stationary over a point on Earth, or whether they should be low flying satellites and a handover from one to the other for the communication link as they passed overhead.

The communication satellites are typified by multiple spot beams. The early satellites had on the directional transmission of their energy- then they came to having it radiated so it would just cover the Earth. Now the newer satellites have spot beams which would cover an area on the Earth - perhaps New York one beam, another one, San Francisco, California, who knows, perhaps even Alamogordo. They were using multiple frequencies and now I get into an arena that becomes a little confusing if you are not working in the communication's business. In fact it is even a little confusing for us who are because there are so many terminologies in terms of bands. Some people - some groups - talk about S-band, K-band, X-band. Others talk about UHF - ultra high frequency - SFH - super high frequency. All denoting the same general things or we can talk in terms of frequencies and even there someone came in and got the bright idea of changing the name from cycles per second to Hertz and that again added confusion. But I will try to talk here in terms of frequencies today and gigahertz which are thousands of millions of cycles per second. The frequencies that were used, oh for the past - almost from the beginning, have been in what is called the S-band region or about the 2 gigahertz region to give you some frame of reference. The commercial communication satellites, the Intelsat series, will operate a little higher in frequency in the 4 to 6 gigahertz region and they are now going up because they do not have enough communication capacity there. They are operating up in what is called the K-band region which is in the 15 gigahertz - about three times higher in frequency, and AT&T are now planning satellites which are up in the 18 to 30 gigahertz region which is about, oh 4 to 5 times higher in frequency -- all with the thrust to get greater communication capacity. The military services have their own frequency allocations and are offering up to 40 gigahertz. Again, going higher and higher in frequency to get more communication capacity capability.

The characteristic of the communication satellite over the others is that they tend to have many antennae. They look like a flying antenna farm because they must satisfy operating at various frequencies and they have many feeds so they can generate many beams to cover many cities, and of course they have a requirement for operating with high reliability, long-term life and are routinely now designed for like ten years life.

As contrasted with the communication satellite area, the application satellites which are again Earth resource, and perhaps weather satellites, they operate at low altitude. They have sensors which are pointing to the Earth and must gather data. They like to get as high a resolution as possible so they fly as low as they can without getting into the Earth's atmosphere. They have very sophisticated sensors on them including high resolution cameras operating at both visual and infrared wavelengths using many bands of wavelengths, and so because of these sensors, they have very high data rates which they must transfer to get the information down. In the past systems, because they had this high data rate, they would have to store this information on tape recorders or some other device, and then when they pass over a ground station, they read it out. As the sensors become more sophisticated, the data rates must become much higher and it is no longer practical to utilize storage onboard and reading out, so NASA is going ahead with a data relay satellite which will operate much as a communication satellite at synchronous altitude, stationary over the Atlantic and Pacific, and which will relay the high data rates from the low flying satellite up to the synchronous data relay and then back to a ground station in the United States.

The third type of space communication system is for planetary missions. This by nature is characterized by long transmission distances and because of that, relatively lower signal levels returned here on Earth and as a result of that, relatively low data rates compared to the Earth-orbiting application satellites. Whereas, on application satellite transmissions, we are talking about hundreds of megabits per second, hundreds of millions bits per second, here we are talking about maybe a 100,000 bits per second for these planetary type applications.

Now, I have highlighted some of the differences. There are obviously some common characteristics which make the communication engineer's job a little easier. I have mentioned that they require high reliability, long life. They also are characterized by having power amplifiers. That component which generates that high power to be transmitted back of traveling wave tubes. They have high gain antennae so as to direct the energy and focus it in a small cone. They all have low noise receivers to detect and amplify very weak signals and they are going to digital signal transmission rather than analog. Now, if I might go back to each of those points - the high reliability long-life. Obviously that is an economic consideration for the communication satellites. The longer life they can design in the satellite, the lower their operating costs and the lower cost they can pass on to the users in terms of cost per channel. As far as planetary missions, long life is obviously mission critical. There was some discussion this morning about Pioneer X and XI, the Pioneer-Jupiter series. If I got the right information before I left, I think Pioneer X is now operating at a distance of about 10 au at this point in time and has been out there about four and a half years. And the way they are operating today, both Pioneer X and XI, it would appear that they will be operating to spec performance for a number of years yet. Pioneer XI will be passing by very close to Saturn and we will be communicating with them as long as they are within distance to reach us. Experience has shown over the years that this high reliability probably exceeds what our expectations were earlier. I mentioned that as late as 1965 we were designing per requirement to meet an objective of like six months. Many programs have experienced much longer life times than were originally designed in the system. I say, for example, the Pioneer VI to IX series that we built there were four of those launched. They were solar orbiters. Two of them I believe went out so the apogee was about the orbit of Mars. The other two went in a little further and their perigee was about the orbit of Venus and they were to explore deep space around the solar system. The first was launched in late 1965 and each one a year later so that the last was in '68. All four of these are still operating and performing all of the mission objectives that they started with back over ten years ago and there is no indication that they won't be operating for several years yet to come.

This is the good news. The bad news is that most of our programs have had to operate because of its availability with these traveling wavetube power amplifiers. These have turned out to be a worrisome and a characteristically weak element in the total communication system. In fact it is the only remaining vacuum tube that we still fly. All the rest of the components are solid state transitor type devices. And industry is working hard to develop solid-state devices - transistors primarily, but also some non-linear diodes for the higher regions to replace these TWTs. They have already been replaced at some of the lower frequencies, in the few hundred megacycles and was mentioned for the GPS - the Global Positioning System which operates at about 1.5 gigahertz. They will be operating with solid-state transistorized power amplifiers. The trend at the higher frequencies - bipolar transistors just do not operate up above 6 to 8 gigahertz so there is a strong development effort to develop other types of transistors, primarily field effect transistors for operating up to maybe 20 giga-hertz and non-linear diodes up above 20, up to perhaps 60 gigahertz.

I mentioned high gain antennae. The obvious need for using higher gain antennae is to focus power because it minimizes the amount of power then that has to be generated to put into the antenna. The general approach, although there are all kinds of antennae, is to lose large parabolic reflectors with either focal point or a casagrain type feed which has a subreflector out front. These are used to generate multiple beams by having multiple feeds in the reflectors and as you position a feed, that determines which direction your beam is pointed. These large antennae are usually of metallic rib construction with a metallic mesh as a reflector for light weight. They are also easily stored and then deployed and so for the lower frequencies up to a few gigahertz, these are routinely used today. I believe ATS-6 used a 30 foot device and it is possible to develop these up to 50 - 60 feet or so. When you get into higher frequencies, maybe above a couple, three gigahertz region, then you need to go to something that has better surface tolerances than a mesh reflector would have. The common approach is to go to a solid aluminum honeycomb which is fairly lightweight, up to maybe 20 gigahertz. And then when you get above this, you get into problems of thermal distortion of aluminum and the common practice is to go to a graphite fiber with an epoxy binding, and the reason for this is this graphite fiber which is primarily developed as I recall for high temperature operation on airplane surfaces. It has a very, very low coefficient of expansion thermally so as it gets alluminated from different size, it does not change the distortion of the antenna. The size constraints, of course, for the solid antennae are either what you can get inside the shroud when you boost these things from launch or if you want to get fairly complicated in terms of deployment, you have to deploy these but generally they range in size from 10 to 20 feet. The Pioneer-Jupiter antenna was about 11 feet in diameter.

On a note that we are reaching the time we have to move back to the afternoon session, so let me just briefly then summarize for you some of the trends that I see happening here in this arena. I have mentioned going to higher and higher frequencies again for primarily the commercial communications, military communications, and for the application satellites because we need to get more and more data transmitted through the limited band which we have available; more beams for point-to-point communication, digital signal transmission rather than analog, primarily because when you use digital data it is relatively free of degradation from noise and as the digital signal degradates, you can reconstruct after you receive it and it also provides a capability for what is called time division multiplex that you can take a number of digital signals, interface them, combine them together into a single signal and get more data that way.

In the applications satellites, the move in the next few years will be toward use of the NASA's new Data Relay. This will operate up to very, very high bit rates and it will provide a higher frequency capability.

In the planetary spacecraft, I note that there is little emphasis there and has been for years in really bringing on any advanced technology. Most of the past missions have been operated with S-band for a number of years. There has been some use recently and some planned use going up to X-band which is relatively low frequency in terms of the other types of communication systems, but I think the thrust there is to try to use pretty much what is available without putting too much development money into it. So this concludes my brief overview here. I again, like I say, it is a pleasure for me to be here with you. I will look forward to seeing you a little later in the conference and thank you very much.

LUNCHEON ADDRESS, 8 OCTOBER 1976

FROM BALLISTIC MISSILES TO SPACE OPERATIONS -
AN EXERCISE IN PERSPECTIVES

Mr. Grant Hansen*

Thank you very much Dr. Steinhoff. You said that just like I wrote it for you. I am certainly honored to be here especially before such a distinguished crowd and to be participating in the International Space Hall of Fame Conference. This is a homecoming for me because I spent several years down in this area in the late 40's and early 50's, and then I have returned here periodically since that time, so I am just delighted to have this chance to be here once again.

In preparation for this, last weekend I took a dictating machine and put into it what I thought I wanted to say and then gave it to my secretary to type. Please let the record show that my secretary types. I told her to take out all the dull stuff. So in conclusion, ladies and gentlemen..... I noticed in your advance information that it said that in this program the speakers would discuss the past, present and future of space. I would like to follow that pattern because I have reached the age where reminiscing is a lot of fun. I enjoy it very much. Winston Churchhill once said, "Hindsight is more accurate but foresight is more helpful." While that statement seems almost axiomatic, I am not sure I accept it completely. Some hindsight gets embellished into the folklore or distorted just a bit to illustrate some speaker's point. Nevertheless, the events of the past must be examined and understood in order to give us the perspective with which we can do our foresight in a more accurate way. I think that this International Space Hall of Fame and the activities that will surround it in the future will be of lasting value in helping to provide us and future generations with that much needed perspective.

When I first started trying to gather up a few thoughts about what I might say speaking in this area, my first impression was that this area was really much stronger in missile achievements than space achievements, but the more I thought about it, the more space activities that I thought of and the more events in the space program that had their roots here that came to my mind. I now have convinced myself that it is most appropriate that the Space Hall of Fame be located here.

NASA has worked hard, as we all know they have, and they should have, to emphasize the fallout from the space program to the benefit of mankind in many areas. One thing that has had much less attention is the fall-in, the ways in which other activities of mankind have contributed to our ability to do the space program and I think this geographic area, the Tularosa Basin, is especially rich in examples of many, many scientific achievements in related fields which have fed directly into the capability that has made the space age possible. Many people who are now prominent in space, including some of you in this room - perhaps all of you, got your early training and experience on rockets, missiles, aircraft or other military systems that were developed or tested in this area. I first came here early in the Flight Test Program for the Nike ground-to-air missile system, and it was a tremendous learning experience for me and it gave me a perspective of what the future might be. I had the privilege of working very closely with the German team that was doing V-2 work at that time, and I had the special

----

privilege in view of my present job of working with the people in my present company, Convair Division of General Dynamics, who were working on the MX-774. The MX-774 later became the Atlas Intercontinental Ballistic Missile and shortly thereafter became the Atlas Space Launch Vehicle, and as you all know, the Atlas has been a real work horse of the space age and it is still going strong. I recently presented a 30-year service pin to a man in our company who told me his entire professional life with General Dynamics has been spent on Atlas. He started working on the project in 1946 and still finds it fascinating because of the wide-range of DOD, NASA, commercial and foreign payloads which have been launched into space by Atlas.

We have launched 427 Atlases to date. Anyway, its predecessor was the MX-774 and I had a story that I'd written down already but these red ribbons might have reminded me of it anyway. I remember one of the MX-774 launches from White Sands that went up a little ways, then plowed in and made a big crater in the ground and it had in it a bright red ribbon-type parachute that was blown all over the area. So those present quickly scavenged those about 1-inch wide parachute ribbons and for years after that, the badge of honor for launch days was a bow tie made out of red MX-774 ribbon.

As I said, I had the privilege of working along side the German team down there in the White Sands launch area blockhouse and there was one telephone in the blockhouse and I was the one that normally answered it. When the party on the other end had an almost unintelligible accent, I would call one of my German associates and he would get on the phone and they would talk away in German and finally the conversation would end with "OK", then they would hang up. So, even then I was very pleased to see we were developing a common language and that our German fellow rocket scientists were beginning to Americanize.

About 1948 I attended an American Rocket Society meeting at the Woman's Club in Las Cruces and heard Wernher von Braun describe how to get a man safely to the moon and back in ten years. It took a little longer but it happened pretty much the way he said it would. Last night I was privileged to attend a joint meeting of the AIAA* and IEEE** at which we heard from Dr. Draper. I just could not help but think of the tremendous change and the tremendous progress that has been made since that meeting which I attended a relatively short 28 years ago in this area and the one last night. It is a special privilege for me to be here today as the immediate past president of the AIAA, and one of the things I want to talk about a little bit is past, present, and future contributions of professional societies to space. But first I would like your indulgence while I reminisce a little more about events that have been in my life in this area.

I would like to start out by telling you a true flying saucer story or I guess nowdays we are more sophisticated and we call them "unidentified flying objects" or "UFOs". But in about 1948 we invited the press for the first time to witness a V-2 launch from inside the blockhouse or the launch control center and in that blockhouse there were just two small slit windows about so wide and about that high out of which one could see the launch. These windows had been subjected to some of this enchanted New Mexico air after a dust storm and were pretty opaque, so I asked my technician to get up in there and clean the windows so people - our visitors - could see the launch well. Well, he was a
----

*American Institute of Aeronautics and Astronautics

**Institute of Electrical and Electronic Engineers

practical joker type and he had in his tool box a green rubber monkey about eight inches high with a suction cup on the end of its arm. So when he finished cleaning the window, he mounted that monkey as if it was leaning against the glass and peering around the corner of the window into the blockhouse. By the time I noticed he had done that, it was too late to do anything about it. The press visitors were coming in there, and so one of the visitors came over right away and looked at that and said, "What's that?" I thought, well anybody can tell it is a little green rubber monkey with a suction cup on the end of its arm but if you want to ask a silly question, you will get a silly answer. So I said, "That is a little green man that we captured from a flying saucer that landed up on the White Sands and we have him under observation." Well, the very next day there was a serious front-page story that a flying saucer from another planet had landed at White Sands in New Mexico and that little green men about a foot high had been captured and were under scientific observation by rocket scientists. Now that was the first time I had ever publicly been called a rocket scientist and I was proud. But you can imagine why I have retained to this day, a pretty healthy skepticism whenever I hear these flying saucer stories reported.

Speaking of flying saucers, I want to digress for just a minute. I have got a lot of funny letters while I was in the Pentagon. This is one of the funnier ones. It is addressed to me as Assistant Secretary for Air Force Research and it is from a man in Connecticut whose name after it says Senior, so I assume there must be a Junior, so I assume this is not a letter from a child. And it had attached to it a picture of a flying saucer. It has got a dimension that is the typical pancake type thing. It has got a dimension 60 feet and it has got some windows in there and a model that says, "China shall rule the Earth" clear. Over here it says, "Operates on four engines perpetual, speed approximately 6,000 miles an hour, scale one inch equals six feet." I will read you a few excerpts from the letter: "Dear Sir: Enclosed you will find a sketch of a new type ship called the flying saucer. I have the knowledge and information needed to build such a ship. It operates on four engines which are perpetual. It has a price and must be constructed to my exact specifications including lights which change color as the ship increases speed. Now we will speak of price...It will not now, or ever, have bomb bay doors to drop bombs on children. It will be obedient to the will of God. It will not make any trips to even the Moon planet. It will travel to other planets only when God wants it to. Now let me tell you a little about life and my name...First, I would like to explain a symbol known in another world. It is 7-look. This symbol is the 7th wonder of the world - birth. Seven is life, L is when one loses his life. He goes into one hole 0 - out of another hole 0 - and only the K - king knows what happens between holes. My name is Jehovah. Jeh in French means I. Hovah is what I have invented that hovers in the heavens. Your president was told not to go to the Moon. (This came in at the time Nixon was in there.) That is why he is in trouble." So, let this speech go down in history as being the talk at which the mystery of Watergate was finally explained. We now know why our president got in trouble.

Sometimes a personal encounter will touch us and affect our whole lives thereafter. I had such an experience out here at White Sands. Radio telemetry has improved tremendously over the last few decades. It is hard to remember that back in 1948 it was pretty crude and much of what we learned about a flight was diagnosed from the vehicles' onboard recorders and from examining the wreckage. And so a standard part of each launch was the recovery party in which three or four people in a 4-wheel drive vehicle would drive long distances over and around the boondocks and search the salvage being directed by a light observation plane overhead. One day I went on such a recovery party and met a tall, thin impressive man wearing military fatigues. I understood he was a Reserve Officer on temporary duty. During our rather lengthy trip, he spoke to me about the greatness of America and the challenges that he foresaw in aviation and missilery and in the exploration and eventual exploitation of space. That man was Charles Lindberg. What a valuable opportunity that was for a 27 year old engineer to share some intimate thoughts from one of our great American heroes and to gain inspiration and confidence that we were on the threshold of a space age. I later had many occasions to talk with Charles Lindberg after I became Assistant Secretary of The Air Force for Research and Development. Few people know of the contributions that he made to the Air Force space program. So I am pleased to be able to mention it here. I noticed in the Dedication Edition of the Alamogordo Daily News on page 26, that there is a picture of Dr. Goddard and some people, but it does not say what people. One of those people standing next to Dr. Goddard and Professor Guggenheim was Charles Lindberg. So he has been involved-- was involved in the space business for a long time and in some ways that never were publicized in relation to his aviation activity.

Last week I had the privilege of hearing Lowell Thomas speak in San Diego. He is well up in his eighties now. He is 84 or 85 and he is still a very useful appearing and vigorous person. He said the one problem with getting older is that everything you say reminds you of something else to say. So by that criterion I must be getting older because as I think about the times with the Air Force and here at Holloman Air Force Base it reminds me of chimpanzees. At the time I went to the Pentagon there was a rising -- there was a crescendo of concern about military spending especially the elimination of anything non-essential and that is still going on of course. There were at that time, perhaps still are, some Air Force bases -- both in the United States and abroad -- for which there really isn't very much useful mission. The air wing for which the base was built has moved away and there is still a lot of people there that are mostly busy mowing each other's lawns and working in the bowling alley, Base Exchange and the Motor Pool. So it immediately becomes apparent that they are not contributing much to defense effectiveness and that that is a way to save some money - you close the base. Well, I soon learned about the national and international political forces which can be brought to bear to prevent a base closing with its intent tenant impact on the community even when that base isn't really needed. The Air Force at that time had some rather extensive activities and supported the technology of military man in space. Some of the research and life sciences was done with chimpanzees. We are all familiar with the first astronaut - Ham, and one of my favorite cartoons ever is the one that shows Ham with the 100 mission Air Force hat and the Kurt LeMay cigar, standing up briefing, and the Mercury astronauts were all sitting down there lined up and he says, "Men, one of the things that you should know is when you get about 85,000 feet you'll have a craving for bananas." It is really a great cartoon and one of the funny parts of it is that one of the later missions, they found that one of the astronauts had some erratic heartbeat and they diagnosed that was due to a deficiency of potassium in his diet and one of the things that has a lot of potassium is bananas. So those guys should have listened to that chimp. Anyway, the Air Force had a chimp breeding colony out here near Alamogordo. Located here, both because of the favorable climatic conditions, and because some of the important research efforts were being conducted at the Holloman Air Force facilities. After the manned orbiting laboratory program unhappily was cancelled, and recognizing the extensive manned space flight research program of NASA, it really did not seem necessary to me that the Air Force should be spending its money in a project to breed and raise chimps for manned space research. I came out here and paid a visit to the colony to make sure that I really understood the activity. There was one chimp out there called Big Sam that had a talent for accurate, long-distance spitting. It was alleged that he could hit a beer can at 10 meters, but I was not told that in time. And after being defiled by Big Sam, I decided that chimp colony really had to go and I thought I was just the guy that could do that because there weren't any chimp Mayors, Congressmen, or Senators to oppose that closing politically. Did I ever find out I was wrong. The entire New Mexico delegation was up in arms and I believe Big Sam was right in there with them. One of my deputies, Dr. Bill Lehmann, and I spent an absolutely inordinate amount of time working on the problem of trying to get those chimps off the Air Force payroll. However, I am happy to say that that story had a happy ending. We finally succeeded in giving it to the Albany Medical College and they continued its operation with some support from the National Institute of Health and I understand it is still going strong just outside of town here.

Now I believe we hear an awful lot about what NASA does, but I believe that the Department of Defense, in general, and the Air Force in particular, are among the unsung heroes of the emerging space age. There are very good reasons for their being unsung. NASA, of course, is a non-military organization and the viability depends upon public support for their space activities. The work they do is open "for all mankind" and they have had and must continue to have a very heavy public information program to stimulate the appropriate public support. But the providing for common defense is an already accepted expenditure of public funds and so a similar campaign is not needed there, and much of the activitiy is sensitive and therefore, militarily classified. So for these reasons, there is much less public information available on the many contributions of the military to today's space achievements and today's capabilities. Much of our present space capability has roots deeply into the contribution of the military, both in space programs themselves and in applicable technology from weapon systems. For example, the Thor, Atlas, and Titan space launch vehicles all came out of the ballistic missile program - just one example. I think it is fair to say that without the military contributions we simply could not be where we are today in space. Conversely, non-military space activities have greatly enhanced the mechanisms of today's national security through the utilization of space research that has been done in other areas, so it is a reciprocal relationship - the military and the non-military of space. The word reciprocal, of course, then reminds me that I wanted to talk about professional societies and something about their past, present, and future activities in space. The whole of any professional society is greater than the sum of its individual parts. There is much more that all of us can do together than all of us can do each individually. In my opinion, four of the principle functions of any professional organization are to learn from each other, to encourage and assist younger professionals, to provide mechanisms for publication and wide dissemination of professional information, and to motivate through prestigious awards giving recognition to those that have excelled in their contributions to the profession.

Now I don't know what the Academy of Motion Picture Arts and Sciences does besides have an annual dinner at which they present Oscars to the winners in the various categories for outstanding performance in their profession. But if that was all that they did, I would still think it is a very worthwhile organization and everyone that is in that profession should belong to it and support it.

What a great thing this International Space Hall of Fame is to help motivate continuing excellence in professional contributions to advancement of our infinite space age. And what a great thing it is that strong support comes from the professional organizations such as the International Academy of Astronautics, the AIAA, the IEEE, and many others. I know there are people here at this conference from the International

Astronautical Federation and the American Astronautic Society, for
example. The space program is richer today because of the national and
international exchange of information between working professionals in
astronautics, through the mechanism provided by the various professional
societies, institutes, academies, and associations, by whatever name.
I think that professional associations have been a very important
contributor of that fall-in to the space program from various other
activities as I mentioned before. My own experience has clearly shown
me that professional associations are often the most rapid and the most
efficient method by which cross-fertilization of technological progress
can take place, not only nationally but internationally, and especially
internationally. Through my activities in AGARD*, I came to have an
appreciation for the fact that a professional association of some kind
can cut across months and years of bureaucracy in getting people communicating to each other in technical matters.

While this cross-fertilization has been important in the past, I
think it is going to be even more important in the future because I
think we all perceive the need for improved efficiency and for cooperation, for marshalling our resources, including people power resources,
and using them efficiently so this ability of our technological
associations has to be enhanced and I think there is much that can be
done. For example, the innovative use of our own brain child, the
satellite, can make our communication of ideas and progress more rapid
and effective. Those of you in the Space Station meeting this morning
heard about wrist-radio communication point to point around the world -
things like that. As I think of that small meeting in, I think about
1948, over in Las Cruces where von Braun was telling us about going to
the Moon, and there were just a few people there, and I think of the
space progress since and where we are going. I do not think it is too
much to think of such an imaginative lecture as that or some of those
we heard this morning being presented simultaneously to gatherings of
professional people worldwide by satellite and large screen displays
stepping immediately across the national and international barriers and
the language barriers. We do not have to do too much more to train our
computers to mechanically give us simultaneous translations so that
everyone can get these thoughts and start cooking them in their minds
at the same time.

With the past roles of technical associations in communicating
among ourselves and providing peer recognition as major activities,
these are certainly important and they should be continued, and they
should be expanded, but there are much greater opportunities that are
available to us. By cooperative joint activities of societies, much
greater achievements are possible. This is happening to some extent now
and it is in this week's activities here in Alamogordo where several
societies are participating. But I think a much greater degree of
related inter-society cooperation is needed for the good of all participating societies and for the good of the profession. The role of our
technical associations must, however, be greatly expanded in the future
to communicate with those outside our profession. We all talk about this
in the various societies, and we say we are preaching to the choir and
all that, but the fact that we keep saying that does not solve the
problem which still needs solving. There must be two-way communication
between people in the space profession and people who are not in the
space profession. We tend to want to make our politicians and the
public understand the space program and we should. But we need to listen
better too. We must gain a much greater sensitivity and understanding of
how we fit into our complex world of energy, and economics, and environment, and ecology, and crime, and the satisfaction of important human
----

*Advisory Group for Aerospace Research and Development

needs such as food, etc. Many of the societies have recently started activities in the public policy area but I think the surface has only been scratched. We have a tremendous power to help shape the future properly if we will harness ourselves effectively to understand what is needed, to really understand what is needed, and how we fit in, and offer our collective knowledge to help satisfy that need. In the NASA outlook for space, they talk about how great the earth's resources' satellites are in surveillance of crops to help improve the production of food to solve the food problem of the world. But that is not going to get accepted outside until we are in a position to prove to people that that is a better investment of resources than buying more fertilizer. If it isn't, we are riding the wrong horse and if it is, why? We should figure that out and preach that message. I think one of the big problems with space people is we take space as an investment in our future. Space is an alternative. If we want that public support, we have to understand what the alternatives are and prove conclusively and sell the fact that it is a superior alternative. I do not think that we are in the future at all in a posture where people are going to do things in space just because you can or because it's there. We have to say what the various alternatives are, make the proper analyses comparing those alternatives, and demonstrate that space is the best way, then it will be accepted and I think we can do that.

Now, it is interesting that we are usually overly optimistic in predicting what we can do in the next couple of years or so but in very long range forecasts, we have nearly always been overly conservative. I think the record shows that. At a ceremony at our General Dynamics/ Convair Astronautics Plant on Kearny Mesa in San Diego in July 1963, 13 years ago, we buried a time capsule to be opened in 100 years. It contained predictions by astronauts, statesmen, educators, industrialists, and scientists -- some of you in this room -- on what they thought astronautics would be doing by the year 2063. Lyndon B. Johnson envisioned regular travel of people and freight between the Earth and the planets. Dr. Bill Pickering foresaw scientific colonies throughout the solar system, man's travel to nearby stars, and exploration of other planetary systems. Our Chairman for the Space Station Sessions today, Dr. Krafft Ehricke, the father of Centaur, his prediction which he has embroidered on a little in his presentation this morning, was that by then the Moon would be man's cosmic front yard and that it would be a spaceport for interplanetary and stellar spaceships and that it would have hotels and hospitals.

The creators of the Buck Roger's comic strip predicted in 1929 that man would go to the Moon and back safely in 500 years. It took only 40 years, so that is an example of how conservative some of the long range predictions can be, but it is also an example of how imaginative some people can be, and perhaps only the science fiction writers have the imagination to really tell us what the potential is of technology and space. Now I cannot imagine a more exciting and challenging time to be alive than now. After all the time that man has been confined near the surface of the Earth, we are just now opening the doors to the heavens around us. We cannot be so egotistical as to believe that in this whole boundless universe, the only life is on this tiny little speck called Earth. In future years we will see in this International Hall of Fame, the enshrinement of those who have had the vision and the ability to make it happen. Will it be you or your associates, or your children? I hope it will be you.

LUNCHEON ADDRESS, OCTOBER 1976

THE ROLE OF SPACE TECHNOLOGY IN SOCIETY

Robert Gervais*

In my address I will try to involve the AAS in this role, specifically the Society, as many of you now have been involved in the leading edge ideas over the years, and specifically the idea of Skylab and Viking to name two, were first identified through the AAS in terms of publications and/or meetings. With that in mind, and considering technology, I am not the first obviously to say this, but each generation tends to think that it has reached the pinnacle of achievement. This is by prelude into what the theme of my talk would be is technology and what we should be doing or might be doing in the future.

Specifically, regarding this pinnacle of achievement and the stagnating, if you will, of technology advancement, a couple of examples many of you know but was interesting reading to me anyway in the last few weeks, that a good idea of this was in the period of Napoleon. Robert Fulton, the American inventor, came to him with the idea that you should use steampower, if you would, steamships for his invasion of England. The French were trying to break the English blockade and having presented this, Napoleon's answer to Fulton was that, "How in God's name are you ever going to buck the tides and currents with a fire under your decks?" and dismissed him. Well, obviously, historically anyway, Napoleon never broke the blockade and never invaded England. Another good example is Simon Newcombe, a British astronomer, I think, around the turn of this century, made the statement that "aerial flight is one of the class of problems with which man will never be able to cope." Unfortunately, for that noted gentleman, the Wright Brothers flew two months later. So, again one always has the stigma involved that technology has advanced about as far as it is going to and therefore we should not continue with technology.

Unfortunately also, that type of thinking has prevailed since Eagle landed on the Moon. We have not been ambitious in our thinking of space as we had formerly been. A good example, Ray and I were just talking about it, when we had met last, -- I remember in 1962 I think it was, I was working on an advanced program then on the contract to NASA defining Man-Martian, at that time, landings. Our first landing, if the pace had kept up that we were going at that time with regard to the Moon landing, etc., the first Man-Martian landing or forecasting at that time was two years from now - 1979. So, again money tempers all things but again one has to continue looking forward. Now, getting back to the theme here.

After Apollo landings, we became very Earth-oriented - not that that's a bad thing, but we were not looking outward. Also, we were and have been forced to become very cost-conscious to justify all our endeavors in terms of cost. Not that that's a bad thing either. On the other hand, it should not inhibit the forward thinking, and again a very good example you have all heard over and over again is: "If Columbus ever had to go through a cost justification to Queen Isabella, obviously discovery of America would have been delayed." However, I think in the last year and a half or so through several people, we have started to think again. Two such examples, I am going to give you several, but two such examples started out. One, in 1968, a fairly good friend of mine
----

*President, American Astronautical Society (AAS); Manager of Energy Systems, McDonnell-Douglas Corp., Santa Monica, California

from A. D. Little Corporation, Peter Glazer, wrote an article addressing
solar satellite power stations, i.e. generating electricity in orbit
with all its advantages and beaming it down to Earth. I quizzed Peter
on that. I said, "Why did you write it? Were you serious?" He said,
"Well, I did it as a lark in a sense, but also I wanted to basically
try something new and see if it would work - test the industry, test
the whole space business." What has happened as a result of that is
the political attractiveness and political in the very broad sense, has
started the people here in the United States and other people, thinking.
We have an energy problem, obviously, and it has become accentuated
since that time with the boycotts and vast depletion of our resources.
We are starting to think along those lines. As an example, something I
just heard two or three days ago in Washington, are terms of a description
of that. I have been an advocate - fine, it is nice to think about
orbital generation, but you have to involve it with the Earth's needs
and the Earth's societial interaction at this time, i.e. for you and I
to use it and for this audience to use it, the way it has to get in here
is to come through utility lines. Looking through the various -- at
least it is the first step -- and to get through social trauma, one
goes through the various planning exercises, etc. As an example, if we
wanted to satisfy 40% of the United States' electrical energy demands
by the year of 2025, which is not too far when you remember it takes 50
years of planning to turn electrical network or electrical generating
system around in its totality - in order to satisfy 40% of the United
States' demands, we need 112 ten-gigawatt stations in geosynchronous
orbit. Not that that's that much of an undertaking really either in the
sense of the technology. The problem is the cost. Just to give you an
idea of what that means, each one of these stations would be 10 x 20
kilometers, have to obviously be built in space, would require the
support of 2 to 500 men in low-Earth orbit and 10 to 20 in geosynchronous orbit. Now when you start thinking of that, the next thing one
might worry about is the environmental issues. For example, which may
or may not go away by the year 2025 or 2000, for example the emissions
of the launch vehicles putting all this into orbit. Of course, the
problems of microwave radiation, both at high and low altitudes --
there are different forms of them but again there are problems there--
something appropriate to your theme this morning of space law, and only
as an example: 112 satellite power-stations in geosynchronous orbit,
beamed down on the United States would occupy 40% of the libration
points over the Western continents. That's a problem of space law in a
sense. What happens if Canada, South America, Mexico, wants their fair
share as an example? One has to get into that a little. I don't know
what the answers are. I am just posing the question. Find new libration
points I guess, is one of the answers.

With regard to supporting such stations, of course, we have the
tremendous space station logistics, housing, and supplying two to 500
men. We also need launch vehicles to push this up into orbit. As I
will mention in a few minutes, there may be other alternatives. Of
course, an effective propulsion system from low-Earth orbit to geosynchronous orbit, specifically probably some form of electrical propulsion.
That is the first step in forward thinking. The next step is Jerry
O'Neill* who many of you know, or at least have heard of, with regard to
his space colony colonization in orbit. Again, you are at the libration
points but maybe not those necessarily of the United States, but on the
other hand, if you want to launch into them you better think about them.
Again, you might have a conflict of space. I won't get into the details
but he is thinking of rendezvousing AL-2 and having a station of L5.

Now, getting into the launch requirements in the latter stages, of
----
*Princeton University

course. One could consider space manufacturing of the requirements, of
construction facilities, etc., and perhaps using raw materials off the
moon. As most of you -- those of you who read Aeronautics and
Astronautics, there is an extensive article in that in this month's
issue. Again, and I talked to Jerry the other day on this, there is a
problem of cost justification. He is trying to cost justify that
system and obviously he is way in the hundreds of billions. One has to
consider, I think, the technical growth. Again, here is the movement
of technology, the technical growth, or technology will enable nations
to grow richer. For example, one of Jerry's associates at Princeton
University did a little exercise several years ago, again on a 100
billion dollar spaceship, but I can relate that to Jerry's orbital
colonization scheme that at the present rate of economic growth, this
is of the world now which technology feeds, Earth's wealth will increase
a thousand times in two hundred years, which isn't hard to visualize by
those of you from the United States, of what has happened in the last
200 years. So, if you play that game, reducing the cost of the project,
you can reduce the cost of this 100 billion dollar investment for this
spaceship in 1968 dollars to roughly 100 million dollars with this
thousand time growth over the next 200 years. Again, it is relative
economics, it is obvious but one has to look at it like that. Let me
put it another way, if our forefathers were on a cost effectiveness
analysis 200 years ago, and did not provide the investment to us, well
here in this case the expansion into the West, obviously we would not be
here today, or at least not in such an auditorium as this.

With this two step process, one, the Earth-power station, got people
thinking. The next step, orbital colonization, has gotten people thinking. Let me take you on a far trip here. Carl Sagan at Cornell, who by
the way was inducted as one of our Directors the other night with me, has
proposed a transformation of the planets such that we can colonize them.
I will give you two examples. One is Venus. Venus has a very harsh
atmosphere. It is loaded with CO_2. Pressure would crush us if we were
sitting on its surface and 900 degrees is not exactly healthy either.
Carl's suggestion is to transform the atmosphere of Venus to become more
inhabitable. Specifically, he is looking at approximately the same size
as Earth, approximately the same distance from the sun as planets go.
So, he is suggesting that we boil the atmosphere, the Venus atmosphere,
with a variety of blue-green algae. Algae love carbon dioxide, which of
course the atmosphere of Venus is, which alone with water would convert
into organic compounds and oxygen. Now as the carbon-dioxide breaks
down into carbon and oxygen, the atmospheric pressure would decline, the
clouds would break up, moisture would rain down, and Venus presumably
would have oceans. If that be the case, then the planet would cool down.
If it took "some time" -- and I don't know what "some time" is -- to
cool off, potentially it could become habitable.

Another idea he has, is Mars. Of course, this is a very topical
issue these days. I just heard the latest reports from JPL on both the
orbiter and the lander. But, what his thoughts are -- again this is Carl
Sagan -- is to put a giant orbiting mirror above the Martian poles.
Basically, it would melt the polar ice caps which would create water and
then carbon dioxide would then turn this barren planet into something
inhabitable. Now, whether this is at all possible, time obviously will
tell. But this is the type of thinking we need, because it spurs our
present day thinking. You just don't leap forward to 25, 500, 200, 300
years. Now to bring all this down into perspective, I took you on a trip
that went from a solar space station or orbiting power station to an
orbiting colony, on to an evolutional transformation of the planets. Now,
that goes back to where we started - people have started thinking "How
do we get there?" Again, it is a question of this investment with
Columbus. We have to invest in the future. Of course, NASA, and, of

course, European people are thinking of **space** industrialization which is either space-based or Earth-oriented.

The Earth-oriented industrialization presents the imagery which would locate various minerals, various navigation aids, etc. A little example of this: About a month and a half ago, I was over at JPL and I saw some data that they were trying to correlate. Specifically, they had mapped an area around Lake Tahoe, California, which is very rich - has been rich, in gold deposits. From the imagery, they obtained from orbit, they correlated the existing gold mines, if you will. In truth, many of them had been worked over but in fact there was correlation. But surprisingly, about 50% more areas were identified that had not been previously known and I am sure, some very smart gold mining company is running down to JPL now trying to figure out where the sites are so they can start up digging again with the price of gold being where it is. So, here is an example of Earth benefits. Now the space-base benefits are, of course, the manufacturing which would have to start in the long run. There are two ways to look at this. One is manufactured items that could be used in space such as building a colony and the other is items which can be brought down here. To tie this in with the American Astronautical Society - I thought I would never get there -- the AAS again has been a very forward looking group over the years and what we have embarked on recently is what we call a "user-developer theme" where we integrate the user of the industrial or commercial needs with the developer being either the government or a private firm like mine. Try to get them defining their requirements and their needs on Day-1, such that together we can go forth. They can tell us what their R&D problems are, what types of manufacturing they need, etc. There are a number of companies who are involved in this. Now, what we have done is formed four groups at the moment. One is space processing which addresses biological preparations which can be made in orbit and brought down - metals and alloys, electronic parts, glass, ceramics and chemicals. We have a navigation committee and its obvious reason is the navigation - both airways, seaways, etc. Communications and Earth-sensing resources, alas, my gold for example. Again, we also provide them a form and these groups are meeting monthly and we are trying to gain more users in... it is pretty hard... If you take an ABC Mining Company here in Alamogordo, the incentive is not there immediately for them to come to Washington or come to wherever and interface with us determining what are their future exploration requirements. As an association, it may work and that is some of the types of things we are trying to do.

A good example of how user's needs need to be identified -- I will get back now to the space-based electrical generation system. It is fine to go off and define the orbital rays, 10 x 20 kilometers or whatever it is, the logistics required to support it, but in fact it has to be used by utilities. You will have to factor into this the utility requirements such things as, if it is going to be used for base-load plant, it has to have a reliability of 90%. When you talk about a reliability of 90%, I doubt if you can get that out of one big 10 gigowatt system. So now you have to start thinking in multiple systems to supply them the reliability. Another requirement they have which falls into the same category is the capability of the transmission lines to accommodate these large pulses of power. It took 50 years and big zillions of dollars, I guess, trillions for sure, to develop the transmission nets in the United States. You are not going to turn that around and they can only carry so much power. So now you are talking about distributed stations - not just one big receiver on the ground, but several. What you also argue for is multiple, smaller satellite systems. Again, here is bringing the user to bear immediately because it affects the requirements of what we are trying to do in space.

Now in conclusion, for two reasons -- you are behind schedule and I have got to reach an airplane, I want to leave with the thought that we have to - in order to progress, we have to keep on thinking. Thank God we are starting to think again, but we have to continue this momentum, and I hope meetings such as this which look backward versus forward will continue such that we can influence the growth of numbers of our population and our industry. I look around this room and I -- as I did in Washington this week -- and I am still the young kid on the block and I am not that young anymore. This is the problem. In order to stimulate future generations to pass on the technology, the interest, the momentum, we have to give them challenging ideas, challenging situations, such that they can become involved. This meeting, hopefully through its publications, will do that and we as a society are endeavoring to do that. Thank you.

THE STATE AND PROMISE OF SPACE

Dr. Thomas O. Paine*

I have heard of St. Thom'as Aquinas but this is a new one. Thank you Ray, it is a great pleasure to be with you at this historic occasion, as the society gets together to review the status of many different fields of space exploration, as we pause for a look backward at the great names that have made the space program possible and from all over the world, whose genius and whose ability to foresee the future has created the program which in the last several decades has moved forward so rapidly. As we meet here tonight history is still being made. The exploration of Mars is proceeding apace with the information coming back from the Viking explorers. Other far ranging spacecraft are hurdling out beyond the orbit of Jupiter for rendezvous with Saturn in several years, and the first man made object is now gone beyond Jupiter and is being transported outside the solar system never to return.

You think about the few short years from the inception of almost wild ideas, about rocket propulsion, Dr. Goddard, Dr. Tsiolkovski, and others. It certainly was a great pleasure to reflect that Herman Oberth was able to come to the Cape and watch the launch of Apollo XI, and l remember how thrilled we all were to have a man whose basic ideas had led to such a great succession of events such as the first launch to the Moon, and to have him present with us. But it is an indication as to how rapidly things can move in science and technology when a national will and national purpose is put behind these things.

Stark Draper said that he'd like to not look forward too much, but I think that tonight is as much an opportunity and occasion to looking forward as it is for looking backwards. In founding this new Hall of Fame, although we honor the pioneers of the past, I think its primary purpose and its primary effect is to inspire the leaders of the future. And as the young people come here and as people from all over the country and all over the world come and thoughtfully study the lives of those who are commemorated here, I think it will be as an inspiration to intensify the commitment that we all have to furthering the purposes of Mankind through bold new ventures in many fields. We, I believe, are at a crossroads in which an entire way of thought of Mankind is being replaced with a new broader vision. I think the last time this happened was during the late middle ages when the explorers who sailed the oceans from the western coast of Europe brought back to Europe a new vision, a new model, a new world view, and it was this global vision which was brought to medieval Europe which I think lead to the great rise of the west.

I think the global vision of Mankind is coming to an end, and we are replacing it with a much more basic three dimensional view of Man's real environment as being this region of the solar system. Perhaps this region of the galaxy where we live. As we look to the future we can see step by step the same kind of progress which has been so dramatic in the last hundred years increasing our reach out into the next hundred years; and perhaps even a bicentenial of the founding of this Space Hall of Fame, will see some dramatic new things indeed. As we look to the next ten years, the Space Shuttle will begin to operate, taking men from the sur-

---

*Administrator of the National Aeronautics and Space Administration March, 1959 to Sept., 1970.

face of the earth up to orbit and back again in a routine, highly reliable and low cost fashion that will indeed open for the first time a new continent of orbital space for men of many nations who are working together on this space program.

I had the pleasure several weeks ago to attend the Space Shuttle roll-out and as the program moves ahead and we begin to actually fly into orbit at the end of this decade, it is possible to begin to envision in the 1980's the building of the first permanent structure in Earth orbit when the Space Shuttle payloads can be assembled and modest space stations begin to evolve. Beyond these space stations lie substantial space research bases, not only in low Earth orbital areas, but in the farther regions up in the geostationary orbit, and of course once operations in the geostationary orbit are achieved, it requires very little energy beyond that to begin to create a true economical, reliable transportation system to the Moon. It will make it possible by the end of the century, the establishment of the first modest research base and the first space stations in orbit around the Moon. Beyond that I am sure we can envision in the next century the first manned trips to Mars, based on the transportation system which we are slowly but steadily establishing. Trips that will be far more economical than any we can envision today with our present crude rockets. Beyond the travels around this part of the solar system of course, lies the possiblity of creating artifical habitats in space. These have been envisioned by all of the space pioneers, indeed some of the first writings of Tsiolkovski envisioned the creation of artifical colonies in space, and he foresaw the ability to use solar energy to provide the type of living accommodations which would make it not only desirable but extremely attractive. Today we are actually studying these in the first early phases and speculating about what kinds of costs, and what kinds of environments, and what kinds of purposes would be served by men living in large communities in space perhaps hundreds of years from now.

The work that is going on to release the tremendous power of nuclear fusion in a controlled way certainly would provide the kind of propulsion that we would need to make space travel routine around this part of our environment. But if we could actually harness nuclear fusion for propulsion and for powering a space colony, I think in a hundred to two-hundred years period from now, it would certainly be possible to begin to propel a space colony of several tens of thousands of souls out across the void of space, perhaps to the near by stars not reaching it necessarily in one generation but taking all the time required with the kind of energy that might be provided. These dreams are still, of course, very far off by today's standards but so are the accomplishments of today when we look back to the work of the pioneers who we honor.

I think it is a very sensible policy on the part of NASA to develop not only the farther out applications and the scientific research in space, but at the same time to bring benefits on an accompanying basis so that as we proceed decade by decade we can get the kind of returns from our space investments that will encourage us to continue. And certainly the word that is going on in the earth resource satellites, communication satellites, navigation, weather, and the basic research that is supporting all of this, is making it increasingly obvious that the returns we are getting for our space investments are very substantial, and extend across many parts of the economy.

So, I think, that we are quite assured that the work which we are honoring here at the International Space Hall of Fame is a beginning and not an end, just as your institution as lovely as it is up on the hill there, is also in its beginning and much work remains to be done, and perhaps as that work is done I think that it will bring to the community and to the world, a great deal of satisfaction and opportunities to see new ways to move forward.

I would like to conclude with a short quotation from one of my favorite poems written by another great visionary, Alfred Lord Tennyson. He also had the ability to look to the future. In one of his poems entitled "Ulysses", he is considering really not just the wonder but the great seaman who returns from the flames of Troy via many trials and tribulations, but he is really speaking of the exploring spirit of the human mind and the human heart. At the end of this poem Ulysses decides to set forth again. Tennyson says that Ulysses calls to his band and says "Come my friends, 'tis not too late to seek a better world. Set forth and sitting well and order smite the sound furoughs, for my purpose holds to sail beyond the vast of all the western stars till I die." This to me is the essence of humanity. We are the intellectual creation, we are the people who through the human intellect have been able for the first time to take life as it is evolved on this planet, and move it across the void to other worlds. It really doesn't make much difference whether we find life on Mars with the Viking expedition or not, if it isn't there now it will be. It's only the question of whether we have to be careful of the ingenious life as we move to enliven that planet in future generations. The purpose of Mankind is something which philosophers can debate, but I think, there is no question that it is to press on and to try to comprehend the grandeur of the Universe, and the place of intelligence in it and this is indeed the higher purpose of the space program.

Thank you very much for your hospitality in having all of us here to your fine community, and good-night.

CLOSING REMARKS

Dr. Charles Stark Draper

Well, I have been an observer of this project even before it was a project because I came out here on another matter and I talked to Ernst and John Stapp and they introduced the notion of this Space Hall of Fame and introduced me to some other nice people, and we did with no great amount of planning, made a little excursion up to Santa Fe and talked to some people. And, indeed, the idea took hold and I would say that beyond doubt I am filled with admiration for Ernst Steinhoff and the people who have worked with him because they have done a job that borders on the miraculous. It's really a miracle and if anybody has seen the details, I have seen the details. He was kind enough, John Stapp went along, to feel that the International Academy of Astronautics might be useful by being an associate in the project, and our particular part was to help out by having some symposium activities like we have at the IAF Congresses and the other meetings that we attend. So you have seen the results and the members of the Academy who have taken part here, I won't try to mention all of them because they're too many of them. I think, a very good job and the three areas that were attacked - the guidance and control which to me was very interesting because I saw perhaps more clearly than ever before an overview of the way in which a region of technology developed and this was a region of technology that had to be developed because you couldn't do these jobs with the biggest engines in the world if you couldn't make the things go where you wanted them to.

So, the discussion of guidance and control, navigation, was certainly excellent and the space law people -- it seemed to me had some very interesting discussions of some old subjects but things that are going to be important in the future, and I think they are beginning to see they're going to be important, and getting around to the place where they are zeroeing in -- if you want to call it that -- on some decisions that have to be made. This was a very, it seems to me, high quality discussion. And then the very complete, I would say, for a meeting like this, looking over science and engineering technology and what these fields of human activities are going to offer to the "space" of the future is opening at least -- or maybe not opening but at least getting a foot in the door of many things that will be much plainer as the years go by. It appears to me that these papers that we've heard and discussions that we've heard produced some mighty fine things to think about. We've heard in the third activity that the Academy fostered these biographies, the stories that tell what some people have really contributed...some people who are among these very few who lead the human race into new paths of importance. We've gotten some good information there, so I think that when you add to these symposia the roundtables that the Academy didn't take the responsibility for but Ernst Steinhoff got up himself - the roundtables, the speakers that he got for the luncheons and for the other social functions and the entertainment that we have been given at the Inn of the Mountain Gods and out on the desert, and various other places has been excellent. So it seems to me in summary, that this has been a week that is on the plus side. It's been well spent, and I for one am very happy about this and I thank you again, Ernst, for the wonderful job that you did, and the Academy is pleased to have been a part of it.

CLOSING REMARKS

Dr. Ernst Steinhoff

Now that the work is done, I think I can do some summarizing here and, I would say, recognize a number of people who have been very helpful in accomplishing extraordinary efforts of a small community to dedicate an International Space Hall of Fame. I think without the encouragement of Dr. Draper himself, I would not have been able to accept the task, and we both had, I would say, an earlier start in mind than we really could get off. I was ready to lay the shovel down and say it can't be done anymore. He told me I'd still do it, and I have to thank all of you who came, in one function or the other, to this program and who kept their promise and came. I had just about two days ago been notified that one of our key speakers could not come because the Secretary of Transportation told him he had to go to Congress and represent the program. He called me up and said, "Ernst, I tried everything. I can't come." I said, "Okay, you send me someone who can give your talk", and he came. You know, of course, who it was -- our luncheon speaker today. I had another case where a friend, the Senior Vice Presiden of one of our greatest aerospace companies, one of the most successful ones, called me - "My corporate office has called me to be up there for deliberations in the week from the 4th through the 9th of October." (I had him as a luncheon speaker.) I asked him, "Can you send me someone who can talk to what you wanted to talk about" -- and he came, George Harter. He was with us and the speaker. You know, if you have such a team to work with, there isn't any problem you can't solve. We are a very, very small office here in Alamogordo. We have handled this with only two volunteers for the last few weeks until I could get more volunteers, and we have handled this generally with two secretaries. I had a few more there in between, but we cannot switch horses or have benefit. I would say the two more we had caused us more time loss than saved. So with this we, in order to bring the proceedings out, had to set the deadline. We did slide the deadline after we talked to the publisher for about five days, and we did get our proceedings out which we were not sure the publisher would publish or not. On the morning of the 2nd of October, Dr. Draper's 75th birthday, I received the first copies, and I could make a very quick decision at the same time.

When I saw the copy, I decided that we would bring instead of five copies, and have the preliminary proceedings dressed in the coat of the final proceedings, with us, I would take a few more so that everyone could see how the final proceedings would look like, except maybe twice the volume. It may require a complete new typesetting of all the first issue to get something which will really look uniform. We could also do the following: have a second volume of the first edition which would contain all the papers which did not come into the first one, some more history of the base, history of the Space Hall of Fame and also who came to the dedication. We have here a number of papers which are devoted to the biographies of the most important persons that came. The commission decided that we probably should have a few more then we could get the biographies written on, and we have done this. I think we did get a good start on staffing our Space Hall with pioneers.

I think that the final version of the proceedings will be a book worth reading. I think it will give and reflect the history -- what has been achieved in the past -- and it will show what we certainly will expect to achieve in the near future -- and will give a hint of what could be achieved if the space effort world wide receives the support which is needed to do the things the world needs done. I think this conference has at least painted the picture of that, and maybe this picture will penetrate wide layers of our population. I have succeeded in interesting the International Space Hall of Fame Foundation into helping me print the final issue and possibly, if necessary, also typeset it to make a real book out of it. They will sell at no benefit whatsoever, the next 1000 copies will support this. We do not know right now what it will cost but I think there will be a number of Americans, when they see what we have done here, who will buy that issue and if we succeed, this issue can be printed in 4000 units at a time, and spread with gradually dropping price to all Americans who come to see the Space Hall. It will be a real cheap book in a very, very expensive coat and with a very expensive preparation.

Maybe we will succeed in having a space textbook. The parents may not see the significance of this book, but I think the youngsters will. I think this was my summary. Thank you. (Remarks of the Editor: We have the space textbook from 6 December 1976, and the New Mexico Department of Education adopted Volume I of the proceeding as textbook for New Mexico Public Schools. We also learned the Part II, which you are reading will be included in this arrangement.

I would like to single out a few people who have helped me, so that the work could be done. I couldn't have done it alone. I think I have done it in a way that cost me the least time to do the job and this was necessary because I had no more time.

I would like to single out a few persons here. One is just walking at me here. I mentioned that I had a few associates who were sticking with me on their own time, and drive some substantial distance to come here to help me, and one of them is, Len Sugerman, who has come here for the last three weeks -- every day pretty much -- and then before we met often to kind of chew the fat on what needed to be done. I intentionally kept myself out of many mass meetings where this was discussed because I didn't have the time, and I think it was too early to enter the activities because much was in flux, and I didn't want to hear over and over again what was in flux. I wanted to know how things really were going and then say what I needed, and stop with that. I think I have done that, and I've kept up to date with everything that did go on and needed to go on, and nothing else.

I have here, Fred Ray. He has prepared with me for many years, while I was in charge, the "Inertial Guidance Test Symposia" and later served Martin Jaenke who is in there too, to continue the tradition in which we had the proceedings on hand on the day when the symposium came, containing the speeches...the manuscripts of the speeches to be given, etc. He has been of great help to organize the gound support with the professional associates, from the AIAA and IEEE, of course, who are also my associates and Dr. Jaenke's associates. Dr. Jaenke himself who prepared in a amazingly short time the paper which he gave. He is retired and was in the middle of building a house, and you know how something like, I think, a paper comes like a ton of bricks on you, but he did it in time. It went into the preliminary proceedings.

I'd like to talk about the two...-staff members of mine. They are both married and have children. They have had to leave their children at a nursery, and there are many, many, many hours where they did not see their children. Susan, would you stand up please. Susan's husband is in Europe at this time, came to work, and worked I would say, literally her head off. She has had problems with her children, every now and then one became sick so she had to stay home a day or so. At the most crucial time, she came down with pneumonia. The doctor said, "Dr. Steinhoff, if Susan gets the other lung inflamed, I may have to put her in the Emergency Ward. He told her to stay home and get over it.

She came the next Monday after two or three days at home. We had to send her home at noon again at the end of the week because the second lung was inflamed. This girl has not quit and the work was done. We wouldn't have any proceedings here today, if she had quit.

Rita, stand up please. She has the same typewriter model she has in the office at home and without me knowing it, she has spent many night hours with Susan reading to her, to get things on the typewriter fast. This was the only way to do it. I think this office team is hard to beat.

I think, we have another person who came to see his father honored. He couldn't come on the day his father was honored so I asked him to be here today, and we'll repeat the honoring because most of you have not heard it. We have had quite a lot of people here but some stayed a day only, or not even a day -- came, gave their paper, and went on again. I would say congressional presentations, all kinds of things that they had to do to come, but they came. Let's say we have many of the people we have seen only two days around here, and a few we have seen all week. We had about 40 representatives from at least 11 countries here, and one country which dropped out on account of currency restrictions which did not permit their scientists to be here. But I have one, who is to be honored, and I think he will be ...I don't know whether he was or may have been in...was he presented, Corocco. Was Corocco presented today? Ok, so we got a presenter for him. They tried every possibility, etc. but it was not possible to get everyone to get enough money to come here.

We had to depend on the personal mobility of all the people who were on the program. There was no way for me to at this time, refund or reimburse anything, and I think this is something everyone should recognize. This is the way, how we work in the Academy and how the Academy has been able to keep people who were once members of the Academy within the Academy, and who continued to work hard. We have many representatives here tody, and you will see them with us, and I will introduce them today. I think with this now, I will close my final remarks, and we now go quickly to get ready for the banquet maybe for one or the other who has spent long hours and midnight oil to recover a few hours of sleep, so that he can be fresh this evening when we all get together, and will be happy about what has been accomplished this week.

For the preparation of Part II of "The Eagle has Returned", Susan LaFlam, Liovy Chavez, and Donna Utsunomiya assisted in getting the book together.

Mrs. LaFlam, whose nusband is currently enrolled at the University of Arizona, School of Engineering, at Tempe, has made several trips down to Alamogordo in order to help complete the book. She has put in many, many hours of typing and listening to tapes in order to accomplish this feat.

Mrs. Chavez and Mrs. Utsunomiya also worked very hard in order to have the finished draft of the book ready by the deadline. Both were able to work long hours on weekends and in the evenings which helped make the completion of the book come to a reality.

I think it will take time for all of this to really sink in and know what we really did here. His own picture may not have to himself reflected it, but if you look at the great picture, it does. With that, I thank you very, very much for all the work and preparation.

During the Space Hall Dedication Conference, space pioneers from 3 nations and their spouses visit the Cloudcroft observatory. Robert Gilruth leading the tour (first right) (Courtsey Col. Sugerman)

On his birthday in 1973 Hermann Oberth explains his theses of space flight 50 years after their 1st publication (courtsey Photo Uecker, Bremen)